M000287308

ETHICS AND EMERGING TECHNOLOGIES

Also by Ronald L. Sandler

CHARACTER AND ENVIRONMENT: A Virtue-Oriented Approach to Environmental Ethics

DESIGNER BIOLOGY: The Ethics of Intensively Engineering Biological and Ecological Systems (*co-edited with John Basl*)

ENVIRONMENTAL JUSTICE AND ENVIRONMENTALISM: The Social Justice Challenge to the Environmental Movement (*co-edited with Phaedra Pezzullo*)

ENVIRONMENTAL VIRTUE ETHICS (*co-edited with Philip Cafaro*)

NANOTECHNOLOGY: The Social and Ethical Issues

THE ETHICS OF SPECIES

VIRTUE ETHICS AND THE ENVIRONMENT (*co-edited with Philip Cafaro*)

ETHICS AND EMERGING TECHNOLOGIES

Edited by

RONALD L. SANDLER
Northeastern University, Boston, USA

Selection and editorial matter © Ronald L. Sandler 2014

Individual chapters © Respective authors 2014

All rights reserved. No reproduction, copy or transmission of this publication may be made without written permission.

No portion of this publication may be reproduced, copied or transmitted save with written permission or in accordance with the provisions of the Copyright, Designs and Patents Act 1988, or under the terms of any licence permitting limited copying issued by the Copyright Licensing Agency, Saffron House, 6–10 Kirby Street, London EC1N 8TS.

Any person who does any unauthorized act in relation to this publication may be liable to criminal prosecution and civil claims for damages.

The authors have asserted their rights to be identified as the authors of this work in accordance with the Copyright, Designs and Patents Act 1988.

First published 2014 by
PALGRAVE MACMILLAN

Palgrave Macmillan in the UK is an imprint of Macmillan Publishers Limited, registered in England, company number 785998, of Houndmills, Basingstoke, Hampshire RG21 6XS.

Palgrave Macmillan in the US is a division of St Martin's Press LLC, 175 Fifth Avenue, New York, NY 10010.

Palgrave Macmillan is the global academic imprint of the above companies and has companies and representatives throughout the world.

Palgrave® and Macmillan® are registered trademarks in the United States, the United Kingdom, Europe and other countries

ISBN: 978–0–230–36702–9 hardback
ISBN: 978–0–230–36703–6 paperback

This book is printed on paper suitable for recycling and made from fully managed and sustained forest sources. Logging, pulping and manufacturing processes are expected to conform to the environmental regulations of the country of origin.

A catalogue record for this book is available from the British Library.

A catalog record for this book is available from the Library of Congress.

To Howard Barry Sandler

CONTENTS

PREFACE

Technology is ubiquitous. It shapes every aspect of human experience and it is the primary driver of social and ecological change. Moreover, the influence of technology on our lives and the rate of technological innovation are increasing. Given this, it is surprising that we spend so little time studying, analyzing, and evaluating new technologies. Occasionally, an issue grabs public attention – for example, the use of human embryonic stem cells in medical research or online file sharing of music and movies. However, these are the exceptions. For the most part, we enthusiastically embrace each new technology and application with little critical reflection on how it will impact our lives and our world. What is more, when an issue raised by an emerging technology is attended to, we often lack the language, concepts, critical skills, and perspectives to thoroughly address it.

The aim of this textbook is to introduce students and other readers to the ethical issues associated with a broad array of emerging technologies, with the goal of helping to develop analytic skills and perspectives for effectively evaluating novel technologies and applications. Toward this end, the features of this book include:

- *Thirty-six chapters authored by leading thinkers on the ethical dimensions of emerging technologies.* Fifteen of the chapters are original to this textbook, written specifically to introduce readers to the technologies and the ethical issues that they raise. The chapters address a diverse range of recent and emerging technologies, including: genetic engineering, sex selection, reproductive assistance technologies, nanomedicine, stem cell research, neurotechnologies, human enhancement technologies, robotics, artificial intelligence, surveillance technologies, virtual reality, geoengineering, synthetic meat, genetically modified crops and animals, synthetic biology, and artificial life.
- *An extensive introduction that proposes a framework for analyzing the ethical dimensions of emerging technologies.* The framework incorporates several critical perspectives from which emerging technologies can be evaluated – for example, power, life-cycle, and form of life perspectives. The perspectives are developed from the work of influential thinkers in the area, several of whose writings also appear in the textbook – for example, Hans Jonas, Langdon Winner, and Arnold Pacey. The Introduction uses examples to illustrate the perspectives, and situates the textbook's readings within the proposed analytic and evaluative framework. In addition to providing a framework for assessing technologies, the Introduction provides an overview of the most prominent themes in the ethics of emerging technologies – for example, safety and risk, liberty/

autonomy, decision-making under uncertainty, justice/equality, hubris/playing God, responsible development, and the relationship between ethics and policy. The Introduction also provides a brief terminological and conceptual primer on ethical theory.

- *Section overviews and chapter summaries.* The section overviews situate the section's chapters with respect to each other, and provide social and ecological context for the technologies discussed in them. The chapter summaries provide brief abstracts of the chapters. They also indicate parts of the Introduction and other readings in the textbook with which the chapter connects.
- *Discussion questions.* At the end of each reading, several questions are suggested to help to stimulate reflection on the issues and ideas discussed in the chapter. The questions also encourage students to consider the technologies in light of the critical perspectives and themes discussed in the Introduction.

STRUCTURE AND USER'S GUIDE

As mentioned in the Preface, the Introduction includes an overview of common themes in the ethics of technology, provides a very brief primer on ethical theory and value terminology, and proposes a set of critical perspectives from which novel technologies can be evaluated. The readings in Part I are general reflections on the ethics of technology, from which several of the critical perspectives presented in the Introduction are developed. The remaining sections are structured around uses of technology (for example, reproductive technologies, therapeutic technologies, agricultural technologies) and particular technology types (for example, synthetic genomics and robotics).

The textbook's structure tracks an approach to teaching and studying ethics and technology that I have found effective:

(1) *Establish cross-cutting themes and perspectives in general* – this is done with the Introduction and the Readings in Part I;
(2) *Focus in detail on issues raised by particular technologies and applications* – this is done by having the sections organized by technology type and application;
(3) *Further develop the critical perspectives and themes in light of what is learned by studying the particular technologies* – this is encouraged by the related readings feature, which connects chapters to the relevant parts of the Introduction and other chapters, as well as by the discussion questions.

Through iterative application of this approach to different technologies, students continually develop their ability to identify the ethically relevant characteristics of emerging technologies, strengthen their analytical and evaluative skills, and deepen their appreciation of the ethical significance of emerging technologies.

An alternative to the approach described above is to begin with readings on particular technologies and applications, and to address the critical perspectives and themes as they arise or become relevant in those readings (and so to use the Introduction in a piecemeal way). The chapter summaries and discussion questions facilitate this approach as well, by indicating for each chapter which parts of the Introduction (and so themes/perspectives) and other readings are related to it.

Yet another way to use this textbook is to focus on the section readings and skip over the Introduction altogether. This may be the preferred way for instructors who already have an established approach to teaching ethics and technology, as well as for instructors newer to teaching the topic who do not find the approach proposed in the Introduction compelling.

So while there is an approach to teaching the ethics of emerging technologies embedded in the structure of this textbook, the diversity and independence of the readings also makes it highly flexible with respect to how it can be used.

ACKNOWLEDGMENTS

This textbook would not have been completed if it were not for the help of John Basl, Colin Pugh, and Jennifer Haskell. John provided extensive advice and feedback on the selection and editing of the chapters, as well as on the Introduction. It was enormously valuable for me to be able to discuss the textbook on a day-to-day basis with someone else who regularly teaches the ethics of emerging technologies. John also helped to formulate the discussion questions for many of the chapters. Gathering permissions and formatting selections for a textbook such as this is perhaps the most difficult aspect of the project (and certainly the most tedious). Colin and Jennifer did much of this dirty work, and I am deeply grateful to them for it.

This project received support from two grants from the U.S. National Science Foundation – Nanotechnology in the Public Interest (SES-0609078) and Center for High-rate Nanomanufacturing (NSE-0425826), for which I thank both the NSF and the principal investigators on the grants, Christopher Bosso and Ahmed Busnaina. I would also like to thank all the people who I worked with at Palgrave Macmillan for their support and assistance throughout the project – Priyanka Gibbons, Melanie Blair, Brendan George, and Ranjan Chaudhuri.

This book is dedicated to my father, Howard Sandler, whose amazement at each new technology never ceases to amaze me. Thanks for everything, Pop.

LIST OF CONTRIBUTORS

Keith Abney

Fritz Allhoff

John Basl

Francoise Baylis

Mark A. Bedau

George Bekey

Nick Bostrom

Maria Bottis

Philip Brey

Philip Cafaro

James P. Collins

Gary Comstock

Inmaculada Melo Martin

Thomas Douglas

Kevin C. Elliott

Lucy Frith

Tamara Garcia

Michele Garfinkle

Marin Gillis

Walter Glannon

Clive Hamilton

Kenneth Himma

Hans Jonas

Leon Kass

Lori Knowles

Ray Kurzweil

S. Matthew Liao

Patrick Lin

Ben A. Minteer

Arnold Pacey

Christopher J. Preston

Michael Ravvin

Jason Robert

Ronald Sandler

Vandana Shiva

Richard Spinello

Jan Stanley

Barry Steinhardt

Robert Streiffer

Paul Thompson

Mark Triant

Jeroen Van Den Hoven

Wendell Wallach

Langdon Winner

1 INTRODUCTION: TECHNOLOGY AND ETHICS

Technology shapes every aspect of human experience. It is the primary driver of social and ecological change. It is a source of power, vulnerability, and inequality. It influences our perspectives and mediates our relationships. Given this, it is surprising that we spend so little time studying, analyzing, and evaluating new technologies. Occasionally, an issue grabs public attention – for example, the use of human embryonic stem cells in medical research or privacy in social networking. However, these are the exceptions. For the most part, we seem to suffer from technological somnambulism (to borrow a term from Langdon Winner [Ch. 4]) – we incorporate new technologies into our lives with little critical reflection on what the impacts will be.

The goal of this textbook is to help students develop linguistic, conceptual, critical, and perspectival resources for thinking carefully about the ethics of new technologies. Toward that end, this Introduction provides an overview on ethics and emerging technologies and suggests an approach to analyzing the ethical dimensions of emerging technologies. Section 1 discusses the significance of technology in human life and culture. Section 2 highlights several prominent themes in the ethics of emerging technologies and proposes a framework for ethical analysis and evaluation based on the themes. Section 3 is a primer on some common ethical theory and value concepts employed in the ethics and technology discourse.

1. TECHNOLOGY IN HUMAN LIFE

When analyzing and evaluating the ethical dimensions of emerging technologies, it is crucial to keep in view the robust interactions between technology and society, as well as the essential role that technology plays in our way of life. The aim of this section is to begin to elucidate these fundamental features of the human–technology relationship.

1.1 Technology and society: beyond technology as a tool

Technology is often conceived of as a tool, something that is developed and used by people to accomplish their goals. It is also often thought to be value neutral – the idea is that the goodness or badness of a technology depends entirely upon the goals or ends for which it is used. There is some truth to this *technology as tool* view, since technology certainly helps people to accomplish things they would not otherwise be capable of doing. In fact, engineering is often described as the creative application of scientific principles to design processes and structures in order to solve problems and overcome barriers. Drugs are engineered to address diseases; crops are engineered to increase food productivity; networking software is engineered to facilitate communication; solar panels are engineered to produce energy; manufacturing processes are engineered to produce reliable products at high volumes and low costs. Moreover, many technologies can be used for both good and bad ends. Synthetic genomics can be used to develop pharmaceuticals or for bioterrorism; GPS technologies can be used to improve supply chain efficiency or to track people without cause and without their knowledge; autonomous military robotics can be used for self-defense or for unjust attacks.

However, the technology as tool view is only part of the story about the complex relationships between technology, society, and ethics. The reason for this is that technologies, in addition to being means to ends, are also complex social phenomena.

Some technologies are encouraged by society through social demand or public funding. This is often the case with medical technologies, for example. Other technologies are opposed or rejected by society (or at least by some members of society). For example, genetically modified crops have been resisted in parts of Europe and Africa, and many countries have passed laws banning human reproductive cloning. Technologies are always implemented in and disseminated through society. Sometimes they help us to solve social problems. For example, vaccines, medical databases and analytical tools help us to respond to disease epidemics and generally improve public health. Sometimes technologies create social problems. For example, many mining technologies have caused tremendous environmental pollution and degradation. In no cases are technologies separate from social context. They are all, always, socially situated. Every instance of technology creation and use is historical. It occurs in a particular place, time, and circumstance (Pacey, Ch. 2). As a result, technologies are in constant interaction with social systems and structures. They are not merely tools to be used by us. *Technology shapes us and our social and ecological world as much as we shape technology.*

Technology shapes the spaces we inhabit. In our homes, businesses, and public buildings almost every aspect of the physical space is structured by technology. This has social implications. It influences who we see and interact with, as well as the conditions under which those interactions occur. Technology also shapes broader spaces – for example, many cities and towns are organized in ways that accommodate themselves to car travel. Entire geographical areas – e.g. the Midwestern United States – have been transformed by technology. Where there were once vast prairies and woodlands, there are now vast farmlands. This impacts who lives there, what they do, how they relate to each other, and what they value. Social interactions and perspectives are structured by the types of places we inhabit, and the places we inhabit are shaped by technology.

Technology also shapes our conceptions of sociability – i.e. how we conceive of social life and what constitutes social relationships. Perhaps the clearest example of this is the impact of information technologies on social interactions. Cell phones, web chatting, social networking, massively multiplayer online gaming, and virtual realities have opened up new forms of social interaction and new types of social relationships. As a result of these technologies, physical proximity is less and less a crucial component of meaningful social interaction. We also are almost always able to reach our friends and family, and plans are made in real time (as opposed to being set in advance). These technologies have extended space and compressed time with respect to social interaction. They have altered how we spend time with people, as well as who we can spend time with. They have transformed social worlds.

Information technologies have also transformed social institutions and organizations. For example, college students and faculty interact with each other and access and exchange information much differently than they did 25 years ago, before email, the internet, personal computers, and PowerPoint. News agencies operate and disseminate information dramatically differently. Government services are accessed and delivered differently. The examples could go on and on.

All of this alters our expectations. We expect to access information quickly and easily. We expect to be able to reach people at any time. We expect government to make information available and to be transparent. We expect that people will be able to learn more about us (and to do so more quickly) than before. We expect to be able to rapidly travel large distances. We also expect to live longer, healthier, more comfortable lives than people have previously. Life expectancy has increased by over a third in technologized nations over the past 100 years, and people expect to be healthy, comfortable and active until the end of life. These are just some of the diverse ways in which technology impacts values and valuing.

Finally, technology shapes our daily lives and activities. From the moment we wake up, we are dealing with technology and moving through a world configured by technology. Many of us spend large amounts of our days looking at a monitor and punching buttons. Others spend it driving vehicles, or using manufacturing technologies to produce still more technologies. Many of us take our recreation in ways that involve technology – e.g. television, video games, off-road vehicles, and geocaching. As already discussed, technology impacts how we interact with each other and the natural environment, as well as our perspectives and expectations. All of us, almost all the time, are interacting with technology in myriad and significant ways. For all of these reasons, technology is much more than a value neutral tool (Pacey, Ch. 2; Winner, Ch. 4).

1.2 The technological animal

Technology is so socially and ecological significant because it is fundamental to and inseparable from our cultural way of life. Our capacity as cultural animals distinguishes us from all other species. Many species, such as starlings and dolphins, have complex communications systems. Many species, such as honey bees and meerkat, have elaborate social systems. Many species, such as elephants and octopi, exhibit social learning and tool use. However, no other species that we are aware of innovates, accumulates, and transmits ideas, information and practices on the scale or at the rate that we do. A thermostat is a far more complex tool than anything found in the nonhuman world – let alone a smart phone or a space shuttle. A university is far more complex than any social structure found in nature – let alone a democratic state or the Catholic Church. Moreover, our social systems and tools change much more rapidly than does anything found in the nonhuman world. For example, the way universities function today – online libraries, smart classrooms, distance learning – is much different from how they functioned prior to widespread personal computing, digitization of information, and the internet. In comparison, the social systems of wolves and the tool use of chimpanzees – two of the most psychologically complex nonhuman species – have changed very little over that time.

We, members of the species *Homo sapiens*, have a characteristic way of going about the world, and that is the cultural way. We have the capacity, far greater than that of any other species, to imagine how the world might otherwise be, to deliberate on whether we ought to try to bring those alternatives about, to devise and implement strategies for realizing those alternatives we judge to be desirable, and to disseminate them (through teaching and learning) if they prove to be successful. Our way of life is characterized by our comparatively large capacity for gathering information, social interactions, moral agency, and technology: *we are the cultural animal*. In fact, our life form is only possible because of our capacity for culture. A human being alone, without social cooperation and without technology, would have difficulty surviving very long in any environment.

The basis for our comparatively large cultural capacity is our biology. The robust psychological and cognitive abilities that make culture possible arise from the features of our brains, and we have the brains that we do (and not the brains of rattlesnakes or chickadees) because of our DNA. Indeed, the DNA that "codes" for brains like ours evolved, in part, because our increasingly large capacity for culture was fitness enhancing under obtaining environmental conditions, and other parts of human biology (such as skull size and language capabilities) co-evolved with the cultural capacities.

Although culture is made possible by our biology, the content of culture is not determined by biology. Cultural diversity is not the product of biological differences between groups of people, but the result of the development of language, social systems, and technology by (relatively) independent populations over time. Cultural evolution is influenced by the environments in which populations live, the resources available to them, and their interactions with other populations (e.g., trade and conflict). Cultural innovation and dissemination is much more rapid than is biological evolution, so since robust culture has emerged, cultural evolution has accounted for most of the changes in the way people live. Indeed, the "ages" of human history – stone, bronze, iron – are often marked by the technologies in use; and the great "revolutions" in human history – agricultural, Copernican, industrial, Darwinian, information – refer to cultural innovations in techno-social and conceptual systems, rather than biological transitions.

Due the pace of technological innovation since the industrial revolution, as well as other factors such as globalization (which is itself enabled by technological innovation), technological change is arguably now the most significant driver of cultural change (Kurzweil, Ch. 26). To be sure, technological innovation is shaped by social systems and ideas (just as social systems and ideas are shaped by technology). But there can be no doubt that technology is restructuring our social, ecological, and personal worlds at an increasingly rapid rate.

2. THEMES IN THE ETHICS OF TECHNOLOGY

As we have seen, technology is inseparable from human life. It structures and mediates our social and ecological worlds. It increasingly drives social and ecological change. However, technology is not one thing. There are different types of technologies and applications. The ethics of emerging technologies is not about whether to have technological innovation at all. It is about which technologies to promote, which to discourage, and how to develop and disseminate them to promote human flourishing in just and ecologically sensitive ways. Determining this requires being able to analyze the quite divergent social and ecological profiles of emerging technologies. We cannot make informed decisions regarding which technologies and applications to encourage, how to optimize their designs (from a social and ecological perspective), or how to regulate them unless we characterize their social, ethical, and ecological dimensions.

In this section, I review prominent themes in the ethics of emerging technologies. Among them are several critical perspectives and concerns that, taken together, constitute a robust set of resources for analyzing and assessing emerging technologies.

2.1 The innovation presumption: liberty, optimism, and inevitability

Ethical evaluation of emerging technologies tends to focus on what might be problematic about them. The reason for this is typically not luddism (i.e. a general opposition to new technologies), but rather that there is a presumption in favor of new technologies. Given this presumption, the question is not 'Why should we pursue or permit this new technology?', but rather 'Are there any good reasons not to develop it?' and 'Are there any concerns that need to be addressed in its development and dissemination?'

There are three considerations that provide the basis for the presumption in favor of invention, adoption, and use of emerging technologies (the *innovation presumption*). The first is *liberty*. This is the idea that people ought to be permitted to do as they like, so long as it is not harmful to others or otherwise socially or ecologically problematic.

The second basis for the innovation presumption is *technological optimism*. Technological innovations have, in general, increased the longevity, health, comfort, and opportunity in the lives of those who have access to them. This is why people are so keen to adopt new technologies. Given this, it seems as if we ought not only allow, but also encourage technological innovation and adoption. The fewer impediments to invention and dissemination, the sooner further technological innovations can improve human lives.

The third basis for the innovation presumption is *technological determinism*. As discussed earlier, technological innovation is crucial to our cultural way of life. Moreover, the rate of technological innovation has continually accelerated, and there are historically very few cases of relinquishment – i.e. societies that have forgone technological innovation. If this is right, then it makes little sense to ask whether we should support or restrict technological innovation, since it is inevitable.

Taken together, appeals to liberty, technological optimism, and technological determinism provide some support for an innovation presumption. However, it is critical to the ethics of emerging technologies that their significance not be overstated, since doing so closes off rather than advances ethical analysis and evaluation.

As indicated above, liberty is not the right to do whatever one likes. Appeals to liberty do not justify human rights violations or ecological degradation, for example. Moreover, governments, particularly those that are democratically elected, are empowered to limit individual choices in order to promote the public good. (There is reasonable disagreement about the extent to which governments ought to have that power.) So, while liberty does support an innovation presumption, it is still necessary to determine whether technological research programs or applications are problematic in ways that justify restrictions or regulations.

Furthermore, the fact that a person has a right to do something does not imply that she ought to do it. People have the right to play massively multiplayer online games for eight hours each day, but doing so is not good for them. In fact, people often have the right to do things that are morally problematic. Parents have the right to install spyware on their children's computers, while telling them that they are not doing so. But it would nevertheless be dishonest, untrusting, and a privacy violation for them to do so (unless special circumstances obtain). Again, liberty only goes so far, and it is not always an overriding consideration when it comes to the ethics of emerging technologies (Kass, Ch. 6; de Melo-Martin, Ch. 7; Liao, Ch. 8).

The view that technological innovation and dissemination improves people's lives also must be qualified. It is true that people live longer, healthier, more comfortable lives today in highly technologized countries than they have at any other point in human history. However, this is not just the product of technology. Ideas, such as democracy and human rights, and social institutions, such as universities and governments, have also played a significant role. Moreover, technological innovations often have had very serious problems and costs that social institutions have had to address. In the United States, for example, an environmental movement, a host of environmental laws, and an Environmental Protection Agency have been needed to address the detrimental ecological and human health effects of technological development – e.g. pollution, resource depletion, and biodiversity loss. Similarly, a labor movement, labor laws, and Labor Department continue to be needed to promote workplace safety and prevent workplace abuses. The United States is not exceptional in these respects; most highly technologized countries have had similar social movements and have institutions with similar responsibilities. It is often only after tremendous effort, sacrifice, and social innovation that the detrimental aspects of technology are addressed.

Moreover, it is not clear that our current levels of consumption are sustainable or just given the finitude of planetary resources and a global population that is now over seven billion (Cafaro, Ch. 28). Strong technological optimists are confident that further technological innovations will help us to address our natural resource challenges, but so far that optimism is unsupported. We have not seen safe and effective solutions to climate change, top soil degradation, desertification, and fresh water shortages, for example. And the three billion people in the world who live on less than $2.50US ppp/day ('ppp' stands for 'purchasing power parity', that is the equivalent of what $2.50 can purchase in the United States) have not benefitted nearly so much from modern technological innovations. In many cases, they have suffered from them – for example, by having their labor and natural resources exploited. These considerations demonstrate that technological innovation and dissemination is not inevitably conducive to human and nonhuman flourishing. Consideration of and responsiveness to the social and ethical challenges of emerging technologies is therefore crucial, even given the very large potential for new technologies to improve human lives.

As with liberty and technological optimism, claims about technological determinism have a kernel of merit but must also be highly qualified. The historical record and a proper understanding of the role of technology in our way of life does support the view that technological innovation and dissemination will continue. However, what the particular innovations will be, who will have access to them, and how they will be used is not at all determined. For example, it may have been largely inevitable that countries with the capacity to do so would begin to develop space programs. But the decision to have the United States Space Program be run by a civilian, scientific organization, rather than by a military one has been crucial to which space technologies have been developed, not to mention the geopolitical implications of their development. Ethically informed policies, regulations, and designs can and do shape the development of emerging technologies.

It should also be noted that ethical concerns do sometimes result in restricted use of technologies. This is the case with genetically modified crops having only limited adoption in Europe, with ozone depleting chemicals being phased out under the Montreal Protocol, and with national prohibitions on human reproductive cloning, for example. Technological innovation and dissemination is not going to be relinquished on a large scale (there are, of course, particular societies that do forgo it to some extent, such as the Amish), but particular technologies often are prohibited or severely restricted. The truth of technological determinism is in historical trajectories and generalities. It in no way undermines the importance of social and ethical evaluation of particular emerging technologies (Winner, Ch. 4; Kass, Ch. 6).

The foregoing shows that, while there might be a presumption in favor of technological innovation and dissemination, it is not nearly as strong as is often supposed. Moreover, it does not at all diminish the importance of thorough ethical evaluation of emerging technologies in order to inform judgments about which technologies to promote, which to discourage, and how to develop and disseminate them.

2.2 Situated technology

As discussed in Section One, technology is always historically situated. It is always located at a time, in a place, and within a set of practices and institutions. As a result, it is not possible to identify the full range of social and ethical issues raised by a technology merely by reflecting on the technology as such, or the distinctive features of the technology, abstracted away from its context. For example, there is no way to determine, just by considering the distinctive features of human enhancement technologies, whether they are likely to exacerbate or diminish social injustices. One must also know about the social structures and systems that will enable or frustrate access to the technologies, as well as what sorts of competitive advantages access to the technologies are likely to impart (Garcia and Sandler, Ch. 17). Similarly, one cannot know how a particular information technology, such as internet browsers, RFID chips or cell phones, might challenge people's privacy without knowing what laws and institutions are (or are not) in place to prevent capture, dissemination, and use of the information (Stanley and Steinhardt, Ch. 18; van den Hoven, Ch. 19). It is, of course, crucial to know what the technologies are, and how they work, but that is not enough. Developing and implementing emerging technologies in ways that are just and sustainable can only be accomplished if the institutional, cultural, and ecological contexts of the technologies are carefully considered (Pacey, Ch. 2).

Attending to the social and ecological situatedness of a technology is also crucial to designing effective technology (Pacey, Ch. 2; van den Hoven, Ch. 19; Spinello, Ch. 20). Consider, for example, the One Laptop Per Child initiative, which had the goal of producing inexpensive laptops (around $100 each) "to empower the children of developing countries to learn" and to "create educational opportunities for the world's poorest children" (http://laptop.org/en/vision/index.shtml). Given this goal, producing a $100 laptop was not sufficient for success. Not only did the laptop need to come close to the target price point, it needed to function well in the conditions in which the children live, which often include unreliable or no access to electricity, unreliable or no internet access, and little if any technical support. As a result, the computers were designed so that they could be charged on alternate power sources, such as car batteries. They have no hard drive and only two internal cables. Their software is open source. They have a long range antenna. And the keyboards are sealed with a rubber membrane to protect them from humidity.

When cultural and ecological context is not adequately attended to, technologies are much more likely to be ineffective or to have detrimental social and ecological impacts (Shiva, Ch. 32). Moreover, it will often be necessary to address aspects of the social context – for example, by providing training or support – into which a technology is being introduced in order for it to be successful (Pacey, Ch. 2). Careful attention to social context might also reveal that the goal a technology aims to accomplish – e.g. educational, public health, or ecological improvement – can be more easily and efficiently accomplished by less technologically sophisticated means, such as implementation of best practices, institutional reform, education, or providing access to already established technologies (Cafaro, Ch. 28; Hamilton, Ch. 29; Thompson, Ch. 34).

2.3 Lifecycle (or Cradle to Grave)

When evaluating an emerging technology or application it is crucial to consider its entire lifecycle, from extraction of the natural resources that are used in its production to where it ends up when it is no longer used. The reason for this is that a technology might appear socially or ecologically benign, or even beneficial, when one focuses only on the use portion of its lifecycle, when in fact it raises significant socially and ecologically issues when one looks at its production or end of life disposal. For example, a comprehensive ethical analysis of cell phones involves considering not only how mobile communication has impacted users and the social institutions (e.g. the workplace) and practices (e.g. social networking) they participate in. It also includes attending to the materials that are used in cell phone production, some of which are relative scarce minerals the control of which has contributed to violent conflicts in parts of Africa. It includes attending to the manufacturing conditions where the phones are assembled, which in some cases have involved human rights violations. It includes attending to the fate of hazardous components when the phones are discarded, which in some cases has involved environmental release and human exposure.

There is nothing distinctive about mobile phones that gives rise to the need for comprehensive lifecycle evaluation. All technologies are made of raw materials that must be extracted, transported, and processed or refined into a usable form. All technologies must be manufactured or constructed, which involves energy and other inputs (e.g. chemicals). All technologies must be transported to consumers. And all used technologies must be disposed of in one way or another. Thus, the use stage of technology constitutes only part of its ecological and social profile. Worker health and safety, environmental impacts of production, greenhouse gases emitted in transportation, effects of extraction activities on local communities, and technological displacement of prior practices must also be considered, for example (Elliott, Ch. 27). Ethanol might burn more cleanly and with fewer greenhouse gas emissions than gasoline; however, this does not tell us anything about the biodiversity losses associated with clearing forests to grow oil palms for ethanol, the impacts on food availability and food prices of using agricultural lands for ethanol inputs rather than food crops, how much energy is used in the production of ethanol (and what the sources and so associated emissions of that energy are), or the effects of ethanol production on farming communities.

The point of the foregoing is not that ethanol and mobile phones are overall objectionable or should be eliminated. It is that optimizing these technologies from a social and ecological perspective (as opposed to a bare technical one) requires evaluating them in a situated way and over their lifecycles. Only then can the full range of challenges and opportunities associated with them be identified, evaluated and addressed. It may be that some emerging technologies are sufficiently risky, ethically objectionable, or otherwise problematic that relinquishment of them is the most justified course of action (Kass, Ch. 6; Hamilton, Ch. 29; Shiva, Ch. 32). But in most cases, the ethical challenge is how to develop them responsibly – i.e. in ways that respect rights, are consistent with principles of justice, and promote human and nonhuman flourishing. Situated lifecycle analysis is crucial to this.

2.4 Power

Technology affords power to those who have access to it. This is perhaps clearest with respect to what we might call *efficient power*, or the power to do or accomplish things. Guns increase the capacity to kill or injure. Washing machines increase the capacity to clean clothes. Steam-shovels increase the capacity to clear land.

Internet access increases the capacity to gather information and communicate with others. Synthetic genomics increases the capacity to modify organisms. Magnetic resonance imaging (MRI) increases the capacity to visualize structures internal to the body. Technology enables individuals, groups of people, and organizations to do things that they would otherwise be capable of doing. It empowers by increasing the scope of our agency (Jonas, Ch. 3). A prominent theme in the ethics of emerging technologies is the need to take responsibility for the power that modern technology provides and to develop ethics appropriate for that power. For example, it is our technologically enabled capacity to impact the natural environment and distant people (spatially and temporally) that makes environmental ethics, global ethics, and future generation ethics so important (Jonas, Ch. 3).

Efficient or material power often translates into *social or political power* (Lin et al., Ch. 23; Shiva, Ch. 32). Social and political power is relational. It is power relative to others within a particular domain or activity. For example, possessing a technologically sophisticated military or nuclear weapon capability empowers a nation in the domain of international negotiations. A more efficient water pump empowers a farmer in the domain of resource competition (particularly if the farmer's neighbors lack the technology). Here is a slightly more detailed example. Among the distinctive feature of digital media is that it can be easily, inexpensively, and reliably copied and disseminated, without loss of quality. It materially empowers consumers to share music and videos online. As a result, their social or political power is increased relative to music distribution companies and record labels. Although record companies have tried to mitigate consumers' power with legal and technological measures (for example, anti-piracy legislation and digital rights management), the power provided to consumers (and musicians) by digital media has caused tremendous change in the music industry (Spinello, Ch. 20).

As the digital media example illustrates, the increase in social and political power provided by a novel technology often comes with a correlative decrease in the power of others. The fact that one farmer can bring water up more quickly or from greater depths than can another farmer disempowers the farmer that does not have access to the technology and so cannot irrigate as reliably or extensively. The fact that Google is able to track individual browsing histories disempowers advertising companies that do not have access to personalized data, and so cannot as effectively target advertisements.

Because technology provides material power that often translates into differential social and political power, a comprehensive assessment of the social and ethical dimensions of an emerging technology requires conducting a *power analysis* of the technology (and constitutes another reason that technology is not value neutral). One must try to determine, given the features of the technology and its situatedness: Who is likely to control and/or have access to the technology? Who is likely to be empowered or disempowered by the technology? How are they likely to be empowered or disempowered? Whose interests are promoted by the technology and whose are not (and whose may be compromised)?

As with lifecycle analysis, the point of a power analysis is not primarily to determine whether a technology should be pursued or permitted at all. It is crucial to identifying how to design technologies and address aspects of their social and ecological context in order to ensure that they contribute to, rather than undermine, justice, autonomy, sustainability, and flourishing.

2.5 Form of life

As we have seen, technology shapes us and our relationships by configuring our social and ecological worlds. As a result, new technologies often involve a change in *form of life*. When we adopt a technology (or have one imposed on us, as is often the case) it provides not only possibilities and power, but also responsibilities, requirements, incentives, perspectives, relationships and constraints (Winner, Ch. 4). This is yet another respect in which technology is value laden, rather than value neutral.

Here is a non-technological example to help illustrate the idea of a form of life: adopting a dog. When a person adopts a dog, she does not merely get a furry four-legged canine. She also adopts a set of responsibilities, to provide care for the dog and to ensure that it does not harm others. As a result, she must organize her life in certain ways – for example, adjusting her schedule so that the dog is not at home alone too long. It also places constraints on her – for example, she cannot live in places that do not allow dogs and must make arrangements for others to take care of the dog if she is travelling. Having a dog is also likely to result in her going to new places (such as dog parks), meeting new people (such as other dog owners), and learning and caring about new things (such as Lyme disease and leash laws). In these and many other ways, the decision to adopt a dog is a decision to adopt a form of life, with economic, lifestyle, relationship, perspectival, responsibility, and opportunity dimensions.

Technology adoption also often has form of life implications, at both the individual and societal levels. A classic example of this is the widespread adoption of cars (Winner, Ch. 4). It required cities to be designed in order to accommodate them, homes constructed with places to park them, and roads laid where people wanted to take them. An infrastructure to support them was also necessary – e.g. refineries, filling stations, repair shops, traffic laws, and licensing systems. Moreover, experiencing the world from a moving automobile is quite different from doing so as a pedestrian. It alters what you perceive, who you interact with (and how you do so), what you are attentive to, and what you care about. The automobile brought with it a form of life, with enormously significant spacial, perspectival, economic, geopolitical, and lifestyle dimensions.

We find the same thing with many more recent and emerging technologies – though of course not always to so profound an extent as automobiles. Mobile computing has changed where and how we can work; it has required a supporting infrastructure (e.g. internet access and electrical outlets); it has increased vulnerability to privacy and security violations; it has altered personal and professional interactions; and it has changed how we take our recreation. The genomics revolution has changed the types of research questions that can be asked, how diagnosis of diseases and illnesses takes place, how health and sickness are conceptualized, the types of treatments that are possible, and the structure of patient–provider interactions and relationships. Genetically engineered organisms in agriculture have promoted a particular type of agricultural practice (industrial, high-input monoculture), increased the power of transnational seed companies, and displaced traditional farming practices, technologies, and traditions (Shiva, Ch. 32).

When a person, community or culture chooses to adopt a new type of technology, or when it is imposed on them by social or economic pressures or authorities, they very often are adopting a new (or modifying a prior) form of life as well. Therefore, when analyzing an emerging technology it is necessary to consider how it might impact such things as how we spend our time, who we interact with (and how we do so), our dependencies and vulnerabilities, what values we attend to (e.g. aesthetic, cultural, efficiency, or economic), and our perspectives more generally. Only then are we able to discuss in an informed way whether the changes in how we live that the technology will bring about are desirable, and how to incorporate them into our lives so that they are so. This applies to everything from whether to join Facebook to whether to genetically enhance one's children (President's Council on Bioethics, Ch. 14).

2.6 Common concerns regarding emerging technologies: extrinsic concerns

Ethicists working on emerging technologies often make a distinction between extrinsic and intrinsic concerns regarding them (Comstock, Ch. 31; Preston, Ch. 36; Bedau and Triant, Ch. 37). *Extrinsic concerns* refer to concerns about possible problematic outcomes or consequences of a technology – for example, that its widespread adoption would result in human health problems, ecological degradation, unjust distribution of risks and benefits, or human rights violations. *Intrinsic concerns* refer to objections to the technology itself, independent of what its impacts might be. For example, some people are opposed to transgenic organisms on the grounds that their creation involves crossing species boundaries; and some people are opposed to embryonic stem cell research on the grounds that it violates the moral status of stem cells or is disrespectful of human life.

Extrinsic and intrinsic concerns regarding particular technologies are addressed at length in the readings. Here I just briefly introduce the most prominent types of concerns, and indicate the chapters in which they are discussed. The primary extrinsic concerns are these:

2.6.1 Environment, Health, and Safety (EHS)

EHS concerns are those to do with the possible negative impacts of technology on human welfare and the nonhuman environment. They are frequently raised in connection with pollutants, as well as biotechnologies. Workplace safety, consumer safety, public health, and ecological integrity (including concerns about biodiversity) all fall within EHS. The negative EHS impacts of an emerging technology are typically unintended and unwanted, though there are exceptions (e.g. bioterrorism). EHS concerns are addressed in chapters on nanomaterials (Elliott, Ch. 27), nanomedicine (Allhoff, Ch. 11), genetically modified crops (Comstock, Ch. 31), synthetic genomics (Garfinkle and Knowles, Ch. 35; Bedau and Triant, Ch. 37), neurotechnologies (Glannon, Ch. 12), global climate change (Cafaro, Ch. 28), geoengineering (Hamilton, Ch. 29), and Robotics (Lin et al., Ch. 23; Wallach, Ch. 24).

2.6.2 Justice, access, and equality

Emerging technologies often distribute their burdens and benefits unequally; they often empower some people (or institutions) while disempowering or disadvantaging others; and there is often differential access to them. They therefore raise significant concerns about justice across a wide range of domains – e.g. economic, political, and social. This is particularly so when the technologies are introduced into contexts that already include unjust inequalities, and when the introduction is done through market mechanisms that favor access for those that are socially and economically advantaged. Justice and access issues, particularly as they pertain to class and gender, are addressed in chapters on cognitive enhancement (Garcia and Sandler, Ch. 17), intellectual property and medicine (Ravvin, Ch. 13; Allhoff, Ch. 11), agricultural biotechnologies (Shiva, Ch. 32), information and computer technologies (Himma and Bottis, Ch. 22), genetic engineering (de Melo-Martin and Gillis, Ch. 9), virtual reality (Brey, Ch. 21), and reproductive technologies (Frith, Ch. 5; de Melo-Martin, Ch. 7).

2.6.3 Individual rights and liberties

EHS considerations include concerns about human health and environmental impacts. Sometimes those impacts constitute rights violations – for example, if they displace or kill people. However, some harms and wrongs are independent from or not reducible to their EHS impacts. For example, forcibly taking someone's property without adequate authority or cause violates their property rights even if it is not harmful to their health. As discussed above, individual liberties are often appealed to in the ethics of technology as a basis for justifying innovation and access. For example, procreative liberty is thought by some to justify access to genetic enhancement and cloning technologies. However, concerns are also often raised regarding the potential for a technology to violate rights or liberties. For example, a primary concern about many information technologies is their potential to compromise the right to privacy, and concerns about violating the right to informed consent often arise in the context of testing new medical technologies. Among the chapters that address rights and liberties issues are those on information technologies (Stanley and Steinhardt, Ch. 18; van den Hoven, Ch. 19), reproductive technologies (Frith, Ch. 5; de Melo-Martin, Ch. 7), and genetic enhancement (Kass, Ch. 6; Liao, Ch. 8).

2.6.4 Autonomy, authenticity, and identity

Because technology confers or constitutes power, and shapes the conditions of human experience (including our perspectives and values), emerging technologies often raise concerns about autonomy, authenticity, and identity. One type of concern is that people can have technologies imposed on them against their will or without their consent. This could be done through authority – for example, if employers were to require employees to take pharmaceuticals in order to improve their performance. Or it could be done through social pressure – for example, if parents feel they must select the genetic traits of their children so that they will be able to "keep up." Another type of concern is that technologies will be used in manipulative ways – for example, when kids are cyber-bullied into doing things they would not otherwise do. Issues related to autonomy and authenticity are discussed in chapters on reproductive technologies (Frith, Ch. 5), stem cell research (de Melo-Martin and Gillis, Ch. 9), and virtual reality (Brey, Ch. 21). A related type of concern is that technologies could so change or alter people that their biographical identity will be disrupted. This concern primarily arises regarding technologies that impact the brain and mental functions, such as neuro-technologies, (Glannon, Ch. 12), enhancement technologies (President's Council on Bioethics, Ch. 14; Bostrom, Ch. 15; Douglas, Ch. 16), and brain–machine integration technologies (Kurzweil, Ch. 26).

2.6.5 Dual use

As mentioned above, problematic outcomes of the development and dissemination of emerging technologies are almost always unintended byproducts (for this reason they are often referred to as unintended effects, secondary effects, or collateral effects). Therefore, when analyzing and evaluating the social and ecological dimensions of a novel technology or application, one must consider unintended impacts, as well as the possibility of unintended uses. Many technologies are *dual use* technologies in that they can be effectively employed in ways other than their intended or designed use. This is frequently the case with drugs. For example, methylphenidate (Ritalin) and modifinil (Provigil) treat ADHD and narcolepsy, respectively, but they are also widely used by healthy people to enhance focus and wakefulness. Restricting a technology to its intended use once it enters the marketplace is difficult, and this needs to inform evaluations of it, particularly as it concerns technology and regulatory design. The dual use challenge of emerging technologies is discussed in chapters on synthetic genomics (Garfinkle and Knowles, Ch. 35; Bedau and Triant, Ch. 37), information technologies (Stanley and Steinhardt, Ch. 18; van den Hoven, Ch. 19), and robotics (Lin et al., Ch. 23; Wallach, Ch. 24).

Extrinsic concerns regarding emerging technologies are often cast in terms of costs and benefits. However, as the forgoing review indicates, there are in fact several varieties of extrinsic concerns. Moreover, they can carry very different normative weight or significance. For example, rights violations are often considered more serious than economic losses. As a result, thorough ethical evaluation of emerging technologies requires not only identifying extrinsic concerns, but also attending to their different normative features or logics.

Because extrinsic concerns are not to do with the technology itself, the same technology can have different extrinsic profiles in different contexts. For example, a technology might perpetuate injustice if it is introduced into an already unjust context that leads to differential access to it, but might be justice neutral if the social context were different. Therefore, addressing extrinsic concerns about emerging technologies often requires both designing the technologies in ways that minimize the likelihood of problematic effects through *value sensitive design* (Elliott, Ch. 27; van den Hoven, Ch., 19; Spinello, Ch. 20) and responding to problematic features of their social context (Sandler and Garcia, Ch. 17).

Furthermore, because the same technology (or type of technology) can have a different extrinsic profile in different circumstances, extrinsic considerations typically do not favor either comprehensive acceptance or comprehensive rejection of an emerging technology. *Discriminatory assessment* across contexts is needed, and extrinsic considerations tend to favor *selective endorsement* of them – i.e. endorsement under these or those conditions or if this or that factor is addressed. Of course, the difficult and important ethical work is determining what those conditions and factors are for a given technology, as well as the extent to which they can be met through technology design and social practices or policies.

2.7 Common concerns regarding emerging technologies: intrinsic concerns

Intrinsic concerns regarding an emerging technology are objections to the technology itself, independent of the outcomes or consequences of its creation and use. Intrinsic concerns have primarily been raised regarding bio- and eco-related technologies, though they are applicable as well to other emerging technologies, such as artificial intelligences. The primary (and often interconnected) intrinsic concerns regarding emerging technologies are these:

2.7.1 Playing God

Research on emerging technologies, particularly genetic technologies, is sometimes described as "playing God." The "playing God" language appears theological, and sometimes it is intended that way. However, it is often used non-theologically as well. The idea that "playing God" language is meant to capture is that there are types of activities that it is simply wrong for people to engage in. With respect to genetic technologies and artificial life, the idea is that people ought not to be trying to create novel species or novel life forms. With respect to human embryonic stem cell research, the idea is that embryonic stem cells are not merely material for us to use. In each case, the idea is that the activity is beyond the purview of what is appropriate for humans to be doing. Different proponents of "playing God" concerns will provide different bases for the claim that certain technologies involve extending our agency to activities that are inappropriate; and, again, some will be theological (e.g. divine prohibition) and some secular (e.g. the moral status of embryos). Among the chapters that discuss "playing God" concerns are those on genetically modified crops (Comstock, Ch. 31), synthetic genomics (Bedau and Triant, Ch. 37), and stem cells (de Melo-Martin and Gillis, Ch. 9).

2.7.2 Hubris

The central idea of hubris-oriented concerns is that those who develop and embrace emerging technologies often overestimate their ability to predict what the technologies will do when they are put into use, as well as their ability to address any problematic effects that might arise (particularly when the technologies are being introduced into complex, dynamic and inadequately understood biological and ecological systems). The hubris concern arises regarding the use of geoengineering to address global climate change (Hamilton, Ch. 29), the release of genetically modified organisms into agricultural systems (Comstock, Ch. 31), reproductive cloning (Kass, Ch. 6), and genetically engineering people for enhancement purposes (President's Council on Bioethics, Ch. 14).

The hubris involved in these activities is thought to make them intrinsically problematic (in addition to the extrinsic EHS concerns involved).

2.7.3 Respecting nature

Bio- and eco-related technologies also often are objected to on the grounds that they destroy naturalness, violate nature, or are disrespectful of the natural. The idea is that there is value or normativity in naturalness – i.e. the biological and ecological world independent of human design and control – that the technologies undermine or violate. For example, human enhancement is sometimes objected to on the grounds that it aims to modify our given, human nature (President's Council on Bioethics, Ch. 14; Liao, Ch. 8; Bostrom, Ch. 15), and synthetic genomics is sometimes objected to on the grounds that the aim is to create organisms without a natural history (Preston, Ch. 36). Naturalness concerns have also been raised about human reproductive cloning (Kass, Ch. 6), ecological restoration (Minteer and Collins, Ch. 30), and geoengineering (Hamilton, Ch. 29).

What these intrinsic concerns or objections have in common is that they are based on features of the technologies themselves. In some cases, intrinsic objections are taken to be overriding – i.e. if there is something intrinsically objectionable about a technology then it ought not to be pursued no matter what. In other cases, intrinsic objections are taken to be defeasible – i.e. they count against the development and use of a technology, but can be overridden by extrinsic considerations. Whether intrinsic objections are taken to be overriding or not often depends upon the type of ethical theory in which the objection is situated – for example, whether rightness of action is understood in terms of outcomes, adhering to rules, or acting virtuously. Thus, the normative significance of intrinsic (and extrinsic) objections depends not only on the type of objection that they are, and whether they are legitimate or reasonable concerns, but also on the ethical system in which they are located (and whether that system is well justified). I briefly discuss different types of ethical systems in Section 3.

2.8   RESPONSIBLE DEVELOPMENT

Everyone – researchers, industry, government, and citizens – is in favor of responsible development of emerging technologies; that is, promoting innovation and commercialization while also addressing legitimate social and ethical concerns. However, there are several factors that make accomplishing responsible development difficult in practice. One is the high rate of technological innovation, which frequently exceeds that of policy and regulatory processes. This difficulty is compounded by the fact that many oversight institutions lack capacity – e.g. they are underfunded or understaffed relative to their regulatory mandates – and often are not as familiar with novel technologies as are those they are regulating. Because of this, soft law and non-legislative resources are crucial to responsible development – for example, product liability, development and promotion of best practices, professional standards and licensing, codes of conduct, and training and education (Garfinkle and Knowles, Ch. 35). However, while these avoid some of the time-lag and information deficit challenges associated with hard regulation, they typically do not have the same authority and sanction behind them (market and liability incentives can be an exception).

A second challenge to responsible development is what might be called *the lure of the technological*. Technological optimism and beliefs about technological determinism favor uncritical adoption of emerging technologies. They also tend to favor technologically-oriented solutions to human health and ecological problems. After all, a technological fix holds the promise of addressing a problem – from high cholesterol to global climate change – without requiring individual sacrifices or significant institutional, cultural, or lifestyle changes (Pacey, Ch. 2; Hamilton, Ch. 29).

Yet another challenge to responsible development is that stakeholders in an emerging technology may have quite different views about what responsible development involves. For example, many non-governmental organizations (NGOs) advocate for premarket regulatory approval for all new industrial chemicals – i.e. the chemicals must be demonstrated to be safe to people and the environment before being used – whereas industry often advocates for post-market monitoring and enforcement instead – i.e. a new chemical can be brought onto market without a rigorous approval process, but if it is shown to be hazardous then it may be pulled from the market and the manufacturer will be liable for damages. Similarly, some researchers, industry and patient advocacy groups favor an expedited process for approving new medical drugs and devices on the grounds that delays to market prevent treatments from getting to those who need them (particularly when the technologies are potentially lifesaving). Others argue that it is more important to extensively demonstrate that a drug or device is safe and effective, even if this results in a longer time to market, since it is crucial that people have confidence in approved technologies and that they not cause further health problems. This issue – how much caution and confidence with which to proceed under conditions of uncertainty – applies to a great many emerging technologies (Bedau and Triant, Ch. 37; Garfinkle and Knowles, Ch. 35; Elliott, Ch. 27; Hamilton, Ch. 29).

Another area in which there are substantive disagreements about what constitutes responsible development is the role of public input in technology policy and regulation. Public consultations and citizen forums on such things as nanotechnology, synthetic genomics, and geoengineering are increasingly common. The motivation is to provide opportunities for citizens to provide input on funding priorities, research goals, and regulatory decisions. However, some argue that such forums are problematic in that many citizens lack an adequate understanding of the technologies, institutions, and political processes at issue. What this and the previous examples indicate is that, in addition to identifying social and ecological opportunities and challenges associated with particular technologies, the ethics of emerging technologies also includes characterizing what responsible development involves: What is the appropriate role of public engagement? How should public funding decisions regarding emerging technologies be made? What standards for safety ought to be met prior to dissemination? How should responsibility for oversight be allocated? And so on.

The rapid pace of technological innovation has led to a widespread call for more anticipatory and agile technology policy and responsible development processes, so that social and ecological issues can be addressed early in technology development (sometimes called *upstream*). Proactive responsible development is contrasted with reactive approaches, in which social and ecological issues are addressed only after they arise. The idea is that we have had sufficient experience with emerging technologies, and have developed sufficient responsible development capacities (e.g. laws, civic organizations, governmental institutions, codes of conduct, and market expectations), that we should now be able to identify potential issues and prevent or mitigate them before they manifest. For example, life cycle analysis is increasingly used by industry to assess the ecological impacts of industrial processes and products to determine where their impacts can be reduced. There is also increased interest in value sensitive design in engineering – i.e. designing technologies in ways that are informed by social and ecological evaluation of them (van den Hoven Ch. 19; Spinello, Ch. 20; Brey, Ch. 21; Elliott, Ch. 27). And regulators and policy makers are employing such means as citizen consultations and mandatory reporting in order to move toward more anticipatory governance.

To be sure, there remain significant challenges to proactive responsible development. Not least of these is the inherent information deficits involved (sometimes referred to as the *Collingridge Dilemma*). On the one hand, the earlier in the development process that one attempts to evaluate the social and ecological aspects of an emerging technology, the less information about the technology, its applications, and its impacts are available. On the other hand, the later one does the evaluation, the more entrenched is the technical design and the more momentum there is to the commercialization process, so the more difficult it is (and fewer opportunities there are) to address social and ecological concerns in the development process. Thus, even (or especially) in proactive responsible development, the issue of how to proceed under conditions of uncertainty and with incomplete information remains crucial.

2.9 ETHICS AND PUBLIC POLICY

A prominent issue in the ethics of emerging technologies is what role peoples' moral or religious concerns ought to play in developing policy and regulation. For example, can a ban on the use of federal funding for embryonic stem cell research be justified on the grounds that many taxpayers believe that such research is disrespectful of human life? Could a broader prohibition on conducting such research be justified?

In democratic societies committed to basic liberties, such as freedom of speech, thought and expression, government is expected to refrain from privileging one worldview over another. This is often referred to as *state neutrality*. Government should not make it more difficult for citizens to live according to one worldview than another – for example, by prohibiting particular religions. It also should not try to promote one worldview over others – for example, by requiring certain ideological commitments in order to hold public office or be part of particular professions. Not all worldviews need to be respected in this way, only those that are reasonable and consistent with basic democratic principles and liberties. Governments do not need to be neutral toward a worldview on which it is permissible to deny basic rights to women, for example. However, for all reasonable worldviews, public policies should not privilege one over others.

The implication of state neutrality for public policy regarding emerging technologies is that it is inappropriate to formulate policy on the basis of moral or theological concerns that are distinctive to particular worldviews. Instead, policies need to be based on fundamental liberal democratic values or else supported by an overlapping consensus among reasonable worldviews. For example, theologically based concerns about violating species boundaries are not a legitimate basis for restricting research on and use of genetically modified organisms (GMOs), since there are many reasonable worldviews on which species boundaries are not ethically significant. Such a ban would therefore violate state neutrality. However, it is permissible to regulate GMOs in order to protect or promote public health, since that is a value that any reasonable worldview would endorse. Moreover, respect for the autonomy of those who believe that cross-species genetic engineering is unethical may favor a policy of labeling foods that contain GMOs. In this way, citizens can make informed judgments about whether to eat them and can live according to their own values.

What the forgoing illustrates is that part of characterizing the ethical profiles of emerging technologies is determining which social and ecological considerations raised by them are policy relevant, and how they are so.

2.10 A FRAMEWORK FOR ETHICAL ANALYSIS OF EMERGING TECHNOLOGIES

Taken together, the critical perspectives and issues discussed in the previous sections (particularly 2.2–2.7) constitute a rich set of resources for analyzing, evaluating, and reflecting upon the personal, social, and ecological dimensions of emerging technologies. Indeed, a fairly comprehensive ethical analysis of an emerging technology can be accomplished by:

A. Identifying any *benefits* the technology might produce (with respect to both human and nonhuman flourishing), including how large the benefits would be and how likely they are to occur.
B. Identifying any *extrinsic concerns* (e.g. EHS, justice-oriented, or rights-based) that the technology may raise, including how likely it is to do so.
C. Conducting a *power analysis* to identify who is empowered and who is disempowered by the technology, as well as how they are empowered or disempowered.
D. Conducting a *form of life analysis* to identify how the technology might restructure the activities in which it is involved, as well as the personal, social, and ecological conditions of our lives.
E. Identifying any *intrinsic concerns* that the technology is likely to raise.
F. Identifying any *alternative approaches* to accomplishing the ends at which the technology aims, including less technologically sophisticated possibilities.

These analyses and issue identifications need to be done over the course of the technology's lifecycle and in a situated way. They need to attend to the distinctive features of the technology, how it differs from prior technologies, and the relevant features of the social and ecological contexts into which it is emerging. They also need to be informed by our experiences with relevantly similar prior technologies (van den Hoven, Ch. 19; Garfinkle and Knowles, Ch. 35; Thompson, Ch. 34).

In many cases, there will be a speculative element to the analyses. When the technologies are not yet fully developed we will not know precisely what their features are or how they work; and when they are not widely disseminated we will not know precisely what their impacts are. As discussed earlier, this information deficit is part of the challenge of responsible development. However, there is a very large difference between informed, well measured anticipation and wild speculation. Good ethical analyses will be as informed as possible about the technology – i.e. what it is, how it works, and who uses it. It will be as informed as possible about the social and ecological context of its development and use – e.g. the relevant regulations, oversight mechanisms, social inequalities, and methods of dissemination. It will be as informed as possible by the study of the social and ethical dimensions of relevant prior technologies, as well as by research on the impacts of the emerging technology (e.g. how sex selection has been used when available, toxicity data on nanomaterials, and public surveys on whether creating artificial life is "playing God").

Conducting these analyses for an emerging technology will provide a profile of the technology's potential to contribute to human flourishing, justice and sustainability, as well as the potential challenges, problems, and costs associated with it. It will also provide the full range of alternatives against which they should be evaluated, including less technologically-oriented ones (Cafaro, Ch. 28; Hamilton, Ch. 29; Thompson, Ch. 34). This, in turn, puts one in a strong position for evaluating the technology, as well as for identifying ways to optimize it (socially and ecologically) through technological design, public policy, and addressing social and cultural factors relevant to its dissemination and use.

3. ETHICAL THEORY AND TERMINOLOGY: A VERY BRIEF PRIMER

Crucial to the ethics of emerging technologies is determining not only which features or impacts of a novel technology are ethically salient, but also how they are salient. As mentioned earlier in this introduction, how a social or ecological concern should be taken into consideration is sensitive to the type of ethical theory that is operative. Ethical theories are systematic accounts of what, why and how things matter, particularly as they relate to deliberations about actions, practices, and policies. For these reasons, the ethics of emerging technologies – both in general and with respect to particular technologies – often involves discussion of ethical theory more generally. The aim of this section is to provide a very brief primer on the elements and types of ethical theories, as well as to introduce some relevant terminology.

3.1 Types of value

The things that matter are the things with value. But there are a variety of different types of value, or ways in which things can matter or have importance. One prominent type of value is *instrumental value*, or the value of something as a means to an end. Novel technologies very often have instrumental value, since they help us to do or accomplish things. They are almost always intended to enable something that is desirable – e.g. human health, security, or economic gain. However, as discussed earlier, they also often have negative unintended byproducts (e.g. pollution), or can be used for detrimental ends (e.g. privacy violations). Thus, both a technology's potential instrumental value and its potential instrumental disvalue are relevant to ethical evaluation of it.

Another prominent type of value is *final value* (or noninstrumental value). Final value, which is also sometimes referred to as intrinsic value, is the value that something has for what it is, or as an end. If something has final value, then its usefulness is not exhaustive of its value. People are commonly thought to have final value. The value of a person is not just to do with how effective they are as a means to ends. Rather, they have value as a person or in virtue of being a person.

Some things have final value because we value them as an end – i.e. they have *subjective final value*. Works of art, landscapes, mementos, religious artifacts, and historical sites are often like this. They have final value because we value them for their beauty, rarity, or history, for example, and not just because they are an effective means to a desired end. Technological accomplishments are frequently valued noninstrumentally – e.g. for their ingenuity, grandeur, or cultural significance. Concerns are also sometimes raised about emerging technologies on the grounds that they compromise something with subjective final value – for example, that a wind turbine energy "farm" would despoil a beautiful (and so valued) landscape or that genetic technologies compromise species purity (which, as discussed earlier, is valued by many people).

Other things have final value in and of themselves, independent of how they are valued by us – i.e. they have *objective final value*. This is typically thought to be the case with people – that is, people have value in virtue of what they are and independent of the valuations of others. Technology usually is not thought to have objective final value, though advances in artificial intelligence and synthetic genomics may require reconsidering this presumption (Basl, Ch. 25). However, technology is often evaluated on the basis of how it would impact entities – e.g. people, species, or nonhuman animals – or states of affairs – e.g. distributions of burdens and benefits – that are thought to have such value. Whether all final value is subjective or some is objective is a contested issue in ethical theory. However, this is not the place to address it. What is important in this context is that there is a difference between instrumental and final (or intrinsic) value, and that there are two possible bases of final value (subjective and objective).

As the foregoing discussion illustrates, there are several varieties of instrumental and final value. Being clear about which types of value are operative is crucial to the ethics of emerging technologies, just as it is in other areas of ethics. The reason is that different types of value have different normative significance. For example, instrumental values often are contingent, replaceable, recreatable, and substitutable, whereas final values often are not; and subjective final values are contingent on the evaluative attitudes of valuers, whereas objective final values are not.

3.2 Types of theories

The idea, discussed earlier, that the same consideration can matter differently depending upon the type of ethical theory in which it is located can be somewhat counterintuitive. Here is an example to illustrate the point. Suppose that the development of a novel medical technology would require extensive testing on nonhuman animals, and that the testing would cause the animals considerable and persistent pain and suffering. Further suppose that nonhuman animals are morally considerable, in that causing them suffering is an ethically relevant consideration. Should the drug's development be supported and allowed to go forward? Here is one possible response: It depends on whether the pain and suffering that is caused by the testing is outweighed by the pain and suffering that the drug would prevent. This is a consequentialist response. The rightness or wrongness of the testing depends upon the balance of the good and bad outcomes that would result (or are expected to result). Thus, it might be permissible to do the testing in a case where the benefits to be gained are sufficiently assured and great (and there is no other way to achieve them with less harm caused), but not in a case where the benefits are not so clear or large. Here is another possible response: Intentionally causing harm to animals as a means to our ends is always wrong. This is a deontological response. The rightness of the practice is determined largely by the features of it – i.e. that it involves using animals in this way – and not by the outcomes. Thus, animal testing is wrong in all cases.

This example shows that knowing that nonhuman animals have final value such that causing them suffering is ethically relevant is not sufficient for ethical evaluation. One must also know how the suffering is relevant to determining what ought to be done. According to *consequentialist* normative theories, what one ought to do, what policy ought to be adopted, or what practices ought to be developed is determined by comparing the outcomes (or expected outcomes) of the different possibilities. The better the outcomes, the more justified is the action, practice, or policy. In *deontological* normative theories, whether an action ought to be done or a policy ought to be adopted is determined by whether it conforms to the operative rules – for example, that it does not violate any human rights or basic principles of justice. In *virtue-oriented* normative theories, an action or policy is justified to the extent that it expresses or hits the target of virtue – e.g. it is compassionate, honest, efficient, and ecologically sensitive. These are the three most prominent types of ethical theories, and they are distinguished by their approaches to evaluation. (There are other types, including atheoretic views.) Consequentialist views prioritize outcomes; deontological views prioritize the features of the action or practice itself; and virtue-oriented views prioritize the character traits that are expressed. This is of course an idealized characterization, and in actuality many ethical theories are hybrids in that they incorporate elements from more than one of them.

Which account of right action and approach to decision making is most justified – i.e. deontological, consequentialism, virtue oriented, some combination, or some other alternative – is another contested issue in ethical theory. Again, it is not an issue that can be addressed here. But it is crucial background. In the discourse on emerging technologies (including in the readings in this textbook), considerations offered for or against a technology will often be situated within a particular type normative theory. For example, a deontological argument against synthetic organisms is developed (Preston, Ch. 36), as is a consequentialist argument for them (Garfinkle and Knowles, Ch. 35). Evaluating them requires not only assessing the arguments, but also considering the broader theoretical frameworks in which they are located.

4. CONCLUSION

Technology continually restructures the conditions of human experience. It shapes our relationships, values, landscapes, and expectations. It alters power relationships. It makes possible new forms of life and displaces previous ones. Moreover, technological innovation and dissemination is accelerating. Perhaps more than ever before, we need to be reflective about how to develop technologies, how to incorporate them into our lives, and how to use them. That is, we need to develop frameworks and resources for evaluating emerging technologies, as well as create spaces and opportunities, both personal and public, for doing so. That is, we need ethics and policies for emerging technologies.

Most of the chapters in this textbook focus on a type of technology or field of application – for example, nanotechnology, synthetic genomics, robotics, therapeutic technologies, agricultural technologies, and enhancement technologies. However, they often employ perspectives, advocate approaches to evaluation, and raise issues that cut across fields and technology types. The aim of this introduction was to highlight those perspectives, approaches and themes, as well as to show how, when taken together, they can provide a robust framework for ethical analysis and evaluation of emerging technologies.

PART

1

GENERAL REFLECTIONS ON ETHICS AND TECHNOLOGY

The readings in this section develop crucial perspectives for ethical analysis and evaluation of emerging technologies. Each of the perspectives – situated (2.2), life-cycle (2.3), power (2.4), and form of life (2.5) – is discussed at length in the Introduction and each plays a prominent role in the ethical reflections on the emerging technologies and applications discussed throughout the textbook.

In "Technology: Practice and Culture," Arnold Pacey draws upon several detailed cases (from snowmobiles in the arctic to water pumps in India) to illustrate the importance of attending to social context when developing and implementing technologies so that they are well fitted to the conditions of their use – i.e. their situated aspects. In doing so he also addresses the need to consider the entire life-cycle of a technology from production, to use and maintenance, to end-of -life disposal.

In "Technology and Responsibility," Hans Jonas emphasizes the relationships between technology, responsibility, and power. Technology increases our power of acting or capacity to bring about change in the world. Therefore, we need a corresponding increase in our understanding of the scope of our responsibility. It is modern technological power that makes the need to develop environmental, global, and future generation ethics so urgent. Moreover, it is necessary to conduct a power analysis on emerging technologies as part of characterizing their ethical profiles.

In "Technologies as Forms of Life," Langdon Winner discusses how technologies define the conditions of human experience by structuring our physical and social worlds. Because of this, the ethics of technology must consider more than just the benefits and costs associated with developing and implementing a technology. It must also consider the ways in which technology reconfigures social institutions, relationships, and possibilities – i.e. our forms of life.

More detailed summaries, as well as a listing of related readings, are located at the start of the chapters.

2 TECHNOLOGY: PRACTICE AND CULTURE[1]

Arnold Pacey

CHAPTER SUMMARY

All technologies, qua technologies, have technical components – the materials, designs, parts, and processes that constitute them, as well as the scientific principles by which they operate. However, all technology is also socially, culturally, ecologically, and politically situated. How effective a technology is in the world, depends not only upon its technical aspects – i.e. on how well it works mechanically, chemically or electrically – but also on how well suited it is to the social and ecological contexts into which it is introduced. In this chapter, Pacey uses several detailed cases studies to emphasize the importance of attending to the cultural and organizational aspects of technology in its design and implementation. In the course of doing so he provides a critique of the view that technology is merely a value neutral tool. He argues that cultural values are embedded within technological design and dissemination.

RELATED READINGS

Introduction: Technology and Society (1.1); Situated Technology (2.2); Lifecycle (2.3); Power (2.4); Responsible Development (2.8)

1. QUESTIONS OF NEUTRALITY

Winter sports in North America gained a new dimension during the 1960s with the introduction of the snowmobile. Ridden like a motorcycle, and having handlebars for steering, this little machine on skis gave people in Canada and the northern United States extra mobility during their long winters. Snowmobile sales doubled annually for a while, and in the boom year of 1970–1 almost half a million were sold. Subsequently the market dropped back, but snowmobiling had established itself, and organized trails branched out from many newly prosperous winter holiday resorts. By 1978, there were several thousand miles of public trails, marked and maintained for snowmobiling, about half in the province of Quebec.

Although other firms had produced small motorized toboggans, the type of snowmobile which achieved this enormous popularity was only really born in

---

[1] This chapter is excerpted from Arnold Pacey (1983) *The Culture of Technology* (Cambridge, MA: MIT Press). It appears here by permission of John Wiley and Sons, Inc.

1959, chiefly on the initiative of Joseph-Armand Bombardier of Valcourt, Quebec.[2] He had experimented with vehicles for travel over snow since the 1920s, and had patented a rubber-and-steel crawler track to drive them. His first commercial success, which enabled his motor repair business to grow into a substantial manufacturing firm, was a machine capable of carrying seven passengers which was on the market from 1936. He had other successes later, but nothing that caught the popular imagination like the little snowmobile of 1959, which other manufacturers were quick to follow up.

However, the use of snowmobiles was not confined to the North American tourist centres. In Sweden, Greenland and the Canadian Arctic, snowmobiles have now become part of the equipment on which many communities depend for their livelihood. In Swedish Lapland they are used for reindeer herding. On Canada's Banks Island they have enabled Eskimo trappers to continue providing their families' cash income from the traditional winter harvest of fox furs.

Such use of the snowmobile by people with markedly different cultures may seem to illustrate an argument very widely advanced in discussions of problems associated with technology. This is the argument which states that technology is culturally, morally and politically neutral – that it provides tools independent of local value-systems which can be used impartially to support quite different kinds of lifestyle.

Thus in the world at large, it is argued that technology is 'essentially amoral, a thing apart from values, an instrument which can be used for good or ill' (Buchanan, 1965, p. 163). So if people in distant countries starve; if infant mortality within the inner cities is persistently high; if we feel threatened by nuclear destruction or more insidiously by the effects of chemical pollution, then all that, it is said, should not be blamed on technology, but on its misuse by politicians, the military, big business and others.

The snowmobile seems the perfect illustration of this argument. Whether used for reindeer herding or for recreation, for ecologically destructive sport, or to earn a basic living, it is the same machine. The engineering principles involved in its operation are universally valid; whether its users are Lapps or Eskimos, Dene (Indian) hunters, Wisconsin sportsmen, Quebecois vacationists, or prospectors from multinational oil companies. And whereas the snowmobile has certainly had a social impact, altering the organization of work in Lapp communities, for example, it has not necessarily influenced basic cultural values. The technology of the snowmobile may thus appear to be something quite independent of the lifestyles of Lapps or Eskimos or Americans.

One look at a modern snowmobile with its fake streamlining and flashy colours suggests another point of view. So does the advertising which portrays virile young men riding the machines with sexy companions, usually blonde and usually riding pillion. The Eskimo who takes a snowmobile on a long expedition in the Arctic quickly discovers more significant discrepancies. With his traditional means of transport, the dog-team and sledge, he could refuel as he went along by hunting for his dogs' food. With the snowmobile he must take an ample supply of fuel and spare parts; he must be skilled at doing his own repairs and

---

[2] See Doyle (1978, pp. 14, 47). On Joseph Armand Bombardier, see Ross (1978, p. 155).

even then he may take a few dogs with him for emergency use if the machine breaks down. A vehicle designed for leisure trips between well-equipped tourist centres presents a completely different set of servicing problems when used for heavier work in more remote areas. One Eskimo 'kept his machine in his tent so it could be warmed up before starting in the morning, and even then was plagued by mechanical failures' (Usher, 1972, p. 173). There are stories of other Eskimos, whose mechanical aptitude is well known, modifying their machines to adapt them better to local use.

So is technology culturally neutral? If we look at the construction of a basic machine and its working principles, the answer seems to be yes. But if we look at the web of human activities surrounding the machine, which include its practical uses, its role as a status symbol, the supply of fuel and spare parts, the organized tourist trails, and the skills of its owners, the answer is clearly no. Looked at in this second way, technology is seen as a part of life, not something that can be kept in a separate compartment. If it is to be of any use, the snowmobile must fit into a pattern of activity which belongs to a particular lifestyle and set of values.

The problem here, as in much public discussion, is that 'technology' has become a catchword with a confusion of different meanings. Correct usage of the word in its original sense seems almost beyond recovery, but consistent distinction between different levels of meaning is both possible and necessary. In medicine, a distinction of the kind required is often made by talking about 'medical practice' when a general term is required, and employing the phrase 'medical science' for the more strictly technical aspects of the subject. Sometimes, references to 'medical practice' only denote the organization necessary to use medical knowledge and

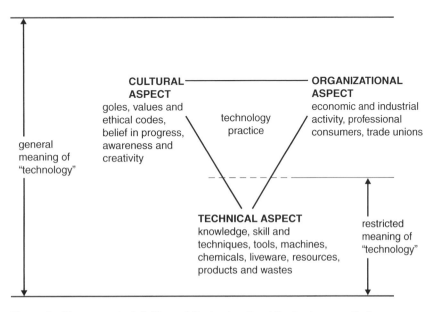

Figure 1   Diagrammatic definitions of "technology" and "technology practice"

skill for treating patients. Sometimes, however, and more usefully, the term refers to the whole activity of medicine, including its basis in technical knowledge, its organization, and its cultural aspects. The latter comprise the doctor's sense of vocation, his personal values and satisfactions, and the ethical code of his profession. Thus 'practice' may be a broad and inclusive concept.

Once this distinction is established, it is clear that although medical practice differs quite markedly from one country to another, medical science consists of knowledge and techniques which are likely to be useful in many countries. It is true that medical science in many western countries is biased by the way that most research is centred on large hospitals. Even so, most of the basic knowledge is widely applicable and relatively independent of local cultures. Similarly, the design of snowmobiles reflects the way technology is practised in an industrialized country – standardized machines are produced which neglect some of the special needs of Eskimos and Lapps. But one can still point to a substratum of knowledge, technique and underlying principle in engineering which has universal validity, and which may be applied anywhere in the world.

We would understand much of this more clearly, I suggest, if the concept of practice were to be used in all branches of technology as it has traditionally been used in medicine. We might then be better able to see which aspects of technology are tied up with cultural values, and which aspects are, in some respects, value-free. We would be better able to appreciate technology as a human activity and as part of life. We might then see it not only as comprising machines, techniques and crisply precise knowledge, but also as involving characteristic patterns of organization and imprecise values.

Medical practice may seem a strange exemplar for the other technologies, distorted as it so often seems to be by the lofty status of the doctor as an expert. But what is striking to anybody more used to engineering is that medicine has at least got concepts and vocabulary which allow vigorous discussion to take place about different ways of serving the community. For example, there are phrases such as 'primary health care' and 'community medicine' which are sometimes emphasized as the kind of medical practice to be encouraged wherever the emphasis on hospital medicine has been pushed too far. There are also some interesting adaptations of the language of medical practice. In parts of Asia, para-medical workers, or para-medics, are now paralleled by 'para-agros' in agriculture, and the Chinese barefoot doctors have inspired the suggestion that barefoot technicians could be recruited to deal with urgent problems in village water supply. But despite these occasional borrowings, discussion about practice in most branches of technology has not progressed very far.

PROBLEMS OF DEFINITION

In defining the concept of technology-practice more precisely, it is necessary to think with some care about its human and social aspect. Those who write about the social relations and social control of technology tend to focus particularly on organization. In particular, their emphasis is on planning and administration, the management of research, systems for regulation of pollution and other abuses, and professional organization among scientists and technologists. These are important topics, but there is a wide range of other human content in technology-practice which such studies often neglect, including personal values and individual experience of technical work.

To bring all these things into a study of technology-practice may seem likely to make it bewilderingly comprehensive. However, by remembering the way in which medical practice has a technical and ethical as well as an organizational element, we can obtain a more orderly view of what technology-practice entails. To many politically-minded people, the *organizational aspect* may seem most crucial. It represents many facets of administration, and public policy; it relates to the activities of designers, engineers, technicians, and production workers, and also concerns the users and consumers of whatever is produced. Many other people, however, identify technology with its *technical aspect*, because that has to do with machines, techniques, knowledge and the essential activity of making things work.

Beyond that, though, there are values which influence the creativity of designers and inventors. These, together with the various beliefs and habits of thinking which are characteristic of technical and scientific activity, can be indicated by talking about an ideological or *cultural aspect* of technology-practice. There is some risk of ambiguity here, because strictly speaking, ideology, organization, technique and tools are all aspects of the culture of a society. But in common speech, culture refers to values, ideas and creative activity, and it is convenient to use the term with this meaning. It is in this sense that the title of this book refers to the cultural aspect of technology-practice.

All these ideas are summarized by Figure 1, in which the whole triangle stands for the concept of technology-practice and the corners represent its organizational, technical and cultural aspects. This diagram is also intended to illustrate how the word technology is sometimes used by people in a restricted sense, and sometimes with a more general meaning. When technology is discussed in the more restricted way, cultural values and organizational factors are regarded as external to it. Technology is then identified entirely with its technical aspects, and the words 'technics' or simply 'technique' might often be more appropriately used. The more general meaning of the word, however, can be equated with technology-practice, which clearly is not value-free and politically neutral, as some people say it should be.

Some formal definitions of technology hover uncertainly between the very general and the more restricted usage. Thus J. K. Galbraith defines technology as 'the systematic application of scientific or other organized knowledge to practical tasks' (Galbraith, 1972, ch. 2). This sounds a fairly narrow definition, but on reading further one finds that Galbraith thinks of technology as an activity involving complex organizations and value-systems. In view of this, other authors have extended Galbraith's wording.

For them a definition which makes explicit the role of people and organizations as well as hardware is one which describes technology as "the application of scientific and other organized knowledge to practical tasks by … ordered systems that involve people and machines" (Naughton, 1979). In most respects, this sums up technology-practice very well. But some branches of technology deal with processes dependent on living organisms. Brewing, sewage treatment and the new biotechnologies are examples. Many people also include aspects of agriculture, nutrition and medicine in their concept of technology. Thus our definition needs to be enlarged further to include 'liveware' as well as hardware; technology-practice is thus the *application of scientific and other knowledge to practical tasks by ordered systems that involve people and organizations, living things and machines.*

This is a definition which to some extent includes science within technology. That is not, of course, the same as saying that science is merely one facet of technology with no purpose of its own. The physicist working on magnetic materials or semiconductors may have an entirely abstract interest in the structure of matter, or in the behaviour of electrons in solids. In that sense, he may think of himself as a pure scientist, with no concern at all for industry and technology. But it is no coincidence that the magnetic materials he works on are precisely those that are used in transformer cores and computer memory devices, and that the semiconductors investigated may be used in microprocessors. The scientist's choice of research subject is inevitably influenced by technological requirements, both through material pressures and also via a climate of opinion about what subjects are worth pursuing. And a great deal of science is like this, with goals that are definitely outside technology-practice, but with a practical function within it.

Given the confusion that surrounds usage of the word 'technology', it is not surprising that there is also confusion about the two adjectives 'technical' and 'technological'. Economists make their own distinction, defining change of technique as a development based on choice from a range of known methods, and technological change as involving fundamentally new discovery or invention. This can lead to a distinctive use of the word 'technical'. However, I shall employ this adjective when I am referring solely to the technical aspects of practice as defined by figure 1. For example, the application of a chemical water treatment to counteract river pollution is described here as a 'technical fix' (not a 'technological fix'). It represents an attempt to solve a problem by means of technique alone, and ignores possible changes in practice that might prevent the dumping of pollutants in the river in the first place.

By contrast, when I discuss developments in the practice of technology which include its organizational aspects, I shall describe these as 'technological developments', indicating that they are not restricted to technical form. The terminology that results from this is usually consistent with everyday usage, though not always with the language of economics.

EXPOSING BACKGROUND VALUES

One problem arising from habitual use of the word technology in its more restricted sense is that some of the wider aspects of technology-practice have come to be entirely forgotten. Thus behind the public debates about resources and the environment, or about world food supplies, there is a tangle of unexamined beliefs and values, and a basic confusion about what technology is for. Even on a practical level, some projects fail to get more than half way to solving the problems they address, and end up as unsatisfactory technical fixes, because important organizational factors have been ignored. Very often the users of equipment and their patterns of organization are largely forgotten.

Part of the aim of this book is to strip away some of the attitudes that restrict our view of technology in order to expose these neglected cultural aspects. With the snowmobile, a first step was to look at different ways in which the use and maintenance of the machine is organized in different communities. This made it clear that a machine designed in response to the values of one culture needed a good deal of effort to make it suit the purposes of another.

A further example concerns the apparently simple hand-pumps used at village wells in India. During a period of drought in the 1960s, large power-driven drilling rigs were brought in to reach water at considerable depths in the ground by means of bore-holes. It was at these new wells that most of the hand-pumps were installed. By 1975 there were some 150,000 of them, but surveys showed that at any one time as many as two-thirds had broken down. New pumps sometimes failed within three or four weeks of installation. Engineers identified several faults, both in the design of the pumps and in standards of manufacture. But although these defects were corrected, pumps continued to go wrong. Eventually it was realized that the breakdowns were not solely an engineering problem. They were also partly an administrative or management issue, in that arrangements for servicing the pumps were not very effective. There was another difficulty, too, because in many villages, nobody felt any personal responsibility for looking after the pumps. It was only when these things were tackled together that pump performance began to improve.

This episode and the way it was handled illustrates very well the importance of an integrated appreciation of technology-practice. A breakthrough only came when all aspects of the administration, maintenance and technical design of the pump were thought out in relation to one another. What at first held up solution of the problem was a view of technology which began and ended with the machine – a view which, in another similar context, has been referred to as tunnel vision in engineering.

Any professional in such a situation is likely to experience his own form of tunnel vision. If a management consultant had been asked about the hand-pumps, he would have seen the administrative failings of the maintenance system very quickly, but might not have recognized that mechanical improvements to the pumps were required. Specialist training inevitably restricts people's approach to problems. But tunnel vision in attitudes to technology extends far beyond those who have had specialized training; it also affects policy-making, and influences popular expectations. People in many walks of life tend to focus on the tangible, technical aspect of any practical problem, and then to think that the extraordinary capabilities of modem technology ought to lead to an appropriate 'fix'. This attitude seems to apply to almost everything from inner city decay to military security, and from pollution to a cure for cancer. But all these issues have a social component. To hope for a technical fix for any of them that does not also involve social and cultural measures is to pursue an illusion.

So it was with the hand-pumps. The technical aspect of the problem was exemplified by poor design and manufacture. There was the organizational difficulty about maintenance. Also important, though, was the cultural aspect of technology as it was practised by the engineers involved. This refers, firstly, to the engineers' way of thinking, and the tunnel vision it led to; secondly, it indicates conflicts of values between highly trained engineers and the relatively uneducated people of the Indian countryside whom the pumps were meant to benefit. The local people probably had exaggerated expectations of the pumps as the products of an all-powerful, alien technology, and did not see them as vulnerable bits of equipment needing care in use and protection from damage; in addition, the local people would have their own views about hygiene and water use.

Many professionals in technology are well aware that the problems they deal with have social implications, but feel uncertainty about how these should be handled. To deal only with the technical detail and leave other aspects on one side is the easier option, and after all, is what they are trained for. With the hand-pump problem, an important step forward came when one of the staff of a local water development unit started looking at the case-histories of individual pump breakdowns. It was then relatively easy for him to pass from a technical review of components which were worn or broken to looking at the social context of each pump. He was struck by the way some pumps had deteriorated but others had not. One well-cared-for pump was locked up during certain hours; another was used by the family of a local official; others in good condition were in places where villagers had mechanical skills and were persistent with improvised repairs. It was these specific details that enabled suggestions to be made about the reorganization of pump maintenance.[3]

A first thought prompted by this is that a training in science and technology tends to focus on general principles, and does not prepare one to look for specifics in quite this way. But the human aspect of technology – its organization and culture – is not easily reduced to general principles, and the investigator with an eye for significant detail may sometimes learn more than the professional with a highly systematic approach.

A second point concerns the way in which the cultural aspect of technology-practice tends to be hidden beneath more obvious and more practical issues. Behind the tangible aspect of the broken hand-pumps lies an administrative problem concerned with maintenance. Behind that lies a problem of political will – the official whose family depended on one of the pumps was somehow well served. Behind that again were a variety of questions concerning cultural values regarding hygiene, attitudes to technology, and the outlook of the professionals involved.

This need to strip away the more obvious features of technology-practice to expose the background values is just as evident with new technology in western countries. Very often concern will be expressed about the health risk of a new device when people are worried about more intangible issues, because health risk is partly a technical question that is easy to discuss openly. A relatively minor technical problem affecting health may thus become a proxy for deeper worries about the way technology is practised which are more difficult to discuss.

An instance of this is the alleged health risks associated with visual display units (VDUs) in computer installations. Careful research has failed to find any real hazard except that operators may suffer eyestrain and fatigue. Yet complaints

---

[3] Heineman (1975), quoted in Pacey (1977).

about more serious problems continue, apparently because they can be discussed seriously with employers while misgivings about the overall systems are more difficult to raise. Thus a negative reaction to new equipment may be expressed in terms of a fear of 'blindness, sterility, etc.', because in our society, this is regarded as a legitimate reason for rejecting it. But to take such fears at face value will often be to ignore deeper, unspoken anxieties about 'deskilling, inability to handle new procedures, loss of control over work' (Damodaran, 1980).

Here, then, is another instance where, beneath the overt technical difficulty there are questions about the organizational aspect of technology – especially the organization of specific tasks. These have political connotations, in that an issue about control over work raises questions about where power lies in the work-place, and perhaps ultimately, where it lies within industrial society. But beyond arguments of that sort, there are even more basic values about creativity in work and the relationship of technology and human need.

In much the same way as concern about health sometimes disguises work-place issues, so the more widely publicized environmental problems may also hide underlying organizational and political questions. C. S. Lewis once remarked that 'Man's power over Nature often turns out to be a power exerted by some men over other men with Nature as its instrument', and a commentator notes that this, 'and not the environmental dilemma as it is usually conceived', is the central issue for technology.[4] As such, it is an issue whose political and social ramifications have been ably analysed by a wide range of authors.[5]

Even this essentially political level of argument can be stripped away to reveal another cultural aspect of technology. If we look at the case made out in favour of almost any major project – a nuclear energy plant, for example – there are nearly always issues concerning political power behind the explicit arguments about tangible benefits and costs. In a nuclear project, these may relate to the power of management over trade unions in electricity utilities; or to prestige of governments and the power of their technical advisers. Yet those who operate these levers of power are able to do so partly because they can exploit deeper values relating to the so-called technological imperative, and to the basic creativity that makes innovation possible. This, I argue, is a central part of the culture of technology, and its analysis occupies several chapters in this book. If these values underlying the technological imperative are understood, we may be able to see that here is a stream of feeling which politicians can certainly manipu-late at times, but which is stronger than their short-term purposes, and often runs away beyond their control.

WORKS CITED

R. A. Buchanan (1965) *Technology and Social Progress* (Oxford: Pergamon Press).

L. Damodaran (1980) 'Health hazards of VDUs? – Chairman's introduction,' conference at Loughborough University of Technology, 11 December 1980.

---

[4] Quoted by Hartley (1980, p. 353).

[5] E.g., Elliott and Elliott (1976).

M. B. Doyle (1978) *An Assessment of the Snowmobile Industry and Sport* (Washington DC: International Snowmobile Industry Association).

D. Elliott and R. Elliott (1976) *The Control of Technology* (London and Winchester: Wykeham).

K. Galbraith (1972) *The New Industrial State*, 2nd British edition (London: Andre Deutsch).

P. Hartley (1980) 'Educating engineers,' *The Ecologist*, 10 (10), December.

C. Heineman (1975) 'Survey of hand-pumps in Vellakovil …,' unpublished report, January.

J. Naughton (1979) 'Introduction: technology and human values,' in *Living with Technology: a Foundation Course* (Milton Keynes: The Open University Press).

A. Pacey (1977) *Hand-pump Maintenance* (London: Intermediate Technology Publications).

A. Ross (1978) *The Risk Takers* (Toronto: Macmillan and the Financial Post).

P. J. Usher (1972) 'The use of snowmobiles for trapping on Banks Island,' *Arctic* (Arctic Institute of North America), 25.

DISCUSSION QUESTIONS

1. In what ways does Pacey believe technology expresses cultural values? Do you think he is right that values are embedded in technology in these ways? Are there other aspects of technology design, dissemination, and use that you believe express cultural values?

2. Consider a particular novel technology, such as social networking or genetic modification. What values do you think are embedded in the technology or expressed in the pursuit of it?

3. Pacey emphasizes the cultural and organization aspects of technology-practice. What does he mean by these? Can you think of any other aspects of the situatedness of technology? That is to ask, what other ways is technology situated besides culturally and organizationally?

4. Pacey is critical of addressing social and ecological challenges with a 'technical fix.' What does he mean by this and why does he think it is problematic? Can you think of any current social or ecological problems for which technical fixes are being used or proposed? Do you agree that the use of technical fixes is problematic? Why or why not?

5. What do you think are the lessons about the relationship between technology and culture that the cases discussed in this chapter illustrate – e.g. the snowmobile and water pump cases? Can you think of other technologies to which these lessons apply?

TECHNOLOGY AND RESPONSIBILITY: REFLECTIONS ON THE NEW TASKS OF ETHICS[1]

Hans Jonas

3

CHAPTER SUMMARY

In this Chapter, Hans Jonas begins from the premise that ethics is concerned with action. Technology significantly increases our scope of agency or power to impact the world through our actions. Therefore, technology requires developing an expanded conception of responsibility. Prior to modern technology, it made sense for ethics to focus on local, immediate, and interpersonal interactions. However, technological advances, particularly since the industrial revolution, have empowered those with access to modern technologies to significantly affect people on the other side of the planet, future generations, and non-human nature; and biotechnologies raise the prospect of our modifying our own genetic natures. We must take responsibility for these new powers, Jonas argues, and develop ethics appropriate for them – global ethics, environmental ethics, future generation ethics, bioethics, as well as an ethic of technology more generally.

RELATED READINGS

Introduction: Power (2.4); Form of Life (2.5); Responsible Development (2.8)

1.

All previous ethics – whether in the form of issuing direct enjoinders to do and not to do certain things, or in the form of defining principles for such enjoinders, or in the form of establishing the ground of obligation for obeying such principles – had these interconnected tacit premises in common: that the human condition, determined by the nature of man and the nature of things, was given once for all; that the human good on that basis was readily determinable; and that the range of human action and therefore responsibility was narrowly circumscribed. It will be the burden of my argument to show that these premises no longer hold, and to reflect on the meaning of this fact for our moral condition. More specifically,

---

[1] This chapter is excerpted from Hans Jonas, "Technology and Responsibility: Reflections on the New Tasks of Ethics," *Social Research: An International Quarterly*, 40 (1): 31–54. It appears here by permission of *Social Research*.

it will be my contention that with certain developments of our powers the *nature of human* action has changed, and since ethics is concerned with action, it should follow that the changed nature of human action calls for a change in ethics as well: this not merely in the sense that new objects of action have added to the case material on which received rules of conduct are to be applied, but in the more radical sense that the qualitatively novel nature of certain of our actions has opened up a whole new dimension of ethical relevance for which there is no precedent in the standards and canons of traditional ethics.

2.

The novel powers I have in mind are, of course, those of modern *technology*. My first point, accordingly, is to ask how this technology affects the nature of our acting, in what ways it makes acting under its dominion *different* from what it has been through the ages. Since throughout those ages man was never without technology, the question involves the human difference of *modern* from previous technology....

3.

[There are several] characteristics of human action [prior to modern technology] which are relevant for a comparison with the state of things today.

1. All dealing with the non-human world, i.e., the whole realm of *techne* (with the exception of medicine), was ethically neutral – in respect both of the object and the subject of such action: in respect of the object, because it impinged but little on the self-sustaining nature of things and thus raised no question of permanent injury to the integrity of its object, the natural order as a whole; and in respect of the agent subject it was ethically neutral because *techne* as an activity conceived itself as a determinate tribute to necessity and not as an indefinite, self-validating advance to mankind's major goal, claiming in its pursuit man's ultimate effort and concern. The real vocation of man lay elsewhere. In brief, action on non-human things did not constitute a sphere of authentic ethical significance.
2. Ethical significance belonged to the direct dealing of man with man, including the dealing with himself: all traditional ethics is *anthropocentric*.
3. For action in this domain, the entity "man" and his basic condition was considered constant in essence and not itself an object of reshaping *techne*.
4. The good and evil about which action had to care lay close to the act, either in the praxis itself or in its immediate reach, and were not a matter for remote planning. This proximity of ends pertained to time as well as space. The effective range of action was small, the time-span of foresight, goal-setting and accountability was short, control of circumstances limited. Proper conduct had its immediate criteria and almost immediate consummation. The long run of consequences beyond was left to chance, fate or providence. Ethics accordingly was of the here and now, of occasions as they arise between men, of the recurrent, typical situations of private and public life. The good man was he

who met these contingencies with virtue and wisdom, cultivating these powers in himself, and for the rest resigning himself to the unknown.

All enjoinders and maxims of traditional ethics, materially different as they may be, show this confinement to the immediate setting of the action. "Love thy neighbor as thyself"; "Do unto others as you would wish them to do unto you"; "Instruct your child in the way of truth"; "Strive for excellence by developing and actualizing the best potentialities of your being *qua* man"; "Subordinate your individual good to the common good"; "Never treat your fellow man as a means only but always *also* as an end in himself" – and so on. Note that in all these maxims the agent and the "other" of his action are sharers of a common present. It is those alive now and in some commerce with me that have a claim on my conduct as it affects them by deed or omission. The ethical universe is composed of contemporaries, and its horizon to the future is confined by the foreseeable span of their lives. Similarly confined is its horizon of place, within which the agent and the other meet as neighbor, friend or foe, as superior and subordinate, weaker and stronger, and in all the other roles in which humans interact with one another. To this proximate range of action all morality was geared....

4.

All this has decisively changed. Modern technology has introduced actions of such novel scale, objects, and consequences that the framework of former ethics can no longer contain them.... To be sure, the old prescriptions of the "neighbor" ethics – of justice, charity, honesty, and so on – still hold in their intimate immediacy for the nearest, day by day sphere of human interaction. But this sphere is overshadowed by a growing realm of collective action where doer, deed, and effect are no longer the same as they were in the proximate sphere, and which by the enormity of its powers forces upon ethics a new dimension of responsibility never dreamt of before.

Take, for instance, as the first major change in the inherited picture, the critical *vulnerability* of nature to man's technological intervention – unsuspected before it began to show itself in damage already done. This discovery, whose shock led to the concept and nascent science of ecology, alters the very concept of ourselves as a causal agency in the larger scheme of things. It brings to light, through the effects, that the nature of human action has *de facto* changed, and that an object of an entirely new order – no less than the whole biosphere of the planet – has been added to what we must be responsible for because of our power over it. And of what surpassing importance an object, dwarfing all previous objects of active man! Nature as a human responsibility is surely a *novum* to be pondered in ethical theory. What kind of obligation is operative in it? Is it more than a utilitarian concern? Is it just prudence that bids us not to kill the goose that lays the golden eggs, or saw off the branch on which we sit? But the "we" that here sits and may fall into the abyss is all future mankind, and the survival of the species is more than a prudential duty of its present members. Insofar as it is the fate of man, as affected by the condition of nature, which makes us care about the preservation of nature, such care admittedly still retains the anthropocentric focus of all classical ethics.

Even so, the difference is great. The containment of nearness and contemporaneity is gone, swept away by the spatial spread and time-span of the cause–effect trains which technological practice sets afoot, even when undertaken for proximate ends. Their irreversibility conjoined to their aggregate magnitude injects another novel factor into the moral equation. To this take their cumulative character: their effects add themselves to one another, and the situation for later acting and being becomes increasingly different from what it was for the initial agent. The cumulative self-propagation of the technological change of the world thus constantly overtakes the conditions of its contributing acts and moves through none but unprecedented situations, for which the lessons of experience are powerless. And not even content with changing its beginning to the point of unrecognizability, the cumulation as such may consume the basis of the whole series, the very condition of itself. All this would have to be co-intended in the will of the single action if this is to be a morally responsible one. Ignorance no longer provides it with an alibi.

Knowledge, under these circumstances, becomes a prime duty beyond anything claimed for it heretofore, and the knowledge must be commensurate with the causal scale of our action. The fact that it cannot really be thus commensurate, i.e., that the predictive knowledge falls behind the technical knowledge which nourishes our power to act, itself assumes ethical importance. Recognition of ignorance becomes the obverse of the duty to know and thus part of the ethics which must govern the ever more necessary self-policing of our out-sized might. No previous ethics had to consider the global condition of human life and the far-off future, even existence, of the race. Their now being an issue demands, in brief, a new conception of duties and rights, for which previous ethics and metaphysics provide not even the principles, let alone a ready doctrine.

And what if the new kind of human action would mean that more than the interest of man alone is to be considered – that our duty extends farther and the anthropocentric confinement of former ethics no longer holds? It is at least not senseless anymore to ask whether the condition of extra-human nature, the biosphere as a whole and in its parts, now subject to our power, has become a human trust and has something of a moral claim on us not only for our ulterior sake but for its own and in its own right. If this were the case it would require quite some rethinking in basic principles of ethics. It would mean to seek not only the human good, but also the good of things extra-human, that is, to extend the recognition of "ends in themselves" beyond the sphere of man and make the human good include the care for them. For such a role of stewardship no previous ethics has prepared us – and the dominant, scientific view of *Nature* even less. Indeed, the latter emphatically denies us all conceptual means to think of Nature as something to be honored, having reduced it to the indifference of necessity and accident, and divested it of any dignity of ends. But still, a silent plea for sparing its integrity seems to issue from the threatened plenitude of the living world. Should we heed this plea, should we grant its claim as sanctioned by the nature of things, or dismiss it as a mere sentiment on our part, which we may indulge as far as we wish and can afford to do? If the former, it would (if taken seriously in its theoretical implications) push the necessary rethinking beyond the doctrine of action, i.e., ethics, into the doctrine of being, i.e., metaphysics, in which all ethics must ultimately be grounded. On this speculative subject I will here say no more than that we should keep ourselves open to the thought that natural science may not tell the whole story about Nature.

5.

Returning to strictly intra-human considerations, there is another ethical aspect to the growth of *techne* as a pursuit beyond the pragmatically limited terms of former times. Then, so we found, *techne* was a measured tribute to necessity, not the road to mankind's chosen goal – a means with a finite measure of adequacy to well-defined proximate ends. Now, *techne* in the form of modern technology has turned into an infinite forward-thrust of the race, its most significant enterprise, in whose permanent, self-transcending advance to ever greater things the vocation of man tends to be seen, and whose success of maximal control over things and himself appears as the consummation of his destiny. Thus the triumph of *homo faber* over his external object means also his triumph in the internal constitution of *homo sapiens*, of whom he used to be a subsidiary part. In other words, technology, apart from its objective works, assumes ethical significance by the central place it now occupies in human purpose. Its cumulative creation, the expanding artificial environment, continuously reinforces the particular powers in man that created it, by compelling their unceasing inventive employment in its management and further advance, and by rewarding them with additional success – which only adds to the relentless claim. This positive feedback of functional necessity and reward – in whose dynamics pride of achievement must not be forgotten – assures the growing ascendancy of one side of man's nature over all the others, and inevitably at their expense. If nothing succeeds like success, nothing also entraps like success. Outshining in prestige and starving in resources whatever else belongs to the fullness of man, the expansion of his power is accompanied by a contraction of his self-conception and being. In the image he entertains of himself – the potent self-formula which determines his actual being as much as it reflects it – man now is evermore the maker of what he has made and the doer of what he can do, and most of all the preparer of what he will be able to do next. But not you or I: it is the aggregate, not the individual doer or deed that matters here; and the indefinite future, rather than the contemporary context of the action, constitutes the relevant horizon of responsibility. This requires imperatives of a new sort. If the realm of making has invaded the space of essential action, then morality must invade the realm of making, from which it had formerly stayed aloof, and must do so in the form of public policy. With issues of such inclusiveness and such lengths of anticipation public policy has never had to deal before. In fact, the changed nature of human action changes the very nature of politics.

For the boundary between "city" and "nature" has been obliterated: the city of men, once an enclave in the non-human world, spreads over the whole of terrestrial nature and usurps its place. The difference between the artificial and the natural has vanished, the natural is swallowed up in the sphere of the artificial, and at the same time the total artifact, the works of man working on and through himself, generates a "nature" of its own, i.e., a necessity with which human freedom has to cope in an entirely new sense. Once it could be said *Fiat justitia, pereat mundus*, "Let justice be done, and may the world perish" – where "world," of course, meant the renewable enclave in the imperishable whole. Not even rhetorically can the like be said anymore when the perishing of the whole through the doings of man – be they just or unjust – has become a real possibility. Issues never legislated on come into the purview of the laws which the total city must give itself so that there will be a world for the generations of man to come.

That there *ought* to be through all future time such a world fit for human habitation, and that it ought in all future time to be inhabited by a mankind worthy of the human name, will be readily affirmed as a general axiom or a persuasive desirability of speculative imagination (as persuasive and as undemonstrable as the proposition that there being a world at all is "better" than there being none): but as a *moral* proposition, namely, a practical *obligation* toward the posterity of a distant future, and a principle of decision in present action, it is quite different from the imperatives of the previous ethics of contemporaneity; and it has entered the moral scene only with our novel powers and range of prescience.

The *presence of man in the world* had been a first and unquestionable given, from which all idea of obligation in human conduct started out. Now it has itself become an *object* of obligation – the obligation namely to ensure the very premise of all obligation, i.e., the *foothold* for a moral universe in the physical world – the existence of mere *candidates* for a moral order....

6.

The new order of human action requires a commensurate ethics of foresight and responsibility, which is as new as are the issues with which it has to deal. We have seen that these are the issues posed by the works of *homo faber* in the age of technology. But among those novel works we haven't mentioned yet the potentially most ominous class. We have considered *techne* only as applied to the non-human realm. But man himself has been added to the objects of technology. *Homo faber* is turning upon himself and gets ready to make over the maker of all the rest. This consummation of his power, which may well portend the overpowering of man, this final imposition of art on nature, calls upon the utter resources of ethical thought, which never before has been faced with elective alternatives to what were considered the definite terms of the human condition.

a.

Take, for instance, the most basic of these "givens," man's mortality. Who ever before had to make up his mind on its desirable and *eligible* measure? There was nothing to choose about the upper limit, the "threescore years and ten, or by reason of strength fourscore." Its inexorable rule was the subject of lament, submission, or vain (not to say foolish) wish-dreams about possible exceptions – strangely enough, almost never of affirmation. The intellectual imagination of a George Bernard Shaw and a Jonathan Swift speculated on the privilege of not having to die, or the curse of not being able to die. (Swift with the latter was the more perspicacious of the two.) Myth and legend toyed with such themes against the acknowledged background of the unalterable, which made the earnest man rather pray "teach us to number our days that we may get a heart of wisdom"

(Psalm 90). Nothing of this was in the realm of doing and effective decision. The question was only how to relate to the stubborn fact.

But lately, the dark cloud of inevitability seems to lift. A practical hope is held out by certain advances in cell biology to prolong, perhaps indefinitely extend the span of life by counteracting biochemical processes of aging. Death no longer appears as a necessity belonging to the nature of life, but as an avoidable, at least in principle tractable and long-delayable, organic malfunction. A perennial yearning of mortal man seems to come nearer fulfillment. And for the first time we have in earnest to ask the question "How desirable is this? How desirable for the individual, and how for the species?" These questions involve the very meaning of our finitude, the attitude toward death, and the general biological significance of the balance of death and procreation. Even prior to such ultimate questions are the more pragmatic ones of who should be eligible for the boon: persons of particular quality and merit? of social eminence? those that can pay for it? everybody? The last would seem the only just course. But it would have to be paid for at the opposite end, at the source. For clearly, on a population-wide scale, the price of extended age must be a proportional slowing of replacement, i.e., a diminished access of new life. The result would be a decreasing proportion of youth in an increasingly aged population. How good or bad would that be for the general condition of man? Would the species gain or lose? And how *right* would it be to preempt the place of youth? Having to die is bound up with having been born: mortality is but the other side of the perennial spring of "natality" (to use Hannah Arendt's term). This had always been ordained; now its meaning has to be pondered in the sphere of decision.

To take the extreme (not that it will ever be obtained): if we abolish death, we must abolish procreation as well, for the latter is life's answer to the former, and so we would have a world of old age with no youth, and of known individuals with no surprises of such that had never been before. But this perhaps is precisely the wisdom in the harsh dispensation of our mortality: that it grants us the eternally renewed promise of the freshness, immediacy and eagerness of youth, together with the supply of otherness as such. There is no substitute for this in the greater accumulation of prolonged experience: it can never recapture the unique privilege of seeing the world for the first time and with new eyes, never relive the wonder which, according to Plato, is the beginning of philosophy, never the curiosity of the child, which rarely enough lives on as thirst for knowledge in the adult, until it wanes there too. This ever renewed beginning, which is only to be had at the price of ever repeated ending, may well be mankind's hope, its safeguard against lapsing into boredom and routine, its chance of retaining the spontaneity of life. Also, the role of the *memento mori* in the individual's life must be considered, and what its attenuation to indefiniteness may do to it. Perhaps a non-negotiable limit to our expected time is necessary for each of us as the incentive to number our days and make them count.

So it could be that what by intent is a philanthropic gift of science to man, the partial granting of his oldest wish – to escape the curse of mortality – turns out to be to the detriment of man. I am not indulging in prediction and, in spite of my noticeable bias, not even in valuation. My point is that already the promised gift raises questions that had never to be asked before in terms of practical choice, and that no principle of former ethics, which took the human constants for granted, is competent to deal with them. And yet they must be dealt with ethically and by principle and not merely by the pressure of interests.

b.

It is similar with all the other, quasi-utopian powers about to be made available by the advances of biomedical science as they are translated into technology. Of these, *behavior control* is much nearer to practical readiness than the still hypothetical prospect I have just been discussing, and the ethical questions it raises are less profound but have a more direct bearing on the moral conception of man. Here again, the new kind of intervention exceeds the old ethical categories. They have not equipped us to rule, for example, on mental control by chemical means or by direct electrical action on the brain via implanted electrodes – undertaken, let us assume, for defensible and even laudable ends. The mixture of beneficial and dangerous potentials is obvious, but the lines are not easy to draw. Relief of mental patients from distressing and disabling symptoms seems unequivocally beneficial. But from the relief of the *patient*, a goal entirely in the tradition of the medical art, there is an easy passage to the relief of *society* from the inconvenience of difficult individual behavior among its members: that is, the passage from medical to social application; and this opens up an indefinite field with grave potentials. The troublesome problems of rule and unruliness in modern mass society make the extension of such control methods to non-medical categories extremely tempting for social management. Numerous questions of human rights and dignity arise. The difficult question of preempting versus enabling care insists on concrete answers. Shall we induce learning attitudes in school children by the mass administration of drugs, circumventing the appeal to autonomous motivation? Shall we overcome aggression by electronic pacification of brain areas? Shall we generate sensations of happiness or pleasure or at least contentment through independent stimulation (or tranquilizing) of the appropriate centers – independent, that is, of the objects of happiness, pleasure, or content and their attainment in personal living and achieving? Candidacies could be multiplied. Business firms might become interested in some of these techniques for performance-increase among their employees.

Regardless of the question of compulsion or consent, and regardless also of the question of undesirable side-effects, each time we thus bypass the human way of dealing with human problems, short-circuiting it by an impersonal mechanism, we have taken away something from the dignity of personal selfhood and advanced a further step on the road from responsible subjects to programmed behavior systems. Social functionalism, important as it is, is only one side of the question. Decisive is the question of what kind of individuals the society is composed of – to make its existence valuable as a whole. Somewhere along the line of increasing social manageability at the price of individual autonomy, the question of the worthwhileness of the whole human enterprise must pose itself. Answering it involves the image of man we entertain. We must think it anew in light of the things we can do to it now and could never do before.

C.

This holds even more with respect to the last object of a technology applied on man himself – the genetic control of future men. This is too wide a subject for cursory treatment. Here I merely point to this most ambitious dream of *homo faber*, summed up in the phrase that man will take his own evolution in hand, with the aim of not just preserving the integrity of the species but of modifying it by improvements of his own design. Whether we have the right to do it, whether we are qualified for that creative role, is the most serious question that can be posed to man finding himself suddenly in possession of such fateful powers. Who will be the image-makers, by what standards, and on the basis of what knowledge? Also, the question of the moral right to experiment on future human beings must be asked. These and similar questions, which demand an answer before we embark on a journey into the unknown, show most vividly how far our powers to act are pushing us beyond the terms of all former ethics.

7.

The ethically relevant common feature in all the examples adduced is what I like to call the inherently "utopian" drift of our actions under the conditions of modern technology, whether it works on non-human or on human nature, and whether the "utopia" at the end of the road be planned or unplanned. By the kind and size of its snowballing effects, technological power propels us into goals of a type that was formerly the preserve of Utopias. To put it differently, technological power has turned what used and ought to be tentative, perhaps enlightening plays of speculative reason into competing blueprints for projects, and in choosing between them we have to choose between extremes of remote effects. The one thing we can really know of them is their extremism as such – that they concern the total condition of nature on our globe and the very kind of creatures that shall, or shall not, populate it. In consequence of the inevitably "utopian" scale of modern technology, the salutary gap between everyday and ultimate issues, between occasions for common prudence and occasions for illuminated wisdom, is steadily closing. Living now constantly in the shadow of unwanted, built-in, automatic utopianism, we are constantly confronted with issues whose positive choice requires supreme wisdom – an impossible situation for man in general, because he does not possess that wisdom, and in particular for contemporary man, who denies the very existence of its object: viz., objective value and truth. We need wisdom most when we believe in it least.

If the new nature of our acting then calls for a new ethics of long-range responsibility, coextensive with the range of our power, it calls in the name of that very responsibility also for a new kind of humility – a humility not like former humility, i.e., owing to the littleness, but owing to the excessive magnitude of our power, which is the excess of our power to act over our power to foresee and our power to evaluate and to judge. In the face of the quasi-eschatological potentials of our technological processes, ignorance of the ultimate implications becomes itself a reason for responsible restraint – as the second best to the possession of wisdom itself.

One other aspect of the required new ethics of responsibility for and to a distant future is worth mentioning: the insufficiency of representative government to meet the new demands on its normal principles and by its normal mechanics. For according to these, only *present* interests make themselves heard and felt and enforce their consideration. It is to them that public agencies are accountable, and this is the way in which concretely the respecting of rights comes about (as distinct from their abstract acknowledgment). But the *future* is not represented, it is not a force that can throw its weight into the scales. The nonexistent has no lobby, and the unborn are powerless. Thus accountability to them has no political reality behind it yet in present decision-making, and when they can make their complaint, then we, the culprits, will no longer be there.

This raises to an ultimate pitch the old question of the power of the wise, or the force of ideas not allied to self-interest, in the body politic. What *force* shall represent the future in the present? However, before *this* question can become earnest in practical terms, the new ethics must find its theory, on which do's and don'ts can be based. That is: before the question of what *force*, comes the question of what *insight* or value-knowledge shall represent the future in the present.

8.

And here is where I get stuck, and where we all get stuck. For the very same movement which put us in possession of the powers that have now to be regulated by norms – the movement of modern knowledge called science – has by a necessary complementarity eroded the foundations from which norms could be derived; it has destroyed the very idea of norm as such. Not, fortunately, the feeling for norm and even for particular norms. But this feeling becomes uncertain of itself when contradicted by alleged knowledge or at least denied all sanction by it. Anyway and always does it have a difficult enough time against the loud clamors of greed and fear. Now it must in addition blush before the frown of superior knowledge, as unfounded and incapable of foundation. First, Nature had been "neutralized" with respect to value, then man himself. Now we shiver in the nakedness of a nihilism in which near-omnipotence is paired with near-emptiness, greatest capacity with knowing least what for. With the apocalyptic pregnancy of our actions, that very knowledge which we lack has become more urgently needed than at any other stage in the adventure of mankind. Alas, urgency is no promise of success. On the contrary, it must be avowed that to seek for wisdom today requires a good measure of unwisdom. The very nature of the age which cries out for an ethical theory makes it suspiciously look like a fool's errand. Yet we have no choice in the matter but to try.

It is a question whether without restoring the category of the sacred, the category most thoroughly destroyed by the scientific enlightenment, we can have an ethics able to cope with the extreme powers which we possess today and constantly increase and are almost compelled to use. Regarding those consequences imminent enough still to hit ourselves, fear can do the job – so often the best substitute for genuine virtue or wisdom. But this means fails us towards the more distant prospects, which here matter the most, especially as the beginnings seem mostly innocent in their smallness. Only awe of the sacred with its unqualified veto is independent of the computations of mundane fear and the solace of uncertainty about distant consequences. But religion as a soul-determining force is no longer there to be summoned to the aid of ethics. The latter must stand on its worldly feet – that is, on reason and its fitness for philosophy. And while of faith it can be said that it either is there or is not, of ethics it holds that it must be there.

It must be there because men act, and ethics is for the ordering of actions and for regulating the power to act. It must be there all the more, then, the greater the powers of acting that are to be regulated; and with their size, the ordering principle must also fit their kind. Thus, novel powers to act require novel ethical rules and perhaps even a new ethics.

"Thou shalt not kill" was enunciated because man has the power to kill and often the occasion and even inclination for it – in short, because killing is actually done. It is only under the *pressure* of real habits of action, and generally of the fact that always action already takes place, without *this* having to be commanded first, that ethics as the ruling of such acting under the standard of the good or the permitted enters the stage. Such a *pressure* emanates from the novel technological powers of man, whose exercise is given with their existence. *If* they really are as novel in kind as here contended, and if by the kind of their potential consequences they really have abolished the moral neutrality which the technical commerce with matter hitherto enjoyed – then their pressure bids to seek for new prescriptions in ethics which are competent to assume their guidance, but which first of all can hold their own theoretically against that very pressure. To the demonstration of those premises this paper was devoted. If they are accepted, then we who make thinking our business have a task to last us for our time. We must do it in time, for since we act anyway we shall have some ethic or other in any case, and without a supreme effort to determine the right one, we may be left with a wrong one by default.

DISCUSSION QUESTIONS

1. Jonas argues that with modern technology "the nature of human action has changed." What does he mean by this? Do you think that he is correct? Why or why not?
2. What new fields or horizons for ethics does Jonas believe modern technology makes necessary? For each of these, why does Jonas believe that traditional ethical injunctions, such as 'love thy neighbor as you love thyself' are inadequate?
3. Do you think any emerging technologies, such as information technologies, virtual reality, or artificial intelligences, could reveal yet another new ethical field or horizon? Why or why not?
4. Jonas seems to believe that new ethical outlooks or principles need to be developed in response to the powers associated with modern technology. Do you agree with this? Or do you think traditional ethics can be extended to provide the necessary guidance in areas such as environmental ethics, global ethics, and technology ethics? Why or why not?
5. What does Jonas mean by 'responsibility'? What do you think is involved in truly taking responsibility for your actions in the modern technologized and globalized world?

TECHNOLOGIES AS FORMS OF LIFE[1]

Langdon Winner

CHAPTER SUMMARY

In this chapter, Langdon Winner discusses what he calls "technological somnambulism." Our capacity and willingness to reflect on the significance of technology and to critically evaluate new technologies lags far behind our capacity for creating and disseminating technologies. As a result, we "willingly sleepwalk through the process of reconstituting the conditions of human existence." Winner suggests several reasons for this somnambulism, including beliefs about the neutrality of technology and technological determinism. He is critical of each of these. Winner is also critical of the near exclusive focus of technological assessment on the positive and negative impacts or effects of a technology. Technologies certainly have impacts, but they also can restructure our physical and social worlds, and so how we live. Winner argues that this understanding of technologies as "forms of life" needs to inform our evaluations and choices regarding technological innovation and adoption.

RELATED READINGS

Introduction: The Innovation Presumption (2.1); Power (2.4); Form of Life (2.5); Responsible Development (2.8)

1. INTRODUCTION

From the early days of manned space travel comes a story that exemplifies what is most fascinating about the human encounter with modern technology. Orbiting the earth aboard *Friendship* 7 in February 1962, astronaut John Glenn noticed something odd. His view of the planet was virtually unique in human experience; only Soviet pilots Yuri Gagarin and Gherman Titov had preceded him in orbital flight. Yet as he watched the continents and oceans moving beneath him, Glenn began to feel that he had seen it all before. Months of simulated space shots in sophisticated training machines and centrifuges had affected his ability to respond. In the words of chronicler Tom Wolfe, "The world demanded awe,

---

[1] This chapter is excerpted from Langdon Winner (1983) 'Technologies as Forms of Life,' in *Epistemology, Methodology and the Social Sciences*, eds. Cohen and Wartofsky (Kluwer Academic Publishers). It appears here by permission of Springer Press.

because this was a voyage through the stars. But he couldn't feel it. The backdrop of the event, the stage, the environment, the true orbit … was not the vast reaches of the universe. It was the simulators. *Who could possibly understand this*?" (Wolfe, 1980, p. 270). Synthetic conditions generated in the training center had begun to seem more "real" than the actual experience.

It is reasonable to suppose that a society thoroughly committed to making artificial realities would have given a great deal of thought to the nature of that commitment. One might expect, for example, that the philosophy of technology would be a topic widely discussed by scholars and technical professionals, a lively field of inquiry often chosen by students at our universities and technical institutes. One might even think that the basic issues in this field would be well defined, its central controversies well worn. However, such is not the case. At this late date in the development of our industrial/technological civilization the most accurate observation to be made about the philosophy of technology is that there really isn't one.

The basic task for a philosophy of technology is to examine critically the nature and significance of artificial aids to human activity. That is its appropriate domain of inquiry, one that sets it apart from, say, the philosophy of science. Yet if one turns to the writings of twentieth-century philosophers, one finds astonishingly little attention given to questions of that kind. The six-volume *Encyclopedia of Philosophy*, a recent compendium of major themes in various traditions of philosophical discourse, contains no entry under the category "technology" (Encyclopedia of Philosophy, 1967). Neither does that work contain enough material under possible alternative headings to enable anyone to piece together an idea of what a philosophy of technology might be.

True, there are some writers who have taken up the topic. The standard bibliography in the philosophy of technology lists well over a thousand books and articles in several languages by nineteenth- and twentieth-century authors (Bibliography of the Philosophy of Technology, 1973). But reading through the material listed shows, in my view, little of enduring substance. The best writing on this theme comes to us from a few powerful thinkers who have encountered the subject in the midst of much broader and ambitious investigations – for example, Karl Marx in the development of his theory of historical materialism or Martin Heidegger as an aspect of his theory of ontology. It may be, in fact, that the philosophy is best seen as a derivative of more fundamental questions. For despite the fact that nobody would deny its importance to an adequate understanding of the human condition, technology has never joined epistemology, metaphysics, esthetics, law, science, and politics as a fully respectable topic for philosophical inquiry.

Engineers have shown little interest in filling this void. Except for airy pronouncements in yearly presidential addresses at various engineering societies, typically ones that celebrate the contributions of a particular technical vocation to the betterment of humankind, engineers appear unaware of any philosophical questions their work might entail. As a way of starting a conversation with my friends in engineering, I sometimes ask, "What are the founding principles of your discipline?" The question is always greeted with puzzlement. Even when I explain what I am after, namely, a coherent account of the nature and significance of the branch of engineering in which they are involved, the question still means nothing to them. The scant few who raise important first questions about their technical professions are usually seen by their colleagues as dangerous cranks and radicals. If Socrates' suggestion that the "unexamined life is not worth living" still holds, it is news to most engineers.[2]

2. TECHNOLOGICAL SOMNAMBULISM

Why is it that the philosophy of technology has never really gotten under way? Why has a culture so firmly based upon countless sophisticated instruments, techniques, and systems remained so steadfast in its reluctance to examine its own foundations? Much of the answer can be found in the astonishing hold the idea of "progress" has exercised on social thought during the industrial age. In the twentieth century it is usually taken for granted that the only reliable sources for improving the human condition stem from new machines, techniques, and chemicals. Even the recurring environmental and social ills that have accompanied technological advancement have rarely dented this faith. It is still a prerequisite that the person running for public office swear his or her unflinching confidence in a positive link between technical development and human well-being and affirm that the next wave of innovations will surely be our salvation.

There is, however, another reason why the philosophy of technology has never gathered much steam. According to conventional views, the human relationship to technical things is too obvious to merit serious reflection. The deceptively reasonable notion that we have inherited from much earlier and less complicated times divides the range of possible concerns about technology into two basic categories: *making* and *use*. In the first of these our attention is drawn to the matter of "how things work" and of "making things work." We tend to think that this is a fascination of certain people in certain occupations, but not for anyone else. "How things work" is the domain of inventors, technicians, engineers, repairmen, and the like who prepare artificial aids to human activity and keep them in good working order. Those not directly involved in the various spheres of "making" are thought to have little interest in or need to know about the materials, principles, or procedures found in those spheres.

What the others do care about, however, are tools and uses. This is understood to be a straightforward matter. Once things have been made, we interact with them on occasion to achieve specific purposes. One picks up a tool, uses it, and puts it down. One picks up a telephone, talks on it, and then does not use it for a time. A person gets on an airplane, flies from point A to point B, and then gets off. The proper interpretation of the meaning of technology in the mode of use seems to be nothing more complicated than an occasional, limited, and nonproblematic interaction.

The language of the notion of "use" also includes standard terms that enable us to interpret technologies in a range of moral contexts. Tools can be "used well or poorly" and for "good or bad purposes"; I can use my knife to slice a loaf of bread or to stab the next person that walks by. Because technological objects and

---

[2] There are, of course, exceptions to this general attitude. See Unger (1982).

processes have a promiscuous utility, they are taken to be fundamentally neutral as regards their moral standing.

The conventional idea of what technology is and what it means, an idea powerfully reinforced by familiar terms used in everyday language, needs to be overcome if a critical philosophy of technology is to move ahead. The crucial weakness of the conventional idea is that it disregards the many ways in which technologies provide structure for human activity. Since, according to accepted wisdom, patterns that take shape in the sphere of "making" are of interest to practitioners alone, and since the very essence of "use" is its occasional, innocuous, nonstructuring occurrence, any further questioning seems irrelevant.[3]

If the experience of modern society shows us anything, however, it is that technologies are not merely aids to human activity, but also powerful forces acting to reshape that activity and its meaning. The introduction of a robot to an industrial workplace not only increases productivity, but often radically changes the process of production, redefining what "work" means in that setting. When a sophisticated new technique or instrument is adopted in medical practice, it transforms not only what doctors do, but also the ways people think about health, sickness, and medical care. Widespread alterations of this kind in techniques of communication, transportation, manufacturing, agriculture, and the like are largely what distinguishes our times from early periods of human history. The kinds of things we are apt to see as "mere" technological entities become much more interesting and problematic if we begin to observe how broadly they are involved in conditions of social and moral life.

It is true that recurring patterns of life's activity (whatever their origins) tend to become unconscious processes taken for granted. Thus, we do not pause to reflect upon how we speak a language as we are doing so or the motions we go through in taking a shower. There is, however, one point at which we may become aware of a pattern taking shape – the very first time we encounter it. An opportunity of that sort occurred several years ago at the conclusion of a class I was teaching. A student came to my office on the day term papers were due and told me his essay would be late. "It crashed this morning," he explained. I immediately interpreted this as a "crash" of the conceptual variety, a flimsy array of arguments and observations that eventually collapses under the weight of its own ponderous absurdity. Indeed, some of my own papers have "crashed" in exactly that manner. But this was not the kind of mishap that had befallen this particular fellow. He went on to explain that his paper had been composed on a computer terminal and that it had been stored in a time-sharing minicomputer. It sometimes happens that the machine "goes down" or "crashes," making everything that happens in and around it stop until the computer can be "brought up," that is, restored to full functioning.

As I listened to the student's explanation, I realized that he was telling me about the facts of a particular form of activity in modern life in which he and others similarly situated were already involved and that I had better get ready for. I remembered J. L. Austin's little essay "A Plea for Excuses" and noticed that the student and I were negotiating one of the boundaries of contemporary moral life – where

---

[3] An excellent corrective to the general thoughtlessness about "making" and "use" is to be found in Mitcham (1978, pp. 229–294).

and how one gives and accepts an excuse in a particular technology-mediated situation (Austin, 1961). He was, in effect, asking me to recognize a new world of parts and pieces and to acknowledge appropriate practices and expectations that hold in that world. From then on, a knowledge of this situation would be included in my understanding of not only "how things work" in that generation of computers, but also how we do things as a consequence, including which rules to follow when the machines break down. Shortly thereafter I got used to computers crashing, disrupting hotel reservations, banking, and other everyday transactions; eventually, my own papers began crashing in this new way.

Some of the moral negotiations that accompany technological change eventually become matters of law. In recent times, for example, a number of activities that employ computers as their operating medium have been legally defined as "crimes." Is unauthorized access to a computerized data base a criminal offense? Given the fact that electronic information is in the strictest sense intangible, under what conditions is it "property" subject to theft? The law has had to stretch and reorient its traditional categories to encompass such problems, creating whole new classes of offenses and offenders.

The ways in which technical devices tend to engender distinctive worlds of their own can be seen in a more familiar case. Picture two men traveling in the same direction along a street on a peaceful, sunny day, one of them afoot and the other driving an automobile. The pedestrian has a certain flexibility of movement: he can pause to look in a shop window, speak to passersby, and reach out to pick a flower from a sidewalk garden. The driver, although he has the potential to move much faster, is constrained by the enclosed space of the automobile, the physical dimensions of the highway, and the rules of the road. His realm is spatially structured by his intended destination, by a periphery of more-or-less irrelevant objects (scenes for occasional side glances), and by more important objects of various kinds – moving and parked cars, bicycles, pedestrians, street signs, etc., that stand in his way. Since the first rule of good driving is to avoid hitting things, the immediate environment of the motorist becomes a field of obstacles.

Imagine a situation in which the two persons are next-door neighbors. The man in the automobile observes his friend strolling along the street and wishes to say hello. He slows down, honks his horn, rolls down the window, sticks out his head, and shouts across the street. More likely than not the pedestrian will be startled or annoyed by the sound of the horn. He looks around to see what's the matter and tries to recognize who can be yelling at him across the way. "Can you come to dinner Saturday night?" the driver calls out over the street noise. "What?" the pedestrian replies, straining to understand. At that moment another car to the rear begins honking to break up the temporary traffic jam. Unable to say anything more, the driver moves on.

What we see here is an automobile collision of sorts, although not one that causes bodily injury. It is a collision between the *world* of the driver and that of the pedestrian. The attempt to extend a greeting and invitation, ordinarily a simple gesture, is complicated by the presence of a technological device and its standard operating conditions. The communication between the two men is shaped by an incompatibility of the form of locomotion known as walking and a much newer one, automobile driving. In cities such as Los Angeles, where the physical landscape and prevailing social habits assume everyone drives a car, the simple act of walking can be cause for alarm. The U.S. Supreme Court decided one case involving a young man who enjoyed taking long walks late at night through the streets of San Diego and was repeatedly arrested by police as a suspicious character. The Court decided in favor of the pedestrian, noting that he had not been engaged in burglary or any other illegal act. Merely traveling by foot is not yet a crime.[4]

Knowing how automobiles are made, how they operate, and how they are used and knowing about traffic laws and urban transportation policies does little to help us understand how automobiles affect the texture of modern life. In such cases a strictly instrumental/functional understanding fails us badly. What is needed is an interpretation of the ways, both obvious and subtle, in which everyday life is transformed by the mediating role of technical devices. In hindsight the situation is clear to everyone. Individual habits, perceptions, concepts of self, ideas of space and time, social relationships, and moral and political boundaries have all been powerfully restructured in the course of modern technological development. What is fascinating about this process is that societies involved in it have quickly altered some of the fundamental terms of human life without appearing to do so. Vast transformations in the structure of our common world have been undertaken with little attention to what those alterations mean. Judgments about technology have been made on narrow grounds, paying attention to such matters as whether a new device serves a particular need, performs more efficiently than its predecessor, makes a profit, or provides a convenient service. Only later does the broader significance of the choice become clear, typically as a series of surprising "side effects" or "secondary consequences." But it seems characteristic of our culture's involvement with technology that we are seldom inclined to examine, discuss, or judge pending innovations with broad, keen awareness of what those changes mean. In the technical realm we repeatedly enter into a series of social contracts, the terms of which are revealed only after the signing.

It may seem that the view I am suggesting is that of technological determinism: the idea that technological innovation is the basic cause of changes in society and that human beings have little choice other than to sit back and watch this ineluctable process unfold. But the concept of determinism is much too strong, far too sweeping in its implications to provide an adequate theory. It does little justice to the genuine choices that arise, in both principle and practice, in the course of technical and social transformation. Being saddled with it is like attempting to describe all instances of sexual intercourse based only on the concept of rape. A more revealing notion, in my view, is that of technological somnambulism. For the interesting puzzle in our times is that we so willingly sleepwalk through the process of reconstituting the conditions of human existence.

---

[4] See Kolender et al. (983). Edward Lawson had been arrested approximately fifteen times on his long walks and refused to provide identification when stopped by the police. Lawson cited his rights guaranteed by the Fourth and Fifth Amendments of the U.S. Constitution. The Court found the California vagrancy statute requiring "credible and reliable" identification to be unconstitutionally vague. See also Mann (1983, pp. 1, 19).

3. BEYOND IMPACTS AND SIDE EFFECTS

Social Scientists have tried to awaken the sleeper by developing methods of technology assessment. The strength of these methods is that they shed light on phenomena that were previously overlooked. But an unfortunate shortcoming of technology assessment is that it tends to see technological change as a "cause" and everything that follows as an "effect" or "impact." The role of the researcher is to identify, observe, and explain these effects. This approach assumes that the causes have already occurred or are bound to do so in the normal course of events. Social research boldly enters the scene to study the "consequences" of the change. After the bulldozer has rolled over us, we can pick ourselves up and carefully measure the treadmarks. Such is the impotent mission of technological "impact" assessment.

A somewhat more farsighted version of technology assessment is sometimes used to predict which changes are likely to happen, the "social impacts of computerization" for example. With these forecasts at its disposal, society is, presumably, better able to chart its course. But, once again, the attitude in which the predictions are offered usually suggests that the "impacts" are going to happen in any case. Assertions of the sort "Computerization will bring about a revolution in the way we educate our children" carry the strong implication that those who will experience the change are obliged simply to endure it. Humans must adapt. That is their destiny. There is no tampering with the source of change, and only minor modifications are possible at the point of impact (perhaps some slight changes in the fashion contour of this year's treadmarks).

But we have already begun to notice another view of technological development, one that transcends the empirical and moral shortcomings of cause-and-effect models. It begins with the recognition that as technologies are being built and put to use, significant alterations in patterns of human activity and human institutions are already taking place. New worlds are being made. There is nothing "secondary" about this phenomenon. It is, in fact, the most important accomplishment of any new technology. The construction of a technical system that involves human beings as operating parts brings a reconstruction of social roles and relationships. Often this is a result of a new system's own operating requirements: it simply will not work unless human behavior changes to suit its form and process. Hence, the very act of using the kinds of machines, techniques, and systems available to us generates patterns of activities and expectations that soon become "second nature." We do indeed "use" telephones, automobiles, electric lights, and computers in the conventional sense of picking them up and putting them down. But our world soon becomes one in which telephony, automobility, electric lighting, and computing are forms of life in the most powerful sense: life would scarcely be thinkable without them.

My choice of the term "forms of life" in this context derives from Ludwig Wittgenstein's elaboration of that concept in *Philosophical Investigations*. In his later writing Wittgenstein sought to overcome an extremely narrow view of the structure of language then popular among philosophers, a view that held language to be primarily a matter of naming things and events. Pointing to the richness and multiplicity of the kinds of expression or "language games" that are a part of everyday speech, Wittgenstein argued that "the speaking of language is a part of an activity, or of a form of life" (Wittgenstein, 1958, p. 11e). He gave a variety of examples – the giving of orders, speculating about events, guessing riddles, making up stories, forming and testing hypotheses, and so forth – to indicate the wide range of language games involved in various "forms of life." Whether he meant to suggest that these are patterns that occur naturally to all human beings or that they are primarily cultural conventions that can change with time and setting is a question open to dispute (Pitkin, 1972, p. 293). For the purposes here, what matters is not the ultimate philosophical status of Wittgenstein's concept but its suggestiveness in helping us to overcome another widespread and extremely narrow conception: our normal understanding of the meaning of technology in human life.

As they become woven into the texture of everyday existence, the devices, techniques, and systems we adopt shed their tool-like qualities to become part of our very humanity. In an important sense we become the beings who work on assembly lines, who talk on telephones, who do our figuring on pocket calculators, who eat processed foods, who clean our homes with powerful chemicals. Of course, working, talking, figuring, eating, cleaning, and such things have been parts of human activity for a very long time. But technological innovations can radically alter these common patterns and on occasion generate entirely new ones, often with surprising results. The role television plays in our society offers some poignant examples. None of those who worked to perfect the technology of television in its early years and few of those who brought television sets into their homes ever intended the device to be employed as the universal babysitter. That, however, has become one of television's most common functions in the modern home. Similarly, if anyone in the 1930s had predicted people would eventually be watching seven hours of television each day, the forecast would have been laughed away as absurd. But recent surveys indicate that we Americans do spend that much time, roughly one-third of our lives, staring at the tube. Those who wish to reassert freedom of choice in the matter sometimes observe, "You can always turn off your TV." In a trivial sense that is true. At least for the time being the on/off button is still included as standard equipment on most sets (perhaps someday it will become optional). But given how central television has become to the content of everyday life, how it has become the accustomed topic of conversation in workplaces, schools, and other social gatherings, it is apparent that television is a phenomenon that, in the larger sense, cannot be "turned off" at all. Deeply insinuated into people's perceptions, thoughts, and behavior, it has become an indelible part of modern culture.

Most changes in the content of everyday life brought on by technology can be recognized as versions of earlier patterns. Parents have always had to entertain and instruct children and to find ways of keeping the little ones out of their hair. Having youngsters watch several hours of television cartoons is, in one way of looking at the matter, merely a new method for handling this age-old task, although the "merely" is of no small significance. It is important to ask, Where, if at all, have modern technologies added *fundamentally new* activities to the range of things human beings do? Where and how have innovations in science and technology begun to alter the very *conditions of life* itself? Is computer programming only a powerful recombination of forms of life known for ages – doing mathematics, listing, sorting, planning, organizing, etc. – or is it something unprecedented? Is industrialized agribusiness simply a renovation of older ways of farming, or does it amount to an entirely new phenomenon?

Certainly, there are some accomplishments of modern technology, manned air flight, for example, that are clearly altogether novel. Flying in airplanes is not just another version of modes of travel previously known; it is something new. Although the hope of humans flying is as old as the myth of Daedalus and Icarus or the angels of the *Old Testament*, it took a certain kind of modern machinery to realize the dream in practice. Even beyond the numerous breakthroughs that have pushed the boundaries of human action, however, lie certain kinds of changes now on the horizon that would amount to a fundamental change in the conditions of human life itself. One such prospect is that of altering human biology through genetic engineering. Another is the founding of permanent settlements in outer space. Both of these possibilities call into question what it means to be human and what constitutes "the human condition."[5] Speculation about such matters is now largely the work of science fiction, whose notorious perversity as a literary genre signals the troubles that lie in wait when we begin thinking about becoming creatures fundamentally different from any the earth has seen. A great many futuristic novels are blatantly technopornographic.

But, on the whole, most of the transformations that occur in the wake of technological innovation are actually variations of very old patterns. Wittgenstein's philosophically conservative maxim "What has to be accepted, the given, is – so one could say – *forms of life*" could well be the guiding rule of a phenomenology of technical practice (Wittgenstein, 1958, p. 226e). For instance, asking a question and awaiting an answer, a form of interaction we all know well, is much the same activity whether it is a person we are confronting or a computer. There are, of course, significant differences between persons and computers (although it is fashionable in some circles to ignore them). Forms of life that we mastered before the coming of the computer shape our expectations as we begin to use the instrument. One strategy of software design, therefore, tries to "humanize" the computers by having them say "Hello" when the user logs in or having them respond with witty remarks when a person makes an error. We carry with us highly structured anticipations about entities that appear to participate, if only minimally, in forms of life and associated language games that are parts of human culture. Those anticipations provide much of the persuasive power of those who prematurely claim great advances in "artificial intelligence" based on narrow but impressive demonstrations of computer performance. But then children have always fantasized that their dolls were alive and talking.

The view of technologies as forms of life I am proposing has its clearest beginnings in the writings of Karl Marx. In Part I of *The German Ideology*, Marx and Engels explain the relationship of human individuality and material conditions of production as follows: "The way in which men produce their means of subsistence depends first of all on the nature of the means of subsistence they actually find in existence and have to reproduce. This mode of production must not be

---

[5] For a thorough discussion of this idea, see Arendt (1958, 1978).

considered simply as being the reproduction of the physical existence of the individuals. Rather it is a definite form of activity of these individuals, a definite form of expressing their life, a definite *mode of life* on their part. As individuals express their life, so they are" (Marx and Engels, 1976, p. 31).

Marx's concept of production here is a very broad and suggestive one. It reveals the total inadequacy of any interpretation that finds social change a mere "side effect" or "impact" of technological innovation. While he clearly points to means of production that sustain life in an immediate, physical sense, Marx's view extends to a general understanding of human development in a world of diverse natural resources, tools, machines, products, and social relations. The notion is clearly not one of occasional human interaction with devices and material conditions that leave individuals unaffected. By changing the shape of material things, Marx observes, we also change ourselves. In this process human beings do not stand at the mercy of a great deterministic punch press that cranks out precisely tailored persons at a certain rate during a given historical period. Instead, the situation Marx describes is one in which individuals are actively involved in the daily creation and recreation, production and reproduction of the world in which they live. Thus, as they employ tools and techniques, work in social labor arrangements, make and consume products, and adapt their behavior to the material conditions they encounter in their natural and artificial environment, individuals realize possibilities for human existence that are inaccessible in more primitive modes of production.

Marx expands upon this idea in "The Chapter on Capital" in the *Grundrisse*. The development of forces of production in history, he argues, holds the promise of the development of a many-sided individuality in all human beings. Capital's unlimited pursuit of wealth leads it to develop the productive powers of labor to a state "where the possession and preservation of general wealth require a lesser labour time of society as a whole, and where the labouring society relates scientifically to the process of its progressive reproduction, its reproduction in constantly greater abundance." This movement toward a general form of wealth "creates the material elements for the development of the rich individuality which is all-sided in its production as in its consumption, and whose labour also therefore appears no longer as labour, but as the full development of activity itself" (Marx, 1973, p. 325).

If one has access to tools and materials of woodworking, a person can develop the human qualities found in the activities of carpentry. If one is able to employ the instruments and techniques of music making, one can become (in that aspect of one's life) a musician. Marx's ideal here, a variety of materialist humanism, anticipates that in a properly structured society under modern conditions of production, people would engage in a very wide range of activities that enrich their individuality along many dimensions. It is that promise which, he argues, the institutions of capitalism thwart and cripple.[6]

As applied to an understanding of technology, the philosophies of Marx and Wittgenstein direct our attention to the fabric of everyday existence. Wittgenstein points to a vast multiplicity of cultural practices that comprise our common world.

---

[6] An interesting discussion of Marx in this respect is Axelos (1976).

Asking us to notice "what we say when," his approach can help us recognize the way language reflects the content of technical practice. It makes sense to ask, for example, how the adoption of digital computers might alter the way people think of their own faculties and activities. If Wittgenstein is correct, we would expect that changes of this kind would appear, sooner or later, in the language people use to talk about themselves. Indeed, it has now become commonplace to hear people say "I need to access your data." "I'm not programmed for that." "We must improve our interface." "The mind is the best computer we have."

Marx, on the other hand, recommends that we see the actions and interactions of everyday life within an enormous tapestry of historical developments. On occasion, as in the chapter on "Machinery and Large-Scale Industry" in *Capital*, his mode of interpretation also includes a place for a more microscopic treatment of specific technologies in human experience (Marx, 1976, ch. 15). But on the whole his theory seeks to explain very large patterns, especially relationships between different social classes, that unfold at each stage in the history of material production. These developments set the stage for people's ability to survive and express themselves, for their ways of being human.

4. RETURN TO MAKING

To invoke Wittgenstein and Marx in this context, however, is not to suggest that either one or both provide a sufficient basis for a critical philosophy of technology. Proposing an attitude in which forms of life must be accepted as "the given," Wittgenstein decides that philosophy "leaves everything as it is" (Wittgenstein, 1958, p. 49e). Although some Wittgensteinians are eager to point out that this position does not necessarily commit the philosopher to conservatism in an economic or political sense, it does seem that as applied to the study of forms of life in the realm of technology, Wittgenstein leaves us with little more than a passive traditionalism. If one hopes to interpret technological phenomena in a way that suggests positive judgments and actions, Wittgensteinian philosophy leaves much to be desired.

In a much different way Marx and Marxism contain the potential for an equally woeful passivity. This mode of understanding places its hope in historical tendencies that promise human emancipation at some point. As forces of production and social relations of production develop and as the proletariat makes its way toward revolution, Marx and his orthodox followers are willing to allow capitalist technology, for example, the factory system, to develop to its farthest extent. Marx and Engels scoffed at the Utopians, anarchists, and romantic critics of industrialism who thought it possible to make moral and political judgments about the course a technological society ought to take and to influence that path through the application of philosophical principles. Following this lead, most Marxists have believed that while capitalism is a target to be attacked, technological expansion is entirely good in itself, something to be encouraged without reservation. In its own way, then, Marxist theory upholds an attitude as nearly lethargic as the Wittgensteinian decision to "leave everything as it is." The famous eleventh thesis on Feuerbach – "The philosophers have only interpreted the world in various ways; the point, however, is to change it" – conceals an important qualification: that judgment, action, and change are ultimately products of history. In its view of technological development Marxism anticipates a history of rapidly evolving material productivity, an inevitable course of events in which attempts to propose moral and political limits have no place. When socialism replaces capitalism, so the promise goes, the machine will finally move into high gear, presumably releasing humankind from its age-old miseries.

Whatever their shortcomings, however, the philosophies of Marx and Wittgenstein share a fruitful insight: the observation that social activity is an ongoing process of world-making. Throughout their lives people come together to renew the fabric of relationships, transactions, and meanings that sustain their common existence. Indeed, if they did not engage in this continuing activity of material and social production, the human world would literally fall apart. All social roles and frameworks – from the most rewarding to the most oppressive – must somehow be restored and reproduced with the rise of the sun each day.

From this point of view, the important question about technology becomes, As we "make things work," what kind of *world* are we making? This suggests that we pay attention not only to the making of physical instruments and processes, although that certainly remains important, but also to the production of psychological, social, and political conditions as a part of any significant technical change. Are we going to design and build circumstances that enlarge possibilities for growth in human freedom, sociability, intelligence, creativity, and self-government? Or are we headed in an altogether different direction?

It is true that not every technological innovation embodies choices of great significance. Some developments are more-or-less innocuous; many create only trivial modifications in how we live. But in general, where there are substantial changes being made in what people are doing and at a substantial investment of social resources, then it always pays to ask in advance about the qualities of the artifacts, institutions, and human experiences currently on the drawing board.

Inquiries of this kind present an important challenge to all disciplines in the social sciences and humanities. Indeed, there are many historians, anthropologists, sociologists, psychologists, and artists whose work sheds light on long-overlooked human dimensions of technology. Even engineers and other technical professionals have much to contribute here when they find courage to go beyond the narrow-gauge categories of their training.

The study of politics offers its own characteristic route into this territory. As the political imagination confronts technologies as forms of life, it should be able to say something about the choices (implicit or explicit) made in the course of technological innovation and the grounds for making those choices wisely. That is a task I take up in the next two chapters. Through technological creation and many other ways as well, we make a world for each other to live in. Much more than we have acknowledged in the past, we must admit our responsibility for what we are making.

WORKS CITED

H. Arendt (1958) *The Human Condition* (Chicago: University of Chicago Press).
H. Arendt (1978) *Willing*, Vol. II of *The Life of the Mind* (New York: Harcourt Brace Jovanovich).

J. L. Austin (1961) *Philosophical Papers* (Oxford: Oxford University Press), 123–153.

Bibliography of the Philosophy of Technology (1973) C. Mitcham and R. Mackey (eds) (Chicago: University of Chicago Press).

Kostas Axelos' Alienation, Praxis and Technē in the Thought of Karl Marx (1976) Translated by R. Bruzina (Austin: University of Texas Press).

The Encyclopedia of Philosophy, 8 Vols., (1967) P. Edwards (editor-in-chief) (New York: Macmillan).

W. Kolender et al. (1983) "Petitioner v. Edward Lawson," Supreme Court Reporter 103: 1855–1867.

J. Mann (1983) 'State Vagrancy Law Voided as Overly Vague,' *Los Angeles Times*, May 3, 1983, 1, 19.

K. Marx (1976) *Capital*, vol. 1, translated by B. Fowkes, with an introduction by E. Mandel (Harmondsworth, England: Penguin Books).

K. Marx and F. Engels (1976) "The German Ideology," in Collected Works, vol. 5 (New York: International Publishers).

C. Mitcham (1978) "Types of Technology," in *Research in Philosophy and Technology*, P. Durbin (ed.) (Greenwich, Conn: JAI Press), 229–294.

H. Pitkin (1972) *Wittgenstein and Justice: On the Significance of Ludwig Wittgenstein for Social and Political Thought* (Berkeley: University of California Press).

S. H. Unger (1982) *Controlling Technology: Ethics and the Responsible Engineer* (New York: Holt, Rinehart and Winston).

L. Wittgenstein (1958) *Philosophical Investigations*, ed. 3, translated by G. E. M. Anscombe, with English and German indexes (New York: Macmillan).

T. Wolfe (1980) *The Right Stuff* (New York: Bantam Books).

DISCUSSION QUESTIONS

1. What does Winner mean by "technological somnambulism"? Do you agree with Winner that, as a culture, we suffer from it?

2. What does Winner mean by "form of life"? In what ways does he think that technologies constitute forms of life?

3. One example that Winner uses of a technology constituting a form of life is automobiles. Do any more recent technologies, such as modern genomics or smart phones, constitute a form of life? If so, in what areas and in what ways do they do so?

4. What is Winner's critique of the view that technologies are just neutral tools that can be used for either good or bad ends? How does his critique (and the examples that he uses in making it) relate to his view that technologies constitute forms of life?

5. Winner believes that the ethics of technology needs to go beyond "impacts and side effects." What does he mean by this? Do you agree with him, why or why not?

PART II

REPRODUCTIVE TECHNOLOGIES

Reproductive technologies are technologies that are used in the context of procreation, such as contraception, reproductive assistance technologies, and pre- and post-natal medical technologies. The selections in this section focus on human reproductive technologies since the innovation of in vitro fertilization in the late 1970s.

In many countries, 1–3% of all pregnancies are now the result of in vitro fertilization. In "Reproductive Technologies: Ethical Debates," Lucy Frith discusses an array of ethical issues associated with widespread use of in vitro fertilization, including whether it should be covered by insurance, whether it is permissible to pay gamete donors, and what should be the fate of unused embryos.

While not yet used on humans, cloning technologies are increasingly employed in reproduction of nonhuman animals. There is no technical reason that the same technologies could not be used for human reproduction, and while laws against human cloning have been passed in some countries, it is not illegal everywhere (including in the United States). In "Preventing a Brave New World," Leon Kass argues against the use of cloning technologies for both human reproduction and therapeutic purposes, based on the possible negative consequences of the technology (extrinsic reasons) as well as the nature of the technology itself (intrinsic reasons).

A reproductive technology that is already in use is sex selection. Increasingly, prospective parents are deciding to choose the sex of their child. In "The Ethics of Sex Selection," Inmaculada de Melo-Martin discusses the cases for and against the use of sex selection technologies. She argues that we ought to be concerned about the way in which the technology expresses and may perpetuate problematic gender expectations.

In the future, it is likely to be possible for parents to select not only the sex of their children, but also other genetic traits. In "Selecting Children: The Ethics of Reproductive Genetic Engineering," S. Matthew Liao evaluates several ethical principles that might be applied to evaluate genetic engineering in general, as well as to guide choices about particular traits.

Each of the technologies discussed in this section – in vitro fertilization, cloning, sex selection, and genetic engineering – restructures choices about reproduction and increases the power or agency of prospective parents with respect to conceiving and influencing the traits of their children. They are ethically controversial in part because they challenge norms and expectations regarding reproduction, family and parenting. They also raise a host of ethical issues specific to the technologies – for example, whether the technologies should be covered by insurance, whether people have a right to reproduce, and whether they will feed into problematic social practices.

More detailed summaries, as well as a listing of related readings, are located at the start of the chapters.

5 REPRODUCTIVE TECHNOLOGIES: ETHICAL DEBATES

Lucy Frith

CHAPTER SUMMARY

In this chapter, Lucy Frith explores the ethical issues associated with reproductive assistance technologies – technologies that aid conception – since the development of in vitro fertilization. She discusses not only whether reproductive assistance technologies are ethically acceptable, but whether people have a right to access to them. These topics intersect with issues such as whether infertility is a disease or condition to be treated, what it means to be a parent, and whether people have a right to reproduction. Frith also addresses ethical issues surrounding the practice of reproductive assistance, including payment for gamete donors and treatment of unused embryos.

RELATED READINGS

Introduction: Power (2.4); Form of Life (2.5); Individual Rights and Liberties (2.6.3)

Other Chapters: Langdon Winner, *Technologies as Forms of Life* (Ch. 4); Leon Kass, *Preventing a Brave New World* (Ch. 6); Inmaculada de Melo-Martin, *The Ethics of Sex Selection* (Ch. 7); S. Matthew Liao, *Selecting Children* (Ch. 8)

1. INTRODUCTION

Reproductive Technologies (RT) have given urgency to some perennial questions: the reasons for wanting children, what it means to be a parent, and the importance of the genetic bond between people. In 1978, scientific developments in embryology and embryo transfer culminated in the birth of the world's first in vitro fertilization baby in Britain. This event prompted extensive ethical debate over the acceptability of reproductive technologies, with opinions ranging from whole-hearted support to condemnation. Fertility treatment is now a multi-million dollar international phenomenon with increasing success rates and new techniques developing all the time. The numbers born from these techniques are also growing, with approximately 1 in every 100 babies born in the US a result of some form of RT (Sunderam et al., 2009). The ethical issues raised by RTs can be divided into two broad categories: (1) arguments for and against the use and development of RTs, and (2) the ethical dilemmas created by specific aspects of RTs. The discussion will largely focus on the Anglo-American debate over RTs, recognizing that these debates have taken different forms in other countries.

2. WHAT ARE REPRODUCTIVE TECHNOLOGIES?

I will use the term reproductive technologies (RTs) as a generic term for techniques that artificially assist conception. The main technique used in RTs is in vitro fertilization (IVF). Eggs (oocytes) are collected from the woman by a process of egg retrieval (this is done by trans vaginal ultrasound guidance) and mixed in a petri dish with sperm. Once fertilization has taken place, the embryo(s) are placed (transferred) into the uterus where the pregnancy can become established. There are other techniques that build on the basic IVF procedure. Intra cytoplasmic sperm injection (ICSI) is used for male factor infertility (where the man has severely impaired or few sperm). The procedure follows IVF, eggs are retrieved and then a single sperm is injected into the inner cellular structure of the egg to enable fertilization. Pre-implantation genetic diagnosis (PGD) also uses IVF procedures: after fertilization occurs genetic tests are performed to determine the presence or absence of a particular gene or chromosome prior to implantation of the embryo in the uterus.

3. THE DEBATE OVER RTs

When the first IVF baby was born in 1978 there was moral outrage, with the clinicians accused of 'playing God' and interfering with something that should be sacrosanct – the bringing of human life into the world. A Roman Catholic view is that RTs are a deviation from normal intercourse and, in separating the unitive and procreative aspects of sexual intercourse, they devalue the reproductive process. To introduce a third person into this process is seen as defiling the sanctity of marriage and the family. Early debates over RTs focused on trying to make a case for why they were necessary medical procedures. This was not only an academic argument: societal approval was needed so that research funding would be forthcoming and the field could be developed into a recognized sub-specialty of gynecology. The importance of infertility as a condition and how it is conceptualized is also important as it has implications for who pays for people's fertility treatment. It is estimated that around only a quarter of employer-provided health plans in the US cover fertility treatment (Hawkins, 2007). In the UK there have been debates over how much treatment people should receive on the National Health Service. At present, around three quarters of treatment is provided privately in the UK, making it one of the most privately provided medical interventions (Woodward, 2008). Thus, infertility treatment is generally not seen as a central aspect of health care provision.

3.1 Should we help the infertile?

RTs are often portrayed as a response to the needs of the infertile (approximately one in ten couples, although figures vary (ASRM, 2012)). This raises the initial question: on what basis should the infertile have their desire for a child met by medicine when there are many other competing demands on health care budgets?

3.1.1 What is infertility?

Robert Winston, a British fertility specialist, said, "Infertility is actually a terrible disease affecting our sexuality and well being" (Winston, 1985). A wide definition of health could be employed to support the claim that infertility is a disease: the World Health Organization (WHO) defines health as a complete sense of well being. It could be argued that infertility treatments are enabling the infertile to function as fully healthy individuals and should be part of health care provision. In response, one might argue that although infertility can be a consequence of certain physical problems, such as blocked fallopian tubes, there is no specific condition of infertility itself which medicine can be called upon to cure. However, many established medical treatments fall into this category. For instance, diabetes is not cured but its unpleasant side effects managed, so this response in itself does not constitute an argument against the merits of medical treatment for infertility.

Conversely, it has been argued that infertility is purely a social problem and one that does not need medical intervention. Infertility is often seen as a problem for the couple, rather than the individual. Couples could have counseling to come to terms with their problem, which is an inability to participate in social customs rather than any specific medical condition. Thus if medicine seeks a role in the treatment of infertility it could be said that this is medicalizing a problem that is predominantly social in origin (see below). One of the difficulties with this view is that it is hard to distinguish clearly between social and medical problems, as the two often have a complex and interactive relationship.

Whether one views infertility as a disease rests heavily on the view one takes as to the importance of having children (Overall, 2012). Professor Edwards, the doctor involved in the birth of the first IVF baby, considers the genetic pressures to have children to be part of the very foundation of our nature, and believes that it is these pressures that lie at the heart of most couples' desire for children. Thus, in order for people to live fulfilled lives, they must have certain basic needs satisfied, and having children is one such need. An alternative view of the desire for a child is that it is socially constructed. Here it is the social pressure to reproduce that creates and determines the desire for a child. This view highlights the existence of pronatalism, an attitude or policy that encourages reproduction and promotes the role of parenthood. Pronatalism particularly affects women who are encouraged to become mothers. In a patriarchal society, true femininity is often equated with childbearing, and motherhood is thereby regarded as a necessary aspect of womanhood. The way in which society pressures women to have children and the focus on genetic relationships can be said to be socially determined ways of constructing our reproductive relationships.

3.1.2 The right to reproduce

A further reason for RTs contends that the infertile have the same right to reproduce as do the fertile. This is an area of debate that has been changed radically by the scientific development of RTs. The right to reproduce, free from interference, is generally viewed as an important basic human right. Enshrined in the United Nations Declaration of Human Rights is the right of "men and women of full age … to marry and found a family" (Article 16). The US Supreme Court has on various occasions supported procreative liberty, both prohibitions on contraceptives (Griswold v. Connecticut 381 U.S. 479 (1965)) and compulsory sterilization have been upheld as unconstitutional (Skinner v. Oklahoma, 316 U.S. 535 (1942). These articles and rulings are usually understood as stipulating what can be called a negative right, that is a right not to have one's reproductive capacities interfered with against one's will.

The infertile, however, in order to exercise their freedom and therefore their right to reproduce, need some assistance. This assistance, it is argued, should be forthcoming, as a couple's interest in reproduction is the same no matter how reproduction takes place. Accordingly, the infertile have to rely on some notion of positive rights that place an obligation on others to provide them with the means they need to reproduce.

John Robertson (1994) has put forward a rights based argument to support the extended use of RTs. He begins by arguing that the concept of procreative liberty should be given primacy when making policy decisions in this area. Procreative liberty is the freedom to decide whether or not to reproduce. At first sight, this appears to be a negative right, not to have one's reproductive capacities interfered with. However, Robertson endorses subsidiary enabling rights to procreation, that is, someone has the right to something if it can be regarded as a prerequisite for procreation. This effectively turns a negative right (not to be interfered with) into a positive right (to have something made available to one). Thus, individuals have a right to RTs, since it enables the infertile to exercise their reproductive rights, and could be said to redress the disparity between the infertile and the fertile.

Robertson's claim is controversial and contested. The feminist philosopher Laura Purdy (1996) argues that Robertson adopts a position that blurs the distinction between negative and positive rights. He seems to infer that the strong right not to have one's reproductive capabilities interfered with implies an equally strong right to reproduce. Embedded in this discussion of reproductive rights is the assumption, made by Robertson, that the issues raised by natural reproduction are similar to those raised by artificial reproduction. Christine Overall (1993), a feminist critic of RTs, argues that the issues raised by the two forms of reproduction are fundamentally different, and correspondingly the justification for the right to use RTs must meet a different burden of proof from the right to be free from reproductive interference.

Hence, it is by no means clear that there is a good argument to support the claim that there should be a positive right to reproduce (that the infertile should have the means made available for them to reproduce) or what the practical (e.g. financial) implications of such a right would be. It is not generally recognized that just because the means are available to achieve some end, people have a right to that means.

3.2 Reproductive choice

RTs broaden the range of reproductive options that are available for both the infertile and, in certain circumstances, the fertile. Any extension of choice is frequently portrayed as desirable, and this is just as true for reproductive choices. Reproductive choice has been one of the main arguments used in favor of allowing developments and new techniques in RTs. Advocates of reproductive autonomy (e.g. Savulescu, 1999; Harris, 2004) endorse the pre-eminence of parental choice in most circumstances. The central claim of this view is that personal reproductive decisions should be free from interference unless they will cause serious harm to others.

There should be a presumption towards liberty unless exercising that liberty harms others (Feinberg, 1973). In other words, the burden of proof should reside with those wishing to restrict choices. Dahl sums up this argument when he says, "each citizen ought to have the right to live his life as he [*sic*] chooses so long as he [*sic*] does not infringe upon the rights of others. The state may interfere with the free choices of its citizens only to prevent serious harm to others" (Dahl, 2007:158).

This argument is sometimes reinforced by claims that reproductive choices are "integral to a person's sense of being" (Jackson, 2007: 48) and any restrictions therefore require even more robust justification than less important choices (Robertson, 1994). Therefore, it might be argued that as reproductive choice is very important – the desire for a child is a fundamental need – allowing people to exercise it is a good in itself and this good outweighs the production of a certain level of harm. In sum, there is a belief that the more important the particular choice, the stronger the case for restricting it has to be.

While some feminists argue that the existence of RTs extends women's procreative choices, since women are free to choose another set of reproductive options, others argue that women do not freely choose to use these procedures but that they respond to pressures that are exerted by society, particularly by their partners. Thus, the very existence of RTs can constrain and influence choices. Here what are presented as new options can quickly become seen as the standard of care that women have to actively refuse. It is argued that to be childless due to infertility is no longer seen as an acceptable option unless women have tried to conceive by using RTs. Even then, if conception does not take place, the feeling of failure and the stress, strain and costly medical treatment can all take their toll.

4. AREAS FOR ETHICAL DEBATE

Despite opposition, IVF is now generally recognized as a therapeutic technique. In this section, I shall consider a selection of current issues raised by the practice of RTs. It is not exhaustive, but gives an impression of the kinds of questions currently being debated.

4.1 Regulation of RTs

The UK was one of the first countries to introduce comprehensive regulation of RTs. The Warnock Committee was set up to consider the regulatory issues raised by RTs. It reported in 1984 and concluded that an Act of Parliament was necessary to regulate this area to ensure that scientific developments were conducted in an ethical way. In 1990, the Human Fertilisation and Embryology Act was passed which set out the parameters for the conduct of RTs) and established the Human Fertilisation and Embryology Authority (HFEA) (Morgan and Lee, 1991. The HFEA's job was to oversee the rapid developments that would take place and formulate ongoing policies. It also had a role as the regulator of clinics, setting good practice guidelines and conducting clinic inspections. The HFEA also keeps a record of all treatments carried out and maintains a register of donors and recipients. In 2010, the Government conducted a widespread review and recommended that the HFEA be abolished (McHale, 2010). Although this review was based almost purely on cost cutting criteria, it prompted a debate over the utility of the HFEA. One prominent clinician argued that clinical practice cannot "be dictated by rules which come from quangos [the HFEA], which do not have the flexibility to move with the variety of problems we deal with in clinical practice" (quoted in Guy, 2011). However, others argued that this area still needs a body like the HFEA to inspect clinics, provide day-to-day guidance and respond quickly to new developments.

In the U.S. the situation is very different, with no overarching legal framework but a reliance on professionals guidelines and state laws. In 2009, a Californian woman gave birth to octuplets after having IVF treatment and this promoted a call for greater regulation of infertility treatment in the US. The American Society of Reproductive Medicine (ASRM) convened a meeting in 2009 to discuss whether more regulation was needed in this area and produced a report (ASRM, 2010) arguing that RTs are, "one of most highly regulated of all medical practices in the United States" and therefore further regulation was not needed.

RTs are regulated at three different levels in the US. First, RTs are overseen at the federal level by the Fertility Clinic Success Rate and Certification Act (FCSRCA), Centers for Disease Control and Prevention (CDC), the Food and Drug Administration (FDA) and the Centers for Medicare and Medicaid Services (CMS). Second, state legislation regulates medical practitioners and includes specific regulations on RTs (in the main relating to insurance coverage). Third, the ASRM and its affiliate, the Society for Assisted Reproductive Technology (SART), offer professional self-regulation through guidelines and codes of conduct for clinics and staff. However, much of this regulation only requires voluntary compliance. For instance, a major plank in professional regulation, the ASRM and SART professional codes and guidelines, are essentially voluntary codes recommending, rather than enforcing, good practice, if membership is rescinded due to non-compliance clinics may still operate.

The multiple-birth rate provides an example of the limited impact of self and voluntary regulation in the US. It is widely acknowledged that the main problem with assisted reproductive technology is multiple pregnancies. This is a phenomenon largely attributable to the number of embryos transferred in a single IVF cycle. The ASRM first issued guidelines on the numbers of embryos to transfer in 1998, recommending no more than 3 embryos for women aged under 35 years, no more than 4 for women aged 35–40, and no more than 5 for women aged over 40 years. Following several revisions, the most recent guidance issued in 2009 (ASRM, 2009a) recommends that for women aged under 35 consideration should be given to transferring one embryo and no more than two.

The ASRM argues that an 80% decrease in the number of triplet births between 1999–2007 demonstrates the success of this 'self-policing' (ASRM, 2010). However, in 2006, transfer of three or more embryos was still common practice in the U.S. As a result, approximately 41% of ART infants born in 2006 were pre-term (thus posing a health risk to these babies), compared with approximately 13% of pre-term births in the general U.S. population, and 49% were born in multiple-birth deliveries, compared with 3% in the general U.S. population. The twin rate was 44%, compared with 3% in the general U.S. population, and the rate of triplets and higher-order multiples was 5%, approximately 25 times higher than the general US population rate (Sunderam et al., 2009). Hence, it is clear that not all clinics are following the guidelines. One study found that at least half of the clinicians surveyed would deviate from ASRM embryo transfer number guidelines in certain situations (Jungheim et al., 2010).

Other areas that could be subject to national regulations are age limits for treatment. For example, there has been debate over whether post-menopausal women should be treated and when it is reasonable to say someone is too old to have children. The welfare of any child produced from RTs is also an area that some have argued needs to be better regulated in the US. Although the welfare of the future child is often held to be of paramount importance, there is considerable disagreement over how such welfare is to be protected, and views on this have changed over the last 20 years. For instance, The Ethics Committee of the ASRM advised in 1994 that the best interests of the child are served when it is born and reared by a heterosexual couple in a stable marriage. In their 2006 guidance they now state that marital status and sexual orientation should not be factors that are automatically seen as detrimental to the welfare of the future child. In 2008, the updated HFEA Act replaced the criteria of the 'need for a father' with 'the need for supportive parenting' when assessing the welfare of the child to determine people's eligibility for treatment. Therefore, what features facilitate the future welfare of children are uncertain and attitudes have changed over time.

4.2 The treatment of embryos

The scientific development of IVF and related techniques has enabled embryos to be created outside the body and this has opened up new areas of debate.

4.2.1 The storage of embryos

In a typical IVF cycle around six embryos will be created. Not all of these will be implanted during treatment. Thus, the couple can either have their spare embryos destroyed, used for research or frozen for future use. If the IVF attempt fails or if future egg retrieval is impossible (due to the woman's age, for example), couples can use these stored embryos for further IVF treatments. There are a number of potential problems with the storage of embryos. How long should embryos be stored? Should there be a time limit or can they be stored indefinitely? Freezing embryos can create problems in the future if there are disputes over ownership, for instance if the couple split up or if one dies. Who should have the decision as to what to do with the embryos? In the UK, embryos are usually stored for a maximum of 10 years.

One response might be to use a detailed consent form setting up the storage terms to preempt these problems, but no consent form, however well worded, can anticipate every possible future circumstance. As the law stands in the UK, gamete (sperm and egg) donors and members of the couple undergoing treatment must give explicitly consent to any use of their gametes or participation in infertility treatment. However, in an area that changes rapidly due to scientific advances, how detailed should the consent be? How can any consent form adequately anticipate all future circumstances?

4.2.2 Embryo relinquishment

How to organize the transfer of stored frozen embryos from the individuals who created them to recipients to use in their fertility treatment is a subject of much debate. Recently, embryo adoption agencies have been established, though this model of embryo relinquishment has been heavily criticized.

There are many ways of organizing embryo relinquishment: it can operate on a continuum from anonymous relinquishment to procedures that have parallels with the adoption of an existing child – i.e. conditional relinquishment. At one end of the scale there is what could be termed a medically orientated model of anonymous embryo donation that would operate in a similar way to anonymous gamete donation. The embryo donors would generally play no part in any decision regarding the selection of recipients of their embryos, as anonymous gamete donors are rarely given a say over who receives their gametes. Those receiving the embryo(s) could get some information about the donors, but this (as is often the case in gamete donation) is generally limited to the physical characteristics and health status of the donor. The identity of the embryo donor would not be made available to the recipient(s) or to the future child.

A conditional embryo relinquishment model lies at the other end of the scale. This model involves allowing those relinquishing their embryos to vet and choose the recipient(s) of their embryos, and negotiate with them possible information exchange, contact, and involvement in the life of the future child. Even if the relinquishing couple chooses not to know the identity of recipients and plans no contact with them, conditional relinquishment programs generally operate as an 'open' non-anonymous system and stipulate that any child born should be able to learn the identity of her or his genetic parents. Therefore, records are kept and the possibility of the child having access to the identifying information about their genetic parents is ensured. Embryo adoption agencies can be seen to operate a form of conditional embryo relinquishment. The first embryo 'adoption' program, Snowflakes® Embryo Adoption, was launched in 1997 by Nightlight Christian Adoptions, a California-based adoption agency (Nightlight, 2012). There are now at least seven agencies in the United States offering an embryo 'adoption' service (Embryoadoption.org, 2012).

In 2009, The Ethics Committee of the ASRM published a report, *Defining Embryo Donation* (ASRM, 2009b), to consider the ethical acceptability of an embryo adoption model of embryo relinquishment. In this they argued that the "application of the term 'adoption' to embryos is inaccurate [and] misleading" (2009b:1818) and concluded that, "the donation of embryos for reproductive purposes is fundamentally a medical procedure intended to result in pregnancy and should be treated as such" (ASRM, 2009b: 1819). The Report goes on to argue that: "applying the procedural requirements of adoption designed to protect existing children to embryos is not ethically justifiable and has the potential for harm" (ASRM, 2009b:1819). The ASRM Ethics Committee's central argument against embryo 'adoption' is that the embryo is not a person and therefore applying the language of adoption to embryo relinquishment is deceptive, "because it reinforces a conceptualization of the embryo as a fully entitled legal being and thus leads to a series of procedures that are not appropriate" (2009b:1818).

However, it can be argued that a particular view of the moral status of the embryo does not lead, automatically, to a particular way of organizing the relinquishing of frozen embryos. A critique of embryo adoption based on the grounds that the embryo is not a person does not completely refute the justification for such programs, as those relinquishing the embryos may wish to organize the practice through an adoption agency. To focus the discussion at the level of the moral status of the embryo leads to the neglect of the experiences of those relinquishing their embryos and debates over how such practices can be best organized and good practice guidelines developed (Frith et al., 2011).

4.3 The donation of gametes

Infertility treatments use donor gametes for patients whose own gametes are not viable. Factors such as abnormal sperm findings, or a woman who is not ovulating but is otherwise able to carry a pregnancy, can indicate the need for donated material. This practice raises a number of ethical issues.

4.3.1 Gamete donor anonymity

One ethical dilemma that is still fiercely debated in assisted reproduction is whether children born by gamete (egg or sperm) donation should be allowed to have information about their gamete donor. The key issue is whether the donor offspring has a right to have access to *identifying* information about their gamete donor. The debate over whether to allow offspring to have *non-identifying* information has been less heated, with many commentators agreeing that non-identifying information should be made available.

The main reason that telling the child how he or she was conceived is thought to be important is that they can then find out information about their donor. The right of the child to have identifying information about their donor has been used by various legislatures to justify policies of non-anonymous gamete donation. The most common reason given for why knowledge of one's genetic origins is thought to be a right is that it is deemed essential for a person's well being. Much of the evidence for the harm caused by not knowing one's origins is drawn from the literature on adoption, but it can be questioned whether this is an accurate comparison. The position of donor offspring within the family differs from that of adoptive children: they have not been separated or removed from their genetic parents and they are often biologically related to one member of the couple. Still, it can be counted that donor offspring have just as much interest in knowing about their origins as adoptees. The absence of information about their genetic parent(s), including the lack of knowledge of their identity, can represent a missing part of their lives. While clearly the analogy between donor offspring and adoptees is not perfect, it appears that these two groups often have similar concerns about knowledge of their genetic identity.

Traditionally, gamete donation was largely practiced anonymously and any offspring would have no or little access to any information, identifying or non-identifying, about their donor. However, the trend is now towards making gamete donation non-anonymous. A number of countries (UK, Victoria Australia and Holland, for example) have changed their laws and stipulated that all gamete donation should be non-anonymous. In 2011, Washington State passed a law that mandated that all donor offspring should be allowed access to the medical history of their donor and identifying information at 18, subject to the donor allowing it. This is an important piece of legislation, as it is the first such law in the U.S., and allows donor offspring access to medical information. In other states this is not guaranteed, as donor medical records may be destroyed or not made available to offspring. However, this legislation only allows the offspring to know the identity of their donor if the donor agrees, so is not as far reaching as the law in the UK for instance, where all donors have to donate anonymously. The Victoria Parliament (Australia) Law Reform Committee (2012) has gone further and recommended retrospective non-anonymity for donors. Offspring of donors who donated under conditions of anonymity (before 1998) can now find out their donors' identity with the proviso of a contact veto. This radical move was justified on the grounds that the welfare and interests of these children are paramount and trump any interests the donor may have.

One major concern regarding non-anonymous gamete donation is that it could adversely affect the numbers of gamete donors, and this is frequently advanced as an argument against such a policy. There are two responses to this contention. First, it is uncertain that even in countries that have recently removed gamete donor anonymity that the fall in donor numbers is solely to do with the removal of anonymity. Second, if non-anonymous donation is the morally right way of organizing gamete donation, then the low donor numbers is the price that has to be paid for a morally sound system.

4.3.2 Payment of donors

It is argued by the HFEA that altruistic donation is to be encouraged and that donors of gametes should not receive any payment over and above a minimal compensation for time and inconvenience. It is thought that any payment could have the detrimental effect of encouraging inappropriate motivation (such as merely seeking financial benefit) on the part of the donors. Excessive inducement to donate gametes could possibly exploit the less well off in society, as it is likely that they would be the ones to respond to financial inducement. This concern reflects the general problem of the ethical acceptability of paying people for their body parts (i.e. organs) or body products (i.e. blood). It could be argued that there are additional ethical problems in the case of selling gametes as the vendors are not only selling body products, but also their genetic material. Conversely, one could argue, as Harris does (Harris, 1985), that not allowing payment is compounding the problems of those who are less well off by depriving them of a source of income. In the UK, donors are paid reasonable expenses. In the US, payment for gametes is allowed, but the ASRM guidance (2007) suggests acceptable limits for egg donors (i.e. amounts above $10,000 are not appropriate), and this is framed partly as compensation for the time, effort and risk the women have to undergo to donate eggs. However, internationally, with the development of the sale of gametes over the internet, donors can sell their wares on the open market to the highest bidder.

4.4 Social egg freezing

A relatively recent development has been advances in the technological ability to freeze eggs (oocytes). It is the age of the oocytes that affects fertility, not the age of the woman, and some women are choosing to freeze their eggs when they are young as 'an insurance policy' against age related fertility decline. In the UK, regulations state that eggs may be frozen for a basic storage period of 10 years. This can be extended in certain circumstances to a total time period of 55 years if a doctor confirms in writing that either gamete provider or the intended recipient is 'prematurely infertile'. Professional organizations have been skeptical about the acceptability of using social egg freezing – that is, freezing one's egg when 'young' to counter-act age related fertility decline (Cutting et al., 2009). The ASRM (2008) has said oocyte cryopreservation is an experimental procedure that should not be offered or marketed as a means to defer reproductive aging, primarily because data relating to clinical outcomes are limited.

The ethical arguments for such social egg freezing include (Goold and Savulescu, 2009):

- Reproductive autonomy – social egg freezing should be allowed as it extends individual's choices
- Redresses reproductive gender equality – women suffer from steeply declining fertility after the age of 35 and although men's' fertility does decline it does not do some completely or so soon
- Egg freezing is allowed for those with iatrogenic infertility – people undergoing cancer treatment have been able to freeze their gametes before treatment that would have rendered them infertile and therefore it is argued that social egg freezing should also be allowed
- Avoids problems of embryo freezing – such as one's partner withdrawing consent, and would also allow a women to reproduce with a partner she met later

On the other hand, there are counter arguments to the acceptability of social egg freezing: it can be said to over-medicalize reproduction, perfectly healthy women would have to undergo egg retrieval techniques. The chances of pregnancy with egg freezing are hard to quantify and the number of frozen eggs needed to achieve a pregnancy is not wa given. Consequently, organizing one's life around the assumption that one could become pregnant later on might be a false hope. Therefore, social egg freezing could result in harm to the women undergoing the procedure. If the procedure is offered, there needs to a robust consent procedure so women can be made aware of: success rates; health risks; storage (cost, regulations); and the psychological aspects of the process (from harvesting, through storage and implanting).

5. CONCLUSION

RTs raise many difficult ethical dilemmas. One of the difficulties in both regulating RTs and considering their ethical implications is the speed at which new techniques and processes are developed. It is imperative that such developments only proceed after careful consideration as to the short- and long-term implications, for those who use them, the child and family that they create, and for society as a whole.

WORKS CITED

American Society for Reproductive Medicine (ASRM) (2010) *Oversight of Assisted Reproductive Technology*, accessed 5th April, 2012 at http://www.asrm.org/uploadedFiles/Content/About_Us/Media_and_Public_Affairs/OversiteOfART%20(2).pdf

American Society for Reproductive Medicine (ASRM) (2009a) 'Guidelines on the number of embryos transferred,' *Fertility & Sterility* 92:1518–19.

American Society for Reproductive Medicine (ASRM) (2009b) 'Defining embryo donation,' *Fertility & Sterility*, 92:1818–9.

American Society for Reproductive Medicine (ASRM) (2008) Essential elements of informed consent for elective oocyte cryopreservation: a Practice Committee opinion, *Fertility & Sterility*, 90:134–5.

American Society for Reproductive Medicine (ASRM) (2007) 'Financial compensation of oocyte donors,' *Fertility & Sterility*, 88:305–309.

R. Cutting, S. Barlow, and R. Anderson (2009) 'Human oocyte cryopreservation: Evidence for practice,' *Human Fertility*, 2 (3):125–136.

E. Dahl (2007) 'The 10 most common objections to sex-selection and why they are far from being conclusive: a Western perspective,' *Reproductive BioMedicine Online*, 14:158–161.

J. Feinberg (1973) *Social Philosophy* (Englewood Cliffs, NJ: Prentice Hall).

L. Frith, E. Blyth, MS. Paul, and R. Berger (2011) 'Conditional Embryo Relinquishment: Choosing to relinquish embryos for family building through an embryo 'adoption' programme,' *Human Reproduction*, 26 (12): 3327–3338.

I. Goold, and J. Savulescu (2009) 'In Favour of Freezing Eggs for Non-Medical Reasons,' *Bioethics*, 23, 1:January.

S. Guy (2011) 'The end of the HFEA,' *Bionews*, 592.

J. Harris (2004) *On Cloning: Thinking in Action* (London: Routledge).

J. Harris (1985) *The Value of Life* (London: Routledge).

J. Hawkins (2007) 'Separating Fact from Fiction: Mandated Insurance Coverage of Infertility Treatments,' *Journal of Law & Policy*, 23:203–227.

E. Jackson (2007) 'Rethinking the pre-conception welfare principle,' in Horsey, K. and Biggs, H. (eds.) *Human Fertilisation and Embryology: Reproducing Regulation* (London: Routledge-Cavendish).

E. Jungheim, L. Ginny, G. Ryan, et al. (2010) 'Embryo transfer practices in the United States: a survey of clinics registered with the Society for Assisted Reproductive Technology,' *Fertility & Sterility*, 94 (4):1432–1436.

J. McHale (2010) 'The bonfire of the regulators,' *British Journal of Nursing*, 19:1124–1125.

D. Morgan, and R. Lee (1991) *Blackstone's Guide to the Human Fertilisation and Embryology Act 1990* (London: Blackstone Press Limited).

Nightlight Christian Adoptions, http://www.nightlight.org/adoption-services/default.aspx, accessed February 2012.

C. Overall (2012) *Why have children? The ethical debate* (Cambridge, MA: MIT Press).

C. Overall (1993) *Human Reproduction: Principles, Practices, Policies* (Toronto: Oxford University Press).

L. Purdy (1996) '*Reproducing persons: issues in feminist bioethics*,' Cornell University Press.

J. Robertson (1994) *Children of Choice: Freedom and the New Reproductive Technologies* (Princeton: Princeton University Press).

J. Savulescu (1999) 'Sex selection: the case for,' *Medical Journal of Australia*, 171:373–375.

S. Sunderam, J. Chang, L. Flowers, et al (2009) 'Assisted Reproductive Technology Surveillance – United States 2006,' MMWR 2009 58 (SS05):1–25, accessed August 10, 2010 at http://www.cdc.gov/mmwr/preview/mmwrhtml/ss5805a1.htm?s_cid=ss5805a1_e).

M. Warnock (1985) *A Question of Life: The Warnock Report* (Oxford: Basil Blackwell).

R. Winston (1985) 'Why we need to experiment,' *The Observer*, 10th February.

P. Woodward (2008) (Press Officer HFEA) Personal communication by email.

DISCUSSION QUESTIONS

1. Should infertility be considered a disease, or a state of unhealthiness?
2. If you were a manager of a health insurance scheme would you cover IVF? What would your arguments be for and against offering coverage?
2. Should insurance plans limit access to infertility treatment (i.e. restrict it to married couples, or those who have not already had children)? Or should it be available to anyone who seeks to have children?
3. Is there a right to have IVF? Or a positive right to the means to have children more generally?
4. How do IVF and other fertility treatments restructure choices about having children?
5. Who is empowered and who is disempowered by IVF and other fertility treatments? And in what ways?
6. Is there anything ethically problematic with freezing gametes or embryos for later use?
7. Should gamete donors be anonymous to any future child born? What arguments do you think best support and counter this proposal?
9. Do you think gamete donors should be paid? Why or why not?

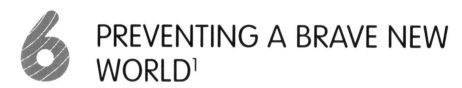

6 PREVENTING A BRAVE NEW WORLD[1]

Leon Kass

CHAPTER SUMMARY

In this chapter, Leon Kass, former chairperson of the United States President's Council on Bioethics, discusses the ethics of human reproductive cloning. Kass begins by critically assessing the presumption that technological innovations in reproduction should be allowed. He then offers several different types of objections, both intrinsic and extrinsic, to reproductive cloning, including: it may have detrimental effects on cloned children and on society; it will open the door to designing children; it expresses and encourages problematic attitudes toward reproduction and parenting; and that it is hubristic. For these reasons, Kass argues that the repugnance response that many people have toward human reproductive cloning is appropriate. He further argues that human reproductive cloning should be prohibited, and that to accomplish this all forms of human cloning, including cloning for therapeutic purposes, need to be banned.

RELATED READINGS

Introduction: Innovation Presumption (2.1); Form of Life (2.5); EHS (2.6.1); Dual Use (2.6.5); Intrinsic Concerns (2.7)

Other Chapters: Lucy Frith, *Reproductive Technologies* (Ch. 5); Inmaculada de Melo-Martin, *The Ethics of Sex Selection* (Ch. 7); S. Matthew Liao, *Selecting Children* (Ch. 8); Inmaculada de Melo-Martin and Marin Gillis, *Ethical Issues in Human Stem Cell Research* (Ch. 9)

1. INTRODUCTION

All contemporary societies are travelling briskly in the same utopian direction. All are wedded to the modern technological project; all march eagerly to the drums of progress and fly proudly the banner of modern science; all sing loudly the Baconian anthem, "Conquer nature, relieve man's estate." Leading the triumphal procession is modern medicine, which is daily becoming ever more powerful in its battle against disease, decay, and death, thanks especially to astonishing

---

[1] This chapter is excerpted from Leon Kass (2001, June 21) 'Preventing a Brave New World,' *The New Republic Online*. It appears here by permission of the author.

achievements in biomedical science and technology – achievements for which we must surely be grateful.

Yet contemplating present and projected advances in genetic and reproductive technologies, in neuroscience and psychopharmacology, and in the development of artificial organs and computer-chip implants for human brains, we now clearly recognize new uses for biotechnical power that soar beyond the traditional medical goals of healing disease and relieving suffering. Human nature itself lies on the operating table, ready for alteration, for eugenic and psychic "enhancement," for wholesale re-design. In leading laboratories, academic and industrial, new creators are confidently amassing their powers and quietly honing their skills, while on the street their evangelists are zealously prophesying a post-human future. For anyone who cares about preserving our humanity, the time has come to pay attention.

Some transforming powers are already here. The Pill. In vitro fertilization. Bottled embryos. Surrogate wombs. Cloning. Genetic screening. Genetic manipulation. Organ harvesting. Mechanical spare parts. Chimeras. Brain implants. Ritalin for the young, Viagra for the old, Prozac for everyone. And, to leave this vale of tears, a little extra morphine accompanied by Muzak.

Years ago Aldous Huxley saw it coming. In his charming but disturbing novel, Brave New World (it appeared in 1932 and is more powerful on each re-reading), he made its meaning strikingly visible for all to see. Unlike other frightening futuristic novels of the past century, such as Orwell's already dated Nineteen Eighty-Four, Huxley shows us a dystopia that goes with, rather than against, the human grain. Indeed, it is animated by our own most humane and progressive aspirations. Following those aspirations to their ultimate realization, Huxley enables us to recognize those less obvious but often more pernicious evils that are inextricably linked to the successful attainment of partial goods.

Huxley depicts human life seven centuries hence, living under the gentle hand of humanitarianism rendered fully competent by genetic manipulation, psychoactive drugs, hypnopaedia, and high-tech amusements. At long last, mankind has succeeded in eliminating disease, aggression, war, anxiety, suffering, guilt, envy, and grief. But this victory comes at the heavy price of homogenization, mediocrity, trivial pursuits, shallow attachments, debased tastes, spurious contentment, and souls without loves or longings. The Brave New World has achieved prosperity, community, stability, and nigh-universal contentment, only to be peopled by creatures of human shape but stunted humanity. They consume, fornicate, take "soma," enjoy "centrifugal bumble-puppy," and operate the machinery that makes it all possible. They do not read, write, think, love, or govern themselves. Art and science, virtue and religion, family and friendship are all *passe*. What matters most is bodily health and immediate gratification: "Never put off till tomorrow the fun you can have today." Brave New Man is so dehumanized that he does not even recognize what has been lost.

In Huxley's novel, everything proceeds under the direction of an omnipotent – albeit benevolent – world state. Yet the dehumanization that he portrays does not really require despotism or external control. To the contrary, precisely because the society of the future will deliver exactly what we most want – health, safety, comfort, plenty, pleasure, peace of mind and length of days – we can reach the same humanly debased condition solely on the basis of free human choice. No need for World Controllers. Just give us the technological imperative, liberal democratic society, compassionate humanitarianism, moral pluralism, and free markets, and we can take ourselves to a Brave New World all by ourselves – and without even deliberately deciding to go. In case you had not noticed, the train has already left the station and is gathering speed, but no one seems to be in charge.

Some among us are delighted, of course, by this state of affairs: some scientists and biotechnologists, their entrepreneurial backers, and a cheering claque of sci-fi enthusiasts, futurologists, and libertarians. There are dreams to be realized, powers to be exercised, honors to be won, and money – big money – to be made. But many of us are worried, and not, as the proponents of the revolution self-servingly claim, because we are either ignorant of science or afraid of the unknown. To the contrary, we can see all too clearly where the train is headed, and we do not like the destination. We can distinguish cleverness about means from wisdom about ends, and we are loath to entrust the future of the race to those who cannot tell the difference. No friend of humanity cheers for a post-human future.

Yet for all our disquiet, we have until now done nothing to prevent it.... Truth be told, it will not be easy for us to do so, and we know it. But rising to the challenge requires recognizing the difficulties. For there are indeed many features of modern life that will conspire to frustrate efforts aimed at the human control of the biomedical project. First, we...believe in technological automatism: where we do not foolishly believe that all innovation is progress, we fatalistically believe that it is inevitable ("If it can be done, it will be done, like it or not"). Second, we believe in freedom: the freedom of scientists to inquire, the freedom of technologists to develop, the freedom of entrepreneurs to invest and to profit, the freedom of private citizens to make use of existing technologies to satisfy any and all personal desires, including the desire to reproduce by whatever means. Third, the biomedical enterprise occupies the moral high ground of compassionate humanitarianism, upholding the supreme values of modern life – cure disease, prolong life, relieve suffering – in competition with which other moral goods rarely stand a chance. ("What the public wants is not to be sick," says James Watson, "and if we help them not to be sick, they'll be on our side.")

There are still other obstacles. Our cultural pluralism and easygoing relativism make it difficult to reach consensus on what we should embrace and what we should oppose; and moral objections to this or that biomedical practice are often facilely dismissed as religious or sectarian. Many people are unwilling to pronounce judgments about what is good or bad, right and wrong, even in matters of great importance, even for themselves – never mind for others or for society as a whole. It does not help that the biomedical project is now deeply entangled with commerce: there are increasingly powerful economic interests in favor of going full steam ahead, and no economic interests in favor of going slow. Since we live in a democracy, moreover, we face political difficulties in gaining a consensus to direct our future, and we have almost no political experience in trying to curtail the development of any new biomedical technology. Finally, and perhaps most troubling, our views of the meaning of our humanity have been so transformed by the scientific-technological approach to the world that we are in danger of forgetting what we have to lose, humanly speaking.

But though the difficulties are real, our situation is far from hopeless. Regarding each of the aforementioned impediments, there is another side to the story.

Though we love our gadgets and believe in progress, we have lost our innocence regarding technology. The environmental movement especially has alerted us to the unintended damage caused by unregulated technological advance, and has taught us how certain dangerous practices can be curbed. Though we favor freedom of inquiry, we recognize that experiments are deeds and not speeches, and we prohibit experimentation on human subjects without their consent, even when cures from disease might be had by unfettered research; and we limit so-called reproductive freedom by proscribing incest, polygamy, and the buying and selling of babies.

Although we esteem medical progress, biomedical institutions have ethics committees that judge research proposals on moral grounds, and, when necessary, uphold the primacy of human freedom and human dignity even over scientific discovery. Our moral pluralism notwithstanding, national commissions and review bodies have sometimes reached moral consensus to recommend limits on permissible scientific research and technological application. On the economic front, the patenting of genes and life forms and the rapid rise of genomic commerce have elicited strong concerns and criticisms, leading even former enthusiasts of the new biology to recoil from the impending commodification of human life. Though we lack political institutions experienced in setting limits on biomedical innovation, federal agencies years ago rejected the development of the plutonium-powered artificial heart, and we have nationally prohibited commercial traffic in organs for transplantation, even though a market would increase the needed supply. In recent years, several American states and many foreign countries have successfully taken political action, making certain practices illegal and placing others under moratoriums (the creation of human embryos solely for research; human germline genetic alteration). Most importantly, the majority of [us] are not yet so degraded or so cynical as to fail to be revolted by the society depicted in Huxley's novel. Though the obstacles to effective action are significant, they offer no excuse for resignation. Besides, it would be disgraceful to concede defeat even before we enter the fray.

Not the least of our difficulties in trying to exercise control over where biology is taking us is the fact that we do not get to decide, once and for all, for or against the destination of a post-human world. The scientific discoveries and the technical powers that will take us there come to us piecemeal, one at a time and seemingly independent from one another, each often attractively introduced as a measure that will "help [us] not to be sick." But sometimes we come to a clear fork in the road where decision is possible, and where we know that our decision will make a world of difference – indeed, it will make a permanently different world. Fortunately, we stand now at the point of such a momentous decision. Events have conspired to provide us with a perfect opportunity to seize the initiative and to gain some control of the biotechnical project. I refer to the prospect of human cloning, a practice absolutely central to Huxley's fictional world. Indeed, creating and manipulating life in the laboratory is the gateway to a Brave New World, not only in fiction but also in fact.

"To clone or not to clone a human being" is no longer a fanciful question. Success in cloning sheep, and also cows, mice, pigs, and goats, makes it perfectly clear that a fateful decision is now at hand: whether we should welcome or even tolerate the cloning of human beings. If recent newspaper reports are to be believed, reputable scientists and physicians have announced their intention to produce the first human clone in the coming year. Their efforts may already be under way.

The media, gawking and titillating as is their wont, have been softening us up for this possibility by turning the bizarre into the familiar. In the four years since the birth of Dolly the cloned sheep, the tone of discussing the prospect of human cloning has gone from "Yuck" to "Oh?" to "Gee whiz" to "Why not?" The sentimentalizers, aided by leading bioethicists, have downplayed talk about eugenically cloning the beautiful and the brawny or the best and the brightest. They have taken instead to defending clonal reproduction for humanitarian or compassionate reasons: to treat infertility in people who are said to "have no other choice," to avoid the risk of severe genetic disease, to "replace" a child who has died. For the sake of these rare benefits, they would have us countenance the entire practice of human cloning, the consequences be damned.

But we dare not be complacent about what is at issue, for the stakes are very high. Human cloning, though partly continuous with previous reproductive technologies, is also something radically new in itself and in its easily foreseeable consequences – especially when coupled with powers for genetic "enhancement" and germline genetic modification that may soon become available, owing to the recently completed Human Genome Project. I exaggerate somewhat, but in the direction of the truth: we are compelled to decide nothing less than whether human procreation is going to remain human, whether children are going to be made to order rather than begotten, and whether we wish to say yes in principle to the road that leads to the dehumanized hell of Brave New World.

2.

What is cloning? Cloning, or asexual reproduction, is the production of individuals who are genetically identical to an already existing individual. The procedure's name is fancy – "somatic cell nuclear transfer" – but its concept is simple. Take a mature but unfertilized egg; remove or deactivate its nucleus; introduce a nucleus obtained from a specialized (somatic) cell of an adult organism. Once the egg begins to divide, transfer the little embryo to a woman's uterus to initiate a pregnancy. Since almost all the hereditary material of a cell is contained within its nucleus, the re-nucleated egg and the individual into which it develops are genetically identical to the organism that was the source of the transferred nucleus.

An unlimited number of genetically identical individuals – the group, as well as each of its members, is called "a clone" – could be produced by nuclear transfer. In principle, any person, male or female, newborn or adult, could be cloned, and in any quantity; and because stored cells can outlive their sources, one may even clone the dead. Since cloning requires no personal involvement on the part of the person whose genetic material is used, it could easily be used to reproduce living or deceased persons without their consent – a threat to reproductive freedom that has received relatively little attention.

Some possible misconceptions need to be avoided. Cloning is not Xeroxing: the clone of Bill Clinton, though his genetic double, would enter the world hairless, toothless, and peeing in his diapers, like any other human infant. But neither is cloning just like natural twinning: the cloned twin will be identical to an older, existing adult; and it will arise not by chance but by deliberate design; and its entire genetic makeup will be pre-selected by its parents and/or scientists. Moreover, the success rate of cloning, at least at first, will probably not be very high: the Scots transferred two hundred seventy-seven adult nuclei into sheep eggs, implanted twenty-nine clonal embryos, and achieved the birth of only one live lamb clone.

For this reason, among others, it is unlikely that, at least for now, the practice would be very popular; and there is little immediate worry of mass-scale production of multicopies. Still, for the tens of thousands of people who sustain more than three hundred assisted-reproduction clinics in the United States and already avail themselves of in vitro fertilization and other techniques, cloning would be an option with virtually no added fuss. Panos Zavos, the Kentucky reproduction specialist who has announced his plans to clone a child, claims that he has already received thousands of e-mailed requests from people eager to clone, despite the known risks of failure and damaged offspring. Should commercial interests develop in "nucleus-banking," as they have in sperm-banking and egg-harvesting; should famous athletes or other celebrities decide to market their DNA the way they now market their autographs and nearly everything else; should techniques of embryo and germline genetic testing and manipulation arrive as anticipated, increasing the use of laboratory assistance in order to obtain "better" babies – should all this come to pass, cloning, if it is permitted, could become more than a marginal practice simply on the basis of free reproductive choice.

What are we to think about this prospect? Nothing good. Indeed, most people are repelled by nearly all aspects of human cloning: the possibility of mass production of human beings, with large clones of look-alikes, compromised in their individuality; the idea of father–son or mother–daughter "twins"; the bizarre prospect of a woman bearing and rearing a genetic copy of herself, her spouse, or even her deceased father or mother; the grotesqueness of conceiving a child as an exact "replacement" for another who has died; the utilitarian creation of embryonic duplicates of oneself, to be frozen away or created when needed to provide homologous tissues or organs for transplantation; the narcissism of those who would clone themselves, and the arrogance of others who think they know who deserves to be cloned; the Frankensteinian hubris to create a human life and increasingly to control its destiny; men playing at being God. Almost no one finds any of the suggested reasons for human cloning compelling, and almost everyone anticipates its possible misuses and abuses. And the popular belief that human cloning cannot be prevented makes the prospect all the more revolting.

Revulsion is not an argument; and some of yesterday's repugnances are today calmly accepted – not always for the better. In some crucial cases, however, repugnance is the emotional expression of deep wisdom, beyond reason's power completely to articulate it. Can anyone really give an argument fully adequate to the horror that is father–daughter incest (even with consent), or bestiality, or the mutilation of a corpse, or the eating of human flesh, or the rape or murder of another human being? Would anybody's failure to give full rational justification for his revulsion at those practices make that revulsion ethically suspect?

I suggest that our repugnance at human cloning belongs in this category. We are repelled by the prospect of cloning human beings not because of the strangeness or the novelty of the undertaking, but because we intuit and we feel, immediately and without argument, the violation of things that we rightfully hold dear. We sense that cloning represents a profound defilement of our given nature as procreative beings, and of the social relations built on this natural ground. We also sense that cloning is a radical form of child abuse. In this age in which everything is held to be permissible so long as it is freely done, and in which our bodies are regarded as mere instruments of our autonomous rational will, repugnance may be the only voice left that speaks up to defend the central core of our humanity. Shallow are the souls that have forgotten how to shudder.

3.

Yet repugnance need not stand naked before the bar of reason. The wisdom of our horror at human cloning can be at least partially articulated, even if this is finally one of those instances about which the heart has its reasons that reason cannot entirely know. I offer four objections to human cloning: that it constitutes unethical experimentation; that it threatens identity and individuality; that it turns procreation into manufacture (especially when understood as the harbinger of manipulations to come); and that it means despotism over children and perversion of parenthood. Please note: I speak only about so-called reproductive cloning, not about the creation of cloned embryos for research. The objections that may be raised against creating (or using) embryos for research are entirely independent of whether the research embryos are produced by cloning. What is radically distinct and radically new is reproductive cloning.

Any attempt to clone a human being would constitute an unethical experiment upon the resulting child-to-be. In all the animal experiments, fewer than two to three percent of all cloning attempts succeeded. Not only are there fetal deaths and stillborn infants, but many of the so-called "successes" are in fact failures. As has only recently become clear, there is a very high incidence of major disabilities and deformities in cloned animals that attain live birth. Cloned cows often have heart and lung problems; cloned mice later develop pathological obesity; other live-born cloned animals fail to reach normal developmental milestones.

The problem, scientists suggest, may lie in the fact that an egg with a new somatic nucleus must re-program itself in a matter of minutes or hours (whereas the nucleus of an unaltered egg has been prepared over months and years). There is thus a greatly increased likelihood of error in translating the genetic instructions, leading to developmental defects some of which will show themselves only much later. (Note also that these induced abnormalities may also affect the stem cells that scientists hope to harvest from cloned embryos. Lousy embryos, lousy stem cells.) Nearly all scientists now agree that attempts to clone human beings carry massive risks of producing unhealthy, abnormal, and malformed children. What are we to do with them? Shall we just discard the ones that fall short of expectations? Considered opinion is today nearly unanimous, even among scientists: attempts at human cloning are irresponsible and unethical. We cannot ethically even get to know whether or not human cloning is feasible.

If it were successful, cloning would create serious issues of identity and individuality. The clone may experience concerns about his distinctive identity not only because he will be, in genotype and in appearance, identical to another human being, but because he may also be twin to the person who is his "father" or his "mother" – if one can still call them that. Unaccountably, people treat as innocent the homey case of intra-familial cloning – the cloning of husband or wife (or single mother). They forget about the unique dangers of mixing the twin relation with the parent–child relation. (For this situation, the relation of contemporaneous twins is no precedent; yet even this less problematic situation teaches us how difficult it is to wrest independence from the being for whom one has the most powerful affinity.) Virtually no parent is going to be able to treat a clone of himself or herself as one treats a child generated by the lottery of sex. What will happen when the adolescent clone of Mommy becomes the spitting image of the woman with whom Daddy once fell in love? In case of divorce, will Mommy still love the clone of Daddy, even though she can no longer stand the sight of Daddy himself?

Most people think about cloning from the point of view of adults choosing to clone. Almost nobody thinks about what it would be like to be the cloned child. Surely his or her new life would constantly be scrutinized in relation to that of the older version. Even in the absence of unusual parental expectations for the clone – say, to live the same life, only without its errors – the child is likely to be ever a curiosity, ever a potential source of *deja vu*. Unlike "normal" identical twins, a cloned individual – copied from whomever – will be saddled with a genotype that has already lived. He will not be fully a surprise to the world: people are likely always to compare his doings in life with those of his alter ego, especially if he is a clone of someone gifted or famous. True, his nurture and his circumstance will be different; genotype is not exactly destiny. But one must also expect parental efforts to shape this new life after the original – or at least to view the child with the original version always firmly in mind. For why else did they clone from the star basketball player, the mathematician, or the beauty queen – or even dear old Dad – in the first place?

Human cloning would also represent a giant step toward the transformation of begetting into making, of procreation into manufacture (literally, "handmade"), a process that has already begun with in vitro fertilization and genetic testing of embryos. With cloning, not only is the process in hand, but the total genetic blueprint of the cloned individual is selected and determined by the human artisans. To be sure, subsequent development is still according to natural processes; and the resulting children will be recognizably human. But we would be taking a major step into making man himself simply another one of the man-made things.

How does begetting differ from making? In natural procreation, human beings come together to give existence to another being that is formed exactly as we were, by what we are – living, hence perishable, hence aspiringly erotic, hence procreative human beings. But in clonal reproduction, and in the more advanced forms of manufacture to which it will lead, we give existence to a being not by what we are but by what we intend and design.

Let me be clear. The problem is not the mere intervention of technique, and the point is not that "nature knows best." The problem is that any child whose being, character, and capacities exist owing to human design does not stand on the same plane as its makers. As with any product of our making, no matter how excellent, the artificer stands above it, not as an equal but as a superior, transcending it by his will and creative prowess. In human cloning, scientists and prospective "parents" adopt a technocratic attitude toward human children: human children become their artifacts. Such an arrangement is profoundly dehumanizing, no matter how good the product.

Procreation dehumanized into manufacture is further degraded by commodification, a virtually inescapable result of allowing baby-making to proceed under the banner of commerce. Genetic and reproductive biotechnology companies are already growth industries, but they will soon go into commercial orbit now that the Human Genome Project has been completed. "Human eggs for sale" is already a big business, masquerading under the pretense of "donation." Newspaper advertisements on elite college campuses offer up to $50,000 for an egg "donor" tall enough to play women's basketball and with SAT scores high enough for admission to Stanford; and to nobody's surprise, at such prices there are many young coeds eager to help shoppers obtain the finest babies money can buy. (The egg and womb-renting entrepreneurs shamelessly proceed on the ancient, disgusting, misogynist premise that most women will give you access to their bodies, if the price is right.) Even before the capacity for human cloning is perfected, established companies will have invested in the harvesting of eggs from ovaries obtained at autopsy or through ovarian surgery, practiced embryonic genetic alteration, and initiated the stockpiling of prospective donor tissues. Through the rental of surrogate-womb services, and through the buying and selling of tissues and embryos priced according to the merit of the donor, the commodification of nascent human life will be unstoppable.

Finally, the practice of human cloning by nuclear transfer – like other anticipated forms of genetically engineering the next generation – would enshrine and aggravate a profound misunderstanding of the meaning of having children and of the parent–child relationship. When a couple normally chooses to procreate, the partners are saying yes to the emergence of new life in its novelty – are saying yes not only to having a child, but also to having whatever child this child turns out to be. In accepting our finitude, in opening ourselves to our replacement, we tacitly confess the limits of our control.

Embracing the future by procreating means precisely that we are relinquishing our grip in the very activity of taking up our own share in what we hope will be the immortality of human life and the human species. This means that our children are not our children: they are not our property, they are not our possessions. Neither are they supposed to live our lives for us, or to live anyone's life but their own. Their genetic distinctiveness and independence are the natural foreshadowing of the deep truth that they have their own, never-before-enacted life to live. Though sprung from a past, they take an uncharted course into the future.

Much mischief is already done by parents who try to live vicariously through their children. Children are sometimes compelled to fulfill the broken dreams of unhappy parents. But whereas most parents normally have hopes for their children, cloning parents will have expectations. In cloning, such overbearing parents will have taken at the start a decisive step that contradicts the entire meaning of the open and forward-looking nature of parent–child relations. The child is given a genotype that has already lived, with full expectation that this blueprint of a past life ought to be controlling the life that is to come. A wanted child now means a child who exists precisely to fulfill parental wants. Like all the more precise eugenic manipulations that will follow in its wake, cloning is thus inherently despotic, for it seeks to make one's children after one's own image (or an image of one's choosing) and their future according to one's will.

Is this hyperbolic? Consider concretely the new realities of responsibility and guilt in the households of the cloned. No longer only the sins of the parents, but also the genetic choices of the parents, will be visited on the children – and beyond the third and fourth generation; and everyone will know who is responsible. No parent will be able to blame nature or the lottery of sex for an unhappy adolescent's big nose, dull wit, musical ineptitude, nervous disposition, or anything else that he hates about himself. Fairly or not, children will hold their cloners responsible for everything, for nature as well as for nurture. And parents, especially the better ones, will be limitlessly liable to guilt. Only the truly despotic souls will sleep the sleep of the innocent.

4.

The defenders of cloning are not wittingly friends of despotism. Quite the contrary. Deaf to most other considerations, they regard themselves mainly as friends of freedom: the freedom of individuals to reproduce, the freedom of scientists and inventors to discover and to devise and to foster "progress" in genetic knowledge and technique, the freedom of entrepreneurs to profit in the market. They want large-scale cloning only for animals, but they wish to preserve cloning as a human option for exercising our "right to reproduce" – our right to have children, and children with "desirable genes." As some point out, under our "right to reproduce" we already practice early forms of unnatural, artificial, and extra-marital reproduction, and we already practice early forms of eugenic choice. For that reason, they argue, cloning is no big deal.

We have here a perfect example of the logic of the slippery slope. The principle of reproductive freedom currently enunciated by the proponents of cloning logically embraces the ethical acceptability of sliding all the way down: to producing children wholly in the laboratory from sperm to term (should it become feasible), and to producing children whose entire genetic makeup will be the product of parental eugenic planning and choice. If reproductive freedom means the right to have a child of one's own choosing by whatever means, then reproductive freedom knows and accepts no limits.

Proponents want us to believe that there are legitimate uses of cloning that can be distinguished from illegitimate uses, but by their own principles no such limits can be found. (Nor could any such limits be enforced in practice: once cloning is permitted, no one ever need discover whom one is cloning and why.) Reproductive freedom, as they understand it, is governed solely by the subjective wishes of the parents-to-be. The sentimentally appealing case of the childless married couple is, on these grounds, indistinguishable from the case of an individual (married or not) who would like to clone someone famous or talented, living or dead. And the principle here endorsed justifies not only cloning but also all future artificial attempts to create (manufacture) "better" or "perfect" babies.

The "perfect baby," of course, is the project not of the infertility doctors, but of the eugenic scientists and their supporters, who, for the time being, are content to hide behind the skirts of the partisans of reproductive freedom and compassion for the infertile. For them, the paramount right is not the so-called right to reproduce, it is what the biologist Bentley Glass called, a quarter of a century ago, "the right of every child to be born with a sound physical and mental constitution, based on a sound genotype ... the inalienable right to a sound heritage." But to secure this right, and to achieve the requisite quality control over new human life, human conception and gestation will need to be brought fully into the bright light of the laboratory, beneath which the child-to-be can be fertilized, nourished, pruned, weeded, watched, inspected, prodded, pinched, cajoled, injected, tested, rated, graded, approved, stamped, wrapped, sealed, and delivered. There is no other way to produce the perfect baby.

If you think that such scenarios require outside coercion or governmental tyranny, you are mistaken. Once it becomes possible, with the aid of human genomics, to produce or to select for what some regard as "better babies" – smarter, prettier, healthier, more athletic – parents will leap at the opportunity to "improve" their offspring. Indeed, not to do so will be socially regarded as a form of child neglect. Those who would ordinarily be opposed to such tinkering will be under enormous pressure to compete on behalf of their as-yet unborn children – just as some now plan almost from their children's birth how to get them into Harvard. Never mind that, lacking a standard of "good" or "better," no one can really know whether any such changes will truly be improvements.

Proponents of cloning urge us to forget about the science-fiction scenarios of laboratory manufacture or multiple-copy clones, and to focus only on the sympathetic cases of infertile couples exercising their reproductive rights. But why, if the single cases are so innocent, should multiplying their performance be so offputting? (Similarly, why do others object to people's making money from that practice if the practice itself is perfectly acceptable?) The so-called science-fiction cases – say, Brave New World – make vivid the meaning of what looks to us, mistakenly, to be benign. They reveal that what looks like compassionate humanitarianism is, in the end, crushing dehumanization.

5.

Whether or not they share my reasons, most people, I think, share my conclusion: that human cloning is unethical in itself and dangerous in its likely consequences, which include the precedent that it will establish for designing our children. Some reach this conclusion for their own good reasons, different from my own: concerns about distributive justice in access to eugenic cloning; worries about the genetic effects of asexual "inbreeding"; aversion to the implicit premise of genetic determinism; objections to the embryonic and fetal wastage that must necessarily accompany the efforts; religious opposition to "man playing God." But never mind why: the overwhelming majority of [us] remain firmly opposed to cloning human beings.

For us, then, the real questions are: What should we do about it? How can we best succeed? These questions should concern everyone eager to secure deliberate human control over the powers that could re-design our humanity, even if cloning is not the issue over which they would choose to make their stand. And the answer to the first question seems pretty plain. What we should do is work to prevent human cloning by making it illegal.

We should aim for a global legal ban, if possible, and for a unilateral national ban at a minimum – and soon, before the fact is upon us. To be sure, legal bans can be violated; but we certainly curtail much mischief by outlawing incest, voluntary servitude, and the buying and selling of organs and babies. To be sure, renegade scientists may secretly undertake to violate such a law, but we can deter them by both criminal sanctions and monetary penalties, as well as by removing any incentive they have to proudly claim credit for their technological bravado.

Such a ban on clonal baby-making will not harm the progress of basic genetic science and technology. On the contrary, it will reassure the public that scientists are happy to proceed without violating the deep ethical norms and intuitions of the human community. It will also protect honorable scientists from a public backlash against the brazen misconduct of the rogues. As many scientists have publicly confessed, free and worthy science probably has much more to fear from a strong public reaction to a cloning fiasco than it does from a cloning ban, provided that the ban is judiciously crafted and vigorously enforced against those who would violate it.

I believe that what we need is an all-out ban on human cloning, including the creation of embryonic clones. I am convinced that all halfway measures will prove to be morally, legally, and strategically flawed, and – most important – that they will not be effective in obtaining the desired result. Anyone truly serious about preventing human reproductive cloning must seek to stop the process from the beginning.

Here's why. Creating cloned human children ("reproductive cloning") necessarily begins by producing cloned human embryos. Yet some scientists favor embryo cloning as a way of obtaining embryos for research or as sources of cells and tissues for the possible benefit of others. (This practice they misleadingly call "therapeutic cloning" rather than the more accurate "cloning for research" or "experimental cloning," so as to obscure the fact that the clone will be "treated" only to exploitation and destruction, and that any potential future beneficiaries and any future "therapies" are at this point purely hypothetical.)

But a few moments of reflection show why an anti-cloning law that permitted the cloning of embryos but criminalized their transfer to produce a child would be a moral blunder. This would be a law that was not merely permissively "pro-choice" but emphatically and prescriptively "anti-life." While permitting the creation of an embryonic life, it would make it a federal offense to try to keep it alive and bring it to birth. Whatever one thinks of the moral status or the ontological status of the human embryo, moral sense and practical wisdom recoil from having the government of the United States on record as requiring the destruction of nascent life and, what is worse, demanding the punishment of those who would act to preserve it by (feloniously!) giving it birth.

But the problem with the approach that targets only reproductive cloning (that is, the transfer of the embryo to a woman's uterus) is not only moral but also legal and strategic. A ban only on reproductive cloning would turn out to be unenforceable. Once cloned embryos were produced and available in laboratories and assisted-reproduction centers, it would be virtually impossible to control what was done with them.

For all these reasons, the only practically effective and legally sound approach is to block human cloning at the start, at the production of the embryo clone. Such a ban can be rightly characterized not as interference with reproductive freedom, nor even as interference with scientific inquiry, but as an attempt to prevent the unhealthy, unsavory, and unwelcome manufacture of and traffic in human clones.

6.

Some scientists, pharmaceutical companies, and bio-entrepreneurs may balk at such a comprehensive restriction. They want to get their hands on those embryos, especially for their stem cells, those pluripotent cells that can in principle be turned into any cells and any tissues in the body, potentially useful for transplantation to repair somatic damage. Embryonic stem cells need not come from cloned embryos, of course; but the scientists say that stem cells obtained from clones could be therapeutically injected into the embryo's adult "twin" without any risk of immunological rejection. It is the promise of rejection-free tissues for transplantation that so far has been the most successful argument in favor of experimental cloning. Yet new discoveries have shown that we can probably obtain the same benefits without embryo cloning. The facts are much different than they were three years ago, and the weight in the debate about cloning for research should shift to reflect the facts.

Numerous recent studies have shown that it is possible to obtain highly potent stem cells from the bodies of children and adults – from the blood, bone marrow, brain, pancreas, and, most recently, fat. Beyond all expectations, these non-embryonic stem cells have been shown to have the capacity to turn into a wide variety of specialized cells and tissues. (At the same time, early human therapeutic efforts with stem cells derived from embryos have produced some horrible results, the cells going wild in their new hosts and producing other tissues in addition to those in need of replacement. If an in vitro embryo is undetectably abnormal – as so often they are – the cells derived from it may also be abnormal.) Since cells derived from our own bodies are more easily and cheaply available than cells harvested from specially manufactured clones, we will almost surely be able to obtain from ourselves any needed homologous transplantable cells and tissues, without the need for egg donors or cloned embryonic copies of ourselves. By pouring our resources into adult stem cell research (or, more accurately, "non-embryonic" stem cell research), we can also avoid the morally and legally vexing issues in embryo research. And more to our present subject, by eschewing the cloning of embryos, we make the cloning of human beings much less likely....

[We] have lived by and prospered under a rosy optimism about scientific and technological progress. The technological imperative has probably served us well, though we should admit that there is no accurate method for weighing benefits and harms. And even when we recognize the unwelcome outcomes of technological advance, we remain confident in our ability to fix all the "bad" consequences – by regulation or by means of still newer and better technologies. Yet there is very good reason for shifting the paradigm, at least regarding those technological interventions into the human body and mind that would surely effect fundamental (and likely irreversible) changes in human nature, basic human relationships, and what it means to be a human being. Here we should not be willing to risk everything in the naive hope that, should things go wrong, we can later set them right again.

But the present danger posed by human cloning is, paradoxically, also a golden opportunity. In a truly unprecedented way, we can strike a blow for the human control of the technological project, for wisdom, for prudence, for human dignity. The prospect of human cloning, so repulsive to contemplate, is the occasion for deciding whether we shall be slaves of unregulated innovation, and ultimately its artifacts, or whether we shall remain free human beings who guide our powers toward the enhancement of human dignity. The humanity of the human future is now in our hands.

DISCUSSION QUESTIONS

1. What are Kass's ethical concerns regarding human reproductive cloning? Do these concerns establish that human reproductive cloning is "unethical in itself"? Do they show that it is "dangerous in its likely consequences"?

2. Do you find the analogy to "Brave New World" illuminating? Is Kass correct that the trajectory of our reproductive and biomedical technologies is in that direction?

3. Would Kass's concerns about reproductive cloning – for example, that it betrays a problematic attitude toward parenting and that it turns procreation into manufacturing – apply as well to intentional genetic modification of embryos?

4. Do you agree that children who are born as a result of human reproductive cloning are likely to be regarded or treated differently by their parents or society than those born from more traditional forms of reproduction? Why or why not?

5. Do you agree that we are adopting novel reproductive technologies without sufficient attention to the ways in which they restructure reproductive practices and parent–child relationships? Should we be concerned about the technologies' impacts on those practice and relationships?

6. How strong is Kass's argument for a ban on human reproductive cloning? Do you agree with him that the ban should extend to therapeutic cloning as well?

THE ETHICS OF SEX SELECTION

Inmaculada de Melo-Martin

CHAPTER SUMMARY

In this chapter, Inmaculada de Melo Martin addresses the ethical issues raised the technologies that enable parents to select the sex of their children. These technologies are rapidly proliferating. She begins by examining arguments in favor of the use of the technologies based on appeals to reproductive liberty and parental autonomy, and argues that they are not decisive. She then discusses both intrinsic and extrinsic concerns regarding the technologies. She argues that while advocates of sex selection have responses to the intrinsic concerns (or concerns about the nature of the technology itself), several of the extrinsic concerns regarding the technologies have merit, including those related to social harms. De Melo-Martin is particularly concerned about the ways in which sex selection technologies might both express and perpetuate problematic understandings and expectations regarding sex and gender.

RELATED READINGS

Introduction: Innovation Presumption (2.1); Extrinsic Concerns (2.6); Intrinsic Concerns (2.7); Types of Theories (3.2)

Other Chapters: Lucy Frith, *Reproductive Technologies* (Ch. 5); Leon Kass, *Preventing a Brave New World* (Ch. 6); S. Matthew Liao, *Selecting Children* (Ch. 8)

1. INTRODUCTION

The selection of children because of their sex is not a new practice (Sen 2003). Indeed, sex-selective abortion and infanticide have long been used as means of sex selection. However, the availability of prenatal diagnostic technologies has made sex selection significantly more common. For instance, several studies have documented the growing imbalance between the numbers of young girls and boys in countries such as India and China due to selective female abortions (Jha et al. 2011, Nie 2011). In China from 2005 to 2010, there were 1.20 male births per one female birth compared to 1.07 in the world (Nie 2011).

Newer technologies such as preimplantation genetic diagnosis (PGD) and sperm sorting can now be used to select the sex of one's children (Karabinus 2009, Ginsburg et al. 2011). Often, selection of sex through these technologies is done for medical reasons, such as diagnosing X-linked disorders, that is, disorders that are determined by the sex chromosomes or by a defective gene on a sex chromosome. But increasingly, these newer technologies are used to select children of a particular sex not because of medical concerns, but for social reasons (Ginsburg et al. 2011). Although, sex selection for medical purposes is usually thought to be ethically permissible, concerns about endorsement of sexist practices, disruption of the sex ratio, or worsening of sexist discrimination has lead the overwhelming majority of countries regulating PGD and sperm sorting to prohibit their use for sex selection for social reasons (BioPolicyWiki 2009). Professional societies and international organizations in their policy documents have also joined the opposition to this practice on similar grounds (ASRM 2004, ACOG 2007, FIGO 2006, Council of Europe 1997). For instance, the International Federation of Gynecology and Obstetrics rejects the use of any medical techniques that would intensify discrimination against either sex (FIGO 2006). Similarly, a 2007 position statement by the American College of Obstetricians and Gynecologists opposes meeting requests for sex selection for personal or family reasons because of the concern that such requests may ultimately support sexist practices (ACOG 2007).

Although all the existing methods of sex selection raise a wide range of ethical, legal, and social concerns, abortion and infanticide are also ethically controversial for other reasons. PGD is less ethically controversial because it involves embryos rather than fetuses or infants, and because the embryos are found outside of a woman's body. Sperm sorting, on the other hand, presents none of the problems related to embryo selection, abortion, or the killing of infants. This method takes advantage of the differences in DNA content between the X and the Y chromosomes to separate the X- and the Y-bearing sperm (Johnson et al. 1993). It is used in combination with intrauterine insemination or in vitro fertilization (IVF). The technology for sperm sorting is patented, but as of early 2012 it is not offered as a clinical practice. Clinical trials on its safety and efficacy are ongoing (Karabinus 2009).

While sperm sorting and PGD are less controversial methods of sex selection, PGD is significantly more effective than sperm sorting in ensuring a child of a desired sex, and the technology is now routinely offered in fertility clinics. Hence, this chapter will focus on the use of PGD as a means of sex selection for social reasons. First, I will briefly describe PGD. Next, I will offer an overview of some of the main ethical arguments offered in favor and against the use of PGD for embryo sex selection for non-medical reasons. In what follows, and unless otherwise stated, I use the term "sex selection" to refer to the practice of using PGD for sex selection for social or cultural reasons.

2. PREIMPLANTATION GENETIC DIAGNOSIS

PGD is used in combination with IVF, a reproductive technology developed to assist infertile couples in having genetically related children (Steptoe and Edwards, 1978). The process usually involves ovarian stimulation with fertility drugs in order to increase the chance of collecting multiple eggs during one of the woman's cycles. Eggs are then retrieved from the woman's ovaries through transvaginal technique involving an ultrasound-guided needle. The retrieved eggs are combined with appropriately prepared sperm on a petri dish containing culture media to enable fertilization. In some cases, a single sperm may be injected directly into the egg using a technique called intracytoplasmic sperm injection. Once fertilization occurs, the fertilized eggs are passed to a special growth medium and allowed to grow for about 48 hours until they consists of six to eight cells. It is at this stage that PGD can be used to select for particular embryos. Once embryos are selected, they are transferred to the woman's body for implantation (Zhao et al., 2011).

PGD was initially developed, as it is most frequently used, to test for chromosomal abnormalities, X-linked diseases such as fragile X syndrome, Duchenne muscular dystrophy, and hemophilia, or single gene disorders such as cystic fibrosis, b-thalassemia, and sickle cell anemia (Simpson 2010, Harper and Sengupta 2012). The first pregnancies obtained after the use of PGD occurred in 1990, when researchers biopsied a single cell at the six- to eight-cell stage to test for an X-linked disorder (Handyside et al. 1990). PGD is usually carried out on day 3 after fertilization in vitro, although biopsies on day 5 or 6 are also used depending on the reasons for PGD (Harper and Sengupta 2012). One or two cells are biopsied at the eight-cell stage and examined to determine whether particular chromosomal or other genetic abnormalities are present. Fluorescent in-situ hybridization, a technique that can detect and localize the presence or absence of specific DNA sequences on chromosomes, is normally used to examine the biopsied cells for chromosomal abnormalities and for sex determination in cases of X-linked disorders. Polymerase chain reaction, a molecular technique that allows the amplification of particular pieces of DNA, is usually employed to determine the existence of single gene disorders. After cells have been examined, only embryos that are thought to be unaffected are transferred into the woman's uterus.

Although PGD is normally used to determine whether embryos are affected by particular genetic conditions, the ability to identify the sex chromosomes permits the sex selection of embryos for non-medical reasons. Indeed, in the United States, the prevalence of sex selection has been increasing, in spite of the discouragement of the practice by the American Society for Reproductive Medicine. In 2008 for instance, PGD for sex selection accounted for about 20% of the use of this technique in non-research cycles (Ginsburg et al. 2011).

3. ARGUMENTS FOR SEX SELECTION FOR SOCIAL REASONS

Proponents of PGD for sex selection have insisted that objections to this practice are unfounded and legal bans are ethically unjustified (Dahl 2003, Dahl 2005b, Harris 2005, McCarthy 2001, Savulescu 1999, Savulescu and Dahl 2000, Robertson 2001, Robertson 2003b). Two arguments are usually offered to defend the ethical and legal legitimacy of sex selection: reproductive freedom and parental autonomy.

3.1 Reproductive freedom

Most of the proponents of sex selection base their defense on the importance of reproductive freedom (Robertson 1994, Savulescu and Dahl 2000, Dahl 2003, Harris 1998). They take reproductive liberty to be well-grounded on moral or constitutional bases (Robertson 1994, Dworkin 1993, Harris 1998, Dahl 2004).

Normally reproductive freedom is associated with the right not to reproduce. Indeed in the US, this right has been explicitly protected on constitutionals grounds in cases such as *Griswold v. Connecticut* and *Roe v. Wade*.

Some have argued, however, that reproductive freedom involves not only the right not to conceive – and thus a right to abortion and contraception – but also the right to reproduce (Robertson 1994). Childbearing and rearing are normally seen by most people as experiences of great significance that are tied to conceptions of a meaningful life. The right to reproduce would presumably include the freedom to reproduce both by coital means and by non-coital ones, such as the use of reproductive technologies (Robertson 1994). To the extent that reproductive liberty is thought to be a basic right, interference with reproductive decisions is seen as morally and legally legitimate only when they clearly and seriously harm others (Robertson 2003b, Dahl 2004, Harris 2005, Savulescu 2001).

Even if one accepts that reproductive liberty entails the right to reproduce, still further argument is needed to show that such a right also involves a right to have a particular child. Some advocates of sex selection have done just that. For instance, Robertson (1994, 1996, 2001, 2003b) argues that whether a particular reproductive activity, such as sex selection, is included within the scope of reproductive liberty is determined by evaluating whether sex is a characteristic that is central to the reproductive decision of the potential parents. If having a child of a particular sex determines whether reproduction will occur, that is, if parents will not reproduce unless they are able to use sex selection, then this activity, Robertson argues, falls under the scope of reproductive freedom.

Proponents of sex selection thus maintain that there is a presumption in favor of reproductive freedom and hence of allowing people to use sex selection. Given this presumption, advocates claim, those who would like to prohibit the practice are morally required to show that sex selection will cause harms, or increase the risk that harms will be caused, to particular individuals, such as the offspring or to society. Proponents also maintain that, at least for Western nations, no evidence has been presented that allowing people to sex select their offspring will produce serious harms (Robertson 2003b, Dahl 2004, Dahl 2005a, Harris 2005, Mamdani 2005, McCarthy 2001, Savulescu 1999).

In response, critics of sex selection might question the appropriateness of the reproductive liberty framework when evaluating the practice. First, even if one agrees that reproductive freedom is a basic right, there is significant disagreement about what reproductive freedom entails (Conly 2005, Pearson 2007, Ryan 1990). Second, there are no compelling reasons to accept that a right to reproduce involves a right to procreate a child of a particular sex. Proponents of sex selection acknowledge that not all reproductive decisions that might be tied to choices about whether to reproduce at all are covered by reproductive freedom (Robertson 2003b). But if this is the case, then claiming that sex selection falls within the scope of procreative liberty because it is tightly linked to decisions about whether to reproduce simply begs the question. If not all reproductive decisions tied to choices about whether to reproduce at all are covered by reproductive freedom, then one must provide reasons for why a particular choice, i.e. sex selection, is so implicated. And if reproductive freedom cannot properly be said to include a right to sex selection, then advocates are not persuasive when they claim that the only legitimate reason to interfere with sex selection is the causing of harm to others.

Moreover, using the framework of reproductive autonomy when discussing sex selection can also be rejected because it illegitimately restricts the range of concerns that are thought to be relevant to the discussion about the morality of this practice and its possible constraints (Murray 2002, Roberts 1995, Ryan 1990). For instance, proponents of sex selection take the preferences of parents for a child of a particular sex as given and as closed to evaluation. To the extent that some parents believe sex selection to be essential to their reproductive decisions, we must take such preferences as unquestionable. But this seems problematic. Clearly, not all preferences are legitimate. People have all kinds of preferences that we don't have any reasons to promote and might have good reasons to discourage: ignorant or greedy preferences and adaptive preferences for example (Anderson 2001, Sen 1973, Nussbaum 2000). Thus, it is important to scrutinize the social context, which helps to both create and give meaning to individuals' preferences such as those related to gender preferences.

More importantly, because the reproductive liberty framework allows only consideration of harms as legitimate for determining whether constraints can be imposed on sex selection, it blocks attention being paid to what are arguably important aspects of the sex selection process. For example, sex selection occurs under the umbrella of medical practice, thus it seems reasonable to ask how it might affect such practice. Arguably, physicians have responsibilities to use their skills, knowledge, and resources to meet the legitimate health needs of society. They assume those responsibilities, at least in part, as a response to the significant degree of autonomy to set and define standards for their work that our societies confer upon medical professionals. Moreover, public money is often used to subsidize medical training and infrastructure, thus enabling physicians to practice their professions. But if an important part of a physician's role involves attending to peoples' mere preferences, then it is not clear that medicine should have the authority and enjoy the privileges now granted to the profession.

3.2 Parental autonomy

A second argument used by some proponents of sex selection appeals to parental autonomy (Robertson 1994). As in the case of reproductive freedom, the principle of parental autonomy enjoys significant support both morally and legally (Green 1997, Scott 2003). Legally, the right of parents to rear their children according to their own beliefs has been established in cases such as *Wisconsin v. Yoder*, in which the United States Supreme Court found that parents' fundamental right to freedom of religion outweighed the state's interest in educating children. Furthermore, most people believe that parents are the best ones to decide how to raise a child. Most also agree that parents should be free to make choices, such as the selection of a partner's characteristics, that will influence a child's initial genetic endowment. Similarly, most support the belief that parents ought to be allowed to decide when to reproduce and under what conditions, even when such choices can have significant influences on a child's social starting point in life. As with reproductive freedom, however, parental autonomy ought not be constrained unless the parents' actions are shown to seriously harm their children or others (Robertson 1994).

As in the case of reproductive freedom, if children born after sex selection can be harmed, then parental autonomy can be legitimately constrained. But, also as in the case of reproductive liberty, one can question the proponents' assumptions about what parental autonomy entails. Indeed, any claims about what parental autonomy might involve presuppose some normative conception of parenting, as well as notions of the rights and responsibilities associated with being a parent and the moral grounds for such rights and responsibilities. But there is no consensus about such matters, and a variety of reasonable positions have been proposed (Austin 2007, Archard and Benatar 2010, Brighouse and Swift 2006, Feinberg 1980, Lafollette 1980). For some, it is because a dependent child must have decisions made for it that a designated parent is entitled to make those decisions and thus parents acquire rights in order to carry out their responsibilities, while others believe that parents' interests in the goods of the parenting project ground parental rights (Austin 2007, Brighouse and Swift 2006). Normative notions underlying discussions about parental autonomy thus need to be spelled out and supported by arguments. But proponents of sex selection tend to simply assume the good of parental autonomy without defending the normative grounds on which their understanding of parental autonomy rests. Given that such normative grounds are contested, a justification is important.

4. ARGUMENTS AGAINST SEX SELECTION

As mentioned earlier, the use of PGD for sex selection involves ethical issues related to the creation and destruction of embryos. I will not be discussing those concerns here, but will focus only on arguments against the practice of sex selection. Such arguments can be classified as intrinsic – arguments that attend to particular aspects of an action other than the consequences that such action might bring about – and consequentialist – arguments that attend to the consequences of the action. I examine these arguments in turn.

4.1 Intrinsic arguments against sex selection

Whether or not sex selection results in harms to offspring or society, some have argued that the practice is impermissible because it treats children merely as means to their parents' ends (Kass 2002, Habermas 2003, Sandel 2007). The choosing of what, all sides of the debate agree, is a morally neutral characteristic (i.e., it is not morally better to be a woman rather than a man and vice versa) that has nothing to do with the well-being of the child, treats children as mere means to the desired ends of the parents. Sex selection, critics contend, threatens our ethical self-understanding and our autonomy (Habermas 2003). Moreover, it disregards the "giftedness" of children (Sandel 2007). Furthermore, sex selection is also inconsistent with a widely shared (Scully et al. 2006a, Scully et al. 2006b) normative understanding of parenting that calls for parents to love whatever child comes along (Herissone-Kelly 2007, McDougall 2005).

These arguments all attend to the particular activity of sex selection rather than to the consequences that it might produce. To the extent that one believes that human beings should not be treated merely as means to someone else's ends or that particular normative conceptions of parenting ought to be respected, then sex selection would be considered morally impermissible.

Advocates of sex selection tend to dismiss these arguments because they reject the claim that selecting children involves treating them as mere means to the parents' ends or that it is opposed to acceptance of whatever child is born (Harris 2007, Robertson 2003a). They allow that children born as a result of sex selection might indeed be used as means to their parents' ends. They reject however the claim that those children are used as *mere* means. They claim that parents will love and cherish those children, will treat them with respect, and promote their flourishing (Green 2007, Robertson 1994). Hence, advocates assert that the practice of sex selection should be ethically and legally permissible.

4.2 Consequentialist arguments against sex selection

Consequentialist arguments focus on the possible negative consequences that might result from the use of sex selection. Such consequences might attend to the effects on children born through this practice, to particular groups of individuals, i.e., women, or to society in general. These arguments are particularly significant because they tend to assume the reproductive freedom framework used by proponents of sex selection. Their goal is to show that even if reproductive liberty and parental autonomy present us with a strong presumption against interference, they do so only to the extent that the practice does not result in serious harms. Critics using consequentialist arguments thus attempt to show that sex selection is indeed harmful.

4.2.1 Harms to children

One of the consequentialist arguments offered by some critics of sex selection is that children born through sex selection might be harmed. Harm can result because parents impose rigid gender expectations on their child and thus limit the opportunities that a child might have (Seavilleklein and Sherwin 2007). What PGD offers is the selection of an embryo carrying XX or XY chromosomes, but what parents choosing sex selection ultimately want is a child of a particular sex and *gender* (Robertson 2001). But, as is well known, chromosomal sex is only an aspect of biological sex, and as the existence of intersexuals, transsexuals, and transvestites shows, it is clearly not a guarantor of gender (Fausto-Sterling 2000). Moreover ample evidence shows that chromosomal sex does not guarantee a particular sexual orientation. Hence, parents who use sex selection might end up having a child who does not conform to their expectations; and their rigid gender expectations for the child might cause harm.

However, talk of harms to children is difficult when considering children born as a result of PGD for sex selection. This is so because, in general, our ethical reasoning assumes a person-affecting conception of harm. According to this conception, an action is wrong because it will harm some future person(s), or make things worse for some future person(s) than they might have been (Parfit 1984).

Thus, when someone argues that a child born after sex selection might be harmed because of psychological or social conditions, one needs to account for the fact that but for the use of PGD for sex selection the child in question would never have existed, and a different child (if any) would have been born. Under a person-affecting conception of morality, then, the child cannot be said to have been harmed, as that particular child would not exist, had it not been for the use of PGD for sex selection.[1]

4.2.2 Harms to women

Another consequentialist argument against sex selection focuses on the negative effects that this practice might have on women's lives and well-being. Some have argued that sex selection is intrinsically sexist because it expresses a judgment that one sex is more valuable than the other (Berkowitz and Snyder 1998). Sex selection practices are then likely to encourage sexism and thus harm women.

However, proponents of sex selection believe that concerns about promotion of sexism are misplaced and thus cannot be said to constitute a harm (Robertson 2001, Dahl 2003, McCarthy 2001, Savulescu 1999, Savulescu and Dahl 2000, Macklin 2010). The reason they give in support of their conclusion is that there is no compelling evidence to believe that parents choosing sex selection are motivated by the sexist belief that one sex is more valuable than, or superior to, the other (Dahl 2003, Savulescu and Dahl 2000, Robertson 2003b, Robertson 2005, McCarthy 2001, Harris 2007). According to advocates of sex selection, a variety of non-sexist motivations can be offered, and are offered by those who choose sex selection, e.g., that raising a girl is different from raising a boy or that having children of different sexes provide different family experiences. Moreover, proponents argue, evidence suggests that most people choosing to use sex selection do so for 'family balancing' – that is, to ensure that they have both boys and girls as part of their families – and thus are willing to have children of both sexes.

These responses, however, seem problematic for several reasons. First, though it is the case that there might be all sorts of non-prejudicial reasons for preferring children of one sex rather than another, this in no way shows that those choosing sex selection actually do not hold motivating prejudicial beliefs. After all, they are choosing for one sex and against another. Second, it is not at all clear that motivating desires for family balancing are non-sexist (Wilkinson 2008). This, at least in part, depends on how sexism is conceptualized. And this constitutes the third problem with the responses offered by proponents of sex selection. Sexism can be conceptualized not only as the belief that one sex is more valuable than the other, but also as beliefs about rigid gender roles thought to be appropriate for boys and girls (Seavilleklein and Sherwin 2007). Few would question the considerable amount of evidence showing that gender expectations have historically been used to limit the life options of men and, particularly, of women. Still today, gender expectations are at least in part responsible for limiting the number of women in a variety of career paths, such as the physical sciences and engineering; they ground

---

[1] It is beyond the scope of this chapter to discuss the non-identity problem and the variety of solutions that have been offered to address it. For discussion of this problem see: Parfit (1984); Brock (1995); Harman (2004); Roberts and Wasserman (2009); Buchanan et al. (2000).

policies that disadvantage women, such as family leaves and tenure and promotion criteria; and are the bases for norms about activities as relevant to a person's well-being as child care, household responsibilities, and caring for the old and the sick. It does not seem particularly farfetched to argue that acceptance of such rigid gender roles – with the blessing of modern science and medicine – is likely to serve to perpetuate the limitation of life choices and to further injustice. If these claims are reasonable, then opponents of sex selection have presented a significant challenge to advocates of this practice.

4.2.3 Social harms

Sexism harms not only women, but all of society as well. Restrictions on women's choices have consequences for the labor force with fewer women having access to traditionally male-dominated fields such as construction management, law, or engineering. Such restrictions result in a loss of talent for a variety of fields as well as in the loss of competitiveness. Similarly, constraints on women's participation in government, the sciences, and business reduce the diversity of voices in important areas and thus limit the chances of finding new solutions for social and scientific problems, as well as our ability to identify and prevent reliance on problematic or biased assumptions that might be shared or go unnoticed within a homogenous group (Economic and Social Council 2010).

Critics of sex selection have also argued that another significant adverse consequence of sex selection is that it can distort the natural sex ratio. They offer the cases of China and India as instances of this effect (Nie 2011, Jha et al. 2011). Imbalance of the sex ratio could lead to more discrimination against women, unrest among men deprived of a partner, and increase in crimes such as rape, incest, trafficking, and kidnapping (Shenfield 2005).

Critics acknowledge that sex ratio imbalances might indeed be a concern in countries such as China or India. However, they believe that skewing of the sex ratio is unlikely to be of concern in Western countries because the pressure to have offspring of a particular sex, i.e., boys, is less prevalent (Robertson 2001, Dahl 2003, McCarthy 2001, Savulescu 1999, Savulescu and Dahl 2000, Macklin 2010). Moreover, distortions of the sex ratio would require that many people use sex selection services and given the costs and risks involved in the use of PGD and IVF, it is unlikely that the demand for these services will rise to problematic levels.

5. CONCLUSION

The use of new reproductive and genetic technologies presents us with difficult ethical choices. One is that of the ethical permissibility of sex selection for social reasons. As we have seen, the arguments of reproductive freedom and parental autonomy are powerful and appealing. They tend to fit many people's intuitions about the importance of reproduction and allowing parents to make decisions on behalf of their children. Nonetheless, the appeal of these arguments might be deceptive. Arguments need to be provided to substantiate claims that sex selection is part of reproductive freedom and an appropriate prerogative of parental autonomy. Moreover, the simplicity of these arguments masks relevant aspects of the discussion. Because interference with reproductive freedom and parental autonomy is legitimate only to prevent serious harms, we are prevented from attending to factors, such as the role and goals of medicine, because such concerns fit uneasily within the understanding of harm used in these arguments. In any case, evaluations of sex selection need to address the legitimate concerns captured by the reproductive liberty and parental autonomy arguments.

Intrinsic arguments against sex selection focus on what is wrong with sex selection independently of the consequences of this practice. These arguments, however, require that one shares assumptions about the normativity of nature or the ethical relevance of reproduction – assumptions that are certainly rejected by proponents of sex selection. They also involve making judgments about when a human being is treated as *mere* means, and reasonable disagreements about these judgments are common.

Consequentialist arguments, on the other hand, involve projections about what might happen if sex selection were to be used by large sections of the population. They also must address difficulties associated with the non-identity problem. Nonetheless, some of the harms said to result from sex selection, such as the promotion of sexist beliefs, are far from speculative. Indeed, there is plenty of evidence about the harms associated with rigid gender roles, which is what grounds the assumptions underlying several consequentialist concerns regarding sex selection.

The preceding discussion has focused on whether sex selection is ethically permissible. A related issue is whether, if sound reasons are offered for the ethical impermissibility of this practice, it should be legally limited or prohibited. Answers to such questions will need to take into account, among other things, the feasibility of regulating sex selection, the degree of interference that such regulation would involve for related reproductive practices, and the effects on the doctor–patient relationship.

WORKS CITED

American College of Obstetricians and Gynecologists (2007) 'ACOG Committee Opinion No. 360: Sex selection,' *American Journal of Obstetrics and Gynecology*, 109 (2 Pt 1): 475–8.

American Society of Reproductive Medicine (2004) 'Sex selection and preimplantation genetic diagnosis,' *Fertility and Sterility*, 82 Suppl 1, S245–8.

E. Anderson (2001) 'Unstrapping the straitjacket of 'preference': A comment on Amartya Sen's contributions to philosophy and economics,' *Economics and Philosophy*, 17(1): 21–38.

D. Archard, and D. Benatar (2010) *Procreation and parenthood: the ethics of bearing and rearing children* (Oxford; New York: Oxford University Press).

M. W. Austin (2007) *Conceptions of parenthood: ethics and the family, Ashgate studies in applied ethics* (Aldershot, England; Burlington, VT: Ashgate).

J. M. Berkowitz, and J. W. Snyder (1998) 'Racism and sexism in medically assisted conception,' *Bioethics*, 12 (1): 25–44.

BioPolicyWiki (2009) 'Preimplantation Genetic Diagnosis,' available at: http://www.biopolicywiki.org/index.php?title=Preimplantation_genetic_diagnosis [accessed August 18, 2011].

H. Brighouse, and A. Swift (2006) 'Parents' rights and the value of the family,' *Ethics*, 117 (1): 80–108.

D. W. Brock (1995) 'The Nonidentity Problem and Genetic Harms – The Case of Wrongful Handicaps,' *Bioethics*, 9(3–4): 269–275.

A. E. Buchanan et al. (2000) *From chance to choice: genetics and justice* (New York: Cambridge University Press).

S. Conly (2005) 'The Right to Procreation: Merits and Limits,' *American Philosophical Quarterly*, 42 (2): 105–115.

Council of Europe (1997) 'Convention for the protection of Human Rights and dignity of the human being with regard to the application of biology and medicine: Convention on Human Rights and Biomedicine.' Available at http://conventions.coe.int/Treaty/Commun/QueVoulezVous.asp?NT=164&CL=ENG'

E. Dahl (2003) 'Procreative liberty: the case for preconception sex selection,' *Reproductive Biomedicine Online*, 7(4): 380–4.

E. Dahl (2004) 'The presumption in favour of liberty: a comment on the HFEA's public consultation on sex selection,' *Reproductive Biomedicine Online*, 8(3): 266–7.

E. Dahl (2005a) 'Preconception gender selection: a threat to the natural sex ratio?' *Reproductive Biomedicine Online*, 10 Suppl 1: 116–8.

E. Dahl (2005b) 'Sex selection: laissez faire or family balancing?' Health Care Analysis, 13(1): 87–90; discussion 91–3.

R. Dworkin (1993) *Life's dominion : an argument about abortion, euthanasia, and individual freedom*, 1st edn, (New York: Knopf).

Economic and Social Council, U. N. (2010) *Achieving gender equality and women's empowerment and strengthening development cooperation* (New York: United Nations. Dept. of Economic and Social Affairs).

A. Fausto-Sterling (2000) *Sexing the body: gender politics and the construction of sexuality*, 1st edn, (New York, NY: Basic Books).

J. Feinberg (1980) 'On the Child's Right to an Open Future' in Aiken, W. and LaFollette, H., eds, *Whose Child?* (Totowa, NJ: Rowman & Littlefield), 124–53.

FIGO Committee for the Ethical Aspects of Human Reproduction and Women's Health (2006) 'Ethical guidelines on sex selection for non-medical purposes. FIGO Committee for the Ethical Aspects of Human Reproduction and Women's Health,' *International Journal of Gynecology and Obstetrics*, 92 (3): 329–30.

E. S. Ginsburg, V. L. Baker, C. Racowsky, E. Wantman, et al. (2011) 'Use of preimplantation genetic diagnosis and preimplantation genetic screening in the United States: a Society for Assisted Reproductive Technology Writing Group paper,' *Fertility and Sterility*, 96 (4): 865–8.

R. M. Green (1997) 'Parental autonomy and the obligation not to harm one's child genetically,' *Journal of Law Medicine & Ethics*, 25(1): 5–15.

R. M. Green (2007) *Babies by design: the ethics of genetic choice* (New Haven: Yale University Press).

J. R. Habermas (2003) *The future of human nature* (Cambridge, UK: Polity).

A. H. Handyside, E. H. Kontogianni, K. Hardy, and R. M. Winston. (1990) 'Pregnancies from biopsied human preimplantation embryos sexed by Y-specific DNA amplification,' *Nature*, 344 (6268): 768–70.

E. Harman (2004) 'Can we harm and benefit in creating?' *Nous*, 89–113.

J. C. Harper, and S. B. Sengupta (2012) 'Preimplantation genetic diagnosis: state of the art 2011,' *The American Journal of Human Genetics*, 131 (2): 175–86.

J. Harris (1998) 'Rights and Reproductive Choice' in Harris, J. and Holm, S., eds, *The Future of Human Reproduction* (Oxford: Oxford University Press), 5–37.

J. Harris (2005) 'Sex selection and regulated hatred,' *Journal of Medical Ethics*, 31(5): 291–4.

J. Harris (2007) *Enhancing evolution: the ethical case for making better people* (Princeton, NJ: Princeton University Press).

P. Herissone-Kelly (2007) 'The "parental love" objection to nonmedical sex selection: deepening the argument,' *Cambridge Quarterly of Healthcare Ethics*, 16 (4): 446–55.

P. Jha, M. A. Kesler, R. Kumar, F. Ram, et al. (2011) 'Trends in selective abortions of girls in India: analysis of nationally representative birth histories from 1990 to 2005 and census data from 1991 to 2011,' *Lancet*, 377 (9781): 1921–8.

L. A. Johnson, G. R. Welch, K. Keyvanfar, A. Dorfmann, et al. (1993) 'Gender preselection in humans? Flow cytometric separation of X and Y spermatozoa for the prevention of X-linked diseases,' *Human Reproduction*, 8 (10): 1733–9.

D. S. Karabinus (2009) 'Flow cytometric sorting of human sperm: MicroSort clinical trial update,' *Theriogenology*, 71(1): 74–9.

L. Kass (2002) *Life, liberty, and the defense of dignity: the challenge for bioethics*, 1st ed. (San Francisco: Encounter Books).

H. Lafollette (1980) 'Licensing Parents,' *Philosophy & Public Affairs*, 9(2): 182–197.

R. Macklin (2010) 'The ethics of sex selection and family balancing,' *Seminars in Reproductive Medicine*, 28 (4): 315–21.

B. Mamdani, (2005) 'In support of sex selection,' *Indian Journal of Medical Ethics*, 2(1): 26–7.

D. McCarthy (2001) 'Why sex selection should be legal,' *Journal of Medical Ethics*, 27(5): 302–7.

R. McDougall (2005) 'Acting parentally: an argument against sex selection,' *Journal of Medical Ethics*, 31 (10): 601–5.

T. H. Murray (2002) 'What are families for? Getting to an ethics of reproductive technology,' *Hastings Center Report*, 32 (3): 41–45.

J. B. Nie (2011) 'Non-medical sex-selective abortion in China: ethical and public policy issues in the context of 40 million missing females,' *British Medical Bulletin*, 98: 7–20.

M. C. Nussbaum (2000) *Women and human development: the capabilities approach* (Cambridge; New York: Cambridge University Press).

D. Parfit (1984) *Reasons and persons* (Oxford: Clarendon Press).

Y. Pearson (2007) 'Storks, cabbage patches, and the right to procreate,' *Journal of Bioethical Inquiry*, 4 (2): 105–115.

D. E. Roberts (1995) 'Social Justice, Procreative Liberty, and the Limits of Liberal Theory: Robertson's "Children of Choice,"' *Law & Social Inquiry*, 20 (4): 1005–1021.

M. A. Roberts, and D. T. Wasserman (2009) *Harming future persons: ethics, genetics and the nonidentity problem* (Dordrecht: Springer).

J. A. Robertson (1994) *Children of choice: freedom and the new reproductive technologies* (Princeton: Princeton University Press).

J. A. Robertson (1996) 'Genetic selection of offspring characteristics,' *Boston Univeristy Law Review*, 76 (3): 421–82.

J. A. Robertson (2001) 'Preconception gender selection,' *American Journal of Bioethics*, 1 (1): 2–9.

J . A. Robertson (2003a) 'Extending preimplantation genetic diagnosis: the ethical debate. Ethical issues in new uses of preimplantation genetic diagnosis,' *Human Reproduction*, 18 (3): 465–71.

J. A. Robertson (2003b) 'Procreative liberty in the era of genomics,' *American Journal of Law and Medicine*, 29 (4): 439–87.

J. A. Robertson (2005) 'Ethics and the future of preimplantation genetic diagnosis,' *Reproductive Biomedicine Online*, 10 Suppl 1: 97–101.

M. Ryan (1990) 'The Argument for Unlimited Procreative Liberty – A Feminist Critique,' *Hastings Center Report*, 20 (4): 6–12.

M. J. Sandel (2007) *The case against perfection: ethics in the age of genetic engineering* (Cambridge, Mass.: Belknap Press of Harvard University Press).

J. Savulescu (1999) 'Sex selection: the case for,' *Medical Journal of Australia*, 171 (7): 373–5.

J. Savulescu (2001) 'In defense of selection for nondisease genes,' *American Journal of Bioethics*, 1(1): 16–9.

J. Savulescu, and E. Dahl (2000) 'Sex selection and preimplantation diagnosis: a response to the Ethics Committee of the American Society of Reproductive Medicine,' *Human Reproduction*, 15 (9): 1879–80.

E. S. Scott (2003) 'Parental Autonomy and Children's Welfare,' *William & Mary Bill of Rights Journal*, 11 (3): 1071–1100.

J. L. Scully, S. Banks, and T. W. Shakespeare (2006a) 'Chance, choice and control: lay debate on prenatal social sex selection,' *Social Science and Medicine*, 63 (1): 21–31.

J. L. Scully, T. Shakespeare, and S. Banks (2006b) 'Gift not commodity? Lay people deliberating social sex selection,' *Sociology of Health and Illness*, 28 (6): 749–67.

V. Seavilleklein, and S. Sherwin (2007) 'The myth of the gendered chromosome: sex selection and the social interest,' *Cambridge Quarterly of Healthcare Ethics*, 16 (1): 7–19.

A. Sen (1973) 'Behavior and the concept of preference,' *Economica*, 41: 241–59.

A. Sen (2003) 'Missing women – revisited,' *BMJ*, 327(7427): 1297–8.

F. Shenfield (2005) 'Procreative liberty, or collective responsibility? Comment on the House of Commons report Human Reproductive Technologies and the Law, and on Dahl's response,' *Reproductive Biomedicine Online*, 11 (2): 155–7.

J. L. Simpson (2010) 'Preimplantation genetic diagnosis at 20 years,' *Prenatal Diagnosis*, 30 (7): 682–95.

S. Wilkinson (2008) 'Sexism, sex selection and 'family balancing',' *Medical Law Review*, 16 (3): 369–89.

DISCUSSION QUESTIONS

1. What are some of the ethical concerns raised by sex selection for social reasons? Do any of these concerns appear particularly pressing?

2. Some have argued that sex selection is a protected activity. What does this mean? What arguments have been proposed to defend such a claim? Are the arguments compelling?

3. Respect for parental autonomy has been used to support the right of couples to select the sex of their future children. What are the strengths and weaknesses of this view?

4. Proponents of sex selection reject the claim that sex selection might promote sexist behavior. What is their argument? Is their understanding of sexism convincing? If not, what are the implications for the strength of their argument?

5. Which, if any, of the intrinsic and which, if any, of the consequentialist arguments against sex selection is more persuasive? Why?

6. Are there any other considerations, besides those discussed in this chapter, that you believe are relevant to the ethics of sex selection?

SELECTING CHILDREN: THE ETHICS OF REPRODUCTIVE GENETIC ENGINEERING[1]

S. Matthew Liao

CHAPTER SUMMARY

Advances in reproductive genetic engineering promise to allow us to select children free of diseases, and also enable us to select children with desirable traits. However, using technologies to design the genome of one's progeny raise a host of ethical concerns, both intrinsic (e.g. concerning the parent–child relationship) and extrinsic (e.g. concerning societal effects). In this chapter, S. Matthew Liao considers two clusters of arguments for the moral permissibility of reproductive genetic engineering, what he calls the Perfectionist View and the Libertarian View; and two clusters of arguments against reproductive genetic engineering, what he calls the Human Nature View and the Motivation View. He argues that an adequate theory of the ethics of reproductive genetic engineering should take into account insights gained from each of these views.

RELATED READINGS

Introduction: Innovation Presumption (2.1); Form of Life (2.5); Individual Rights and Liberties (2.6.3); Intrinsic Concerns (2.7)

Other Chapters: Leon Kass, *Preventing Brave New World* (Ch. 6); Inmaculada de Melo-Martin, *The Ethics of Sex Selection* (Ch. 7); President's Council on Bioethics, *Beyond Therapy* (Ch. 14); Nick Bostrom, *Why I Want to be a Posthuman When I Grow Up* (Ch. 15)

1. INTRODUCTION

Advances in genetic engineering have already made it possible to select the sex of one's child with great accuracy and screen for the susceptibility to serious genetic diseases (Liao 2005b). Soon, it may be possible to choose a child's attributes such as height, hair and eye color, as well as other physical characteristics, and, perhaps

---

[1] This chapter is excerpted from S. Matthew Liao (2008) 'Selecting Children: The Ethics of Reproductive Genetic Engineering,' *Philosophy Compass* (3): 1–19. It appears here by permission of the author.

even personality and intelligence. In light of these possibilities, how should we think about the ethics of genetically engineering children?

There is already a vast literature devoted to the ethics of genetic engineering. For example, much has been written about the potential of genetic engineering to exacerbate social inequality and about the various problems inherent in genetic research – such as the ethical quandary of conducting research on children without their consent and the potential pressures on women in particular of accepting these new technologies. My aim in this paper is therefore not to be comprehensive. I shall concentrate on some recent philosophical arguments for and against genetic engineering. Also, I would like to focus here on certain kinds of genetic engineering. In particular, I am interested in the ethics of genetically engineering human beings. So I shall set aside ethical issues regarding the genetic engineering of non-human animals (as well as human–animal chimeras) and non-animal entities such as crops. Furthermore, a distinction is often made between *somatic* and *germline* genetic engineering. Somatic engineering targets the genes in specific organs and tissues of the body of a single existing person without affecting genes in their eggs or sperm. Germline engineering targets the genes in eggs, sperm or very early embryos. My concern here will be with the ethical issues surrounding germline engineering. Moreover, within the subset of germline engineering, it is useful to distinguish between *reproductive* and *research/therapeutic* genetic engineering. The former is concerned with using genetic engineering to select and create new beings, whereas the latter is concerned with using genetic engineering for research purposes and/or for developing treatments for diseases. My interest here will primarily be with reproductive genetic engineering.

Finally, it is useful to distinguish between *modification* and *selection*. Modification involves using genetic engineering to alter certain genes in an embryo or a gamete. For example, at present, genetic engineering involves putting the desired "new" gene into a virus-like organism, which is then allowed to enter the target cell (of an embryo or a gamete in the case of germline engineering) and insert the new gene into the cell along with the "old" genes. Selection involves selecting particular pairs of gametes for fertilization or a particular embryo for implantation. In the future, it may be possible to design genes to create entire gametes or embryos for selection. (For this reason, in this paper, I am using the term 'genetic engineering' broadly to cover both modification and selection).

The distinction between modification and selection is relevant because in the case of reproductive modification, it could be argued that the resulting being could later complain that he or she could have been different if the modification (either in the embryo or in the gametes) had not taken place. Reproductive modification therefore raises issues relating to identity and autonomy (Liao 2005b; Habermas 2003, 40–41). Unlike reproductive modification, reproductive selection does not involve these issues. The reason is that the being that actually exists could not later complain that he or she could have been different; this is because he or she would not have existed if a different being had been selected. In the following, I am primarily interested in ethical issues that arise out of reproductive selection, although, as we shall see, reproductive modification will be more pertinent to some of the views I shall discuss.

I shall evaluate two clusters of arguments for the moral permissibility of reproductive selection, what I call the Perfectionist View and the Libertarian View; and two clusters of arguments against reproductive selection, what I call the Human Nature View and the Motivation View. As I shall point out, many of these views are not mutually exclusive.

2. THE PERFECTIONIST VIEW

To begin, I shall examine what might be called

> *The Perfectionist View*: Given a choice between selecting a being that will have the best chance of having the best life and a different being that will not have the best chance of having the best life, it is morally obligatory to select the former.

The notion of 'best chance' can be understood here in terms of the expected value of an outcome multiplied by the probability of its occurrence. The notion of a 'best life' can be understood as a life with the most well-being. There are ongoing controversies about whether to understand well-being in terms of an objective list (Griffin 1996, 29–30), informed desires (Griffin 1986; Darwall 2002), or states of pleasure and pain (Feldman, 2006). However, The Perfectionist View can be neutral with respect to these different accounts of well-being. It can make the minimal assumption that there are things that make a life go better or worse. For example, on any plausible account of well-being, it seems that chronic pain would be something that would make a life go worse.

The chief advocate of the Perfectionist View has been Julian Savulescu (Savulescu 2001). Here it is useful to distinguish between a stronger version of the Perfectionist View, according to which there is an all-things-considered, absolute, obligation to engage in selection if the beings selected will have the best chance of having the best life; and a weaker version, according to which there is a prima facie obligation to engage in selection if the beings selected will have the best chance of having the best life. Savulescu defends the weaker version, which he calls Procreative Beneficence:

> [C]ouples (or single reproducers) should select the child, of the possible children they could have, who is expected to have the best life, or at least as good a life as the others, based on the relevant, available information (Savulescu 2001, 415).

As Savulescu explains, the term 'should' can be understood here as a reason for having a prima facie obligation and is not just a supererogatory reason (Savulescu 2001, 415).

What are some reasons in favour of the Perfectionist View? Following Derek Parfit, Savulescu suggests that the following examples support the Perfectionist View (Parfit 1984, Part IV; Savulescu 2001, 417). For instance, suppose that a woman has rubella. If she conceives now, she will have a blind and deaf child. If she waits three months, she will conceive a different but healthy child. According to Savulescu, she should choose to wait until her rubella has passed. Or, suppose that a couple is having in vitro fertilization (IVF) in an attempt to have a child

(Savulescu 2001, 418). The process produces two embryos. A battery of tests for common diseases is performed, and it is found that Embryo A has no abnormalities on the tests performed; and Embryo B has no abnormalities on the tests performed except that its genetic profile reveals it has a predisposition to developing asthma. Which embryo should be implanted? According to Savulescu, Embryo B has nothing to be said in its favour over A and something against it. Therefore, Embryo A should be implanted, because A has the better chance of having a better life than B.

A purported advantage of the Perfectionist View is that it may have the resources to argue against the selection of disabled children. I say 'purported' because some people believe that it should be morally permissible to select disabled children (at least in some cases). On a certain version of the Perfectionist View (with a particular account of well-being according to which a deaf child would not have the best chance of having the best life), there would be a prima facie obligation against selecting such a child if it is possible to select a non-deaf child instead.

As far as I can see, the Perfectionist View has two potential weaknesses. First, it appears to have certain counterintuitive implications. In particular, suppose that being male or being white or being tall or being heterosexual will provide the best chance of having the best life, the Perfectionist View implies that there will be a prima facie obligation to select children who are male, white, tall or heterosexual.

Advocates of the Perfectionist View might try to respond to this worry by arguing that sexism, racism and homophobia are social prejudices. Given this, so the argument goes, they require social, not biological, solutions. But, arguably, such a response only has limited force. Imagine a non-racist, non-sexist, non-homophobic society, in which, as it happens, being male or being white or being heterosexual still provides the best chance of having the best life. It seems that the Perfectionist View would have to accept that there is a prima facie obligation to select children who are male or white or heterosexual. Indeed, consider our society in which being tall gives one a better chance of having the best life. Arguably, there is no entrenched prejudice – at least not in the same manner as racism or sexism or homophobia – against those who are short. The Perfectionist View, it seems, would have to accept that there is a prima facie obligation to select children who are tall. This seems counterintuitive, because it is difficult to see how there could be an obligation, even a prima facie one, to select children who are tall.

Secondly, the examples used to support the Perfectionist View do not seem to work. To recall, in one case, a woman has rubella, and if she conceives now, she will have a blind and deaf child. In another case, one embryo has a predisposition to developing asthma. It seems that in both cases, the choice is between selecting a healthy child and a possibly sick child. As such, it may well be that, in these kinds of cases, one should select the healthy child. However, the Perfectionist View requires that one select a child who will have the best chance of having the best life, even if the choice is between two healthy children. It is more difficult to see why there would be a prima facie obligation to do this. Consider this case.

The Wine Capacity Case: Suppose that a couple is having in vitro fertilization (IVF) in an attempt to have a child. The process produces two embryos. A battery of tests for common diseases is performed, and it is found that Embryo A has no abnormalities on the tests performed, except that its genetic profile reveals that it has a predisposition to be able to enjoy super fine wine; and Embryo B has no abnormalities on the tests performed except its genetic profile reveals it has a predisposition to be able to enjoy very fine wine, but not super fine wine. Let us assume that being able to enjoy super fine wine over very fine wine gives Embryo A a better chance to have the best life. Which embryo should be implanted?

On the Perfectionist View, Embryo B has nothing to be said in its favour over A, while Embryo A has something to be said in its favour over B. So, on the Perfectionist View, other things being equal, there is a prima facie obligation to implant Embryo A. But it is difficult to see how there could be an obligation, even a prima facie one, to choose A over B in this case. Advocates of the Perfectionist View might reply that given that it is agreed that Embryo A will have a better chance of having a better life than Embryo B, surely, there is *some* reason to prefer Embryo A to B. However, it seems that the notion of reason here is ambiguous. It could be a reason that grounds a prima facie obligation, that is, something that, other things being equal, one ought to do; or a supererogatory reason, that is, something that is good to do but that there is no obligation to do. To illustrate this distinction, consider the following:

The Blind Lady Case: Suppose that you have the option of helping a blind lady to cross a street or helping the blind lady to cross the street and getting her a free ice cream. The blind lady loves ice cream and would be very happy if she had ice cream. Someone just across the street is giving away ice cream for free, and it is no trouble for you to get her some.

There is certainly a reason to help the lady to cross the street. In addition, there may also be a reason to help get ice cream for the lady, since it would make her very happy. But arguably, the two reasons are not the same. The reason to help the lady cross the street looks to be a prima facie obligation, that is, something that, other things being equal, one ought to do. However, the reason to help get ice cream for the lady looks at best to be a supererogatory reason, that is, something that is good to do but that there is no obligation to do. In the Wine Capacity Case, it seems that the reason to prefer Embryo A to B is also something that is good to do but that there is no obligation to do. If this is right, one can grant that there is a reason to prefer Embryo A to B without accepting that this is a reason grounding a prima facie obligation. If so, the Wine Capacity Case does not seem to support the Perfectionist View.

3. THE LIBERTARIAN VIEW

Another set of arguments for the moral permissibility of reproductive selection is what might be called the Libertarian View. Some versions of the Libertarian View are fairly permissive. According to one such version,

The Permissive Libertarian View: It is morally permissible to engage in the selection of any beings.

John Robertson, who has defended a version of such a view, argues that "procreative liberty should enjoy presumptive primacy when conflicts about its exercise arise ... " (Robertson 1994, 24). More recently, Nicholas Agar, who has also advocated a version of such a view, which he has called Liberal Eugenics, argues that

> liberal eugenicists propose that [genetic technologies] be used to dramatically enlarge reproductive choice. Prospective parents may ask genetic engineers to introduce into their embryos combinations of genes that correspond with their particular conception of the good life. Yet they will acknowledge the right of their fellow citizens to make completely different eugenic choices. No one will be forced to clone themselves or to genetically engineer their embryos (Agar 2004, 6).

There are also less permissive versions of the Libertarian View. According to one such version,

The Life Worth Living (LWL) Libertarian View: It is morally permissible to engage in selection if the beings selected can have a life worth living.

The notion of "a life worth living" comes from Parfit (Parfit 1984, Part IV). It means something like a life that contains some positive values, and the amount of negatives values this life contains, e.g. pain, is not sufficient to negate the positive values that this life has. Jonathan Glover, who has defended such a Libertarian View, what he calls the "zero line view," asks,

> Can it be right to bring a child into the world so long as we expect the child to have a quality of life at least at the zero line just above that 'very terrible' level? (Glover 2006, 52)

Glover argues that if, all things considered, a child is glad to be alive, "how can it be that we owed it to the child to prevent his or her life?" (Glover 2006, 57)

As I mentioned previously, the Libertarian View is based on the principle of procreative liberty. In liberal societies, there is a presumption in favor of procreative liberty. Given this, in order to justify its infringement, significant moral reasons must be presented.

Many people would reject the Permissive Libertarian View on the ground that it is too permissive. For example, suppose that an individual wants to select a child whose life will be full of pain and suffering. On the Permissive Libertarian View, it seems that this individual would be morally permitted to do so. For many though, this implication seems counterintuitive. Accordingly, they would reject the Permissive Libertarian View.

Unlike the Permissive Libertarian View, the LWL Libertarian View does not have this implication, since it requires that the being selected has a life worth living, and arguably, a life full of pain and suffering is not one that is worth living. Still, other people may reject the LWL Libertarian View on the ground that it permits the selection of some disabled children, since some of them can have lives worth living. For these people, the LWL Libertarian View would still be too permissive.

As I mentioned earlier, Glover has argued that if, all things considered, a child is glad to be alive, 'how can it be that we owed it to the child to prevent his or her life?' In support of Glover's point, it seems correct that we cannot owe it to the child to prevent him or her from existing when the child can have a life worth living, since the child would not have existed otherwise. But perhaps one can have a (prima facie) obligation not to bring such a child into existence in other ways.

One possibility, following Parfit, is that bringing a child who will merely have a life worth living into existence is an instance of harmless wrongdoing (Parfit 1984, Part IV). That is, it is a wrong even though no one is harmed. Another possibility is that bringing such a child into existence involves wronging a type instead of a token (Kumar 2003). That is, it wrongs the child as a type, whoever the individual, who comes to instantiate the child as a token, may be.

But exactly what grounds the wrongness of bringing into existence a child who has merely a life worth living? As we have seen, advocates of the Perfectionist View have a ready answer to this question: The act is prima facie wrong because one has a prima facie obligation not to bring into existence a child who will not have the best chance of having the best life, if it is possible to bring into existence a different child who will have a better chance of having the best life. As I have suggested earlier though, there may be problems with the Perfectionist View. In light of this, one might hold instead this view.

> *The Sufficientarian View*: Given a choice between selecting a being that will have a decent chance to have a sufficiently decent life and a different being that will not have a decent chance to have a sufficiently decent life, there is a prima facie obligation to select the former (Kamm 1992).

As with the Perfectionist View, the Sufficientarian View also requires an account of well-being. In addition, a plausible Sufficientarian View would need to provide a plausible account of what constituted a 'decent chance' and a 'sufficiently decent' life. As with providing an account of what constitutes 'the best chance' and 'the best life,' this too is a difficult task, but I shall assume that it can be done.

Like the Perfectionist View, it seems that the Sufficientarian View could also have the implication that there is a prima facie obligation to select children who are male, white, tall, or heterosexual, if it turns out that the only way to have a decent chance to have a sufficiently decent life is to be male, white, tall or heterosexual. However, unlike the Perfectionist View, it seems that the Sufficientarian View can explain why there is not a prima facie obligation to select Embryo A over B in the Wine Capacity Case, because both embryos have decent chances to have sufficiently decent lives.

Also, on the Sufficientarian View, it seems that a child who has merely a life worth living would not have a decent chance to have a sufficiently decent life. If so, on the Sufficientarian View, there would be a prima facie obligation not to bring such a child into existence if it is possible to bring into existence another child who has a decent chance of having a sufficiently decent life.

This is a quick sketch of the Sufficientarian View and how one might be able to employ it to answer Glover's challenge. No doubt more can be said regarding it. But I shall move on now to consider some arguments against the permissibility of reproductive selection.

4. THE HUMAN NATURE VIEW

Many people believe that reproductive genetic engineering is impermissible because it interferes with nature, in particular, human nature. For example, Michael Sandel suggests that a problem with genetic engineering is that it represents "a Promethean aspiration to remake nature, including human nature, to serve our purposes and satisfy our desires ..." (Sandel 2007; Sandel 2004).

Similarly, Francis Fukuyama points out that the deepest fear that people have regarding new biotechnology is that

> "biotechnology will cause us in some way to lose our humanity – that is, some essential quality that has always underpinned our sense of who we are and where we are going ..." (Fukuyama 2002, 101).

And, Leon Kass, the former chairman of the President's Council on Bioethics, writes,

> To turn a man into a cockroach – as we don't need Kafka to show us – would be dehumanizing. To try to turn a man into more than a man might be so as well. We need more than generalized appreciation for nature's gifts. We need a particular regard and respect for the special gift that is our own given nature (Kass 2003, 20).

These objections all rest on the idea that there is something special about human nature which grounds the impermissibility of genetic engineering. But, exactly what is special about human nature? These writers do not explain. Fukuyama, for example, says only that "when we strip all of a person's contingent and accidental characteristics away, there remains some essential human quality underneath that is worthy of a certain minimal level of respect – call it Factor X" (Fukuyama 2002, 149). He then goes on to say that "there is no simple answer to the question, What is Factor X?" (Fukuyama 2002, 149). Or, Kass asks, "What is disquieting about our attempts to improve upon human nature, or even our own particular instance of it?" (Kass 2003, 17) His response is that "It is difficult to put this disquiet into words. We are in an area where initial repugnances are hard to translate into sound moral arguments" (Kass 2003, 17).

In the following, I shall attempt to provide some possible explanations regarding the specialness of human nature, and I shall consider their implications for the permissibility of genetic engineering.

To begin, let me briefly mention a view that I will not discuss at any great length here:

> *The Interfering with Nature View*: It is morally impermissible to interfere with human nature, because this is interfering with nature, and it is morally impermissible to interfere with nature.

This view resonates with many people who are worried about the prospects of reproductive genetic engineering, but among other things, this view – at least in its unqualified form – implies (implausibly) that providing vaccination, offering pain relief to women in labor, and so on, are impermissible, since these acts interfere with nature. Below I consider versions of the Human Nature View that may be more defensible.

Also, it is important to note that the Human Nature View may be more applicable to cases of reproductive modification rather than selection. The reason is that typically one can *interfere* with human nature only when human nature is already there. If reproductive selection involves, for example, selecting various genes to produce entire gametes or embryos, it is difficult to see how a Human Nature View could be relevant to this kind of selection. Hence, I shall primarily be discussing reproductive modification rather than selection here.

Moreover, those who wish to argue that genetic engineering can undermine the specialness of human nature may be making this task more difficult than necessary. The reason is that they seem to believe that by appealing to the specialness of human nature, they could show that genetic engineering (at least in respect to enhancement) is absolutely morally impermissible. But such an absolutist position is too strong. Consider

> *The End of the World Case*: The world is about to end, and the only chance of saving the world is genetically to enhance a group of human beings so that they can save the world.

In such a case, it seems that using genetic engineering to enhance human nature could be permissible. Or, consider a less extreme example:

> *The Really Great Net-Benefit Case*: A very minor genetic enhancement to human nature can greatly improve human societies.

If the net benefit is indeed really great, perhaps genetically enhancing human nature in such circumstances could also be permissible. These two cases therefore suggest that appealing to the specialness of human nature does not give one an absolute prohibition against using genetic engineering to alter human nature. In fact, as we shall see, the Human Nature View may in fact provide very few constraints against reproductive genetic engineering.

One possible explanation regarding the specialness of human nature is that the human species is something special and something that ought to be treated in a certain way. As the late Bernard Williams said

> "there are certain respects in which creatures are treated in one way rather than another simply because they belong to a certain category, the human species" (Williams 2008).

Call this the Human Species View.

According to Williams, one ought to treat the human species in a special way because it is ours:

> Suppose we accept that there is no question of human beings and their activities being important or failing to be so from a cosmic point of view. That does not mean that there is no point of view from which they are important. There is certainly one point of view from which they are important, namely ours ... (Williams 2008).

Williams is suggesting that the fact that the human species is our species may provide a reason for treating it in a special way. Following Peter Singer though, it might be thought that this kind of partiality to our own species is just speciesism and is unjustified, much like racism and sexism are unjustified (Singer 1993). However, partiality per se is not necessarily unjustified, as for example, partiality to one's spouse or children. Moreover, Williams argues that partiality to the human species is justified, because, unlike racism or sexism,

> "It's a human being" does seem to operate as a reason, but it does not seem to be helped out by some further reach of supposedly more relevant reasons, of the kind which in the other cases of prejudice turned out to be rationalizations (Williams 2008).

The Human Species View would not rule as impermissible using genetic engineering to select children with certain hair or eye color or children with disability, since these children would all still belong to the human species. As such, it might be thought that the Human Species View provides very few constraints on reproductive genetic engineering. However, advocates of the Human Species View could argue that the constraint that it does provide is very important.

Another possible explanation regarding the specialness of human nature is the following: There is a long line of philosophers who have argued that the specialness of human nature lies in moral agency (e.g. Kant, Rawls, and Scanlon). By moral agency, I mean the capacity to act in light of moral reasons. Among those who share this idea, there is a lively debate about whether what grounds the specialness of human nature is actual, potential, or the genetic basis for, moral agency (Liao 2011; Liao 2012). For our purpose, we can set this debate aside, and call this *the Moral Agency View*.

The value of moral agency is, I think, not in dispute. As an instrumental good, it enables human moral agents to live together. As an intrinsic good, arguably, it provides the basis of human moral status. The Moral Agency View provides a slightly different constraint than the Human Species View. In particular, it would rule as impermissible the deliberate creation of beings without moral agency that would have otherwise had moral agency. Hence, it would rule as impermissible the deliberate creation of children who lacked moral agency if these children would have otherwise had moral agency.

At the same time, as with the Human Species View, the Moral Agency View would not rule as impermissible using genetic engineering to select children with certain hair or eye color, since these children would still have moral agency. Also, the Moral Agency View would also not rule as impermissible the selection of children with disability as long as these children have moral agency. Hence, it might be thought that the Moral Agency View also provides very few constraints on reproductive genetic engineering. Again though, its advocates could argue that the constraint that it does provide is very important.

A third possible explanation regarding the specialness of human nature comes from the idea that human beings flourish in a certain way. For example, human beings flourish by having deep personal relationships, knowledge, and active and passive pleasures (Griffin 1986; Liao 2006). It might be argued that the fact that human beings flourish in a certain way can place some constraints on human reproductive genetic engineering. For example, given that human beings flourish by having deep personal relationships, genetically creating a human being who can have a good life without deep personal relationships would change too much the meaning of what constitutes a flourishing human life as it is currently understood. Or, as Larry Temkin argues, given that mortality is part of human life, genetically creating a human being who will be immortal would also change too much the meaning of what constitutes a flourishing human life (Temkin, 2008). Call this *the Human Flourishing View.*

Allen Buchanan has argued that the Human Flourishing View cannot constrain genetic engineering because it can only tell us that if we have a certain nature, then that should be taken into account when we try to determine what our good is, not that we should persist with that nature and the constraints that it imposes (Buchanan, 2009). Buchanan offers the following analogy:

> if we are limited to a particular canvas, we can only create a painting that fits within its boundaries and we should take that into account in deciding what to paint – on it. But if we have the option of using a different canvas, then there will be other possibilities (and other limitations, as well), if we choose to paint on it. Recognizing that a given canvas limits the artistic good we can achieve does not imply that we should refrain from changing canvases; on the contrary, it suggests that we should at least consider using a different one, if we can (Buchanan, 2009).

Buchanan is surely right that recognizing that a set of goods places certain constraints on us does not mean that we are therefore not permitted to choose a different set of goods which will then have a different set of constraints.

But the Human Flourishing View may have an important implication that is not captured by Buchanan's analogy. If what you care about is making improvements/enhancements within a certain set of constraints as dictated by a certain set of goods, then, arguably, you would not be making improvements/enhancements *for that particular set of goods* if you choose a different set of goods with a different set of constraints (Sandel 2007). Consider the following: Suppose that you are trying to improve your marathon running time. Running a marathon is a good that comes with certain constraints, e.g., that you should physically run the 26.2 miles. It is certainly within the constraint of the good that you purchase better running shoes in order to improve your time. However, suppose you choose to use a car to complete the marathon. In doing so, it seems that you would no longer be improving your marathon time. The reason is that a constraint on the good of running a marathon is that you physically run the marathon, and in using a car, you would no longer be physically running a marathon.

Similarly, suppose your aim is to improve human flourishing for a particular being. But to achieve this aim, you choose to make a being who does not need to flourish according to the constraints set by existing ways of human flourishing. For example, you selected a being that does not need to have deep personal relationships in order to have a good life. Even if this being will indeed have a better life than any human being, on the Human Flourishing View, arguably, you are not improving human flourishing for this being, since this being does not require human flourishing in order to have a good life.

The Human Flourishing View seems compatible with the selection of children with disability as long as these children can still have flourishing human lives. Also, the Human Flourishing View appears to be compatible with the selection of children with certain hair or eye color, since these children can still have flourishing human lives. As such, it might also be thought that the Human Flourishing View provides very few constraints on reproductive genetic engineering, though, again, its advocates could argue that the constraint that it does provide is very important.

5. THE MOTIVATION VIEW

In addition to the Human Nature View, there is a complementary view against the permissibility of reproductive selection, which can be called

> *The Motivation View*: It is not morally permissible to engage in selection if one does not have the appropriate motivation.

Sandel is a prominent spokesperson for a version of this view. Sandel has a number of objections against enhancement, whether by genetic engineering, drugs, or extensive training, but he argues that the deepest objection to it is the desire for mastery that it expresses. Focusing on the attempt of parents to enhance their children, Sandel says,

> the deepest moral objection to enhancement lies less in the perfection it seeks than the human disposition it expresses and promotes. The problem is not that parents usurp the autonomy of a child they design. The problem is in the hubris of the designing parents, in their drive to master the mystery of birth. Even if this disposition does not make parents tyrants to their children, it disfigures the relation between parent and child, and deprives the parent of the humility and enlarged human sympathies that an openness to the unbidden can cultivate (Sandel 2007, 46).

It is worth noting here that Sandel's primary target is using genetic modifications for enhancement purposes. He believes that it may be permissible and even obligatory to treat illnesses by genetic modification, drugs, or training. As he says, "medical intervention to cure or prevent illness… does not desecrate nature but honors it. Healing sickness or injury does not override a child's natural capacities but permits them to flourish" (Sandel 2004, 57). In the context of reproductive selection, we can therefore call a view such as Sandel's

The Hubristic Motivation View: It is not morally permissible to engage in selection if one has a hubristic motivation to control reproduction in enhancement cases.

According to Sandel, "parents bent on enhancing their children are more likely to overreach, to express and entrench attitudes at odds with the norm of unconditional love" (Sandel 2007, 49). Sandel accepts that parents must "shape and direct the development of their children" (Sandel 2007, 49). However, following William F. May, Sandel argues that there should be a balance between "accepting love," which "affirms the being of the child"; and "transforming love," which "seeks the well-being of the child" (Sandel 2007, 50; May 2005).

The Hubristic Motivation View seems too strong. It implies that whenever someone seeks to control reproduction in enhancement cases, one is acting wrongly. Yet, we do not think that a couple who plays Mozart to their still-in-the-womb offspring in order to enhance the offspring's appreciation for music is necessarily acting wrongly. Accordingly, here may be a more plausible version of the Motivation View:

The Weak Hubristic Motivation View: It is not morally permissible to engage in selection if one has a strongly hubristic motivation to control reproduction in enhancement cases or if one has inappropriate motivations; and if one only has a weak motivation to love a selected child for the child's own sake (Liao 2005a).

Such a view would permit couples who love their child for the child's own sake to play some Mozart to their still-in-the-womb offspring in order to enhance the offspring's appreciation for music. It would deem as impermissible a case in which the parents play Mozart to their still-in-the-womb offspring all day long with the intent of enhancing the offspring's appreciation for music so that the offspring may become a musical genius, but in which the parents care very little, if at all, about the welfare of the offspring. On the Weak Hubristic Motivation View, it would not be impermissible for a parent to select a disabled child, if the parent were prepared to love the child for its own sake.

As I mentioned at the outset, many of the views discussed are not mutually exclusive. So, for example, one could combine the Weak Hubristic Motivation View with a Perfectionist View. One would obtain the following:

Given a choice between selecting a being that will have the best chance of having the best life and a different being who has a lesser chance of having the best life, it is morally obligatory to select the former; only if one does not have a strongly hubristic motivation to control reproduction in enhancement cases or only if one does not have inappropriate motivations; and only if one does not have a weak motivation to love a selected child for the child's own sake.

Still, the Perfectionist View and the Motivation View are distinct, because someone could, for example, believe that her obligation is to select beings that will have the best chance of having the best life, irrespective of whether she is motivated to love the child or not.

6. CONCLUSION

Being able to use genetic engineering in reproduction has the potential to transform human lives. Not only does it promise to allow us to select children free of diseases, it can also enable us to select children with desirable traits. In this paper, I identified several factors that should be considered in the ethics of genetically engineering children: the best chance of having the best life for a child; reproductive liberty; life worth living; the chance of having a sufficiently decent life for a child; human species; moral agency; human flourishing; and appropriate motivation. Although I have not defended any particular principle for guiding decision-making about genetically engineering children, I have indicated the some significant strengths and weakness of the different approaches. Ultimately, it is prospective parents with access to the technology who will need to consider these factors and principles and decide whether, to what extent, in what ways, and for what traits they will choose the genetic traits of their children.

WORKS CITED

N. Agar (2004) *Liberal Eugenics: In Defense of Human Enhancement* (Oxford: Blackwell).

K. Birch (2005) 'Beneficence, Determinism and Justice: An Engagement with the Argument for the Genetic Selection of Intelligence,' *Bioethics* 19: 12–28.

N. Bostrom (2003) 'Human Genetic Enhancements: A Transhumanist Perspective,' *Journal of Value Inquiry*, 37 (4): 493–506.

A. Buchanan (2009) 'Human Nature and Enhancement,' *Bioethics* 23 (3): 141–150.

A. Buchanan, D. Brock, N. Daniels, and D. Wikler (2000) *From Chance to Choice* (Cambridge: Cambridge University Press).

S. Darwall (2002) *Welfare and Rational Care* (Princeton, NJ: Princeton University Press).

I. De Melo-Martin (2004) 'On Our Obligation to Select the Best Children: A Reply to Savulescu,' *Bioethics* 18: 72–83.

F. Feldman (2006) *Pleasure and the Good Life* (USA: Oxford University Press).

F. Fukuyama (2002) *Our Posthuman Future* (New York: Farrar, Straus and Giroux), 101.

J. Glover (2006) *Choosing Children: The Ethical Dilemmas of Genetic Intervention* (Oxford: Oxford University Press).

J. Griffin (1996) *Value Judgement: Improving Our Ethical Beliefs* (Oxford: Clarendon Press).

J. Griffin (1986) *Well-Being* (Oxford: Oxford University Press).

J. Habermas (2003) *The Future of Human Nature* (Cambridge: Polity Press).

F. Kamm (1992) *Creation and Abortion: A Study in Moral and Legal Philosophy* (Oxford: Oxford University Press).

F. Kamm (2005) 'Is There a Problem with Enhancement?' *American Journal of Bioethics* 5 (3): 5–14.

I. Kant (1996) *The Metaphysics of Morals* (1996) Translated by M. Gregor (Cambridge: Cambridge University Press).

L. Kass (2003) 'Ageless Bodies, Happy Souls: Biotechnology and the Pursuit of Perfection,' *The New Atlantis*, 1: 9–28.

R. Kumar (2003) 'Who Can Be Wronged?' *Philosophy and Public Affairs*, 31 (2): 99–118.

S. M. Liao (2005a) 'Are 'Ex Ante' Enhancements Always Permissible?' *American Journal of Bioethics*, 5 (3): 23–25.

S. M. Liao (2005b) 'The Ethics of Using Genetic Engineering for Sex Selection,' *Journal of Medical Ethics*, 31: 116–18.

S. M. Liao (2006) 'The Right of Children to Be Loved,' *Journal of Political Philosophy* 14 (4): 420–40.

S. M. Liao (2010) 'The Basis of Human Moral Status,' *Journal of Moral Philosophy*, 7 (2): 159–79.

S. M. Liao (2012) 'The Genetic Account of Moral Status: A Defence,' *Journal of Moral Philosophy*, 9 (2): 265–77.

W. F. May (2005) 'The President's Council on Bioethics: My Take on Some of Its Deliberations,' *Perspectives in Biology and Medicine*, 48: 230–31.

R. Norman (1996) 'Interfering with Nature,' *Journal of Applied Philosophy*, 13 (1): 1–11.

D. Parfit (1984) *Reasons and Persons* (Oxford: Oxford University Press).

M. Parker (2007) 'The Best Possible Child,' *Journal of Medical Ethics*, 33: 279–83.

J. Rawls (1971) *A Theory of Justice* (Oxford: Oxford University Press).

J. A. Robertson (1994) *Children of Choice: Freedom and the New Reproductive Technologies* (Princeton: Princeton University Press).

M. Sandel (2004) 'The Case against Perfection,' *Atlantic Monthly*, 293 (3): 51–62.

M. Sandel (2007) *The Case against Perfection: Ethics in the Age of Genetic Engineering* (Cambridge, MA: Harvard University Press).

J. Savulescu (2007) 'In Defence of Procreative Beneficence,' *Journal of Medical Ethics* 33: 284–88.

J. Savulescu (2001) 'Procreative Beneficence: Why We Should Select the Best Children,' *Bioethics*, 15 (5/6): 413–26.

J. Savulescu, and G. Kahane (under review) '*Procreative Beneficence and Disability: Is There a Moral Obligation to Create Children with the Best Chance of the Best Life?*'

T. Scanlon (1998) *What We Owe to Each Other*, (Cambridge, MA: Belknap Press).

M. Sheehan (forthcoming) 'Making Sense of the Immorality of Unnaturalness,' *Cambridge Quarterly of Healthcare Ethics*.

P. Singer (1993) *Practical Ethics* 2nd edn (Cambridge: Cambridge University Press).

L. Temkin (2008) 'Is Living Longer Living Better?' *Journal of Applied Philosophy*, 25 (3): 193–210.

D. Wasserman (2005) 'The Nonidentity Problem, Disability, and the Role Morality of Prospective Parents,' *Ethics*, 116 (1): 132–52.

B. Williams (2008) 'The Human Prejudice,' in *Philosophy as a Humanistic Discipline*, (ed.) A. W. Moore, 135–54 (Princeton, NJ: Princeton University Press).

DISCUSSION QUESTIONS

1. If you could select traits for your children, which traits would you select? Are any traits off limits? Why or why not?
2. How might genetic engineering alter attitudes toward having children and parenting practices?
3. What problems are there in trying to determine which traits provide the best chance of the best life?
4. What do you think marks the difference between a therapeutic intervention and an enhancement?
5. This article discusses the idea that various principles might be combined to create a more plausible view about the permissibility of enhancement. How do you think these principles should be combined or used to make decisions about genetic engineering?
6. What might be some unintended effects of the widespread practice of genetically engineering children? Would the effects be good or bad? How likely are they to occur?
7. How might genetic selection technologies restructure parent–child relationships and choices about reproduction?

PART III

BIOMEDICAL AND THERAPEUTIC TECHNOLOGIES

Biomedical and therapeutic technologies are technologies that are developed and used to improve physical and mental health and wellbeing. Biomedical technologies encompass diagnostics, pharmaceuticals, medical devices, treatments/ therapies, medical equipment, information management technologies, and much more. There are several powerful incentives that drive biomedical innovation. There primary one is that most people want to live healthier, longer, more comfortable lives. They do not want to be sick, injured, in pain, or deceased, and are willing to spend significant amounts of resources on avoiding these. In the United States, for example, health care expenditures accounted for 17.6% of GDP in 2010. This was by far the highest percentage of any nation in the world (the Netherlands was next at 12% of GDP), but the average for all OECD countries was still a quite substantial 9.5%. This means that nearly one in ten dollars was spent on trying to improving human health in those countries. Thus, there is an enormous economic incentive for developing effective, novel biomedical technologies. Moreover, nations have a public mandate to foster biomedical innovation, develop their medical infrastructure (including their research infrastructure), and expand their medical sectors more generally through public funding and intellectual property systems, for example.

The first four chapters of this section address recent and emerging biomedical technologies that have raised significant ethical concerns. In "Ethical Issues in Human Stem Cell Research," Inmaculada de Melo-Martin and Marin Gillis begin by reviewing the potential therapeutic value of embryonic stem cells, which are cells derived from embryos that can differentiate into any type of cell or tissue in the human body. They then discuss a range of ethical issues associated with stem cell research and stem cell based therapies, including the moral status of embryos, safety, obtaining ova, and justice and access.

In "Crossing Species Boundaries," Jason Robert and Francoise Baylis address the use of chimeras in biomedical research. Chimeras are organisms that have cells from multiple types of species or else have genetic material from multiple species within each of their cells. Robert and Baylis focus in particular on whether there is anything intrinsically ethically problematic with chimera creation and use in virtue of the mixing of genetic material or cells across species. Their approach to the issue involves extended discussion of what species are and whether there are fixed species boundaries.

In "The Coming Era of Nanomedicine," Fritz Allhoff identifies and discusses several possible concerns regarding emerging nanoscale science and engineering based medical technologies. Nanotechnologies take advantage of the distinctive properties and functionalities that materials have at the nanoscale (one billionth of a meter), and it is expected that nanoscale science and engineering will lead to advances in virtual all areas of medical technologies. In his chapter, Allhoff focuses on the potential privacy, safety, and justice issues that some nanotechnology based diagnostics and therapies might raise.

In "Psychopharmacology and Functional Neurosurgery: Manipulating Memory, Thought, and Mood," Walter Glannon discusses the social and ethical issues associated with two neurotherapies, pharmacological memory manipulation and neurosurgical deep-brain stimulation. Memory manipulation is used to treat severe cases of post traumatic stress disorder (PTSD), for example, while deep brain stimulation is used to treat a range of neuropsychiatric disorders including obsessive-compulsive disorder and depression. Due to the ways in which these technologies intervene into the brain and affect thought, mood, and personality, they raise a complex set of issues associated with personal identity, authenticity, and individual responsibility.

In the final chapter of this section, "Incentivizing Access and Innovation for Essential Medicines: A Survey of the Problem and Proposed Solutions," Michael Ravvin discusses the limitations of the market-oriented incentive and intellectual property system that drives biomedical innovation. In particular, he is concerned about the lack of incentives to develop and provide access to pharmaceuticals that would benefit poor people in less developed countries. Ravvin reviews several features of the current system that leads to the problem of access and "orphaned diseases" before evaluating several possible strategies for addressing them.

In each of the chapters, the discussion is not primarily about whether the technology should be developed at all. Instead, it is oriented toward how to develop rapidly emerging and potentially powerful human health technologies in safe, just, and ethically responsible ways.

More detailed summaries, as well as a listing of related readings, are located at the start of the chapters.

ETHICAL ISSUES IN HUMAN STEM CELL RESEARCH: EMBRYOS AND BEYOND[1]

Inmaculada de Melo-Martin and Marin Gillis

CHAPTER SUMMARY

In this chapter, Inmaculada de Melo-Martin and Marin Gillis begin by presenting some background on embryonic stem cells, including on why their distinctive feature – the capacity to differentiate into any type of cell in the body – is thought to hold so much biomedical research and therapeutic potential. They then explore the ethical issues raised by the use of stem cells in research and treatment. Among the issues they discuss are: the moral status of the embryos that are destroyed in the derivation of stem cells; safety and informed consent by patients who undergo stem cell based therapies; obtaining ova from women to create embryos for research; and justice concerns related to the inequality in access to novel medical technologies.

RELATED READINGS

Introduction: EHS (2.6.1); Justice, Access, and Equality (2.6.2); Individual Rights and Liberties (2.6.3); Intrinsic Concerns (2.7)

Other Chapters: Lucy Frith, *Reproductive Technologies* (Ch. 5); Leon Kass, *A Brave New World* (Ch. 6)

1. INTRODUCTION

We live in the era of new biotechnological advances. Discussion of the social, legal, ethical, and scientific aspects of genetic therapy, *in vitro* fertilization, genetically engineered food, or cloning, appear everywhere, from prestigious scientific journals, to television programs and the tabloids. In a world where the Human Genome Project hoards millions in public and private monies and thousands of scientists, where infertility seems rampant, and where the search for the perfect

---

[1] This chapter is excerpted from Inmaculada de Melo-Martin and Marin Gillis (2010) 'Ethical Issues in Human Stem Cell Research: Embryos and Beyond,' in *Technology and Human Values: Essential Readings*, ed. C. Hanks (Wiley-Blackwell). It appears here by permission of the authors.

human baby occupies people's imagination, one might expect to find this focus on biotechnology quite normal and welcomed.

Not only have these discussions captured the public imagination and the interests of scientists, they seem also to have swept many of the members of the bioethics profession away from more mundane issues, such as questions of access to health care, or just distribution of medical resources. Lately, the issue of human stem cell research seems to be the new kid on the block.

The purpose of this work is to present some of the ethical issues involved with research using embryonic stem cells. First, we will briefly give an account of the scientific and medical background surrounding stem cell research. Next, we will offer an overview of some of the main ethical concerns that have been presented in relation to this kind of research.

1.1 Properties and types of stem cells

All stem cells have three general properties: unlike muscle or nerve cells, which do not normally replicate themselves, stem cells are capable of dividing and renewing themselves for long periods of time; they are unspecialized, that is, stem cells do not have any tissue-specific structures that allow them to perform specialized functions; finally, stem cells can give rise to specialized cell types.

Human stem cells are classified depending on their source.[2] hES are derived from embryos that develop from eggs that have been donated by couples undergoing *in vitro* fertilization. The stem cells are obtained from the inner cell mass of embryos that are typically 3–5 days old. hES are isolated by transferring this inner cell mass in a laboratory dish with culture medium. The inner surface of the culture dish is usually coated with a feeder layer made of mouse embryonic skin cells that have been treated so that they cannot divide. Because these systems of support run the risk of cross-transfer of animal pathogens from the animal feeder to the hES cells, thus compromising later clinical application, scientists have recently begun to devise other ways to grow embryonic stem cells (Amit et al., 2003; Lim and Bodnar, 2002; Richards et al., 2002) . After cells proliferate they are removed and plated into fresh culture dishes. This process of replating can be repeated for many months to yield millions of embryonic stem cells. Human embryonic stem cells can thus grow indefinitely in vitro in the primitive embryonic stage while retaining their pluripotentiality or the ability to differentiate into somatic and extraembryonic cell types.

Adult stem cells, considered to be multipotent, are undifferentiated cells that can be found among differentiated cells in a tissue or organ (Filip, Mokry, and Hruska, 2003; Preston, 2003; Vats et al., 20020). They can renew themselves and can differentiate to yield the major specialized cell type of that tissue or organ. Adult stem cells are important cells in maintaining the integrity of tissues like skin, bone and blood. Adult stem cells include haematopoietic stem cells, bone marrow stromal (mesenchymal) stem cells, neural stem cells, dermal (keratinocyte) stem cells, and fetal cord blood stem cells.

---

[2]   See, for example, NIH (2002); Hadjantonakis and Papaioannou (2001).

Until recently, scientists believed that organ specific stem cells were lineage restricted. Recent work has called this idea into question, and proposed that adult stem cells may have much wider differentiation capabilities. Researchers have showed that bone marrow derived cells could target and differentiate into muscle, liver, kidney, cardiomyocytes, neural cell lineages, and gut (Preston, 2003; Orlic et al., 2001; Poulsom, 2001; Alison, 2000; Ferrari, 1998; Eglitis and Mezey, 1997).

1.2 Implications for biomedicine

Although the understanding of human stem cells is at this time limited, it appears that investigations with these cells might have a widespread impact on biomedical research. Some of the areas where stem cell research can yield important results are described in this section.

a. *Human developmental biology.* Studies of human embryonic stem cells can offer information about the genetic processes that help build a human body and can help us to obtain information of the complex events that occur during human development. A better understanding of the genetic and molecular aspects of cell division and differentiation can result in knowledge of medical conditions such as cancer and birth defects. Embryonic stem cells can also be used to identify and study environmental toxins that could cause abnormalities in the differentiation and division of cells.[3] Additionally research on stem cells could help us understand how tissues regenerate.

b. *Testing of new drugs.* Research on stem cells can allow us to develop normal lines of cells that represent specific tissues and organs. This would permit testing of new and already existing drugs and prevent cases of unanticipated liver toxicity that can be fatal.[4] Because testing of new drugs is usually performed on non-human animals, research on stem cells could result in a reduction of such practices.

c. *Regenerative medicine.* Embryonic stem cell research has the potential to provide us with renewable sources of replacement cells and tissues to treat diseases such as Parkinson's, Alzheimer's, spinal cord injury, cancer, athero-sclerosis, burns, stroke, diabetes, heart disease, rheumatoid arthritis, and oste-oarthritis. Moreover, research on stem cells can provide us with a supply of organs for transplantation that could solve the problems of unavailability that we face today.[5] Some believe that this research may lead to the formation of tissue banks to repair or replace damaged body parts (Hall, 2003).

d. *Organ replacement.* In April 2006, for example, 91,846 Americans who are in want of an organ transplant have been registered on UNOS (United Network for Organ Sharing). It is estimated that in the US thousands of people die each year for want of a transplant (United Network of Organ Sharing, 2006).

---

[3] See, for example, Hamasaki et al. (2003); Peault et al. (2002).
[4] See, for example, Rich (2003); Margolin (2002).
[5] See, for example, He et al., (2000); Ostenfeld and Svendsen (2003); Lakatos and Franklin (2002); Brehm, Zeus, and Strauer (2002).

Thus, a successful regenerative medicine would meet an important demand in itself, and consequently bring about important medical benefits. In addition, its application would prevent what some believe to be great harms to others, given existing protocols for transplant research and therapy. These include the cultivating of animals, either genetically modified and/or cloned, for the purpose of providing organs for humans (i.e. xenotransplantation). Also, it would greatly diminish if not obliterate the demand for traffic in organs and thus the questionable means of appropriating human organs that have been documented in China and India (Rothman, 1999; Scheper-Hughes, 2000; The Bellagio Task Force). Further, the ethical and legal debate about the sale of organs is potentially abated.

e. *Somatic cell nuclear transfer (SCNT)*. One of the most serious medical problems occurring in transplant therapies results from immunological incompatibilities between the donor and the recipient. Currently, host immune responses, which can only be overcome by administering long-term immunosuppressive drug therapy, have frustrated cell replacement therapies using allogeneic hES cells (Semb, 2005). Thus, the possible generation of stem cells derived from nuclear transfer embryonic stem cells (NT-ESC) has the potential to avoid immunorejection.

2. ETHICAL ISSUES

In many cases, groups within the scientific community, recognizing the promise of stem cell research, argue that they should be allowed to pursue these investigations. Nonetheless, the possible medical applications that stem cell research might produce are only part of the equation in the discussion of embryonic stem cells. Another part of the equation comes with the ethical implications of such research. In most cases, the debate on stem cell research centers around the question of whether the ethical consequences of this research are such that they require its prohibition, strict government regulation, or local institutional guidelines. In this section we describe some of the ethical issues raised by human stem cell research. Some of the concerns, such as the moral status of the embryo and the use of ova, are specific to investigations with embryonic stem cells. Also, these issues arise whether the human embryonic stem cells are derived from embryos created by the union of egg and sperm or whether the embryos result from nuclear transfer. Other ethical problems, such as issues about justice and safety, affect both embryonic and adult stem cell research.

2.1 Moral status of the embryo

Any research that has to do with the human embryo is bound to be controversial. Consequently, the case of hES cells is not different from other types of research that require the manipulation of human embryos. The main reason for the debate in this case is that, presently, the methods available to derive stem cells from embryos involve their destruction. Thus, the issue of the ontological and moral status of this entity is relevant.[6]

---

[6] See, for example, Warren (1997).

The issue of the ontological status of the embryo concerns the kind of entity embryos are. Embryos can be considered individual organisms, biological human beings, or persons. Although some authors have suggested that there is no difference between the concept of personhood and that of a biological being, many philosophers and theologians have suggested that the criteria for personhood are more demanding that those simply required for being a biological human. Some of the conditions than have been suggested as required for personhood, include self-consciousness, the capacity to feel pain, capacity to act on reasons, capacity to communicate with others using a language, capacity to act freely, and rationality.[7] Granting that an embryo is a person usually gives the embryo a more significant status than merely saying that it is an individual organism.

Inquiries into the moral status of the embryo attempt to discern whether embryos are the kinds of entities that ought to be given moral and legal rights. Having moral status is to qualify under a range of moral protections. Thus, if human embryos have full moral status, then they possess the same rights as human beings who have been born.

Opponents of hES cell research often argue that early human embryos have the full status of persons, and thus the destruction of embryos during research is equated with human sacrifice in order to obtain scientific knowledge. They argue that because the line between the human and the non-human is appropriately drawn at conception, embryos ought to be considered persons and therefore be attributed full moral status. As humans we have moral value simply because we have a human genotype, no matter what the stage of development might be.[8] Given that present ethical and legal regulations prohibit the sacrifice of human life for the sake of knowledge this would mean that the destruction of embryos in this case would not be morally justifiable. Stem cell research would, on these grounds, need to be prohibited because it violates the embryo's rights (Meyer, 2002; Meilaender, 2001; Doerflinger, 1999).

Some authors who oppose embryonic stem cell research might concede that the embryo is not a person, but that it is a potential person and this is a significant moral property. Because of this, killing an embryo to obtain stern cells must be justified to an extent comparable to the justification required for killing a person.[9] Given that most people would not justify the killing of a person for the sake of knowledge, then, even when they are not persons, the killing of embryos for that purpose would be equally unjustified.

On the other extreme of the debate, proponents of hES cell research argue that the moral status of an early human embryo is equivalent to any other cell in the human body. At this point of development, embryonic cells are too unspecialized to be a unique entity. Proponents of this view usually argue that the appropriate line between the human and the non-human must be drawn at birth rather than at conception or at some other point of development. They would recognize that the embryo is certainly an individual organism and biologically human, but it is

---

[7] See, for example, Warren (1973).

[8] See, for example, Marquis (1989).

[9] See Noonan (1970, pp. 51–59).

not human in an ontologically significant sense, and therefore it does not have a significant moral status. Hence, there is no ethical need for any protection or regulation in the use of embryonic stem cells for medical research purposes.[10]

In between these views is the so-called "middle" or "third way", which regards embryos as neither persons nor property. It sees them as being entities worthy of a *profound respect*. Embryos may thus have instrumental value while at the same time they are not simply human tissue. They are special and they cannot be treated or used in any way one might wish.[11] In this context, the protection demanded for an embryo is not absolute and might be weighed against the benefit of research purposes.[12]

This position recognizes the important differences between human embryos and fully developed human beings while at the same time giving weight to the unique relationship they have with each other. Proponents of this view tend to draw the line between the human and the non-human at some point between conception and birth. Thus the embryo has no significant moral status although the fetus might acquire full moral status at some stage of fetal development.

Some in this group argue that we already create spare embryos and sometimes discard them in *in vitro* fertilization practices, and that therefore, nothing is lost and much is gained if, rather than discarding those embryos, we can use them for research that would benefit humanity (Outka, 2002). This position cannot automatically resolve the problem of regulation of embryo research. Although it does reject outright prohibition of such investigations, or the destruction of embryos as non-problematic, proponents of this middle view need to establish parameters that would balance the interests of embryos with those of scientists and other human beings.

Because the destruction of embryos is regarded by many in the popular media and the research world as virtually the only pressing ethical issue in stem cell technology, many believe that if there were a way around the destruction of embryos, all ethical issues, or at least the most pressing and controversial ones, would dissolve.[13] Indeed, significant time and resources have been devoted to develop new ways of obtaining embryonic stem cells that do not necessitate the destruction of the embryo. These beliefs notwithstanding, there are other important ethical issues in stem cell research beyond the embryo.[14]

2.2 Safety concerns

Another ethical issue related to stem cell research (with both embryonic and adult stem cells) is related to the use of these cells for medical purposes. Some scientists argue that the procedures required for implanting cells into the human body put patients at risk (Marr et al., 2002; Smaglik, 1999). Similarly, researchers have questions about whether these cells will grow normally once inserted into the human body or whether they might become cancerous (Solter and Gearhart,

---

[10] See, for example, Warren (1997).

[11] See, for example, Green (2001); Lauritzen (2001) Meyer and Nelson (2001); Shanner (2001); Steinbock (2000).

[12] See, for example, NRC (2002); McLaren (2001); McGee and Caplan (1999); Strong (1997).

[13] See, for example, Murray (2005).

[14] See, for example, Chung et al. (2005); Meissner and Jaenisch (2005).

1999). Also concerns about immunoreaction from transplant recipients are an issue with this kind of research.

Proponents of stem cell research reply to these arguments maintaining that at this point, given the knowledge, or the lack of it, that we have, it would be unethical to attempt stem cell transplantation or similar techniques on a human being. However, if safety is the issue, it can be proposed that more research on animals be done, and more investigation completed to establish its safety and effectiveness for humans, before we proceed to use this technique on human beings. Thus, opposition to stem cell research on the basis of safety would fail as an argument against this research per se.

Proponents could also argue that once we have reasonable beliefs that regenerative medicine, for example, is safe and effective, it would then be ethically justifiable to use it on human beings who give consent to these techniques. The difficulty with this argument, however, is that at least until we have research on human subjects, the issue of safety and efficiency cannot be established. But, if this is so, some critics might argue that given a situation of uncertainty it is unclear how patients can give full informed consent to the use of these techniques.

Shortly after the Nuremberg trials, which presented horrifying accounts of Nazi experimentations on unwilling human subjects, the issue of informed consent began to receive attention.[15] The first sentence of the Nuremberg Codes states that the voluntary consent of human subjects in research is absolutely essential. At Helsinki in 1964, the World Medical Association made consent of patients and subjects a central requirement of ethical research.[16] Since then virtually all prominent medical and research codes as well as institutional rules of ethics dictate that physicians and investigators obtain the free informed consent of patients and subjects prior to any substantial intervention. Procedures for free informed consent have several functions such as the protection of patients and subjects from harm or the promotion of medical responsibility in interactions with patients and subjects. Their more fundamental goal is, however, to enable autonomous choices.

The received approach to the definition of informed consent has been to specify the elements of the concept. Legal, regulatory, medical, psychological, and philosophical literature tend to analyze informed consent in terms of the following elements[17]: (1) disclosure, (2) understanding, (3) voluntariness, and (4) competence. Thus, one gives free informed consent to an intervention if and only if one is competent to act, receives a thorough disclosure about the procedure, understands the disclosure, acts voluntarily, and consents to the intervention. Disclosure refers to the necessity of professionals passing on information to decision-makers and possible risk victims. The professionals' perspectives, opinions, and recommendations are usually essential for a sound decision. They are obligated to disclose a core set of information, including (1) those facts or descriptions that patients usually consider material in deciding whether to consent to or refuse the intervention, (2) information the doctors believe to be material, (3) the

---

[15]   See, for example, Faden and Beauchamp (1986).

[16]   See, for example, Katz (1972).

[17]   See, for example, de Melo-Martin (1997, Ch. 4).

professionals' recommendations, (4) the purpose of seeking consent, and (5) the nature and limits of consent as an act of authorization.

Understanding may be the most important component for free informed consent. It requires professionals to help potential risk victims overcome illness, irrationality, immaturity, distorted information, or other factors that can limit their grasp of the situation to which they have the right to give or withhold consent. Thus people understand if they have acquired pertinent information and justified, relevant beliefs about the nature and impacts of their actions. This understanding need not be complete, because a substantial grasp of central facts is generally sufficient. Normally, diagnoses, prognoses, the nature and purpose of the intervention, alternatives, risks, benefits, and recommendations are essential. Patients or subjects also need to share an understanding with professionals about the terms of the authorization before proceeding. Unless agreement exists (about the crucial features of what the patients authorize) there can be no assurance that they have made autonomous decisions. Thus, even if doctor and patients use a word such as 'ovulation induction', their interpretations could be totally different if standard medical definitions have no meaning for the patients.

Another element of informed consent is voluntariness, or being free to act in giving consent. It requires that the subjects act in a way that is free from manipulation and coercion by other persons. Coercion occurs if and only if one person intentionally uses a credible and serious threat of harm or force to control another. Manipulation is convincing people to do what one wants by means other than direct coercion or rational persuasion. One important form of manipulation in health care is informational manipulation, a deliberate act of handling information that alters patients' understanding of the situation and motivates them to do what the agent of influence plans. Withholding evidence and misleading exaggerations of benefits are instances of manipulation inconsistent with voluntary decisions. The way in which doctors present information by tone of voice, by framing information positively ('the therapy is successful most of the time') rather than negatively ('the therapy fails in forty percent of the cases'), for example, can manipulate patients' perceptions and, therefore, affect understanding.

The criterion of competence refers to the patients' abilities to perform a task. Thus patients or subjects are competent if they have the ability to understand the material information, to make a judgment about the evidence in light of their values, to intend a certain outcome, and to freely communicate their wishes to the professionals.

Given these elements that scholars recognize as necessary for informed consent, to claim that patients can give free informed consent to regenerative medicine, for example, might be questionable. If patients are ignorant of the possible risks involved in these kinds of treatments, then they cannot give genuinely informed consent. Lack of information may seriously hinder people's abilities to make informed choices. A possible way to solve this problem, however, is to inform patients of the uncertainty. It is nevertheless unclear whether people really understand what a situation of uncertainty means for evaluation of risks and benefits.

2.3 Obtaining ova

The use of research embryos also raises ethical concerns related to the interests of the women whose eggs would be used to make the embryos. There are two potential sources of embryonic stem cells: discarded embryos used in infertility treatments and created embryos for the purpose of research. Given that obtaining eggs from women is far more burdensome and risky than obtaining sperm, the interests of women merit special consideration.

According to empirical evidence, risks to women undergoing IVF treatment vary from simple nausea to death.[18] For example, the hormones that doctors use to stimulate the ovaries are associated with numerous side effects. Some studies assert that ovulation induction may be a risk factor for certain types of hormone-dependent cancers. Researchers have associated excessive estrogen secretion with ovarian and breast carcinoma, and gonadotropin secretion with ovarian cancer. A substantial body of experimental, clinical, and epidemiological evidence indicates that hormones play a major role in the development of several human cancers. The ability of hormones to stimulate cell division in certain organs, such as the breast, endometrium, and the ovary, may lead (following repeated cell divisions) to the accumulation of random genetic errors that ultimately produce cancer. Hormone-related cancers account for more than 30% of all newly diagnosed female cancers in the United States. Hence any technique (like IVF) – that relies on massive doses of hormones to obtain ova – may be quite dangerous for women.

The ovarian hyperstimulation syndrome (OHSS) is another possible iatrogenic (caused by medical treatment) consequence of ovulation induction. Women with the severe form of OHSS may suffer renal impairment, liver dysfunction, thromboembolic phenomena, shock, and even death. The incidence of moderate and severe OHSS in IVF treatment ranges from 3% to 4%.

The procedures that doctors normally use to obtain women's eggs, i.e., laparoscopy and ultrasound-guided oocyte retrieval also pose risks. Although there are no accurate statistical data about the hazards associated with these two procedures, risks related to these technologies include postoperative infections, punctures of an internal organ, hemorrhages, ovarian trauma, and intrapelvic adhesions.

If the eggs have been obtained in order to solve a reproductive problem, then, we can say that the potential benefits – the birth of a child – might outweigh the potential risks. Thus, those concerned with issues of pressure and coercion of women might see the use of embryos already created for fertility purposes as less problematic. However, a problem appears when we consider the situation of women who are asked to donate excess embryos for stem cell research. Because of the pressure that women facing reproductive problems suffer in our society, it is not unreasonable to believe that women undergoing IVF treatment might feel coerced to produce more embryos than those necessary for implantation. Moreover, in cases where women might be asked to donate eggs for research purposes other problems arise. Here women would be properly considered research subjects. However, these women would not directly experience any direct benefit from this procedure – such as the birth of a child. Similarly, they would experience neither direct nor indirect health benefits from donating their eggs. Furthermore, women presumably could not be compensated for the risks imposed on them by these procedures. Present regulations recommend that it be illegal to sell human embryos because the practice of selling human embryos would be contrary to the respect owed to them. But it can be argued

---

[18]   See, for example, de Melo-Martin (1997, Ch. 4).

that if such practice ought to be illegal, then consistency would require that the selling of eggs with the sole intention of creating embryos for research be illegal.[19]

The need to obtain embryos for research purposes thus might increase the risk of coercion and exploitation for all women. Exploitation of low-income women would be a clear reality in a system where gametes, embryos, or wombs can be bought and sold. Problems of social and racial discrimination in our society make these concerns even more present and serious. Poor, migrant, refugee, or ethnic minority women might be used as producers of eggs.

Furthermore, in a society where eggs and embryos are treated as properties that can be bought, sold, or rented, problems of commodification arise. "Commodification" refers to the association of a thing or a practice with attitudes and behaviours that accompany typical market transactions (Altman, 1991).[20] Because eggs and embryos have an intimate connection to personhood, their commodification could contribute to a diminishing sense of human personhood on an individual level, and might damage commitments to human flourishing at the societal level (Holland, 2001).

The moral objections to the commodification of women's reproductive material echo the most vocal moral objections people often have to human cloning, the patenting of organisms or organic processes in whole or in part, surrogacy, and eugenics. In these activities, it is acceptable to treat some human bodies, some body parts, and some non-human living things in the same manner that manufactured objects are treated. The ethos that accepts such commodification risks fostering the view that the value of some living things, like fertile women, is the same as that of laundry detergent and toasters.[21] This objection combines deontological and consequentialist concerns. On the one hand, there is something inherently valuable about a living thing such that it is wrong to instrumentalize it. On the other hand, thinking of eggs and embryos as commodities shapes our understanding of what it means to be human. It can contribute to a lack of respect for human dignity or to diminish respect for human life.

Also relevant is the fact that potential egg and embryo donors are encouraged to make such donations to medical research and therapy altruistically. The possibility of saving another person's life (or at least ease his or her suffering in some way) is an act represented as so intrinsically valuable that it would be diminished if the donor were compensated for it. However, as some scholars have pointed out, there is a significant tension between the altruism that individuals exhibit by donating their tissue for research and the current patent system, which encourages companies to stake lucrative property claims in that research (Knowles, 1999; Spar, 2004). Indeed, others argue even more strongly that donation of eggs

---

[19] See, for example, Baylis (2000).

[20] See, for example, Raddin (1991); Raddin (1989).

[21] Statement of the American Humane Association, on behalf of American Society for the Prevention of Cruelty to Animals, Animal Protection Institute, Committee for Humane Legislation, and Massachusetts Society for the Prevention of Cruelty to Animals: "It troubles us that animal patenting reduces the animal kingdom to the same level as laundry detergent and toasters. Animals are not objects" (TAPRA '89 Hearings, 288).

and embryos as a gift both masks and legitimizes what is actually the extension of commodification. On the one hand, we have donors who believe that they are demonstrating altruism, but on the other we find biotechnology firms and researchers using the discourse of profit (Dickenson, 2001; Dodds, 2004; Holland and Davis, 2001). To call for the supplier to be altruistic when there is no similar call placed on those who would profit greatly from the sale of therapies made from the tissue can give way to exploitative practices.[22]

2.4 Justice

Another ethical issue resulting from the debate on stem cell research is related to our duties and obligations to promote social justice.[23] Stem cell research and the possible medical application resulting from it raise issues of social justice for several reasons. First, existing inequalities in access to healthcare mean that new technologies affect people quite differently according to ability to pay. Medical resources are not distributed equally and people do not have equal access to new medical procedures. That means that those with insurance will likely be beneficiaries of the new advances in medicine and technology. But not everybody in our society has adequate insurance or insurance at all. Approximately fifty million people at any given time during the year have not insurance at all. There is a fundamental inequality in the options that people have, and medical benefits resulting from stem cell research do nothing to narrow or solve the gap between them.

Second, even if the problem of access is resolved as an economic problem, this might not resolve the problem of inequalities and injustices present in our social context.[24] Some evaluations and reports on medical technologies (mostly European) have considered "economic issues" – but it seems only to the extent that the reports note that there is a problem with some people not having the ability to pay (Hoedemaekers, ten Have, and Chadwick, 1997). One response to these concerns has been the recommendation that medical technologies and procedures resulting from stem cell research be available to everyone. However, while following this recommendation would go some way toward addressing issues of social justice, it would not fully and adequately respond to those concerns. For example, we might say that government programs will provide access for those who do not have insurance. This might not solve the problems posed by other factors such as distrust of the healthcare system by certain groups, the role of religious beliefs in directing healthcare decisions, lack of access to information, and lack of education. Each of these factors might prevent people from having access to even free and public healthcare programs. Medical research practices such as involuntary sterilization, hysterectomy, mastectomy, and the Tuskegee study, also show that distrust of government health services might appear reasonable. A conception of medical technologies and practices that ignores these issues when evaluating the implementation of such techniques will then be deficient.

---

[22] See, for example, Sample (2003).
[23] See, for example, Rawls (1971).
[24] See, for example, de Melo-Martin and Hanks (2001).

Also, given the fact that many healthcare problems are related to where and how one lives (namely, the relative level of dangers posed by the immediate environment of one's home or work), and that these are differentially distributed throughout the population in ways which track economic status (and also race and sex), then providing the access to diagnosis and treatment alone will not solve problems of inequality. Thus, focusing exclusively on the benefits or risks of stem cell research blinds us to the importance of the environment for preventing or curing diseases.

Furthermore, even if stem cell research were fully privately funded, problems of social justice would still arise for several reasons. First, there may be complications due to stem cell procedures when used as medical techniques. Treatment of such complications may result in costs for the public system. Second, societal interdependencies and professional contracts create and enhance doctors' abilities to use the results of stem cell research. They employ tools and technologies developed in part through societal resources. Also, public money supports physicians through learning, because virtually no student, even in private schools, pays for the full costs of their education; taxes or donations usually supplement that cost.

3. CONCLUSION

Stem cell research is still very new. The potential benefits related to this research have created increasing expectations. However, ethical challenges need also be considered when we are evaluating such benefits. In this work, we have called attention to several ethical issues that deserve our attention in the assessment of stem cell research: the problem of the moral status of the embryo, the issue of safety for patients, concerns about coercion and exploitation of women in order to obtain eggs for research purposes, and issues of social justice related to access to the benefits of new medical technologies and procedures. A complete assessment of stem cell research cannot ignore these ethical issues.

WORKS CITED

M. R. Alison, et al. (2000) 'Hepatocytes from non-hepatic adult stem cells,' *Nature*, 406: 257.

S. Altman (1991) 'Commodifying Experience,' *Southern California Law Review*, 65: 293–340, 293.

M. Amit, et al. (2003) 'Human Feeder Layers for Human Embryonic Stem Cells,' *Biological Reproduction*, 68 (6): 2150–6.

The Bellagio Task Force Report on Transplantation, Bodily Integrity, and the International Traffic in Organs Available at www.icrc.org/Web/eng/siteeng0.nsf/iwpList302/ 87DC95FCA3C3D63EC1256B66005B3F6C (accessed April 2006).

F. Baylis (2000) 'Our Cells/ Ourselves: Creating Human Embryos for Stem Cell Research,' *Women's Health Issues*, 10 (3): 140–5.

M. Brehm, T. Zeus, and B. E. Strauer (2002) 'Stem cells – clinical application and perspectives, *Herz*, 27 (7): 611–20.

L. S. Cahill (2000) 'Social ethics of embryo and stem cell research,' *Women's Health Issues*, 10: 131–5.

Y. Chung, I. Klimanskaya, S. Becker, et al. (2005) 'Embryonic and extraembryonic stem cell lines derived from single mouse blastomeres,' *Nature* (Oct 16).

I. de Melo-Martin (1997) *Making Babies. Biomedical Technologies, Reproductive Ethics, and Public Policy* (Dordrecht: Kluwer University Press).

I. de Melo-Martin, and C. Hanks (2001) 'Genetic technologies and women: the importance of context,' *Bulletin of Science, Technology & Society*, 21 (5): 354–60.

D. Dickenson (2001) 'Property and women's alienation from their own labour,' *Bioethics* 15: 204–17.

S. Dodds (2004) 'Women, com- modification, and embryonic stem cell research' in *Stem Cell Research: Biomedical Ethics Reviews*, J. Humber and R. Almeder, (eds) (Towota, New Jersey: Humana Press).

R. M. Doerflinger (1999) 'The ethics of funding embryonic stem cell research: a Catholic viewpoint,' *Kennedy Institute of Ethics Journal*, 9: 137–50.

M. A. Eglitis, and E. Mezey (1997) 'Hematopoietic cells differentiate into both microglia and macroglia in the brains of adult mice,' *Proceedings of the National Academy of Sciences* USA 94: 4080–5.

R. Faden, and T. Beauchamp (1986) *A History and Theory of Informed Consent* (New York: Oxford University Press).

G. Ferrari, et al. (1998) 'Muscle regeneration by bone marrow-derived myogenic progenitors,' *Science*, 279: 1528–30.

S. Filip, J. Mokry, and I. Hruska (2003) 'Adult stem cells and their importance in cell therapy,' *Folia Biologica*, 49 (1): 9–14.

R. Green (2001) *The Human Embryo Research Debates: Bioethics in the Vortex of Controversy* (New York: Oxford University Press).

A-K. Hadjantonakis, V. E. Papaioannou (2001) 'The stem cells of early embryos,' *Differentiation*, 68: 159–166.

S. Hall (2003) 'The Recycled Generation,' *New York Times Magazine* (January): 30.

T. Hamasaki et al. (2003) 'Neuronal cell migration for the developmental formation of the mammalian striatum,' *Brain Research. Brain Research Reviews*, 41 (1): 1–12.

Q. He et al. (2003) 'Embryonic stem cells: new possible therapy for degenerative diseases that affect elderly people,' *Journals of Gerontology. Series A, Biological Sciences and Medical Sciences*, 58 (3): 279–87.

R. Hoedemaekers, H. ten Have, R. Chadwick (1997) 'Genetic screening: a comparative analysis of three recent reports,' *Journal of Medical Ethics* 23 (3): 135–41.

S. Holland (2001) 'Contested commodities at both ends of life: buying and selling gametes, embryos, and body tissues,' *Kennedy Institute of Ethics Journal*, 11 (3): 263–84.

S. Holland, and D. Davis (eds) (2001) 'Special Issue: Who's afraid of commodification,' *Kennedy Institute of Ethics Journal*, 11.

J. Katz (1972) *Experimenting with Human Beings* (New York: Russell Sage Foundation).

L. Knowles (1999) 'Property, patents, progeny,' *Hastings Center Report*, 2: 38–40.

A. Lakatos, and R. J. Franklin (2002) 'Transplant mediated repair of the central nervous system: an imminent solution?' *Current Opinion Neurology*, 15 (6): 701–5.

P. Lauritzen (2001) 'Neither person nor property: embryo research and the status of the early embryo,' *America*, (March 26).

J. W. Lim, and A. Bodnar (2002) 'Proteome analysis of conditioned medium from mouse embryonic fibroblast feeder layers which support the growth of human embryonic stem cells,' *Proteomics*, 2 (9): 1187–203.

K. Margolin (2002) 'High dose chemotherapy and stem cell support in the treatment of germ cell cancer,' *Journal of Urology*, 169 (4): 1229–33.

D. Marquis (1989) 'Why Abortion Is Immoral,' *Journal of Philosophy*, 86 (4): 470–86.

K. A. Marr, et al. (2002) 'Invasive aspergillosis in allogeneic stem cell transplant recipients: changes in epidemiology and risk factors,' *Blood*, 100 (13): 4358–66.

G. McGee, and A. Caplan (1999) 'The ethics and politics of small sacrifices in stem cell research,' *Kennedy Institute of Ethics Journal*, 9 (2): 151–8.

A. McLaren (2001) Ethical and social considerations of stem cell research, *Nature* 414: 129–31.

G. Meilaender (2001) 'The point of a ban: or how to think about stem cell research,' *Hastings Center Report*, 31 (1): 9–16.

A. Meissner, and R. Jaenisch (2005) 'Generation of nuclear transfer-derived pluripotent ES cells from cloned Cdx2-deficient blastocysts,' *Nature*, (Oct 16).

J. R. Meyer (2002) 'Human embryonic stem cells and respect for life,' *Journal of Medical Ethics*, 26: 166–70.

M. J. Meyer, and L. J. Nelson (2001) 'Respecting what we destroy: reflections on human embryo research,' *Hastings Center Report*, 31: 16–23.

T. Murray (2005) 'Will new ways of creating stem cells dodge the objections?' *Hastings Center Report* 32 (1): 8–9.

National Institutes of Health (2002) 'Stem Cells: A Primer, Bethesda, Maryland: National Institutes of Health,' available at www.nih.gov/news/stemcell/primer.htm (accessed April 6, 2003).

National Research Council and the Institute of Medicine (2002) '*Stem Cell and the Future of Regenerative Medicine*,' (Washington, DC: National Academy Press).

J. Noonan (1970) 'An almost absolute value in history,' in *The Morality of Abortion: Legal and Historical Perspectives*, (ed.) J. Noonan (Cambridge, MA: Harvard University Press).

D. Orlic et al. (2001) 'Bone marrow cells regenerate infarcted myocardium,' *Nature* 410: 701–4.

T. Ostenfeld, and C. N. Svendsen (2003) 'Recent advances in stem cell neurobiology,' *Advances and Technical Standards in Neurosurgery*, 28: 3–89.

G. Outka (2002) 'The ethics of human stem cell research,' *Kennedy Institute of Ethics Journal* 12 (2): 175–214.

B. Peault, E. Oberlin, and M. Tavian (2002) 'Emergence of hematopoietic stem cells in the human embryo,' *Comptes Rendus Biology* 325 (10): 1021–6.

R. Poulsom, et al. (2001) 'Bone marrow contributes to renal parenchymal turnover and regeneration,' *Journal of Pathology*, 195: 229–35.

S. L. Preston (2003) 'The new stem cell biology: something for everyone,' *Molecular Pathology*, 56 (2): 86–96.

M. J. Raddin (1991) 'Reflections on objectification,' *Southern California Law Review*, 65: 341–54.

M. Raddin (1989) 'Justice and the market domain,' in R. Pennock and J. Chapman (eds), *Markets and Justice* (New York: New York University Press).

J. Rawls (1971*) A Theory of Justice* (Cambridge, MA: Harvard University Press).

D. Resnik (2001) 'Regulating the Market for Human Eggs,' *Bioethics*, 15: 1–25.

I. N. Rich (2003) 'In vitro hematotoxicity testing in drug development: a review of past, present and future applications,' *Current Opinion Drug Discovery Development*, 6 (1): 100–9.

M. Richards, et al. (2002) 'Human feeders support prolonged undifferentiated growth of human inner cell masses and embryonic stem cells,' *Nature Biotechnology*, 20 (9): 933–6.

D. Rothman (1999) 'The international organ traffic,' in *Moral Issues in a Global Perspective* (ed.) C. Koggel (Peterborough, Ontario: Broadview Press): 611–18.

R. Sample (2003) *Exploitation: What It Is and Why It's Wrong* (Lanham, MD: Rowman and Littlefield).

N. Scheper-Hughes (2000) 'The global traffic in human organs,' *Current Anthropology*, 41 (April): 2–19.

H. Semb (2005) 'Human embryonic stem cells: origin, properties and applications,' *APMIS*, 113 (11–12): 743–50.

L. Shanner (2001) 'Stem cell terminology: practical, theological and ethical implications,' *Health Law Review Papers* (September 2): 62–6.

P. Smaglik (1999) 'Promise and problems loom for stem cell gene therapy,' *Scientist*, 13:14–15.

D. Solter, and J. Gearhart (1999) 'Putting stem cells to work,' *Science*, 283 (5407): 1468–70.

D. Spar (2004) 'The business of stem cells,' *New England Journal of Medicine*, 351: 211–13.

B. Steinbock (2000) 'What does 'respect for embryos' mean in the context of stem cell research?' *Women's Health Issues*, 10 (3): 127–30.

C. Strong (1997) 'The moral status of preembryos, embryos, fetuses, and infants,' *Journal of Medical Philosophy*, 22: 457–78.

United Network of Organ Sharing, a US non- profit organization and clearing house. Available at www.unos.org (accessed April 2006).

A. Vats, et al. (2002) 'Stem cells: sources and applications,' *Clinical Otolaryngology*, 27 (4): 227–32.

M. Warren (1973) 'On the Moral and Legal Status of Abortion,' *The Monist*, 57 (1): 43–61.

M. Warren (1997) *Moral Status: Obligations to Persons and Other Living Things* (New York: Oxford University Press).

A. Werthheimer (1999) *Exploitation* (Princeton, NJ: Princeton University Press).

DISCUSSION QUESTIONS

1. In what sense, if any, can an embryo be considered a "person"? In what senses are embryos not properly considered to be "persons"? Given your answers to these questions, is it appropriate to regard embryos as having moral and legal rights?

2. Is destroying an embryo in the context of biomedical research intrinsically problematic? Why or why not?

3. The authors raise concerns about the difficulties in meeting the standard of free informed consent for novel stem cell based therapies. Do you think these difficulties are particularly significant in the case of stem cells? Or are they common to all novel (or even familiar) advanced therapies?

4. What considerations might be offered in support of allowing the sale of human eggs and embryos? Do you share the authors' concerns about the possible commodification of human body parts, particularly those related to reproduction and personhood? Why or why not?

5. Do you think it is just for a society to spend scarce medical resources on developing cutting edge new technologies if many people in the society lack access to basic medical care? Why or why not?

6. Overall, are the legitimate ethical concerns associated with embryonic stem cell research such that the research should not be permitted at all? Or are the ethical concerns such that they could be addressed in the development and use of stem cell based therapies?

10 CROSSING SPECIES BOUNDARIES[1]

Jason Robert and Françoise Baylis

CHAPTER SUMMARY

In this chapter, Jason Robert and Francoise Baylis critically examine the morality of crossing species boundaries in the context of research that involves combining human and nonhuman animals at the genetic or cellular level. They begin by discussing the notion of species identity, with a focus on the presumed fixity of species boundaries, as well as the general biological and philosophical problem of defining species. Against this backdrop, they survey and criticize earlier efforts to justify a prohibition on crossing species boundaries in the creation of novel beings. They do not attempt to establish the immorality of crossing species boundaries, but do conclude with some thoughts about such crossings, alluding to the notion of moral confusion regarding social and ethical obligations to novel interspecies beings.

RELATED READINGS

Introduction: Intrinsic Concerns (2.7)

Other Chapters: Gary Comstock, *Ethics and Genetically Modified Foods* (Ch. 31); Christopher J. Preston, *Evolution and the Deep Past: Responses to Synthetic Biology* (Ch. 36); Mark A. Bedau and Mark Triant, *Social and Ethical Implications of Creating Artificial Cells* (Ch. 37)

1. INTRODUCTION

Crossing species boundaries in weird and wondrous ways has long interested the scientific community but has only recently captured the popular imagination beyond the realm of science fiction. Consider, for instance, the print and pictorial publicity surrounding the growth of a human ear on the back of a mouse;[2] the plight of Alba, artist Eduardo Kac's green-fluorescent-protein bunny stranded in

---

[1] This chapter is excerpted from Jason Robert and Francoise Baylis (2003) 'Crossing Species Boundaries,' *American Journal of Bioethics* 3: 1–13. It appears here by permission of Taylor and Francis.

[2] See, e.g., Mooney and Mikos (1999); and the *Scientific American Frontiers* coverage of "The Bionic Body," available from: http://www.pbs.org/saf/1107/features/body.htm. See also Bianco and Robey (2001).

Paris;[3] the birth announcement in *Nature* of ANDi, the first transgenic primate;[4] and, most recently, the growth of pigs' teeth in rat intestines[5] and miniature human kidneys in mice.[6]

But, bizarrely, these innovations that focus on discrete functions and organs are almost passé. As part of the project of harnessing the therapeutic potential of human stem cell research, researchers are now involved in creating novel inter-species whole organisms that are unique cellular and genetic admixtures (DeWitt 2002). A human-to-animal embryonic chimera is a being produced through the addition of human cellular material (such as pluripotent or restricted stem cells) to a nonhuman blastocyst or embryo. To give but four examples of relevant works in progress, Snyder and colleagues at Harvard have transplanted human neural stem cells into the forebrain of a developing bonnet monkey in order to assess stem cell function in development (Ourednik et al. 2001); human embryonic stem cells have been inserted into young chick embryos by Benvenisty and colleagues at the Hebrew University of Jerusalem (Goldstein et al. 2002); and most recently it has been reported that human genetic material has been transferred into rabbit eggs by Sheng (Dennis 2002), while Weissman and colleagues at Stanford University and StemCells, Inc., have created a mouse with a significant proportion of human stem cells in its brain (Krieger 2002).

Human-to-animal embryonic chimeras are only one sort of novel creature currently being produced or contemplated. Others include: *human-to-animal fetal or adult chimeras* created by grafting human cellular material to late-stage nonhuman fetuses or to postnatal nonhuman creatures; *human-to-human embryonic, fetal, or adult chimeras* created by inserting or grafting exogenous human cellular material to human embryos, fetuses, or adults (e.g., the human recipient of a human organ transplant, or human stem cell therapy); animal-to-human embryonic, fetal, or adult chimeras created by inserting or grafting nonhuman cellular material to human embryos, fetuses, or adults (e.g., the recipient of a xenotransplant); animal-to-animal embryonic, fetal, or adult chimeras generated from nonhuman cellular material whether within or between species (excepting human beings); *nuclear-cytoplasmic hybrids*, the offspring of two animals of different species, created by inserting a nucleus into an enucleated ovum (these might be intraspecies, such as sheep–sheep; or interspecies, such as sheep–goat; and, if interspecies, might be created with human or nonhuman material); *interspecies hybrids* created by fertilizing an ovum from an animal of one species with a sperm from an animal of another (e.g., a mule, the offspring of a he-ass and a mare);

---

[3] See the bibliography of media coverage at http:// www.ekac.org/transartbiblio.html.

[4] A sample headline from *The Independent* (London): "How a Glowing Monkey Will Help Cure Disease" (12 January 2001; available from: http://www.independent.co.uk/news/ science/how-a-glowing-monkey-will-help-cure-disease-700310.html). See also Chan et al. (2001); and Harris (2001).

[5] "Scientists Grow Pig Teeth in Rat Intestines" is available from: http://www.laurushealth. com/HealthNews/reuters/ NewsStory0926200224.htm. See also Young et al. (2002).

[6] "Human Kidneys Grown in Mice" is available from: http://news.bbc.co.uk/2/hi/ health/2595397.stm. See also Dekel et al. (2003).

and *transgenic organisms* created by otherwise combining genetic material across species boundaries.

For this paper, in which we elucidate and explore the concept of species identity and the ethics of crossing species boundaries, we focus narrowly on the creation of interspecies chimeras involving human cellular material – the most recent of the transgressive interspecies creations. Our primary focus is on human-to-animal *embryonic* chimeras, about which there is scant ethical literature, though the scientific literature is burgeoning.

Is there anything ethically wrong with research that involves the creation of human-to-animal embryonic chimeras? A number of scientists answer this question with a resounding "no." They argue, plausibly, that human stem cell proliferation, (trans)differentiation, and tumorigenicity must be studied in early embryonic environments. For obvious ethical reasons, such research cannot be carried out in human embryos. Thus, assuming the research must be done, it must be done in nonhuman embryos – thereby creating human-to-animal embryonic chimeras. Other scientists are less sanguine about the merits of such research. Along with numerous commentators, they are quite sensitive to the ethical conundrum posed by the creation of certain novel beings from human cellular material, and their reaction to such research tends to be ethically and emotionally charged. But what grounds this response to the creation of certain kinds of part-human beings? In this paper we make a first pass at answering this question. We critically examine what we take to be the underlying worries about crossing species boundaries by referring to the creation of certain kinds of novel beings involving human cellular or genetic material. In turn, we highlight the limitations of each of these arguments. We then briefly hint at an alternative objection to the creation of certain novel beings that presumes a strong desire to avoid introducing moral confusion as regards the moral status of the novel being. In particular we explore the strong interest in avoiding any practice that would lead us to doubt the claim that humanness is a necessary (if not sufficient) condition for full moral standing.

2. SPECIES IDENTITY

Despite significant scientific unease with the notion of *species identity*, commonplace among biologists and commentators are the assumption that species have particular identities and the belief that the boundaries between species are fixed rather than fluid, established by nature rather than by social negotiation. Witness the ease with which biologists claim that a genome sequence of some organism – yeast, worm, human – represents the identity of that species, its blueprint or, alternatively, instruction set. As we argue below, such claims mask deep conceptual difficulties regarding the relationship between these putatively representative species-specific genomes and the individual members of a species.

The ideas that natural barriers exist between divergent species and that scientists might someday be able to cross such boundaries experimentally fuelled debates in the 1960s and 1970s about the use of recombinant DNA technology (e.g., Krimsky 1982). There were those who anticipated the possibility of research involving the crossing of species boundaries and who considered this a laudable scientific goal. They tried to show that fixed species identities and fixed boundaries between species are illusory. In contrast, those most critical of crossing species boundaries argued that there were fixed natural boundaries between species that should not be breached.

At present the prevailing view appears to be that species identity is fixed and that species boundaries are inappropriate objects of human transgression. The idea of fixed species identities and boundaries is an odd one, though, inasmuch as the creation of plant-to-plant[7] and animal-to-animal hybrids, either artificially or in nature, does not foster such a vehement response as the prospective creation of interspecies combinations involving human beings – no one sees rhododendrons or mules (or for that matter goat–sheep, or geep) as particularly monstrous (Dixon 1984). This suggests that the only species whose identity is generally deemed genuinely "fixed" is the human species. But, what is a *species* such that protecting its identity should be perceived by some to be a scientific, political, or moral imperative? This and similar questions about the nature of species and of species identities are important to address in the context of genetics and genomics research (Ereshefsky 1992; Claridge, Dawah, and Wilson 1997; Wilson 1999b).

Human beings (and perhaps other creatures) intuitively recognize species in the world, and cross-cultural comparative research suggests that people around the globe tend to carve up the natural world in significantly similar ways (Atran 1999). There is, however, no one authoritative definition of species. Biologists typically make do with a plurality of species concepts, invoking one or the other depending on the particular explanatory or investigative context.

One stock conception, propounded by Dobzhansky (1950) and Mayr (1940), among others, is the *biological species concept* according to which species are defined in terms of reproductive isolation, or lack of genetic exchange. …The *evolutionary species concept* advanced by G. G. Simpson and E. O. Wiley, emphasizes continuity of populations over geological time: "a species is a single lineage of ancestral descendant populations of organisms which maintains its identity from other such lineages and which has its own evolutionary tendencies and historical fate" (Wiley 1978, 18; see also Simpson 1961). …On the *homeostatic property cluster* view of species, advocated in different ways by Boyd (1999), Griffiths (1999), and Wilson (1999a), a species is characterized by a cluster of properties (traits, say) no one of which, and no specific set of which, must be exhibited by any individual member of that species, but some set of which must be possessed by all individual members of that species.

To these definitions of species many more can be added: at present, there are somewhere between nine and twenty-two definitions of species in the biological literature.[8] Of these, there is no one species concept that is universally compelling.

---

[7] A possible exception is the creation of genetically modified crops. But here the arguments are based on human health and safety concerns, as well as on political opposition to monopolistic business practices, rather than on concern for the essential identity of plant species.

[8] Kitcher (1984) and Hull (1999) each discuss nine concepts. Mayden (1997) discusses twenty-two.

3. MORAL UNREST WITH CROSSING SPECIES BOUNDARIES

As against what was once commonly presumed, there would appear to be no such thing as fixed species identities. This fact of biology, however, in no way undermines the reality that fixed species exist independently as moral constructs. That is, notwithstanding the claim that biologically species are fluid, people believe that species identities and boundaries are indeed fixed and in fact make everyday moral decisions on the basis of this belief. (There is here an analogy to the recent debate around the concept of race. It is argued that race is a biologically meaningless category, and yet this in no way undermines the reality that fixed races exist independently as social constructs and they continue to function, for good or, more likely, ill, as a moral category.) This gap between science and morality requires critical attention.

Scientifically, there might be no such thing as fixed species identities or boundaries. Morally, however, we rely on the notion of fixed species identities and boundaries in the way we live our lives and treat other creatures, whether in decisions about what we eat or what we patent. Interestingly, there is dramatically little appreciation of this tension in the literature, leading us to suspect that (secular) concern over breaching species boundaries is in fact concern about something else, something that has been mistakenly characterized in the essentialist terms surveyed above. But, in a sense, this is to be expected. While a major impact of the human genome project has been to show us quite clearly how similar we human beings are to each other and to other species, the fact remains that human beings are much more than DNA and moreover, as we have witnessed throughout the ages, membership within the human community depends on more than DNA. Consider, for example, the not-so-distant past in which individual human beings of a certain race, creed, gender, or sexual orientation were denied moral standing as members of the human community. By appealing to our common humanity, ethical analysis and social activism helped to identify and redress what are now widely seen as past wrongs.

Although in our recent history we have been able to broaden our understanding of what counts as human, it would appear that the possible permeability of species boundaries is not open to public debate insofar as novel part-human beings are concerned. Indeed, the standard public-policy response to any possible breach of human species boundaries is to reflexively introduce moratoriums and prohibitions.[9]

But why should this be so? Indeed, why should there be *any* ethical debate about the prospect of crossing species boundaries between human and nonhuman animals? After all, hybrids occur naturally, and there is a significant amount of gene flow between species in nature.[10] Moreover, there is as yet no adequate biological

---

[9] See, for example, s6(2)(b) Infertility (Medical Procedures) Act 1984 (Victoria, Australia); s3(2)(a)–(b) and s3(3)(b) Human Fertilisation and Embryology Act 1990 (United Kingdom); and Article 25 Bill containing rules relating to the use of gametes and embryos (Embryo Bill), September 2000 (the Netherlands). See also Annas, Andrews, and Isasi (2002).

[10] A particularly well-documented example of gene flow between species is Darwin's finches in the Galapagos Islands. For a recent account, see Grant and Grant (2002).

(or moral) account of the distinctiveness of the species *Homo sapiens* serving to capture all and only those creatures of human beings born. As we have seen, neither essentialism (essential sameness, genetic or otherwise) nor universality can function as appropriate guides in establishing the unique identity of *Homo sapiens*. Consequently, no extant species concept justifies the erection of the fixed boundaries between human beings and nonhumans that are required to make breaching those boundaries morally problematic. Despite this, belief in a fixed, unique, human species identity persists, as do moral objections to any attempt to cross the human species boundary – whatever that might be.

According to some, crossing species boundaries is about human beings playing God and in so doing challenging the very existence of God as infallible, all-powerful, and all-knowing. There are, for instance, those who believe that God is perfect and so too are all His creations. This view, coupled with the religious doctrine that the world is complete, suggests that our world is perfect. In turn, perfection requires that our world already contains all possible creatures. The creation of new creatures – hybrids or chimeras – would confirm that there are possible creatures that are not currently found in the world, in which case "the world cannot be perfect; therefore God, who made the world, cannot be perfect; but God, by definition is perfect; therefore God could not exist" (Morriss 1997, 279).[11] This view of the world, as perfect and complete, grounds one sort of opposition to the creation of human-to-animal chimeras.

As it happens, however, many do not believe in such a God and so do not believe it is wrong to "play God." Indeed, some would argue further that not only is it *not* wrong to play God, but rather this is exactly what God enjoins us to do. Proponents of this view maintain that God "left the world in a state of imperfection so that we become His partners" – his co-creators (Breitowitz 2002, 327).

Others maintain that combining human genes or cells with those of nonhuman animals is not so much about challenging God's existence, knowledge, or power, as it is about recognizing this activity as inherently unnatural, perverse, and so offensive. Here the underlying philosophy is one of repugnance. To quote Kass (1998), repugnance

> revolts against the excesses of human willfulness, warning us not to transgress what is unspeakably profound. Indeed in this age in which … our given human nature no longer commands respect … repugnance may be the only voice left that speaks up to defend the central core of humanity. (19)

For many, the mainstay of the argument against transgressing species "boundaries" is a widely felt reaction of "instinctive hostility" (Harris 1998, 177) commonly known as the "yuck factor." But in important respects repugnance is an inchoate emotive objection to the creation of novel beings that requires considerable defense. If claims about repugnance are to have any moral force, the intuitions captured by the "yuck" response must be clarified. In the debate about the ethics of creating novel beings that are part human, it is not enough to register

---

[11]   Note that Morriss does not subscribe to such a position.

one's intuitions. Rather, we need to be able to clearly identify and critically examine these intuitions, recognizing all the while that they derive "from antecedent commitment to categories that are themselves subject to dispute" (Stout 2001, 158).

A plausible "thin" explanation for the intuitive "yuck" response is that the creation of interspecies creatures from human materials evokes the idea of bestiality – an act widely regarded as a moral abomination because of its degrading character. Sexual intimacy between human and nonhuman animals typically is prohibited in law and custom, and some, no doubt, reason from the prohibition on the erotic mixing of human and nonhuman animals to a prohibition on the biotechnological mixing of human and nonhuman cellular or genetic material. There are important differences, however. In the first instance the revulsion is directed toward the shepherd who lusts after his flock and acts in a way that makes him seem (or actually be) less human (Stout 2001, 152). In the second instance the revulsion is with the purposeful creation of a being that is neither uncontroversially human nor uncontroversially nonhuman.

A more robust explanation for the instinctive and intense revulsion at the creation of human-to-animal beings (and perhaps some animal-to-human beings) can be drawn from Douglas's work on taboos (1966). Douglas suggests that taboos stem from conceptual boundaries. Human beings attach considerable symbolic importance to classificatory systems and actively shun anomalous practices that threaten cherished conceptual boundaries. This explains the existence of well-entrenched taboos, in a number of domains, against mixing things from distinct categories or having objects/actions fall outside any established classification system. Classic examples include the Western response to bisexuality (you can't be both heterosexual and homosexual) and intersexuality. Intersexuality falls outside the "legitimate" (and exclusive) categories of male and female, and for this reason intersex persons have been carved to fit into the existing categories (Dreger 2000). Human-to-animal chimeras, for instance, are neither clearly animal nor clearly human. They obscure the classification system (and concomitant social structure) in such a way as to constitute an unacceptable threat to valuable and valued conceptual, social, and moral boundaries that set human beings apart from all other creatures. Following Stout, who follows Douglas, we might thus consider human-to-animal chimeras to be an abomination. They are anomalous in that they "combine characteristics uniquely identified with separate kinds of things, or at least fail to fall unambiguously into any recognized class." Moreover, the anomaly is loaded with social significance in that interspecies hybrids and chimeras made with human materials "straddle the line between *us* and *them*" (Stout 2001, 148). As such, these beings threaten our social identity, our unambiguous status as human beings.

But what makes for unambiguous humanness? Where is the sharp line that makes for the transgression, the abomination? According to Stout, the line must be both sharp and socially significant if trespassing across it is to generate a sense of abomination: "An abomination, then, is anomalous or ambiguous with respect to some system of concepts. And the repugnance it causes depends on such factors as the presence, sharpness, and social significance of conceptual distinctions" (Stout 2001, 148). As we have seen, though, there is no biological sharp line:

we have no biological account of unambiguous humanness, whether in terms of necessary and sufficient conditions or of homeostatic property clusters. Thus it would appear that in this instance abomination is a social and moral construct.

Transformative technologies, such as those involved in creating interspecies beings from human material, threaten to break down the social dividing line between human beings and nonhumans. Any offspring generated through the pairing of two human beings is by natural necessity – reproductive, genetic, and developmental necessity – a human. But biology now offers the prospect of generating offspring through less usual means; for instance, by transferring nuclear DNA from one cell into an enucleated egg. Where the nuclear DNA and the enucleated egg (with its mitochondrial DNA) derive from organisms of different species, the potential emerges to create an interspecies nuclear-cytoplasmic hybrid.

In 1998 the American firm Advanced Cell Technology (ACT) disclosed that it had created a hybrid embryo by fusing human nuclei with enucleated cow oocytes. The goal of the research was to create and isolate human embryonic stem cells. But if the technology actually works (and there is some doubt about this) there would be the potential to create animal–human hybrids (ACT 1998; Marshall 1998; Wade 1998). Any being created in this way would have DNA 99% identical with that of the adult from whom the human nucleus was taken; the remaining 1% of DNA (i.e., mitochondrial DNA) would come from the enucleated animal oocyte. Is the hybrid thus created simply part-human and part-nonhuman animal? Or is it unequivocally human or unequivocally animal (see Loike and Tendler 2002)? These are neither spurious nor trivial questions. Consider, for example, the relatively recent practice in the United States of classifying octoroons (persons with one-eighth negro blood; the offspring of a quadroon and a white person) as black. By analogy, perhaps 1% animal DNA (i.e., mitochondrial DNA) makes for an animal.[12]

A more complicated creature to classify would be a human-to-animal chimera created by adding human stem cells to a nonhuman animal embryo. It has recently been suggested that human stem cells should be injected into mice embryos (blastocysts) to test their pluripotency (Dewitt 2002). If the cells were to survive and were indeed pluripotent, they could contribute to the formation of every tissue. Any animal born following this research would be a chimera – a being with a mixture of (at least) two kinds of cells. Or, according to others, it would be just a mouse with a few human cells. But what if those cells are in the brain, or the gonads (Weissman 2002)? What if the chimeric mouse has human sperm? And what if that mouse were to mate with a chimeric mouse with human eggs?

All of this to say that when faced with the prospect of not knowing whether a creature before us is human and therefore entitled to all of the rights typically conferred on human beings, we are, as a people, baffled.

One could argue further that we are not only baffled but indeed fearful. Hybrids and chimeras made from human beings represent a metaphysical threat to our self-image. This fear can be explained in both historical and contemporary

---

[12] Mitochondrial DNA is not insignificant DNA. Like nuclear DNA it codes for functions.

terms. Until the end of the eighteenth century the dominant Western worldview rested on the idea of the Great Chain of Being. The world was believed to be an ordered and hierarchical place with God at the top, followed by angels, human beings, and various classes of animals on down through to plants and other lesser living matter (Lovejoy 1970; see also Morriss 1997). On this worldview human beings occupied a privileged place between the angels and all nonhuman animals. In more recent times, though the idea of the Great Chain of Being has crumbled, the reigning worldview is still that human beings are superior to animals by virtue of the human capacity for reason and language. Hybrids and chimeras made from human materials blur the fragile boundary between human beings and "unreasoning animals," particularly when one considers the possibility of creating "reasoning" nonhuman animals (Krieger 2002). But is protecting one's privileged place in the world solid grounds on which to claim that hybrid- or chimera-making is intrinsically or even instrumentally unethical?

4. MORAL CONFUSION

Taking into consideration the conceptual morass of species-talk, the lack of consensus about the existence of God and His role in Creation, healthy skepticism about the "yuck" response, and confusion and fear about obscuring, blurring, or breaching boundaries, the question remains as to why there should be any ethical debate over crossing species boundaries. We offer the following musings as the beginnings of a plausible answer, the moral weight of which is yet to be assessed.

All things considered, the engineering of creatures that are part-human and part-nonhuman animal is objectionable because the existence of such beings would introduce inexorable moral confusion in our existing relationships with nonhuman animals and in our future relationships with part-human hybrids and chimeras. The moral status of nonhuman animals, unlike that of human beings, invariably depends in part on features other than species membership, such as the intention with which the animal came into being. With human beings the intention with which one is created is irrelevant to one's moral status. In principle it does not matter whether one is created as an heir, a future companion to an aging parent, a sibling for an only child, or a possible tissue donor for a family member. In the case of human beings, moral status is categorical insofar as humanness is generally considered a necessary condition for moral standing. In the case of nonhuman animals, though, moral status is contingent on the will of regnant human beings. There are different moral obligations, dependent on social convention, that govern our behavior toward individual nonhuman animals depending upon whether they are bred or captured for food (e.g., cattle), for labor (e.g., oxen for subsistence farming), for research (e.g., lab animals), for sport (e.g., hunting), for companionship (e.g., pets), for investment (e.g., breeding and racing), for education (e.g., zoo animals), or whether they are simply cohabitants of this planet. In addition, further moral distinctions are sometimes drawn between "higher" and "lower" animals, cute and ugly animals, useful animals and pests, all of which add to the complexity of human relationships with nonhuman animals.

These two frameworks for attributing moral status are clearly incommensurable. One framework relies almost exclusively on species membership in *Homo sapiens* as such, while the other relies primarily on the will and intention of powerful "others" who claim and exercise the right to confer moral status on themselves and other creatures. For example, though some (including ourselves) will argue that the biological term *human* should not be conflated with the moral term *person*, others will insist that all human beings have an inviolable moral right to life simply by virtue of being human. In sharp contrast, a nonhuman animal's "right to life" depends entirely upon the will of some or many human beings, and this determination typically will be informed by myriad considerations.

It follows that hybrids and chimeras made from human materials are threatening insofar as there is no clear way of understanding (or even imagining) our moral obligations to these beings – which is hardly surprising given that we are still debating our moral obligations to some among us who are undeniably biologically human, as well as our moral obligations to a range of nonhuman animals. If we breach the clear (but fragile) *moral* demarcation line between human and nonhuman animals, the ramifications are considerable, not only in terms of sorting out our obligations to these new beings but also in terms of having to revisit some of our current patterns of behavior toward certain human and nonhuman animals.[13] As others have observed (e.g., Thomas 1983), the separateness of humanity is precarious and easily lost; hence the need for tightly guarded boundaries.

Indeed, asking – let alone answering – a question about the moral status of part-human interspecies hybrids and chimeras threatens the social fabric in untold ways; countless social institutions, structures, and practices depend upon the moral distinction drawn between human and nonhuman animals. Therefore, to protect the privileged place of human animals in the hierarchy of being, it is of value to embrace (folk) essentialism about species identities and thus effectively trump scientific quibbles over species and over the species status of novel beings. The notion that species identity can be a fluid construct is rejected, and instead a belief in fixed species boundaries that ought not to be transgressed is advocated.

An obvious objection to this hypothesis is that, at least in the West, there is already considerable confusion and lack of consensus about the moral status of human embryos and fetuses, patients in a persistent vegetative state, sociopaths, nonhuman primates, intelligent computers, and cyborgs. Given the already considerable confusion that exists concerning the moral status of this range of beings, there is little at risk in adding to the confusion by creating novel beings across species boundaries. Arguably, the current situation is already so morally confused that an argument about the need to "avoid muddying the waters further" hardly holds sway.[14]

---

[13] Animal-rights advocates might object to the creation of part-human hybrids on the grounds that this constitutes inappropriate treatment of animals solely to further human interests. Obviously, proponents of such a perspective will not typically have a prior commitment to the uniqueness and "dignity" of human beings. For this reason we do not pursue this narrative here.

[14] This objection was raised for us by Vaughan Black.

From another tack, others might object that confusion about the moral status of beings is not new. There was a time when many whom we in the West now recognize as undeniably human – for example, women and blacks – were not accorded this moral status. We were able to resolve this moral "confusion" (ongoing social discrimination notwithstanding) and can be trusted to do the same with the novel beings we create.

Both of these points are accurate but in important respects irrelevant. Our point is not that the creation of interspecies hybrids and chimeras adds a huge increment of moral confusion, nor that there has never been confusion about the moral status of particular kinds of beings, but rather that the creation of novel beings that are part-human and part-nonhuman animal is sufficiently threatening to the social order that for many this is sufficient reason to prohibit any crossing of species boundaries involving human beings. To do otherwise is to have to confront the possibility that humanness is neither necessary nor sufficient for personhood (the term typically used to denote a being with full moral standing, for which many – if not most – believe that humanness is at least a necessary condition).

In the debate about the ethics of crossing species boundaries the pivotal question is: Do we shore up or challenge our current social and moral categories? Moreover, do we entertain or preclude the possibility that humanness is not a necessary condition for being granted full moral rights? How we resolve these questions will be important not only in determining the moral status and social identity of those beings with whom we currently coexist (about whom there is still confusion and debate), but also for those beings we are on the cusp of creating. Given the social significance of the transgression we contemplate embracing, it behooves us to do this conceptual work now, not when the issue is even more complex – that is, once novel part-human beings walk among us.

5. CONCLUSION

To this point we have not argued that the creation of interspecies hybrids or chimeras from human materials should be forbidden or embraced. We have taken no stance at all on this particular issue. Rather, we have sketched the complexity and indeterminacy of the moral and scientific terrain, and we have highlighted the fact that despite scientists' and philosophers' inability to precisely define *species*, and thereby to demarcate species identities and boundaries, the putative fixity of putative species boundaries remains firmly lodged in popular consciousness and informs the view that there is an obligation to protect and preserve the integrity of human beings and *the* human genome. We have also shown that the arguments against crossing species boundaries and creating novel part-human beings (including interspecies hybrids or chimeras from human materials), though many and varied, are largely unsatisfactory. Our own hypothesis is that the issue at the heart of the matter is the threat of inexorable moral confusion.

With all this said and done, in closing we offer the following more general critique of the debate about transgressing species boundaries in creating part-human beings. The argument, insofar as there is one, runs something like this:

species identities are fixed, not fluid; but just in case, prohibiting the transgression of species boundaries is a scientific, political, and moral imperative. The scientific imperative is prudential, in recognition of the inability to anticipate the possibly dire consequences for the species *Homo sapiens* of building these novel beings. The political imperative is also prudential, but here the concern is to preserve and protect valued social institutions that presume pragmatically clear boundaries between human and nonhuman animals. The moral imperative stems from a prior obligation to better delineate moral commitments to both human beings and animals before undertaking the creation of new creatures for whom there is no apparent a priori moral status.

As we have attempted to show, this argument against transgressing species boundaries is flawed. The first premise is not categorically true – there is every reason to doubt the view that species identity is fixed. Further, the scientific, political, and moral objections sketched above require substantial elaboration. In our view the most plausible objection to the creation of novel interspecies creatures rests on the notion of moral confusion – about which considerably more remains to be said.

ACKNOWLEDGMENTS

An early version of this paper was presented at the 2001 meeting of the American Society for Bioethics and Humanities, Nashville, Tenn. A significantly revised version of this paper was presented at the 2002 meeting of the International Association of Bioethics, Brasilia, Brazil. We are grateful to both audiences for engaging our ideas. Additional thanks are owed to Brad Abernethy, Vaughn Black, Fern Brunger, Josephine Johnston, Jane Maienschein, Robert Perlman, and Janet Rossant for helpful comments on interim drafts. Research for this paper was funded in part by grants from the Canadian Institutes of Health Research (CIHR) made independently to JSR and FB, a grant from the CIHR Institute of Genetics to JSR, and a grant from the Stem Cell Network (a member of the Networks of Centers of Excellence program) to FB.

WORKS CITED

Advanced Cell Technology (1998) 'Advanced Cell Technology announces use of nuclear replacement technology for successful generation of human embryonic stem cells,' Press release, 12 November (available at http:// www.advancedcell. com/pr_11–12–1998.html).

B. Allen (1997) 'The chimpanzee's tool,' *Common Knowledge* 6 (2): 34–54.

G. J. Annas, L. B. Andrews, and R. M. Isasi (2002) 'Protecting the endangered human: Toward an international treaty prohibiting cloning and inheritable alterations,' *American Journal of Law & Medicine*, 28: 151–78.

S. Atran (1999) 'The universal primacy of generic species in folk biological taxonomy: Implications for human biological, cultural, and scientific evolution,' in *Species: New interdisciplinary essays*, (ed.) R. A. Wilson, 231–61 (Cambridge, MA: MIT Press).

P. Bianco, and P. G. Robey (2001) 'Stem cells in tissue engineering,' *Nature*, 414: 118–21.

R. Boyd (1999) 'Homeostasis, species, and higher taxa,' in *Species: New interdisciplinary essays*, (ed.) R. A. Wilson, 141–85. (Cambridge, MA: MIT Press).

Y. Breitowitz (2002) 'What's so bad about human cloning?' *Kennedy Institute of Ethics Journal* 12: 325–41.

A. Campbell, K. G. Glass, and L. C. Charland (1998) 'Describing our "humanness": Can genetic science alter what it means to be "human"?' *Science and Engineering Ethics* 4: 413–26.

A. W. S. Chan, K. Y. Chong, C. Martinovich, C. Simerly, and G. Schatten (2001) 'Transgenic monkeys produced by retroviral gene transfer into mature oocytes,' *Science* 291: 309–12.

M. F. Claridge, H. A. Dawah, and M. R. Wilson (eds.) (1997) *Species: The units of biodiversity* (London: Chapman and Hall).

J. M. Claverie (2001) 'What if there are only 30,000 human genes?' *Science*, 291: 1255–57.

B. Dekel, T. Burakova, F. D. Arditti et al. (2003) 'Human and porcine early kidney precursors as a new source for transplantation,' *Nature Medicine*, 9: 53–60.

C. Dennis (2002) 'China: Stem cells rise in the East,' *Nature*, 419: 334–36.

N. DeWitt (2002) 'Biologists divided over proposal to create human-mouse embryos,' *Nature*, 420: 255.

B. Dixon (1984) 'Engineering chimeras for Noah's Ark,' *Hastings Center Report*, 10: 10–12.

T. Dobzhansky (1950) 'Mendelian populations and their evolution,' *American Naturalist*, 84: 401–18.

W. F. Doolittle (1999) 'Lateral genomics,' *Trends in Genetics*, 15 (12): M5–M8.

M. Douglas (1966) *Purity and danger* (London: Routledge and Kegan Paul).

A. D. Dreger (2000) *Hermaphrodites and the medical invention of sex* (Cambridge: Harvard University Press).

L. Eisenberg (1972) 'The *human* nature of human nature,' *Science*, 176: 123–28.

W. Enard, P. Khaitovich, J. Klose, et al. (2002) 'Intraand interspecific variation in primate gene expression patterns,' *Science*, 296: 340–43.

M. Ereshefsky (ed.) (1992) *The units of evolution: Essays on the nature of species* (Cambridge, MA: MIT Press).

R. S. Goldstein, M. Drukker, B. E. Reubinoff, and N. Benvenisty (2002) 'Integration and differentiation of human embryonic stem cells transplanted to the chick embryo,' *Developmental Dynamics*. 225: 80–86.

P. R. Grant, and B. R. Grant (2002) 'Unpredictable evolution in a 30-year study of Darwin's finches,' *Science*, 296: 633–35.

P. E. Griffiths (1999) 'Squaring the circle: Natural kinds with historical essences,' in *Species: New interdisciplinary essays*, (ed.) R. A. Wilson, 209–28 (Cambridge, MA: MIT Press).

P.E. Griffiths (2002) 'What is innateness?' in *The Monist* 85: 70–85.

J. Harris (1998) *Clones, genes, and immortality: Ethics and the genetic revolution* (New York: Oxford University Press).

R. Harris (2001) 'Little green primates,' *Current Biology*, 11: R78-R79.

D. L. Hull (1986) 'On human nature,' in *Proceedings of the Biennial Meeting of the Philosophy of Science Association* 2: 3–13.

D. L. Hull (1999) 'On the plurality of species: Questioning the party line,' in *Species: New interdisciplinary essays*, (ed.) R. A. Wilson, 23–48 (Cambridge, MA: MIT Press).

L. J. Kass (1998) 'The wisdom of repugnance,' in *The ethics of human cloning*, by L. J. Kass and J. Q. Wilson, 3–59 (Washington, DC: AEI Press).

P. Kitcher (1984) 'Species,' *Philosophy of Science*, 51: 308–33.

L. M. Krieger (2002) 'Scientists put a bit of man into a mouse,' *Mercury News*, 8 December (available at http://www.bayarea.com/mld/mercurynews/4698610.html).

S. Krimsky (1982) *Genetic alchemy: The social history of the recombinant DNA controversy* (Cambridge, MA: MIT Press)

R. C. Lewontin (1992) 'The dream of the human genome,' *New York Review of Books*, 28 May, 31–40.

E. A. Lloyd (1994) 'Normality and variation: The Human Genome Project and the ideal human type,' in *Are genes us? The social consequences of the new genetics*, (ed.) C. F. Cranor, 99–112 (New Brunswick, NJ: Rutgers University Press).

J. D. Loike, and M. D. Tendler (2002) 'Revisiting the definition of *Homo sapiens*,' *Kennedy Institute of Ethics Journal*, 12: 343–50.

A. C. Love (2003) 'Evolutionary morphology, innovation, and the synthesis of evolutionary and developmental biology,' *Biology & Philosophy*, 18: 309–345.

A. O. Lovejoy (1970) *The great chain of being: A Study of the history of an idea* (Cambridge, MA: Harvard University Press).

J. Marks (2002) *What it means to be 98% chimpanzee: Apes, human beings, and their genes* (Berkeley: University of California Press).

E. Marshall (1998) 'Claim of human-cow embryo greeted with skepticism,' *Science*, 282: 1390–91.

R. L. Mayden (1997) 'A hierarchy of species concepts: The denouement in the saga of the species problem,' in *Species: The units of biodiversity*, (ed.) M. F. Claridge, H. A. Dawah, and M. R. Wilson, 381–424 (London: Chapman and Hall).

E. Mayr (1940) 'Speciation phenomena in birds,' *American Naturalist*, 74: 249–78.

E. Mayr (1959) 'Typological versus populational thinking,' in *Evolution and the Diversity of Life*, E. Mayr, 26–29 (Cambridge: Harvard University Press).

D. J. Mooney, and A. G. Mikos (1999) 'Growing new organs,' *Scientific American*, 280: 38–43.

P. Morriss (1997) 'Blurred boundaries,' *Inquiry* 40: 259–90.

M. V. Olson, and A. Varki (2003) 'Sequencing the chimpanzee genome: Insights into human evolution and disease,' *Nature Reviews Genetics*, 4: 20–28.

V. Ourednik, J. Ourednik, J. D. Flax, et al. (2001) 'Segregation of human neural stem cells in the developing primate forebrain,' *Science* 293: 1820–24.

S. Oyama (2000) *The ontogeny of information: Developmental systems and evolution* (Durham: Duke University Press).

R. Plomin, J. C. Defries, I. W. Craig, P. McGufan, and J. Kagan, (eds.) (2002) *Behavioral genetics in the postgenomic era.* (Washington: American Psychological Association).

J. S. Robert (1998) 'Illich, education, and the Human Genome Project: Reflections on paradoxical counterproductivity,' *Bulletin of Science, Technology, and Society,* 18: 228–39.

G. G. Simpson (1961) *Principles of animal taxonomy* (New York: Columbia University Press).

E. Sober (1980) 'Evolution, population thinking, and essentialism,' *Philosophy of Science,* 47: 350–83.

J. Stout (2001) *Ethics after Babel: The languages of morals and their discontents* (Boston: Beacon Books) Reprint, in expanded form and with a new postscript, (1988) (Princeton: Princeton University Press).

A. I. Tauber, and S. Sarkar (1992) 'The Human Genome Project: Has blind reductionism gone too far?' *Perspectives in Biology and Medicine* 35: 220–35.

K. Thomas (1983) *Man and the natural world: Changing attitudes in England, 1500–1800* (London: Allen Lane).

R. Trigg (1988) *Ideas of human nature: An historical introduction* (Oxford, U.K.: Basil Blackwell).

N. Wade (1998) 'Researchers claim embryonic cell mix of human and cow,' *New York Times,* 12 November, A1. (available at http://query.nytimes.com/ search/ article-page.html?res~9C04E3D71731F931A25752C1A96E 958260.

N. Wade (2002) 'Scientist reveals genome secret: It's his,' *New York Times,* 27 April. (available at http:// www.nytimes.com/2002/04/27/science/27GENO. html).

I. Weissman (2002) 'Stem cells: Scientific, medical, and political issues,' *New England Journal of Medicine,* 346: 1576–79.

E. O. Wiley (1978) 'The evolutionary species concept reconsidered,' *Systematic Zoology* 27: 17–26.

M. B. Williams (1992) 'Species: Current usages,' in *Keywords in evolutionary biology,* (ed.) E. F. Keller and E. A. Lloyd, 318–23 (Cambridge: Harvard University Press).

R. A. Wilson (1999a) 'Realism, essence, and kind: Resuscitating species essentialism?' in *Species: New interdisciplinary essays,* (ed.) R. A. Wilson, 187–207 (Cambridge, MA: MIT Press).

R. A. Wilson (1999b) *Species: New interdisciplinary essays* (Cambridge: MIT Press).

C. S. Young, S. Terada, J. P. Vacanti, et al. (2002) 'Tissue engineering of complex tooth structures on biodegradable polymer scaffolds,' *Journal of Dental Research* 81: 695–700.

DISCUSSION QUESTIONS

1. The authors argue that there are no fixed species boundaries. What exactly do the mean by this? Do you find their arguments against fixed species boundaries compelling?

2. What is the difference between a biological basis for categories like species and race and a conventional basis for such categories? In what respects is species a biological category and in what respects is it a conventional

category? Do you think the same thing holds for racial categories? Why or why not?

3. The authors suggest that it does not matter for ethics whether a way of grouping people or organisms is biologically based, socially constructed, or both. Do you agree with this? Can you think of biological categories that are ethically insignificant and socially constructed categories that are ethically significant?

4. The authors are critical of intrinsic arguments against creating chimeras – i.e. arguments that it involves inappropriate crossing of species boundaries. What are their reasons for this? Do you find those reasons compelling? Why or why not?

5. The authors raise concerns about the moral confusion that chimeras might produce. What is this 'moral confusion'? Why do they think it is problematic? Do you agree that there will be moral confusion created by chimeras and that this is something to worry about?

6. Can you think of any possible objections to creating human–animal chimeras that were not discussed by the authors?

7. Would you support a ban or moratorium on the use of chimeras in research? Why or why not?

11 THE COMING ERA OF NANOMEDICINE[1]

Fritz Allhoff

CHAPTER SUMMARY

In this chapter, Fritz Allhoff presents some general background on nanomedicine, focusing in particular on the investment that is being made in this emerging field by pharmaceutical companies. The bulk of the essay, however, consists of explorations of two areas in which the impacts of nanomedicine are likely to be most significant: (1) diagnostics and medical records; (2) treatment (including surgery and drug delivery). For each of these areas, Allhoff surveys some of the social and ethical issues that are likely to arise from the applications – for example, privacy and justice concerns. Allhoff also raises some concerns about the trajectory of nanomedical development that are related to its being guided almost entirely by market incentives.

RELATED READINGS

Introduction: Extrinsic Concerns (2.6)

Related Chapters: Michael Ravvin, *Incentivizing Access and Innovation for Essential Medicines* (Ch. 13); Jeroen van den Hoven, *Nanotechnology and Privacy* (Ch. 19).

1. INTRODUCTION

Nanotechnology has been hailed as the "next Industrial Revolution" (National Science and Technology Council 2000) and promises to have substantial impacts into many areas of our lives. What, though, is nanotechnology? A common definition, and one that is good enough for our purposes, comes from the U.S. National Nanotechnology Initiative: "nanotechnology is the understanding and control of matter at dimensions of roughly 1 to 100 nanometers, where unique phenomena enable novel applications" (NNI 2007). This definition suggests two necessary (and jointly sufficient) conditions for nanotechnology. The first is an issue of *scale*: nanotechnology is concerned with things of a certain size. 'Nano-' (from the Greek *nannos*, "very short man") means one billionth, and

---

[1] This paper was published, in modified form, as Fritz Allhoff (2009) 'The Coming Era of Nanomedicine,' *The American Journal of Bioethics*, 11: 3–11. It appears here by permission of the author.

in nanotechnology the relevant billionth is that of a meter. Nanometers are the relevant scales for the size of atoms; for example, a hydrogen atom is about a quarter of a nanometer in diameter. The second issue has to do with that of *novelty*: nanotechnology does not *just* deal with small things, but rather must deal with them in a way that takes advantage of some properties that are manifest *because* of the nanoscale.

Applications of nanotechnology to medicine are already underway and offer tremendous promise; these applications often go under the moniker of 'nanomedicine' or, more generally, 'bionanotechnology'. In this essay, I will present some general background on nanomedicine, particularly focusing on some of the investment that is being made in this emerging field. The bulk of the essay, though, will consist in explorations of two areas in which the impacts of nanomedicine are likely to be most significant: diagnostics and medical records and treatment, including surgery and drug delivery.[2] In each of these sections I will survey some of the ethical and social issues that are likely to arise in these areas.

2. THE RISE OF NANOMEDICINE

Drug companies are always striving to increase the success rate of their products, as well as to decrease research and development (R&D) costs, including time to development. Getting a new drug to market is an extremely daunting task: the economic cost can be as high as $800M (Dimasi *et al.* 2003); the time to market is usually 10–15 years; and only one of every 8,000 compounds initially screened for drug development ultimately makes it to final clinical use (Bawa and Johnson 2009). Annual R&D investment by drug companies has climbed from $1B in 1975 to $40B in 2003, though the number of new drugs approved per year has not increased at all; it has stayed at a relatively constant 20–30 approvals per year (Sussman and Kelly 2003). (New drugs account for only approximately 25% of the approvals, with the other 75% coming from reformulations or combinations of already-approved drugs.) And, due to the high costs, only about 30% of new drugs are recovering their R&D costs. International pressures are mounting on U.S. pharmaceutical companies as well, especially with production being increased in low-cost countries like India and China, coupled with the expiration of some American patents (Bawa and Johnson 2009).

None of this is to lament the plights of pharmaceutical companies. The biggest ones – e.g., Johnson and Johnson, Pfizer, Bayer, and GlaxoSmithKline – have annual revenues at or close to $50B, and all have under $10B/year investments in R&D. Even factoring in total expenses, these biggest companies have annual profits of over $10B each (MedAdNews 2007). But of course those companies always want to become more profitable, and nanotechnology offers promise in this regard:

> Nanotechnology not only offers potential to address [the above] challenging issues but it can also provide significant value to pharma portfolios. Nanotechnology can

---

[2] Some of that discussion will be adapted from Allhoff (2007).

enhance the drug discovery process via miniaturization, automation, speed and the reliability of assays. It will also result in reducing the cost of drug discovery, design and development and will result in the faster introduction of new cost-effective products to the market. For example, nanotechnology can be applied to current micro-array technologies, exponentially increasing the hit rate for promising compounds that can be screened for each target in the pipeline. Inexpensive and higher throughput DNA sequencers based on nanotechnology can reduce the time for both drug discovery and diagnostics. It is clear that nanotechnology-related advances represent a great opportunity for the drug industry as a whole (Bawa and Johnson 2009, p. 212).

As expected, pharmaceutical companies are already investing in nanotechnology. Analysts have predicted that, by 2014 the market for pharmaceutical applications of nanotechnology will be close to $18B annually (Hunt 2004); another report indicates that the U.S. demand for medical products incorporating nanotechnology will increase over 17% per year to $53B in 2011 and $110B in 2016 (Bawa and Johnson 2009; Freedonia Group 2007).

The worry is that all of this investment and interest in nanotechnology, particularly insofar as it is motivated by a race to secure profitable patents, will lead to a neglect of important ethical issues. In one of the few significant contributions to discussion of the ethical issues in nanomedicine, Raj Bawa and Summer Johnson point out, correctly, that the time for such reflection is not once these technologies come to market, but before R&D even begins (Bawa and Johnson, pp. 212–213). The principal ethical issues will have to do with safety, especially toxicity; some of these issues have to do with the nanomaterials involved in nanomedicine, which will interact with complex biological systems and have been poorly studied thus far. It is quite likely that some of these applications will have unwelcome results in their hosts, at least some of which might not be predicted ahead of time.

3. DIAGNOSTICS AND MEDICAL RECORDS

Consider the trajectory of any particular health remediation. Effectively, there are two central steps: first, health care professionals have to assess the health of their patients – both through diagnostics and through the use of pre-existing medical records – and, second, they have to choose some treatment plan to address the patients' health issues. This is surely an oversimplified view of medicine, particularly as it fails to include any of medicine's social aspects (e.g., interaction with patients). But, for our purposes, consideration of this two-stage process will suffice: figure out what is wrong with the patient, and then fix it. There are all sorts of diagnostic tools available to the medical community, ranging from simple patient interviews and examinations and including more sophisticated imaging tools like computerized tomography (CT), magnetic resonance imaging (MRI), and positron emission tomography (PET). Other diagnostic tools include blood work, DNA analysis, urine analysis, x-rays, and so on. And there are all sorts of ways to gain access to patients' medical histories. Again, simple patient interviews, while limited, can be effective, and medical records, prepared by other health care professionals, can offer a wealth of salient information. Once the diagnosis has been made, treatment can proceed; whether through, for example, surgery or drugs.

Many diagnostic techniques are limited, particularly given certain maladies that health care professionals would like to be able to effectively diagnose at early stages. Nanotechnology promises to address some of these limitations. Cancer, for example, is one arena in which nanotechnology is likely to have the biggest impact, and this impact will come both in terms of improved diagnostics and improved treatment. Many cancer cells have a protein, epidermal growth factor receptor (EGFR), distributed on the outside of their membranes; non-cancer cells have much less of this protein. By attaching gold nanoparticles to an antibody for EGFR (anti-EGFR), researchers have been able to bind the nanoparticles to the cancer cells (El-Sayed *et al.* 2006). Once bound, the cancer cells manifest different light scattering and absorption spectra than benign cells (El-Sayed *et al.* 2005). Pathologists can thereafter use these results to identify malignant cells in biopsy samples.

This is just one example, but it has three noteworthy features: cost, speed, and effectiveness. Given some other traditional diagnostics that have been available, this one offers an improvement in all three regards. Furthermore, the improvements along these dimensions enabled by nanotechnology can be generalized beyond just cancer. To wit, if nanoparticles can differentially bind to something – by which I mean that they bind more readily to that thing than to everything else – then it will be easier to identify. And improved diagnostics lead to improved treatments, which lead to better health outcomes.

With respect to the ethical issues related to diagnostics, concerns center around toxicity, particularly when the applications are utilized *in vivo* or within the body of a patient (as opposed to *in vitro*). Conventional diagnostic mechanisms, however, manifest the same structural features as nano-diagnostics; there do not seem to be any substantially new issues raised in this regard. Consider, for example, x-rays, which use electromagnetic radiation to generate images that can be used for medical diagnostics. But radiation, absorbed in large dosages, is carcinogenic, so health care professionals have to be judicious in their application thereof. The radiation outputs for x-rays are reasonably well-understood, as are their toxicities in regards to human biology. And when considering treatment options, health care professionals must consider these toxicities, as well as the benefits of this diagnostic mechanism (in comparison with other options). Nano-diagnostics admit of a similar deliberative model, even if some of the risks are, at present, less well-understood.

Another possibility is "lab-on-a-chip" technologies, self-contained diagnostic tools in, for example, pill form, which could detect "cells, fluids or even molecules that predict or indicate disease states" (Bawa and Johnson 2009, p. 218). These devices could also provide real-time monitoring of various biometric indicators, such as blood glucose levels, which might be useful to a diabetic (Bawa and Johnson 2009, p. 218). Again, the ethical issues could have to do with toxicity, depending on how these chips are deployed. They might be kept inside the body indefinitely or permanently, which could be different from some of the binding diagnostics that, if administered *in vivo*, would ideally leave the body soon thereafter (e.g., if the cancer cells to which they bound were destroyed). Certainly clinical trials will be important, but it could be hard to get the sort of data that would be relevant for long-term effects. For example, perhaps the diagnostics only manifest toxicity some distant years in the future and only after they have been degraded; this information might not even be available for decades.

Depending on how information from "lab on a chip" tools is stored and shared, it could give rise to privacy issues: nano-sized chips could be implanted into individuals and serve as repositories for medical information, thus enabling quick access by health care professionals. Strictly speaking, this does not seem to be a diagnostic function, but one of medical record-keeping. In other words, these chips are "passive" in the sense that they are merely storing information. In the previous paragraph, though, I characterized chips that, in some sense, are "active": they could actually determine what is going on in the body and then make that informational available. This distinction is irrelevant ethically insofar as there are privacy issues regardless, but the extent of information that is accessible will vary depending on whether a technology is merely diagnostic or also contains detailed patient information. In terms of diagnostics then, the focus should probably be on the active applications, but the passive ones, in virtue of their greater informational potential, give rise to greater privacy concerns. However, privacy concerns might be mitigated by the fact that, in all likelihood, the chips would merely contain some identification number which would allow access to patients' records through some medical database; the chip itself would not actually contain any medical information, in which case there is no privacy worry. (Even the identification number could require authentication or decryption as further protections and to prevent the possibility of tracking and surveillance.)

Even though data may be protected to some extent, additional privacy issues arise due to the sort of infrastructure required to realize the benefits of these record-keeping technologies. These technologies require databases of patient information. But what are these databases and how are they supposed to work? First, we do not presently have the infrastructure to support this sort of data-basing. The ideal is supposed to be something like the following: imagine that someone from California happens to be traveling to New York on business, when she falls seriously ill and is rushed, unconscious, to the emergency room. Maybe even, in the rush to get her to the emergency room, her identification is lost. The emergency room staff can simply scan her triceps, plug her identification number into the database and learn who she is, what previous conditions she has, what drugs she is allergic to, and so on. Treatment can then ensue given this access to her history and records. But, of course, no such database exists. And several issues exist for its implementation.

First, it would be a huge endeavor and, presumably, very expensive. We would have to think about whether such a project would be worthwhile particularly given that, at least in many cases, there are fairly straightforward ways of getting access to the relevant medical information (e.g., by asking the patient or by using identification to contact a relative). To be sure, this will not always work (e.g., with patients who are unconscious and lack relatives and/or identification), but we would have to think about how many cases could not be handled by more conventional means and whether it would be a worthwhile investment of resources to develop this new system. Second, even if the chips themselves did not pose any substantial privacy worries, the database itself might. If all the medical information really went on some sort of national server, then it could be hacked. If it were retained locally, but remotely accessible (as it would have to be), the same worries would apply. Third, even if the privacy issues were mitigated and gave rise to improved outcomes, there could be ethical issues insofar as those outcomes would only be available to citizens of wealthier countries; as with other disparities, questions of distributive justice arise.

There are some other aspects of nanomedicine that are promising and probably without any significant ethical worries. For example, "quantum dots have been used as an alternative to conventional dyes as contrast agents due to their high excitability and ability to emit light more brightly and over long periods of time" (Bawa and Johnson 2009, p. 218). If we can use quantum dots, especially *in vitro*, to allow for more sensitive detection, then this is a useful application with minimal ethical baggage. As mentioned above, *in vivo* uses raise issues about toxicity, but there are certainly benign applications of nanomedicine that should be pursued if they increase our diagnostic abilities. These other issues mentioned above, regarding toxicity and privacy, are quite likely superable, though ethical attention needs to be paid to how we move forward. Having now offered discussion of how nanomedicine could affect diagnostics and medical record-keeping, let us now move on to treatment.

4. TREATMENT

In this section, I will consider how nanomedicine can be used to improve treatment options, and I propose to draw a distinction between two different kinds of treatment: surgery and drugs. 'Surgery' comes from the Greek χειρουργική and through the Latin *chirurgiae* which is often translated as "hand work". Many cardiac techniques, for example, have to be done manually (literally, by hand) in the sense that a surgeon has to physically intervene on the heart of a patient. Some of these techniques can now be performed remotely, automatically, and so on, but the basic unifier is direct intervention on a physical system by a physician or proxy. Surgery can be contrasted with drug treatment, by which I mean the administration of some pharmacological substance that is prescribed for remediation of some physical ailment; this latter part of the definition is to distinguish drug *treatment* from other uses of drugs (e.g., recreational).

Nano-surgery – by which I mean surgical applications of nanotechnology – enables techniques that are more precise and less damaging than traditional ones; let me offer a few examples. First, a Japanese group has performed surgery on living cells using atomic force microscopy with a nanoneedle (6–8 μm in length and 200–300 nm in diameter) (Obatya *et al.* 2005). This needle was able to penetrate both cellular and nuclear membranes, and the thinness of the needle prevented fatal damage to those cells. In addition to ultra-precise and safe surgical needles, laser surgery at the nanoscale is also possible: femtosecond near-infrared (NIR) laser pulses can be used to perform surgery on nanoscale structures inside living cells and tissues without damaging them (Tirlapur and König 2003). Because the energy for these pulses is so high, they do not destroy the tissue by heat – as conventional lasers would – but rather vaporize the tissue, preventing necrosis of adjacent tissue. I already discussed above that gold nanoparticles can be used for cancer diagnostics insofar as they can be attached to anti-EGFR which would then bind to EGFR; once bound, cancer cells manifest different light scattering and absorption spectra than benign cells, thus leading to diagnostic possibilities. But this technology can be used for cancer treatment as well as diagnostics: since the gold nanoparticles differentially absorb light, laser ablation can be used to destroy the attached cancer cells without harming adjacent cells.

What these applications have in common is that they allow for direct intervention at the cellular level. While some traditional laser technologies, for example, have allowed for precision, the precision offered by surgery at the nanoscale is unprecedented. It is not just the promise of being able to act on individual cells, but even being able to act *within* cells without damaging them. Contrast some of this precision offered by nano-surgery with some more traditional surgical techniques. Just to take an extreme example, consider lobotomies, especially as were practiced in the first half of the 20th century: these procedures were effectively carried out by inserting ice picks through the patient's eye socket, and the objective was to sever connections to and from the prefrontal cortex in the hopes of treating a wide range of mental disorders. Independently of whatever other ethical concerns attached to lobotomies – which have fallen out of practice since the introduction of anti-psychotics such as chlorpromazine – these procedures were extremely imprecise. Even given the dark history of lobotomies, one of its highlights was the invention of more precise surgical devices; for example, António Egas Moniz's introduction of the leucotome, for which he won the Nobel Prize in 1949. However surgeries are practiced, precision is always important, by which I mean roughly that surgeons want to be able to access the damaged area of the patient's body without simultaneously compromising anything not damaged. The gains in precision from ice picks to nano-surgery are measured in multiple orders of magnitude, and even the gains in precision from recent surgical advances to nano-surgery could be a full order of magnitude.

So what are the worries? As was alluded to above and will be discussed more in what follows, the principal one is that these surgical techniques carry risks, and that, in the excitement to rush them to market, those risks will not be adequately explored or assessed. For now, though, it is worth noticing that there is nothing special about nano-surgery in this regard: whatever stance we otherwise adopt towards risk is equally transferrable to this context.

The symmetry of risks between traditional technologies and nanotechnologies also applies to the use of nanoparticles; as discussed above, these might play a role in various cancer treatments. Of particular concern is the toxicity from nanoparticles that might be used, as well as other safety concerns. Whatever is to be said about these risks, though, it hardly follows that similar concerns do not attach to more traditional approaches. Consider, for example, chemotherapy, which uses cytotoxic drugs to treat cancer. The downside of chemotherapy is that these drugs are toxic to benign cells as well as to malignant ones, and there are side effects such as immunosuppression, nausea, vomiting, and so on. When physicians are prescribing chemotherapy, they therefore have to think about these risks and whether the risks are justified. But whether the treatment option involves nanoparticles or not, this basic calculus is unchanged: physicians must choose the treatment option that offers the best prognosis. Toxicity or side effects count against these outcomes, and improved health counts in favor of them. Obviously, there are epistemic obstacles to such forecasting, and physicians must be apprised of the relevant toxicity and side effect data, but there is nothing unique to nano-medicine in this regard.

The point that I want to make is that there are *already* risks in treatments and there is no good reason to think that the risks of nano-surgery are any higher than the risks characteristic of conventional medicine. In fact, there are good reasons to think that the risks of nano-surgery are actually lower and that the benefits are higher. For example, compare laser ablation of malignant cells to chemotherapy. Chemotherapy, as mentioned above, is a hard process and one with a lot of costs to the patient (including physical and psychological). If nanomedicine allows us, unlike chemotherapy, to destroy the malignant cells without harming the rest of the organism, it is a definite improvement. Are the nanoparticles that would be utilized toxic? Do they pass out of the body after the treatment? Even supposing that the answers are yes and no, respectively, it is quite probable that these new treatments would be improvements over the old. That is not to say that due diligence is not required, but I suspect that, in the end, there will not be that much to fear, at least at the appropriate comparative level.

Turning now to drug delivery, there are myriad advantages that nanomedicine will be able to confer. Per the discussion in above, there is already a tremendous interest from the pharmaceutical companies in terms of incorporating nanotechnology into their product lines, and that interest is the primary driving force in the rise of nanomedicine; this is not to say that the above-discussed nano-surgical techniques are not impressive, just that they are probably not as profitable. Let me go through three traditional challenges in drug delivery, and explain how nanotechnology can mitigate them.

First, consider absorption of drugs: when drugs are released into the body, they need to be absorbed as opposed to passed through. Nanotechnology can facilitate a reduction in the size of drugs (or at least their delivery mechanisms) and therefore, an increase in their surface-to-volume ratios. The basic idea here is that it is the surfaces of materials that are most reactive and that, by increasing surface-to-volume ratios, greater reactivity is achieved. As the body absorbs drugs, it has to act on the surfaces of those drugs and, by having more surface area per unit volume, the drugs can be dissolved faster, or even be rendered soluble at all. Speed of absorption is of critical importance to the success of drugs, and nanotechnology will make a difference in this regard. (Also, it is worth noting that nanomedicine may obviate the need for oral (or other) administrations of drugs in some cases and allow for topical administration: because the drugs will be smaller, they will be more readily absorbed transdermally. Insofar as some patients would prefer this method of administration, it offers another advantage.)

Second, it is not always the case that fast absorption is ideal since, in some cases, it would be better if the drug were released slowly over time. Consider, for example, time-release vitamins: because vitamins B and C are water soluble, they quickly flush from the body if not administered in some time-release manner. If the options were taking a vitamin every couple of hours or else taking one that is slowly released over a longer period of time, there are obvious advantages to the latter. Nanotechnology could be used to create better time-release capacities insofar as it could allow for smaller apertures through which the pharmacological molecules would dissipate. In other words, nanotechnology could help to create lattices with openings through which, for example, single molecules would pass. If only single molecules could pass at any given time from the delivery system (e.g., the capsule), then it would take a longer time to disperse the drug supply and, depending on the application, this could lead to more effective treatment.

Third, because nanotechnology will allow for the engineering of smaller drugs, the drugs might be able to traverse various membranes or other biological barriers that had previously restricted their usefulness. For example, consider the blood-brain barrier, which is a membrane that restricts the passage of various chemical substances (and other microscopic entities, like bacteria) from the bloodstream into the neural tissue. Nanoparticles are able to pass this barrier, thus opening up new possibilities for treatment of psychiatric disorders, brain injuries, or even the administration of neural anesthetic (In-Pharmatechnologist. com 2005). In this case, nanomedicine is not just improving existing treatment options, but perhaps even creating new ones.

Again tabling issues of distributive justice, the principal concern with bringing nanotechnology to bear on drug delivery has to do with toxicity and other risks: we simply do not know how these technologies will interact with the body, and there could be negative consequences. As with the discussion of nano-surgery, I am inclined to think that the benefits conferred by the application of nanotechnology to drug delivery outweigh the risks, though the risks in drug delivery are probably greater than the risks with nano-surgery. It is worth reiterating, though, that risks pertaining to delivery are not unique to nanomedicine. Consider, for example, the case of Jesse Gelsinger, who died in a gene therapy trial (Philipkowski 1999). Gelsinger had ornithine transcarbamylase deficiency: he lacked a gene that would allow him to break down ammonia (a natural byproduct of protein metabolism). An attempt to deliver this gene through adenoviruses was made, and Gelsinger suffered an immunoreaction that led to multiple organ failure and brain death. Whether talking about vectors for genetic interventions or nanoparticles, we surely have to think carefully about toxicity, immunoreactions, and other safety concerns; the point is merely that these issues are not unique to nanomedicine.

5. MOVING FORWARD

In this last section, let me try to tie together various themes that have been developed in the preceding three and, in particular, make some comments about the future of nanomedicine. Throughout, I have indicated skepticism about whether nanomedicine raises any new ethical or social issues, or at least ones that have not already been manifest with existing technologies. This skepticism applies to various other areas of nanoethics as well, though perhaps more so to medicine. The issues with nanomedicine seem to be, at most, risks (e.g., toxicity and safety), distributive justice, and privacy, and in fairly standard ways. Of course these considerations ought to be taken into account, but they should always be taken into account and nothing inherent to nanomedicine makes us think differently about them.

What does make nanomedicine interesting, at least from an ethical perspective, is its extreme profitability. As mentioned previously, pharmaceutical companies make tens of billions a dollar on year in profits, and this leads to a different motivational scheme than exists in other applications of nanotechnology. For example, consider nanotechnology and the developing world. The developing world, practically by definition, simply does not have tremendous amounts of money. Staying with pharmaceuticals, a primary ethical concern is that drugs critical for health in the developing world just are not developed because they would not be profitable; consider, for example, the billions of dollars spent in the US on erectile dysfunction drugs as against the lack of investment for lifesaving anti-malarial medication. This is not to say that medicines created for the developing world cannot be made profitable for pharmaceutical companies,[3] but it has a long way to go to rival the profitability in high GDP nations.

Still, on the whole, nanomedicine is likely to be one of the most – if not *the* most – profitable application of nanotechnology and, furthermore, one that is going to be primarily pursued by pharmaceutical companies committed to making profits. These features give rise to ethical concerns insofar as they are harbingers for market-first, ethics-last mantras. Imagine that Johnson and Johnson is competing with Pfizer to bring out the next greatest drug; a multi-year patent and billions of dollars hang in the balance. There is a lot of pressure to be first. And, potentially, a lot to be gained by cutting corners along the way, whether during pharmacological development, clinical trials, complete disclosure, or whatever. Organizations like the FDA will have to be extremely vigilant but, of course, they always have to be vigilant. The profits are there to be had whether the drugs incorporate nanotechnology or not so, in that sense, this is nothing new. The only point is that these pressures are likely to be more significant in nanomedicine than many or all other applications of nanotechnology.

Moving forward, we must develop an engagement between nanomedicine's promise and its ethical and social implications. Thus far there has been very little academic or public work done on these issues. Surely this needs to be remedied. To have the pharmaceutical companies developing products without an existing forum to discuss the potential effects of those products – including toxicity and other risks – is not good. Again, some of those effects might not be manifest for many years as we simply do not have long-term research about how nanotechnology interacts with the body. The promise is certainly high, but it should be negotiated clearly and carefully with attention to ethical and social implications and an accompanying discourse.

WORKS CITED

F. Allhoff (2007) 'On the Autonomy and Justification of Nanoethics', *Nanoethics*, I (3): 185–210.

F. Allhoff (2009) 'The Coming Era of Nanomedicine', *The American Journal of Bioethics*, IX (10): 3–11.

R. Bawa and S. Johnson (2009) 'Emerging Issues in Nanomedicine and Ethics', *Nanotechnology & Society: Current and Emerging Ethical Issues*, F. Allhoff and P. Lin (eds.). Springer, Dordrecht.

---

[3] The Health Impact Fund, for example, seeks to incentivize private pharmaceutical companies to invest in global public health. See Hollis and Pogge (2008).

J. DiMasi, R. Hansen, and H. G. Grabowski (2003) 'The Price of Innovation: New Estimates of Drug Development Costs', *Journal of Health Economics*, 22: 151–185.

I. El-Sayed, H. Xiaohua, and M. A. El-Sayed (2006) 'Selective Laser Photo-Thermal Therapy of Epithelial Carcinoma Using Anti-EGFR Antibody Conjugated Gold Nanoparticles', *Cancer Letters* 239 (1): 129–35.

I. El-Sayed, X. Huang, and M. A. El-Sayed (2005) 'Surface Plasmon Resonance Scattering and Absorption of Anti-EGFR Antibody Conjugated Gold Nanoparticles in Cancer Diagnostics: Applications in Oral Cancer', *Nano Letters* 5 (5): 829–834.

A. Hollis and T. Pogge (2008) *The Health Impact Fund: Making New Medicines Accessible for All* (Incentives for Global Health). Available at http://www.yale.edu/macmillan/igh/ (accessed January 28, 2009).

W. H. Hunt (2004) 'Nanomaterials: Nomenclature, Novelty, and Necessity', *Journal of Materials* 56: 13–19.

I. Obataya, C. Nakamura, S. Han, *et al.* (2005) 'Nanoscale Operation of a Living Cell Using an Atomic Force Microscope with a Nanoneedle', *Nano Letters* 5 (1): 27–30.

National Nanotechnology Initiative. "What is Nanotechnology" Available at http://www.nano.gov/html/facts/whatIsNano.html (accessed July 16, 2007).

National Science and Technology Council's Committee on Technology (2000), "National Nanotechnology Initiative: Leading to the Next Industrial Revolution" Available at http://clinton4.nara.gov/media/pdf/nni.pdf (accessed November 5, 2008).

"Nanotechnology to Revolutionize Drug Delivery" (2005), *Pharmatechnologist.com*. Available at http://www.in-pharmatechnologist.com/Materials-Formulation/Nanotechnology-to-revolutionise-drug-delivery (accessed August 27, 2008).

K. Philipkowski (1999) 'Another Change for Gene Therapy', *Wired*. Available online at http://www.wired.com/science/discoveries/news/1999/10/31613 (accessed August 16, 2007).

N. L. Sussman, and J. H. Kelly (2003) 'Saving Time and Money in Drug Discovery: A Pre-emptive Approach', *Business Briefings: Future Drug Discovery* (London: Business Briefings, Ltd.).

The Freedonia Group, Inc. (2001). *Nanotechnology in Healthcare*. Cleveland, OH: Freedonia.

'The Top 50 Pharmaceutical Companies Charts & Lists' (2007), *MedAdNews* 13 (9).

U. K. Tirlapur, and K. König (2003) 'Femtosecond Near-Infrared Laser Pulses as a Versatile Non-Invasive Tool for Intra-Tissue Nanoprocessing in Plants Without Compromising Viability', *The Plant Journal* 31 (2): 365–374.

DISCUSSION QUESTIONS

1. Is the author correct that the issues associated with nanomedicine are "at most, risks (e.g., toxicity and safety), distributive justice, and privacy, and in fairly

standard ways"? Or does nanomedice raise additional issues, or nonstandard versions of these issues?

2. How might nanomedicine restructure medical practices and organizations? What are the implications of these changes likely to be?

3. How might nanomedicine alter relationships, including power relationships, such as those between patients and medical providers?

4. Might nanomedicine lead to new expectations for healthfulness, as well as new conceptions or ways of thinking about health and disease?

5. Is there any reason to think that current frameworks for addressing the ethical concerns associated with the development of drugs and devices are inadequate to address those raised by nanomedicines?

6. Do you agree with the article that relying primarily on market-based innovation incentives for nanomedicine is ethically problematic? Why or why not?

PSYCHOPHARMACOLOGY AND FUNCTIONAL NEUROSURGERY: MANIPULATING MEMORY, THOUGHT, AND MOOD

Walter Glannon

CHAPTER SUMMARY

Recent years have seen rapid advancement in therapeutic neurotechnologies, or technologies that intervene in the brain to improve physical and mental health. In this chapter, Walter Glannon discusses the ethical issues associated with two highly scrutinized neurotechnologies, pharmacological memory manipulation and neurosurgical deep-brain stimulation to manipulate thought and mood. Because of the ways in which these technologies intervene into the brain and can affect thought, mood, and personality, they raise a complex set of issues associated with personal identity, authenticity, and individual responsibility. Glannon begins the chapter with an accessible explanation of the technologies, including how they work and what their applications are, before moving on to evaluate the ethical concerns that have been raised about them. He argues that because the neuropsychiatric disorders that they treat can be so debilitating, there is often a positive benefit–risk ratio to using them. However, the risks are nevertheless substantial, and merit careful attention in the use of the technologies.

RELATED READINGS

Introduction: Innovation Presumption (2.1); Intrinsic Concerns (2.6); Extrinsic Concerns (2.7)

Other Chapters: President's Council on Bioethics, *Beyond Therapy* (Ch. 14); Thomas Douglas, *Moral Enhancement* (Ch. 16)

1. INTRODUCTION

Advances in neuroscience have made it possible to intervene in the brain and alter a range of cognitive, conative, and affective functions. Psychotropic drugs can modulate neurotransmitters and neural circuits and restore normal brain functions for people with neurological disorders such as Parkinson's disease and psychiatric disorders such as schizophrenia. These drugs may also raise these functions above normal levels. For example, the drug methylphenidate (Ritalin) is prescribed therapeutically for attention deficit/hyperactivity disorder (ADHD). It is also used by students and researchers who do not have this disorder to enhance concentration and perform better on exams and write more successful grant applications. Functional neurosurgery is a different type of intervention in the brain. It treats neurological and psychiatric disorders by electrically stimulating and modulating brain circuits associated with motor control, thought, motivation, and mood. This technique can also elevate mood to a euphoric level in some individuals with depression.

In this chapter, I describe and discuss two of the most contentious issues in clinical neuroscience: memory manipulation through psychopharmacology; and manipulation of thought and mood through the functional neurosurgical technique of deep-brain stimulation (DBS). I will consider the use of these interventions for both therapy and enhancement. The chapter is divided into two general parts. In the first part, I examine the use of drugs to dampen the emotional content of the pathological memories in people with posttraumatic stress disorder (PTSD), to erase or prevent unpleasant memories, and to enhance the capacity to store and retrieve long-term memory. In the second part, I examine the use of DBS for drug-resistant obsessive-compulsive disorder and depression. Electrical stimulation of the neural circuits that mediate cognitive and affective states can relieve the symptoms of these psychiatric disorders. Yet because this intervention may have significant side-effects in a number of cases, we need to consider the trade-offs when restoration of some mental capacities comes at the cost of impairing others. In addition, I explore the possibility of using functional neurosurgery to enhance mood beyond a normal level and whether this would be in a person's best interests. Consideration of the potential benefits and risks of psychopharmacology and functional neurosurgery will go some way toward determining whether or to what extent manipulation of memory, thought, and mood can be medically and ethically justified.

2. DAMPENING PATHOLOGICAL MEMORY

Individuals who have experienced traumatic events such as rape, automobile accidents, or military combat can develop memories of these events that become embedded in their minds. Indeed, they can become so embedded that they recur beyond one's conscious control and result in pathological responses to stimuli. PTSD is characterized by intrusive thoughts, flashbacks, nightmares, and other adverse psychological effects that constitute an affected person's emotionally charged unconscious memory. These responses occur when a stimulus triggers a recall that makes an individual believe that he or she has returned to the threatening environment in which they experienced the event. Biochemically, the initial event triggers the release of epinephrine (adrenaline) from the adrenal gland, which then activates norepinephrine (noradrenaline) in the brain. Receptors for this hormone within the amygdala cause the event to become encoded as an unconscious memory. The flashbacks and nightmares induce continued production of high levels of adrenaline and noradrenaline, which ensure that the memory is firmly stored in the amygdala. This brain structure, which lies deep within the medial temporal lobe, regulates primitive emotions such as fear and is necessary to form memories of threatening events. It is thus critical for survival. For this reason, memories formed and stored in the amygdala are extremely difficult to erase or weaken. Unconscious memories of traumatic experiences are distinct from conscious episodic memories of non-traumatic experiences. The latter have emotional content but are not so negatively charged. Autobiographical memories are a species of episodic memories about the person who has them. These memories are formed in the hippocampus, which is also located within the medial temporal lobe. The function of the hippocampus overlaps to some extent with that of the amygdala in that both regulate not only knowing *when* we experienced an event but also the feeling of *what it is like* to recall it.

Pathological memories of traumatic events can interfere with our ability to adapt to the environment. Cognitive-behavior therapy and virtual reality programs that help individuals cope with perceived fearful or threatening stimuli are among the non-pharmacological interventions used to treat PTSD. Extinction training is another. It consists in the introduction and withdrawal of stimuli that reinforce certain behaviors and has been used to weaken the emotional content of traumatic memories in some people. Unfortunately, in severe cases of PTSD this and other psychological interventions have been only marginally effective. The problem is in not being able to rid oneself of a memory whose triggers are outside of one's conscious awareness and cognitive control. Emotionally charged memories are better remembered than memories that are emotionally neutral, and recall of traumatic experiences is more emotionally charged than recall of pleasant experiences (McGaugh, 2003).

The beta-adrenergic receptor antagonist, or "beta-blocker," propranolol has been used in a number of experimental studies involving subjects who have been exposed to traumatic events. This drug interferes with the release of adrenaline and noradrenaline and their effects on receptors in the amygdala. Through this mechanism, propranolol blocks the formation and consolidation of emotionally charged memories. In these studies, propranolol and similarly acting drugs modulated people's emotional response to traumatic memories. These interventions do not erase the memory but separate its emotional content from the memory itself. One of the studies was conducted by psychiatrist Roger Pitman and colleagues at the Massachusetts General Hospital. Half of a group of people admitted to the emergency room shortly after automobile accidents were given propranolol, and half were given a placebo. One month later, those who had taken the drug had fewer flashbacks and nightmares and more moderate responses to stimuli resembling the initial traumatic event than those who had taken the placebo (Pitman et al., 2002; Kolber, 2011). This manipulation of emotionally charged memory could have therapeutic effects for a significant number of people. In particular, it could benefit many soldiers returning from military combat with PTSD by relieving the symptoms of the disorder.

If the emotional content of a memory could be dampened pharmacologically, then one could recall a traumatic or fearful event without re-experiencing the trauma or fear. One could become emotionally disengaged from the context in which one experienced the event. Yet researchers and clinicians need to be careful in manipulating memory in this way. There are extensive projections between the amygdala and hippocampus, and considerable overlap between the memory functions they mediate. Both of these structures influence how one recalls the events in one's life. Weakening the content of unconscious emotionally charged memories might also weaken the content of conscious episodic memories. It may take the sting out of recalling a traumatic event; but it could also remove some of the positive emotional aspects of one's autobiographical memories.

Research in cognitive neuroscience has shown that recollection of personal episodes from the past depends on activation and reactivation of latent emotional associations with these episodes. This process appears to involve distinct but interacting memory systems. If the amygdala and hippocampus function interdependently in enabling these associations, then weakening the emotional response to unconscious pathological memory might also weaken the emotional response to conscious episodic memory. A network consisting of the amygdala and hippocampus supports emotional arousal (whether the event is exciting or calming), while a network consisting of the prefrontal cortex and hippocampus supports emotional valence (whether the event is positive or negative) (LaBar and Cabeza, 2006). Because the hippocampus is active in both networks and both types of emotion, it is possible that altering one circuit could alter the other and that altering emotional arousal could alter emotional valence. Since emotion is critical to the meaningful recall of experience, this type of memory modification could impair one's ability to make sense of one's past.

Losing some of the content of one's autobiography may be an acceptable price to pay for modulating an emotionally charged memory, but only if that memory has a significant negative impact on one's experience and quality of life. Weakening or neutralizing the emotional content of memory could impair our ability to situate experienced events within a meaningful first-person narrative. We might be left with the capacity to recall particular events without being able to understand how they relate to each other or how they shape our autobiographies. Removing the emotional content of memory from the context in which we experienced the recalled event could disrupt this structure and the narrative unity and continuity of our lives. Researchers and clinicians administering drugs to manipulate traumatic memories, as well as individuals taking them, thus need to carefully weigh the potential benefits against the risks of this type of intervention in the brain.

3. ERASING MEMORIES

In 2003, the U.S. President's Council on Bioethics expressed the concern that pharmacological modification of memory could have untoward consequences in our lives (Staff Working Paper, 2003). Among other things, the Council warned that blocking or dampening emotionally charged memories could erode much of our moral sensitivity. It could remove psychological constraints on doing things we would not otherwise do. Given the essential role of memory in personal identity and moral responsibility, memory modification could reshape us in ways that might not be beneficial or benign. The Council was concerned not only with manipulating pathological memories but also unpleasant episodic memories. Such concern would be warranted if memory manipulation caused us to lose the capacity for the moral emotions of shame and regret. These emotions are essential to our capacity to take responsibility for our actions and their consequences, as well as our practices of holding people morally and legally responsible. Certain drugs might enable us to eradicate memories of our moral failures. If eradicating an unpleasant memory resulted in the loss of moral emotions, then it could have negative effects on the normative content of one's thought and behavior. It could impair one's capacity to respond to and act in accord with moral reasons.

Yet there may be some instances in which erasing an unpleasant memory could benefit a person without weakening the influence of moral emotions on his or her thought and behavior. If it did not harm anyone, then this action could be a defensible form of therapy. Suppose that a young professor gives a lecture in which he makes a glaring factual error or fails miserably in responding to questions from the audience. He is embarrassed by the experience and internalizes it to such a degree that it weakens his self-esteem. In experiments with rats, researchers have shown that a chemical called ZIP can neutralize the action of the protein kinase PKMzeta (Sacktor, 2011). This protein plays an important role in the consolidation and storage of long-term memories by increasing the strength of synapses that regulate memory. In the experiments, the blocking action of ZIP caused the rats to forget a range of learned behaviors (but see Volk et al., 2013). Although it has not been used in humans, theoretically it is possible that a drug consisting of ZIP could block PKMzeta and prevent the formation and storage of a memory. It is also possible that the drug could remove an already formed memory from storage in the brain. If such a drug could erase the professor's unpleasant episodic memory at the moment when he was recalling it, then taking the drug could eliminate his persistent feeling of embarrassment and insecurity and make him better off psychologically. Manipulating the memory in this way would not necessarily undermine the professor's capacity for emotions and responsiveness to moral reasons. It would not incline him to perform harmful acts that he would not otherwise perform because it would not make him fail to respect the rights, interests, and needs of others.

Erasing a memory of a wrongful or harmful act that one performed may desensitize one to an appropriate emotional response to future circumstances. In this respect, memory manipulation would be objectionable. But this would not be a likely consequence of erasing an unpleasant memory of an episode in which one did nothing harmful or wrongful. There would be nothing objectionable about an action that involved deleting one unpleasant episode from one's autobiography for a therapeutic reason. It would not undermine or weaken one's moral sensitivity. Nor would it substantially alter the unity and continuity of the set of life experiences that formed one's identity. Erasing the memory would be editing a small part of one's life narrative while retaining the integrity of the whole.

Some might object that memory erasure would undermine our authenticity. According to philosopher Charles Taylor, authenticity means that it is up to each of us as a human being to find our own way in the world, to flourish, and be true to one's self. Taylor says: "If I am not true to myself, I miss the point of my life. I miss what being human is for me" (Taylor, 1991, p. 29). We come to have authentic selves by identifying with the mental states that issue in our actions. This identification results from a process of critical reflection on our desires, beliefs, intentions, and emotions. It is through this reflection that we reinforce or reject these states as the springs of our actions. Insofar as agency is a necessary condition of selfhood, being an authentic agent is necessary to have an authentic self. Altering the process of critical reflection and the mental states that result from it with psychopharmacology presumably would alienate us from our true selves. We could not identify with these altered mental states and would not have authentic selves because something alien to us would be the agent of change. Leon Kass and other members of the President's Council raise this as a possible consequence of manipulating the mind in general and memory in particular; "As the power to transform our native powers increases both in magnitude and refinement, so does the possibility for 'self-alienation' – for losing, confounding, or abandoning our identity" (Kass, 2003, p. 294).

But it does not necessarily follow that the voluntary use of a drug to dampen or erase a memory would make us inauthentic. If a person with the capacity for critical self-reflection and prudential reasoning freely decides to take a memory-modifying drug, then he or she is the agent of any change in his or her mental states. The drug is merely the means through which the person effects the change. If a person has the capacity to weigh the reasons for and against manipulating memory and to act on these reasons, and if this is what he or she desires, intends and freely decides to do, then the change would be of his or her own doing. Erasing a memory can thus be consistent with one's authenticity by reflecting the set of experiences that one wants to constitute one's self.

4. ENHANCING LONG-TERM MEMORY

Earlier, I noted that PKMzeta plays an important role in the formation and storage of long-term memory. This is distinct from short-term forms of memory such as working memory, which consists in retaining and processing information for a matter of seconds. Drugs that could increase levels of this protein kinase could expedite the process of long-term potentiation (LTP), which is associated with the plasticity of synapses. Increasing LTP with PKMzeta or a class of drugs called ampakines may stimulate synaptic connections and strengthen signal transmission between neurons involved in regulating long-term memory. The effect could be an increase in the brain's capacity to form more memories and store them for longer periods. These could be semantic memories of facts as well as episodic memories of events, and an increase in both types of memory could improve a range of cognitive tasks by making more information available to us. Because they would be used by people with normal memory functions, the drugs would be a form of enhancement and therefore distinct from the therapeutic use of drugs to retard memory loss in people with Alzheimer's disease and other dementias.

However, it is also possible that storing and retrieving more memories of remote facts and events could have the opposite effect, impairing the capacity to process information efficiently and execute immediate cognitive tasks. It is unclear how memory-enhancing drugs could weed out memory of trivial facts from memory of useful facts. It is also unclear whether retrieval of more stored memories would affect the capacity to form new memories, especially given the established neuroscientific view that we must forget a certain amount of facts and events in order to learn new ones. Altering memory mechanisms that regulate the balance between learning and forgetting could allow too much information in the brain, which could interfere with the capacity for learning. Some neuroscientists have argued that we can train our brains to handle increased informational demands. For at least some people, though, an excess of information could overwhelm their brains and impair their capacity for reasoning and decision-making.

There are cases of maladaptive or pathological behavior associated with an excess of episodic memory. Consider the case of 45-year-old Jill Price. At 14, she noticed that she had exceptional autobiographical memory. Most of these memories were intrusive and filled with trivial details. She notes that much of her recall is beyond her control: "I don't make any effort to call memories up; they just fill my mind. In fact, they're not under my conscious control, and as much as I'd like to, I can't stop them" (Price, 2008, p. 2). Price emphasizes the importance of forgetting in her following remarks: "Whereas people generally create narratives of their lives that are fashioned by a process of selective remembering and an enormous amount of forgetting, and continuously re-crafting that narrative through the course of life, I have not been able to do so" (p. 6). Further: "I've come to understand that there is a real value in being able to forget a great deal about our lives." Instead, "I remember all the clutter" (p. 45).

Price has been diagnosed with hyperthymestic syndrome. The excess recall of past experiences interferes with the ability to focus on the present and imagine the future. It can also cause impairment in other types of memory. She has below-normal capacity to memorize facts. This suggests that too much recall can interfere with the capacity to form new memories, learn new cognitive skills, and process information. Neuroimaging has shown that the regions regulating formation and retrieval of long-term memory, particularly in the hippocampal formation and prefrontal cortex, are enlarged and hyperactive in Jill Price's brain. Ordinarily, there is a balance between these regions and other brain regions regulating short-term memory. In her case, too much retrieval of long-term memory comes at the cost of impairment in short-term memory. Although Jill Price's syndrome is rare, it provides a cautionary tale suggesting that increasing storage and retrieval of memories may be more of a burden than a benefit. Drugs designed to "enhance" some memory systems could have the opposite effect on them and have more general deleterious effects on the mind as a whole. The full range of our cognitive capacities could be diminished by our ability to recall more facts and events that had no purpose for us. Constructing a narrative requires the ability to project oneself into the future. This requires one to forget a considerable amount of information about the past. Before we try to pharmacologically manipulate memory systems, we need to consider how this might affect the neural and psychological capacities that regulate the different functions, content, and meaning of different types of memory.

5. FUNCTIONAL NEUROSURGERY TO MODULATE THOUGHT AND MOOD

Unlike structural neurosurgery to resect a tumor or clip an aneurysm, functional neurosurgery aims to restore brain functions associated with motor control, motivation, thought, and mood. It does this by modulating overactive or underactive neural circuits underlying these physical and mental capacities. The most widely used form of functional neurosurgery is deep-brain stimulation (DBS). Since the late 1990s, this technique has been used to treat neurological disorders such as Parkinson's disease. More recently, it has been used to treat psychiatric disorders such as depression and obsessive-compulsive disorder. In DBS, one or more electrodes are surgically implanted in a particular region of the brain. The electrodes are connected by wires ("leads") to a battery-powered pacemaker implanted immediately below the collarbone. The pacemaker can be turned on constantly or intermittently to activate the electrodes and influence function in the targeted brain regions. There are significant differences between psychotropic drugs and DBS regarding their effects on the brain and mind. DBS is used for patients whose conditions are resistant to psychopharmacology. In addition, unlike drugs DBS targets specific structures in neural circuits. This helps to offset indiscriminate actions of medications and their side-effects. It also helps to elucidate and monitor in real-time the areas of the brain responsible for the symptoms that characterize these disorders. DBS is usually performed not by neurosurgeons alone but by teams consisting of neurosurgeons, neurologists, psychiatrists, and anesthesiologists.

DBS can be life-transforming for many patients, resolving debilitating symptoms and enabling them to regain control of their lives. Yet its effects may include alteration of psychological properties critical to personality, identity, and agency. DBS may cause symptoms that patients cannot control, despite its purpose to enable patients to regain control of their thought and behavior. Still, because this technique is used for conditions that are resistant to drugs, and because it is effective in resolving symptoms in many cases, the potential of adverse effects must be weighed against the severity of these conditions and the potential benefit and improvement in quality of life. Neuromodulation techniques have been used experimentally for a number of disorders of the brain and mind. I will focus on its use as therapy for the psychiatric disorders of obsessive-compulsive disorder (OCD) and severe depression. These uses pertain to the regulation of disordered thought and mood. I will also explore how DBS can enhance mood in some depressed individuals when the electrical current is increased beyond a certain level.

6. MODULATING THOUGHT

The first psychiatric disorder to be treated with DBS was OCD. A study conducted in 1999 by Swiss and Belgian researchers showed positive results in a small number of patients with a severe form of the disorder (Nuttin et al., 2003). Since then, a greater number of people with OCD have been treated with DBS. Results of studies indicate that one-third of those with the disorder treated with DBS have gone into remission, and roughly two-thirds are living independently. The technique is able to restore normal functioning in the neural circuits regulating the interpretation of information from external stimuli and in turn restore a normal response to the environment. OCD is characterized by obsessions or compulsions, or both. Obsessions are recurrent and persistent intrusive thoughts, impulses, or images. Compulsions are repetitive behaviors that a person feels driven to perform in response to an obsession or rigid application of rules. Individuals with OCD feel that they must think certain thoughts or do certain things. They do not want to have these mental states and often fight against them. Many retain the rational capacity for insight into the disorder and the capacity to give informed consent to treatments such as DBS. Whether one has these capacities and whether the treatments are effective depend on the severity of the disorder.

DBS is an alternative to the ablative neurosurgical procedures of capsulotomy and cingulotomy for severe OCD. These consist in making small lesions in the internal capsule or anterior cingulate region of the brain, which regulates the perception of conflict and has a critical role in reasoning. Although these procedures are treatments of last resort, they are safer than the frontal lobotomies of the past, which many labeled pejoratively as "psychosurgery." DBS involves fewer risks and thus is preferable to lesioning. This is because the effects of removing brain tissue are permanent. In contrast, adverse effects of stimulating the brain can be reversed in most cases by turning off the stimulator, adjusting the level or frequency of the electrical current, or surgically removing the electrodes. Nevertheless, there have been some serious neurological and psychological outcomes of DBS for OCD, some of which might not be reversible. Findings from one randomized controlled study indicated that stimulating the targeted brain region could reduce the symptoms of OCD. But of the eight patients in the active arm of the study, one had an intracerebral hemorrhage, two had infections from the surgical incision, and others experienced twenty-three adverse effects from the stimulator associated with mood. Most of the psychiatric symptoms involved hypomania, a state characterized by persistently elevated or irritable mood (Mallet, et al., 2008). These symptoms probably resulted from the unintended overstimulation of brain structures mediating emotions. Altered affective states might not be as serious as bleeding in the brain. But they are medically and ethically significant and need to be considered by physicians providing DBS for OCD patients with the decisional capacity to consent to undergo it. As the results of the study just mentioned illustrate, the procedure may achieve the goal of reducing the obsessions and compulsions only to cause other symptoms. These may reduce the net benefit and result in net harm for some patients.

Concern about side-effects notwithstanding, OCD is a psychopathology that causes considerable suffering and may involve a risk of suicide the longer one has to live with the disorder. Any adverse effects of electrically stimulating the brain must be weighed against the poor quality of life resulting from a disorder that has not responded to other therapies and the potential benefit from neuromodulation. Some might question whether any alteration in mental states associated with personality change would benefit a patient. But the obsessions and compulsions that make up a significant part of the lives of people with OCD are not mental states with which they identify. The pathological beliefs in this disorder can cause disunity and discontinuity in an affected person's mental states, disrupt their identity, and result in a personality that is undesirable for the subject. Relieving the symptoms can restore the unity of one's mental states, the integrity of the self, and thus one's identity. The desired personality is not the one associated with the disorder but the one that would be restored when the symptoms resolve. Addressing the question of whether DBS would alter the personality of his patient with OCD, "Herr Z.," psychiatrist Michael Schormann commented: "Patients don't see their obsessions as part of their personality. They see them as something imposed on them, that they yearn to be rid of" (cited by Abbott, 2005, p. 19). Neurologist Matthis Synofzik and psychiatrist Thomas Schlaepfer point out that "the ethically decisive question is not whether DBS alters personality or not, but whether it does so in a good or bad way from the patient's very own perspective" (Synofzik and Schlaepfer, 2008, p. 1514). Whether modulating the brain benefits a person depends not only on improvement according to objective neurological measures but also, and more importantly, according to subjective measures involving the individual's state of mind and quality of life.

A major concern about functional neurosurgery is that there is no way of knowing which adverse effects of this intervention would be temporary or permanent. The probability of memory impairment, mania, or personality changes may be low. Yet it is the magnitude of any deleterious alteration of the person's psyche, if the probability were actualized, that generates ethical concern. We cannot assume that the technique will result in complete remission of symptoms for every person who undergoes it and not have any adverse effects. Nevertheless, the disabling obsessions and compulsions, suffering, and the potential for substantial remission of symptoms give the procedure a favorable benefit-risk ratio and justify it as therapy for severe OCD.

7. MODULATING AND ENHANCING MOOD

A study published in 2005 showed that DBS modulated hyperactivity in the subgenual cingulate region of the brain and resulted in relief of symptoms in six patients with major depression whose condition failed to respond to drug treatment (Mayberg et al., 2005). In a more recent study published in 2008, 12 of 20 severely depressed patients undergoing DBS targeting the subcallosal cingulate gyrus experienced significant improvement, effectively going into remission for one year (Lozano et al., 2008). Another study published in the same year showed that stimulation of an underactive nucleus accumbens in the brain's reward system relieved anhedonia (inability to experience pleasure) without side-effects in some patients with treatment-resistant depression (Schlaepfer et al., 2008).

To say that depressed patients *feel* better after DBS does not always mean that they *are* better. Overstimulation of circuits in the reward system may cause hypomania or even mania in some patients. These states may seem preferable to anhedonia, since feelings such as euphoria intuitively are more desirable than the inability to experience pleasure. But the fact that persistent euphoria often correlates with irrational and self-harming behavior shows that hypomania and mania are not desirable or beneficial states. Instead, the heightened mood can harm people by impairing their capacity to reason. Neuroscientists have a better understanding of the neural circuitry underlying Parkinson's disease and OCD than the circuitry underlying depression. There is also greater variation in symptoms among people with depression than those with other neuropsychiatric disorders. The changes induced by stimulating the brain of a depressed person are poorly understood and many produce different effects among them. So, caution in the selection and monitoring of patients undergoing DBS is especially warranted for those with depression.

People undergoing functional neurosurgery to treat depression are exposed to the same risks as those undergoing it for other disorders of the brain and mind. As with OCD, the suffering, poor quality of life, and risk of suicide associated with the condition must be weighed against the potential benefits and risks of DBS. Patients with severe depression and the physicians treating them need to balance the risks *with* DBS against poor quality of life and the risk of suicide *without* DBS. Provided that patients have the capacity to understand the potential positive and negative outcomes of this intervention in the brain, they can consent to undergo it. Still, because DBS can cause changes in cognition and mood, some may question whether a patient undergoing DBS can retain the decisional capacity to consent to continue or discontinue the technique once it has been initiated. This would have to be determined by a psychiatric consultation on a case-by-case basis. Any changes in the patient's mental states would have to be substantial to undermine decisional capacity. Subtle changes would not necessarily have this effect, in which case there would be a presumption of continued capacity. With careful patient selection and monitoring of symptoms while undergoing the procedure, the benefits of DBS can outweigh the risks for those being treated for severe treatment-resistant depression and justify its use for this purpose.

A more controversial effect of functional neurosurgery is the enhancement of mood above normal therapeutic levels. Suppose that a person has combined generalized anxiety and major depression, neither of which has responded to drug therapy. He undergoes DBS targeting the nucleus accumbens to elevate his mood. Because this brain region projects to the amygdala, the stimulation influences the latter structure and modulates his heightened fear response to stimuli. He is less anxious and feels better after several sessions. Indeed, he likes feeling this way so much that he asks his psychiatrist to increase the level of stimulation so that he can constantly be in a good mood. But the increased voltage overstimulates his brain's reward system and causes him to feel "almost too good." (This example is based on a case discussed by [Synofzik, Schlaepfer, and Fins, 2012]). Fortunately, he retains insight into his conditions and becomes concerned that the increased voltage might induce an overwhelming state of euphoria. This would cause him to lose his ability to reason and control his behavior. So, he asks the psychiatrist to reduce the voltage to the initial level that resolved his dysphoria and restored a normal level of mood.

Modulating underactive or overactive neural circuits implicated in anxiety or depression can restore the fear response and the capacity for pleasurable experiences to normal levels. As the case just cited illustrates, it may also be possible for DBS to elevate mood to a level where one is in a constant euphoric state. But very high levels of mood (feeling "too good") can interfere with the ability to reason and incline one to become locked into maladaptive or pathological behavior. This characterizes people with addictions, whose behavior is driven by a dysfunctional reward system in their brains. There may be optimal levels of mood and the neurotransmitter dopamine in the reward system that regulates it. These levels can influence cognitive functions associated with reasoning and decision-making. A certain level of mood and motivation is necessary for one to adapt to the environment by enabling one to choose and act in accord with reasons. But the range of one's choices and actions can contract if the reward system in the brain is manipulated to raise mood to a euphoric level. There are limits to the extent to which elevating mood can benefit a person. Exceeding these limits can cause a presumed enhancing effect to become a debilitating one.

8. CONCLUSION

Neuroscience has provided us with knowledge of the brain mechanisms mediating memory, thought, and mood. It has also provided us with knowledge of how manipulating these mechanisms pharmacologically or surgically can alter certain regions of the brain and these states of mind. These interventions can be used as therapy to restore normal brain and mental functions or to enhance them. One possible future therapy would be the use of a prosthesis to replace a damaged hippocampus, enabling people with this damage to form new memories (Berger 2012). But I have focused on actual interventions in the brain in this chapter and have emphasized two general points. First, there is interaction between the cognitive functions associated with memory and thought and the affective functions associated with emotion. Medical professionals who provide, and individuals who take, psychotropic drugs or undergo functional neurosurgery need to exercise due diligence in these actions. In many cases, these interventions have a favorable benefit–risk ratio for people suffering from neuropsychiatric disorders. Yet it is possible that manipulating some neural networks mediating some mental capacities could influence others in ways that might not be beneficial for their recipients. Second, any attempt to enhance mood needs to be informed by the idea that there may be optimal levels of mood and mood-regulating neural networks and neurotransmitters that promote adaptive behavior. Elevating mood above this level by increasing the amount of the neurotransmitter dopamine and over-activating the brain's reward system through DBS may result in maladaptive or pathological behavior. These actual and possible ways of altering the brain and mind should make us carefully consider the consequences before we pharmacologically or surgically tinker with them.

WORKS CITED

A. Abbott (2005) 'Deep in Thought,' *Nature*, 436: 18–19.

T. Berger et al. (2012) 'A Hippocampal Cognitive Prosthesis: Multi-Input, Multi-Output Nonlinear Modeling and VLSI Implementation,' *IEEE Transactions on Neural System and Rehabilitation Engineering*, 2: 198–211.

L. Kass (2003) *Beyond Therapy: Biotechnology and the Pursuit of Happiness* (New York: Harper Collins), 294.

A. Kolber (2011) 'Give Memory-Altering Drugs a Chance,' *Nature* 476: 275–276.

K. LaBar, and R. Cabeza (2006) 'Cognitive Neuroscience of Emotional Memory,' *Nature Reviews Neuroscience*, 7: 54–64.

A. Lozano, et al. (2008) 'Subcallosal Cingulate Gyrus Deep-Brain Stimulation for Treatment-Resistant Depression,' *Biological Psychiatry*, 64: 461–467.

L. Mallet, et al. (2008) 'Subthalamic Nucleus Stimulation in Severe Obsessive-Compulsive Disorder,' *New England Journal of Medicine*, 359: 2121–2134.

H. Mayberg, et al. (2005) 'Deep-Brain Stimulation for Treatment-Resistant Depression,' *Neuron* 45: 651–660.

J. McGaugh (2003) *Memory and Emotion: The Making of Lasting Memories* (New York: Columbia University Press).

B. Nuttin, et al. (2003) 'Long-Term Electrical Capsular Stimulation in Patients with Obsessive-Compulsive Disorder,' *Neurosurgery* 52: 1263–1274.

R. Pitman et al. (2002) 'Pilot Study of Secondary Prevention of Posttraumatic Stress Disorder with Propranolol,' *Biological Psychiatry* 51: 189–192.

J. Price (2008) *The Woman Who Can't Forget: A Memoir* (New York: Free Press).

T. Sacktor (2011) 'How Does PKMzeta Maintain Long-Term Memory?' *Nature Reviews Neuroscience*, 12: 9–15.

T. Schlaepfer, et al. (2008) 'Deep-Brain Stimulation to Reward Circuitry Alleviates Anhedonia in Refractory Major Depression,' *Neuropsychopharmacology*, 33: 368–377.

M. Synofzik, T. Schlaepfer, and J. Fins (2012) 'How Happy Is Too Happy? Euphoria, Neuroethics, and Deep-Brain Stimulation of the Nucleus Accumbens,' *American Journal of Bioethics-Neuroscience*, 3 (1): 30–36.

M. Synofzik, and T. Schlaepfer (2008) 'Stimulating Personality: Ethical Concern for Deep Brain Stimulation in Psychiatric Patients and for Enhancement Purposes,' *Biotechnology Journal*, 3: 1511–1520.

C. Taylor (1991) *The Ethics of Authenticity* (Cambridge, MA: Harvard University Press).

U.S. President's Council on Bioethics, Staff Working Paper (2003) 'Better Memories? The Promise and Perils of Pharmacological Intervention,' March (available at http://www.bioethics.gov/transcripts/mar03.html).

L. Volk, et al. (2013) 'PKM-Zeta Is Not Required for Hippocampal Synaptic Plasticity, Learning and Memory,' *Nature*, 493: 420–423.

DISCUSSION QUESTIONS

1. Would there be anything morally objectionable about dampening emotional memory if doing so had no adverse side-effects? Would it violate some essential property of our human nature?

2. Taking a drug to erase or weaken the emotional content of a traumatic memory of a rape or other criminal offence could preclude a person from testifying against the offender in a court of law. Should the victim have the right to take the drug in these cases?

3. Even if it is permissible to delete one or a few selected unpleasant memories, is there a limit beyond which deleting memories would become objectionable?

4. Would deep-brain stimulation (DBS) to treat a severe neurological or psychiatric disorder be justified if it resulted in a radical change in a person's identity?

5. Would it be permissible to offer DBS to modify the behavior of criminal offenders as an alternative to incarceration?

6. Does the fact that the stimulating device in a person's brain regulates his or her behavior mean that the device and not the person is the source or author of his or her actions? Does it undermine individual responsibility?

7. If a person with a psychiatric disorder lacks the decisional capacity to consent to DBS as treatment, is it permissible for a family member or other surrogate to consent to it on their behalf?

8. Should a patient have the right to request DBS to enhance their mood above a normal level? What would the responsibility of the medical practitioner be in these cases?

10. Assuming that someone deliberately chooses to erase all their past memories, are they still the same person? Is this choice authentic?

11. What are the possible unintended (or dual) uses for pharmacological memory erasing, memory dampening, and memory enhancing? Are any of them likely to have detrimental consequences?

12. Many of the concerns associated with neurotherapeutics are related to the continuity of biographical or narrative identity. What is the significance of this continuity and how important is it to retain it?

INCENTIVIZING ACCESS AND INNOVATION FOR ESSENTIAL MEDICINES: A SURVEY OF THE PROBLEM AND PROPOSED SOLUTIONS[1]

Michael Ravvin

CHAPTER SUMMARY

In this chapter, Michael Ravvin discusses the ways in which the existing intellectual property system discourages the innovation of, and access to, essential medicines for the poor in developing countries. He begins by surveying the features of the existing pharmaceutical patent system that give rise to the problems. He then offers critical analysis of some reform proposals. He argues that some existing mechanisms intended to mitigate the limitations of the current pharmaceutical patent system are inadequate and perhaps even counterproductive over the long term; while others are often inefficient and limited in scope. He believes that approaches that offer reward for successful pharmaceutical innovations are the most promising mechanism for overcoming the barriers to access and benefits for the poor in developing countries. He discusses two such approaches in particular, the use of priority review vouchers and the Health Impact Fund.

RELATED READINGS

Introduction: Justice, Access, and Equality (2.6.2)

Other Chapters: Fritz Allhoff, *The Coming Era of Nanomedicine* (Ch. 11); Kenneth Himma and Maria Bottis, *The Digital Divide* (Ch. 22)

---

[1] This chapter is excerpted from Michael Ravvin (2008) 'Incentivizing Access and Innovation for Essential Medicines: A Survey of the Problem and Proposed Solutions,' *Public Health Ethics* (1): 110–123. It appears here by permission of Oxford University Press.

1. INTRODUCTION

Millions of the world's poor die every year from preventable, curable or treatable illnesses. This is partly due to the existing intellectual property regime, which discourages the innovation of, and access to, essential medicines for the poor in developing countries. There are two crucial areas in which the current system must be improved. The first is access. Under the current pharmaceutical patent system, the market exclusivity guaranteed to patented drugs makes them inaccessible to the poor due to high prices. Drugs that are available to the poor through philanthropic and other efforts often face challenges of distribution. Second, it is essential to devise a mechanism that incentivizes innovation of pharmaceuticals for diseases that predominantly afflict the poor in developing countries. Currently, potential innovators have no financial incentive to invest in R&D for poor country diseases, since those who need these medicines will not be able to afford the monopoly prices that are necessary to recover costs. A successful proposal to reform the existing system must address the challenges of access and innovation. This article provides a survey of the problems with the existing pharmaceutical patent system and a critical analysis of some recent proposals to remedy them. I will conclude with some thoughts on the most promising directions for reform, and the challenges that remain.

2. PROBLEMS IN THE CURRENT SYSTEM

The existing pharmaceutical patent regime is defined primarily by the Agreement on Trade-Related Aspects of Intellectual Property Rights (TRIPS), signed at the end of the Uruguay Round of WTO negotiations in 1995. This agreement governs nearly all aspects of intellectual property in international trading and the domestic law of states parties. TRIPS requires all WTO member states to adhere to strict patent protection laws for pharmaceuticals that guarantee at least 20 years of market exclusivity for patented drugs.

Prior to TRIPS, different countries had different patent laws, which often reflected their level of development and social goals. Developed countries typically had very robust patent laws, providing strong protection for monopoly manufacturing and sale of patented products. In developing countries, looser patent protection allowed generic drug manufacturers to provide safe and affordable generic versions of drugs still under patent protection in developed countries. Many countries had no patent protection for pharmaceuticals at all. In the absence of patent protection and alternative reward mechanisms, innovators had no way to recover costs and make profits on pharmaceuticals developed for diseases that predominantly afflict the developing world. It is therefore no surprise that little R&D was devoted to poor country diseases. Nonetheless, poor countries had some access to cheap and reliable medicines that were under patent protection.

This generic industry was effectively shut down in 2005, when the 10-year compliance window for TRIPS closed for all but the least developed countries. WTO members were required to bring their domestic patent laws up to TRIPS standards, effectively universalizing the strong patent protection favored in developed countries. The provisions of this treaty have been supplemented by bilateral TRIPS-plus agreements that often extend the effective life of patents well beyond

20 years. The result is further limits on access to essential medicines. Due to the rapid mutability of many infectious diseases, by the time a drug becomes available for generic production it may no longer be effective against its target disease.

The existing patent system was designed to incentivize investments in research and development (R&D) by guaranteeing innovators some period of protected monopoly during which they could recover R&D costs. Without this guarantee, potential innovators have little incentive to assume the risks of R&D. This is certainly the case with pharmaceuticals, where the current estimated average cost of development is over $800 million per drug brought to market (DiMasi et al., 2003). Though monopolies are generally discouraged in market economies due to the inefficiencies they create, the monopoly marketing rights guaranteed by patents are commonly believed to be essential to encourage socially valuable innovation. Unfortunately, however, this system has proved woefully inadequate for the pharmaceutical market, especially for the poor in developing countries.[2]

The monopoly patent system may be a relatively efficient way to incentivize innovation in fields such as popular music and software. However, the pharmaceutical market differs in three important ways (Hollis, 2005a: 4–5). First, there are strong reasons to doubt whether the monopoly price of a patented drug is a reliable indicator of its social value due to informational asymmetries in the drug market. In normal markets, consumers decide how much they are willing to pay for products that they choose. In the drug market, however, consumers do not have the information necessary to decide what drugs they need, and so do not normally choose the drugs they take. Insurance companies or the government usually pay most of the cost of drugs, so consumers are less sensitive to price. Doctors who prescribe drugs are also not sensitive to price since they are not paying, either. Meanwhile, doctors are susceptible to influence from drug manufacturers, who may persuade them to prescribe more expensive drugs, since it comes at no additional cost to those doctors. Insurers and governments, in most cases the actual purchasers of drugs, are not in a position to know what patients really need. Thus, drug prices are determined by many factors besides what consumers would pay if they had perfect information.

Despite these informational asymmetries, we may still feel that there is no better alternative than the price system for determining the value of pharmaceuticals. Indeed, nearly all commercial products have primarily subjective value to those who consume them, which is reflected in price. However, the second important difference between the pharmaceutical market and other markets is that pharmaceuticals are developed for a specific purpose (the one for which they receive marketing approval) and their impact can be objectively measured. This makes pharmaceuticals candidates for a system of valuation that does not depend on the price they are able to command, but rather on their ability to achieve a particular health impact (Hollis, 2005b: 13).

The final important difference is the moral consequences of inadequate access and incentives for innovation. Whereas the consequences of lack of access to music and software are negligible, millions of people die every year due to lack of

---

[2] For discussion of the advantages and disadvantages of the patent system for pharmaceuticals, see Hollis (2008).

essential medicines. This is an unacceptable consequence of the existing patent regime. We therefore have a strong moral obligation to find a solution to the problems caused by the existing system.

Alternatives to the price system play a role in the most promising proposals to reform the current system, as will be discussed later. In the meantime, the structural characteristics of the existing pharmaceutical patent system have created a series of problems. The evaluation of proposals to reform the current system must be based on their ability to resolve the following issues.[3]

2.1 High prices

The monopoly marketing rights guaranteed under TRIPS allow drug companies to charge extremely inflated prices for patented drugs, often 400 per cent or more above competitive market prices (Baker, 2004: 2). Due to guaranteed monopolies, drug companies are able to make more profits by selling drugs at very high prices to fewer consumers in wealthy markets than they would by selling the same drugs at a lower price to a broader market. In order to prevent parallel imports, drug companies do not charge differential prices in countries that cannot sustain the monopoly prices. The result is that patented drugs remain outside the financial reach of the poor. Furthermore, inflated prices force those who can afford monopoly prices to overpay for the R&D that patent protection is supposedly designed to incentivize.

2.2 Neglected diseases

As a result of pricing problems, pharmaceutical R&D is devoted almost exclusively to the diseases prevalent in affluent countries, where there is a large market for expensive drugs. Pharmaceutical innovators have no incentive to invest in R&D for diseases that predominantly afflict the poor in developing countries, who cannot support high monopoly prices. These neglected diseases are primarily those that have been identified by the WHO as Type II and Type III diseases, including malaria, tuberculosis, Chagas' disease, African sleeping sickness, other tropical diseases and HIV/AIDS, which claims most of its victims in sub-Saharan Africa (WHO, 2006a: 16). It is often claimed that only 10 per cent of health R&D is devoted to diseases that cause 90 per cent of the global burden of disease (GBD). This imbalance is reflected in the types of drugs brought to market. Of the 1556 new drugs approved for commercial sale from 1975–2004, only 21 – barely more than 1 per cent – were for neglected tropical diseases (Trouiller et al., 2002: 2189; Chirac and Torreele, 2006: 1560). It is clear that 'intellectual property rights are an important incentive for the development of new health-care products. [However,] this incentive alone does not meet the need for the development of new products to fight diseases where the potential paying market is small or uncertain' (WHO, 2007: 4).

---

[3] The following list of problems is drawn from (Pogge, 2008b, pp.: 5–6) and (Hollis, 2005a, pp. 4–10).

2.3 Deadweight losses

Deadweight losses occur when people are willing and able to pay more than the marginal cost of production for a drug, and yet are not willing or able to afford the patent price. If these patients cannot purchase the drugs, drug companies make less money than they could have by charging a lower price (Hollis, 2005a: 7–8). Deadweight losses are estimated to exceed $100 billion per year by 2013, and are therefore at least equal to the total amount of private sector spending on R&D. For every dollar spent by pharmaceutical companies on R&D, one dollar is lost in deadweight losses (Baker, 2004: 8). This enormous waste comes primarily at the expense of the poor in developing countries, who are not able to purchase essential medicines. However, like other inefficiencies in the current system, these losses also come at the expense of the affluent, who pay higher prices to compensate for deadweight losses that would not exist under a more efficient system.

2.4 Bias toward symptom relief

Under the current system, pharmaceutical company profits depend on the number of treatments sold, not necessarily on the impact of those treatments. Vaccines and other preventative treatments are the least lucrative drugs, as they prevent future spending on treatments for the same disease, and are most often purchased by governments and other large buyers for low prices. Curative medicines, which remove diseases from the patient's body, are also unattractive because they, too, limit potential future revenues. Drugs that treat the symptoms of a disease, leaving the disease in the body and necessitating the continual use of pharmaceutical treatments, deliver the greatest profits. Thus the unfortunate result of the existing patent system is that pharmaceutical companies are incentivized to develop medicines that deliver lower health impact to rich and poor patients alike (Pogge, 2008b: 4).

2.5 Duplicative drugs

Monopoly pricing gives pharmaceutical innovators strong incentives to invest in R&D to invent drugs that are sufficiently different from existing pharmaceuticals to be eligible for a separate patent, but which provide little or no additional therapeutic qualities (Hollis, 2007a:6). These 'me-too' drugs are particularly lucrative for the sort of blockbuster drugs common in affluent countries for conditions such as high cholesterol and erectile dysfunction. According to the US Food and Drug Administration, over 77 per cent of the drugs approved from 1990–2004 were duplicative, rather than breakthrough drugs (USFDA, 2005). This race to duplication wastes billions of dollars of R&D and produces little substantive improvement in the GBD.

2.6 Counterfeiting

Large markups for patented drugs, high demand for these drugs at cheaper prices in the developing world and lack of access to affordable, reliable generics encourages the illegal manufacture and sale of patented pharmaceuticals. Many counterfeit drugs are ineffective or dangerous. Even if they are faithful copies of the original, they cost the innovator lost revenues, a cost that is passed on to other consumers. This reduces incentives to innovate, creates additional waste and presents a real danger to consumers of counterfeit drugs (Hollis, 2005a: 8; Pogge, 2008b: 5).

2.7 Excessive marketing

Due to high markups and the strong incentives for me-too drugs, there is enormous waste on excessive marketing that targets doctors and consumers directly. The pharmaceutical industry currently spends as much on marketing as it does on R&D (Baker, 2004: 8). This marketing can influence doctors to prescribe medicines that their patients do not really need, or more expensive treatments under patent protection when cheaper and equally effective generic treatments are available. Marketing can also influence consumers to take medicines they do not really need for diseases they don't really have (Hollis, 2005a: 9; Pogge, 2008b: 5).

2.8 Last mile problem

Even when drugs for developing world diseases are available, the current system does not provide any incentive to make sure that these medicines achieve their maximum impact. This is often due to the challenges involved in the final stages of the distribution of medicines, known as the last mile (Pogge, 2008b: 5). The final distribution mechanisms for drugs determine whether the patients who need the drugs receive them on time and in sufficient quantity, and that they are properly administered to achieve maximum effectiveness. The challenges along the last mile in developing countries are significant due to poor health and transportation infrastructures. Any proposal to reform the existing patent system must overcome the challenges of the last mile in order to achieve a significant health impact.

3. PUSH VERSUS PULL MECHANISMS

Most existing efforts to incentivize innovation for neglected diseases and to provide affordable access to the resulting drugs fall in the category of push mechanisms. Push mechanisms reduce the cost of research by providing some or all of the funding for R&D up front. The most common form of push programs is research grants, where researchers are paid by governments or other funding sources for research that is determined to be socially valuable. The research conducted by the US National Institutes of Health, as well as a great deal of R&D conducted at universities through government or other grants, are examples of this sort of push funding. This has historically been the primary way to enable research for socially desirable but unprofitable products. The existence of publicly funded research is itself an indication of the failure of the patent system to encourage many types of socially valuable R&D.

A second common form of push funding is public–private partnerships (PPPs), in which public or non- profit institutions subsidize research by private firms on the understanding that the resulting products will be made available at prices affordable to the poor. There are currently 60–80 PPPs in the global health field. Examples include the International AIDS Vaccine Initiative, the Medicines for Malaria Venture, the Global Alliance for Tuberculosis Drug Development and the Drugs for Neglected Diseases Initiative (Johnston and Wasunna, 2007: S26). Overall, the amount of publicly subsidized or supported R&D in the USA is roughly equal to the amount of private R&D (Baker, 2004: 12).

The advantage of push funding is the elimination of R&D risk. In particular cases, this can be done at relatively low cost because the sponsors can determine the amount of R&D they wish to subsidize. Push mechanisms are popular with research institutes, as they provide a steady stream of funding. Push mechanisms can also be useful for encouraging basic research, the anticipated benefits of which may come too far in the future for patent protection to provide a reasonable assurance of profit (Hollis, 2007c: 5). Despite these advantages, push funding has four important drawbacks.

First, it is inefficient because it funds a great deal of unsuccessful R&D at public expense.[4] Funding is linked to ongoing research, and the end of the research brings with it the end of the funding. Push programs therefore incentivize the pursuit of a continuing stream of research grants, rather than the swift production of viable products (Hollis, 2007b: 10). Since the costs of R&D are covered regardless of the success of the research and because grants are an essential source of revenue, push programs encourage potential innovators to continue research into projects that have a high likelihood of failure, causing enormous waste (Schwartz and Hsu, 2007: 26). The chances of failure are increased by the fact that often only one team of researchers is working on a particular innovation at a time, and each additional team multiplies the costs of research without any assurance of a corresponding multiplication of returns. Lacking competition, researchers have fewer incentives to work quickly and cost-effectively toward early success (Pogge, 2008a: 242).

Second, the selection of projects for push funding is made on the basis of inadequate and often biased information. Decisions about where to allocate push funding must be made before the outcome of the research is known. Research targets can be influenced by political factors, so that research is not necessarily targeted toward innovations that will have the greatest health impact (Baker, 2004: 13). The selection of funding recipients is also open to political manipulation and bias. Even when funding recipients are chosen with the best intentions, due to information asymmetries between donors and innovators donors may not be able to accurately determine which projects are most likely to lead to successful innovations (Hollis, 2007a: 78–79; Johnston and Wasunna, 2007: S26; Pogge, 2008a: 242). It is also difficult to determine the proper amount of money that should be invested in research, because the costs entailed in the R&D process and the value of the innovation are hard to anticipate before research begins. Thus, funding allocations are likely to reflect the personal interests of the sponsors, rather than calculations of efficiency (Hollis, 2007b: 11).

Third, push funding often subsidizes R&D that ultimately leads to drugs sold at high monopoly prices. Public funding normally covers only early phases of research. However, the most expensive part of pharmaceutical R&D is the late-stage trials and testing that precede market approval. As a result,

---

[4] For an extended discussion of the inefficiency of push mechanisms compared to pull mechanisms, see Schwartz and Hsu (2007).

publicly funded researchers sometimes turn over innovative research to private pharmaceutical companies to bring prospective medicines to market. In many of these cases, the pharmaceutical firm, whose goal is to maximize profits, holds the patent rights to the new product, and sells the patented medicines at high prices (Hollis, 2006: 128; Schwartz and Hsu, 2007: 4). This entails a double loss for the public, which has paid to fund early research and then pays again for high-priced drugs sold under a market monopoly, which are also inaccessible to the poor.

Finally, push programs are neither general nor stable over the long term. Publicly funded grants must be frequently reapproved, and are often terminated. Philanthropic support for research may dissolve as sponsors' priorities change. Since push funding subsidizes predetermined research targets, financial support will shift together with the interests and sympathies of funders, denying potential innovators a reliable source of general R&D funding to incentivize innovation into medicines that innovators believe are most likely to yield results.

Pull mechanisms, on the other hand, are designed to incentivize innovation by rewarding successful innovators through enhanced profits or some other form of reward for the achievement of a socially valuable product. The existing patent system is itself an example of a pull mechanism, which promises a market monopoly for patented medicines. Though the patent system is flawed, it has proven effective at stimulating innovation for markets that can afford monopoly pricing. The challenge in overcoming problems of access and innovation for the developing world is to devise a system that replicates this reward structure, but is able to encourage innovation of drugs for neglected diseases, reduce waste and ensure access to affordable essential medicines. There are a number of existing proposals for pull mechanisms, three of which will be discussed in more detail later in this essay.

Publicly funded pull programs are a significant departure from the way in which pharmaceutical innovation for the developing world has traditionally been done, and therefore such programs are often met with skepticism by governments and potential innovators alike. However, given the poor record of existing programs, there is a strong reason to seek a better alternative. Successful pull programs must meet three important conditions. First, the basis for eligibility for rewards must be clearly specified far in advance, so that potential innovators understand the goal they are working toward. Second, the size of the reward must be sufficiently large to incentivize innovation despite the risk of failure. Third, these programs must be a sustainable and reliable source of rewards for pharmaceutical innovators over the long term. Pull mechanisms that adequately meet these conditions offer substantial advantages in terms of efficiency over push mechanisms.

Pull mechanisms are more efficient than push mechanisms because they do not pay for failed research and they encourage innovators to work quickly and cost-effectively toward early success (Hollis, 2006: 128; Pogge, 2008a: 241). A pull mechanism specifies a prize that would be awarded to an innovator who successfully brings to market a drug that met certain conditions, and may specify that only the first successful innovation will be rewarded. Sponsors would not have to pay for research that did not lead to a successful product, and would not have to pay for the duplicative efforts of firms that failed to succeed first. Potential innovators would accept the costs and risks of R&D, hoping to win the reward. Pull mechanisms are also able to overcome the informational asymmetries of push mechanisms by taking advantage of the internal assessment of potential innovators. Firms that believe they stand a good chance of being the first to achieve the research goal would undertake the R&D, while those that feel they are not likely to succeed will stay out of the race.

The risks involved mean that the size of the reward would have to be considerably larger than what a single firm would be paid under a push program, enough to balance a firm's risk of failure. This risk is twofold, as research efforts may fail because the innovator is unable to achieve the goal at all, or because some other innovator succeeds first (Pogge, 2008b: 8). However, because pull mechanisms pay only for successful innovations and can set a low market price for innovated pharmaceuticals as a condition of receiving rewards, the cost of treatments per individual can be substantially lower than through push mechanisms. Pull mechanisms are therefore more likely to lead to successful innovations, and the innovations they incentivize are likely to be more widely accessible to their target populations (Schwartz and Hsu, 2007: 27). Increased health impact per dollar means that pull programs are considerably more efficient than push mechanisms.

Although publicly funded pull programs are a relatively new idea, they have the potential to gain broad political support from taxpayers and pharmaceutical firms alike. Pull mechanisms align the interests of profit-seeking innovators with those of society, which seeks efficient pharmaceutical innovation and affordable medicines (Hollis, 2006: 128). Pharmaceutical firms are competing on terms similar to the existing patent system that is familiar to them. Citizens are more likely to support programs that do not subsidize failed research. By relying primarily on private risk and entrepreneurial innovation, pull mechanisms fit better with the spirit of private enterprise and replicate the advantages of the market system (Pogge, 2008a: 243). Because they reward only successful innovation and can stipulate the conditions for rewards in advance, including the sale price of the drug, well-designed pull mechanisms are more likely to increase access to essential medicines and incentivize innovation for neglected diseases. I now turn to consider [two] recently proposed pull mechanisms.

4. [TWO] PROPOSED PULL MECHANISMS

4.1 Priority Review Vouchers (PRV)

PRVs were initially proposed by Ridley, Grabowski and Moe in 2006 (Ridley et al., 2006). The proposal caught the attention of US legislators, and under the sponsorship of Senator Sam Brownback (R-KS), it was included as the Elimination of Neglected Diseases Amendment in the FDA Amendments Act, which was signed into law on 27 September 2007 (Food and Drug Administration Amendments Act of 2007). Under this scheme, a pharmaceutical company that obtains approval for a drug or vaccine against a specified neglected disease will receive a voucher for priority FDA review of another pharmaceutical. By expediting the FDA review process, the voucher could reduce the time required to gain FDA approval of the second drug by up to one year. The additional profit that a pharmaceutical innovator could earn from this additional year of market exclusivity is estimated at more than $300 million for a blockbuster drug (Ridley et al., 2006: 315). Vouchers can also be sold to other firms. In either case, the increased revenues from the voucher would offset the R&D costs of the development of the drug targeted at a neglected disease.

As a pure pull mechanism, the PRV is highly efficient. It does not pay for unsuccessful research, and it does not even require direct payment for successful research. Even the costs associated with expedited FDA review would be paid by the innovator, and would likely constitute only a very small fraction of the resulting profits. The plan can therefore be implemented at no additional cost to consumers or taxpayers. Further, by leaving the list of targeted diseases broad, PRVs allow innovators to determine which drugs to pursue based on an internal evaluation of the likelihood of success.

In addition to efficiency, PRVs claim to achieve two main benefits: 'faster access to blockbuster drugs in developed countries, and faster access to cures for infectious diseases in developing countries' (Ridley et al., 2006: 315). This claim exposes the shortcomings of PRVs. While they may speed up access to medicines for developing country diseases for those able to afford them, they are unlikely to incentivize a significant amount of innovative R&D into developing country diseases or increase access among the global poor to the products of that R&D. In cases where a drug for a developing country disease is in late stages of development, the innovator will rush to complete the necessary steps to receive product approval from the FDA and the accompanying PRV for another product. However, if the projected profits that will result from the expedited approval of the second drug are estimated at $300 million in the case of a blockbuster drug – an exceptional circumstance – then in most cases the additional profits resulting from the voucher will not be enough to incentivize large-scale exploratory R&D into medicines for developing country diseases. Thus the PRV is more likely to reward firms for medicines already in late stages of development – medicines that would likely be brought to market in any case – rather than to incentivize the innovation of a great number of medicines for developing country diseases that would not otherwise have been developed.

There is also little reason to believe that once drugs eligible for reward under the PRV scheme receive market approval from the FDA they will be widely accessible to the global poor. The original voucher proposal included a stipulation that innovators forgo patent rights for neglected disease drugs in order to receive vouchers (Ridley et al., 2006: 312). Unfortunately, this condition is not included in the version that was actually implemented. Given the relatively small reward, and the absence of conditionality on the price of the innovative medicine, drug companies seeking to maximize profits may prefer to sell innovative treatments for developing country diseases to more limited affluent markets at very high prices. Thus the PRV does little to ensure that the innovative medicines that result from the scheme will actually be available at affordable prices to the vast majority of those who need them. Since these drugs are likely to be sold through the traditional patent system, PRVs are also unable to mitigate the problems of deadweight losses and counterfeiting. Since rewards are not linked to the actual effectiveness of the treatment, PRVs are unlikely to overcome the bias toward symptom relief. Finally, since the necessary threshold for receiving the reward of the PRV is only the achievement of market approval for a neglected disease drug, rather than any actual health impact of the drug, PRVs do nothing to overcome the challenge of the last mile.

However, PRVs can claim one important advantage. Of the proposals considered here, it is the only one to actually have been passed into law. Though the health impact of PRVs is likely to be minimal, the political achievement is highly significant. PRVs were able to achieve broad support by appealing to the interests of all stakeholders, including political leaders, pharmaceutical companies and global health advocates, allowing the proposal to be implemented in remarkably little time. In this respect, PRVs serve as an important example for future reform efforts.

4.2 Health Impact Fund (HIF)

The HIF is designed to offer a more general incentive to pharmaceutical innovation and distribution (Hollis, 2007c, 2008; Pogge, 2008a). Under this plan, instead of receiving profits from selling patented drugs at high monopoly prices, innovators could opt to register any newly patented medicine with the HIF, which would provide a guaranteed payment stream in proportion to the incremental impact of the innovative drug on GBD during its first 10–12 years on the market, corresponding to the average effective market life of a patented drug. Innovative firms could forgo monopoly prices in one of two ways. They could grant an open license to generic producers, or the innovator could maintain exclusive production of the drug, but all sales revenues would be subtracted from reward payments. Innovators whose medicines have a significant impact on GBD would receive a return on their R&D investment through the rewards provided by the HIF, rather than through sales at high prices.

High prices will be avoided because innovators rewarded by the HIF will either have to make their products generically available, or have sales revenues deducted from rewards. In either case, prices for drugs will fall to near the marginal cost of production or perhaps even below. Low prices will be directly incentivized because rewards will be dependent on the actual impact of the drugs, and so innovators will seek to make sure that their products are widely accessible to all who need them. The reduction of drug prices to competitive levels will eliminate the problem of deadweight losses. Since innovations will only be rewarded for their incremental health impact over existing medicines, there will be no incentive to develop duplicative drugs that do not achieve some substantial improvement over existing alternatives. Low prices and the absence of me-too drugs will discourage excessive marketing.

The HIF would evaluate innovative pharmaceuticals only once they have received marketing approval on the basis of an objective measurement of their actual health impact. A candidate measure is the Quality-Adjusted Life Year (QALY), variations of which are currently used to gauge the health impact of pharmaceuticals by national health systems in Australia, Canada, UK and the USA, as well as by private insurance companies (Hollis, 2007a: 83). The rate paid per QALY would be automatically determined through a division of the fund proportionally according to the number of QALYs gained by each drug registered with the HIF (Hollis, 2008). The use of this general, objective measure would prevent the need to determine detailed specifications for eligible drugs in advance. The HIF therefore achieves important advantages of efficiency, because the HIF will pay only for the actual performance of medicines. The HIF does not need to speculate about the probability of success of R&D or the costs involved, so there is less risk that the HIF will pay too much or too little for important research. (Hollis, 2007b: 26). Additionally, the use of a QALY-based measurement will reduce the bias toward symptom relief, as innovators will seek to develop medicines that have the greatest health impact.

A general fund for incentivizing socially valuable innovations that employs an objective measure of health impact would also avoid the sort of political and bureaucratic meddling. Under the HIF, there is no need to make decisions about which diseases to target. Instead, there is a general reward for any patentable pharmaceutical innovation proportional to its impact on global health. This leaves little room for lobbying on behalf of particular diseases, and will result in the allocation of R&D resources by potential innovators to those areas where a significant alleviation of GBD can be most effectively achieved, taking maximum advantage of the internal information about the capacities of innovative firms.

The HIF is also more capable of incentivizing innovation for neglected diseases, and does not need to rely on a committee's decisions on research priorities. Under the HIF, pharmaceutical companies are rewarded only for incremental health impact. Therefore they will target diseases that cause the most suffering and death, and for which effective and accessible treatments do not exist but are likely to be found. Since there is no need to determine specifications for eligibility in advance, the HIF need not limit itself to vaccines. Innovators will prefer to develop cures and vaccines, because for the same disease a cure or vaccine will be valued at a higher QALY level than a drug that only treats symptoms. However, the HIF can also provide incentives when there is potential for a drug that is superior to the *status quo*, but a vaccine or cure remains elusive. The HIF also does much to overcome the challenges of the last mile. Since companies will be rewarded based on the actual health impact of their innovations, they will have strong incentives to not only make sure their medicines are cheaply available, but that they are effectively distributed and administered to those who need them.

The HIF is also politically attractive. It is an optional complement to the existing patent system; while not threatening existing sources of profits, the HIF offers pharmaceutical companies a new source of revenue from the development of drugs for diseases that primarily afflict the poor. Pharmaceutical companies therefore will be able to support this proposal while meeting their fiduciary obligation to shareholders to maximize profits. The HIF thereby aligns the interests of the poor, who desperately need essential medicines, with those of pharmaceutical companies, whose support is probably necessary for any reform proposal to succeed.

Pharmaceutical companies are likely to prefer the HIF because it is general, rather than disease-specific, and allows companies to apply their comparative advantage to find the most cost-effective ways to reduce GBD without worrying about whether or not their innovations will satisfactorily meet predetermined criteria. Because the HIF offers a general reward and is backed by international funding commitments, it is more stable over the long term, encouraging pharmaceutical companies to invest in R&D toward whatever is most likely to achieve effective outcomes. Since the HIF is scalable, it is viable at a range of funding levels. Over time, if the HIF proves to be an effective mechanism for incentivizing innovation and access, countries will devote to it a higher percentage of their GNI, increasing the amount of innovation that it can incentivize (Hollis, 2008). The HIF appeals to the interests of political leaders and stakeholders, making it a politically realistic option. Since the greatest beneficiaries of the HIF will be the poor in developing countries, these advantages are a boon for them, too.

By offering a politically realistic mechanism for expanding access to essential medicines and incentivizing innovation for diseases primarily afflicting the poor in developing countries, the HIF offers the most promising reform of the existing patent system. However, proponents note that there remain some significant obstacles to the implementation of the HIF (Hollis, 2007c: 17–22; Hollis, 2008; Pogge, 2008a: 253–256). First, the mechanism relies on three measurements that must be further specified: the measure of the actual incremental health impact of a particular drug on the real and projected disease burden, the accurate and consistent quantification of the health impact of a particular drug in terms of QALYs, and the measure of the appropriate reward amount per QALY. For the HIF to be effective and fair, it must be able to devise a consistent and reliable measure of these three variables.[5] Second, the HIF must design mechanisms to avoid gaming by pharmaceutical companies, who may attempt to exaggerate the health impact of a new drug in clinical tests and in the field. Third, the HIF must establish strong funding commitments from states over the long term in order to provide reliable assurances to potential innovators. Finally, alternative mechanisms must be created for orphan diseases that are unlikely to benefit from the innovation incentivized by the HIF.

5. CONCLUSION

The pharmaceutical patent system has contributed significantly to a global health crisis that continues to deteriorate. Because this crisis is partially the result of policies instituted primarily by affluent countries, and results in so much avoidable human suffering, we have a strong moral duty to seek alternatives. Any successful reform of the current system must increase access to essential medicines among the poor and incentivize innovation of new medicines for diseases that predominantly afflict developing countries. So long as the existing patent system links rewards for innovation only to the price system, a solution to this crisis is unlikely to be found. However, it is possible to create a mechanism that will reward innovative firms based on the actual health impact of their innovations, rather than the high prices that patented drugs can fetch. To succeed, such a mechanism must expand access and innovation, but it must also be efficient, sustainable and politically feasible. The best option for achieving these goals appears to be the Health Impact Fund. It is hoped that further development of this proposal will yield a viable and successful complement to the existing pharmaceutical patent system.

---

[5] Hollis (2008) addresses these issues. Additional complexities of these measurements are discussed in Selgelid (2008).

ACKNOWLEDGEMENTS

I am grateful to Thomas Pogge, Aidan Hollis, Matt Peterson, Kieran Donaghue, Anna Ravvin and the editors for helpful comments on an earlier draft.

WORKS CITED

D. Baker (2004) 'Financing Drug Research: What Are the Issues?' *Center for Economic and Policy Research*, (available at http://www.cepr.net/documents/publications/intellectual_property_2004_09.pdf, accessed April 14th 2008).

E. Berndt, R. Glennerster, M. Kremer, J. Lee, R. Levine, G. Weizsacker, and H. Williams (2007) 'Advance Market Commitments for Vaccines Against Neglected Diseases: Estimating Costs and Effectiveness,' *Health Economics*, 16: 491–511.

Center for Global Development (CGD) (2005) *Making Markets for Vaccines: Ideas to Action.* (Washington, DC: Center for Global Development), (available at http://www.cgdev.org/doc/books/vaccine/Making-Markets-complete.pdf, accessed April 14, 2008).

P. Chirac, and E. Torreele (2006) 'Global Framework on Essential Health R&D,' *The Lancet*, 367: 1560–1561.

P. Danzon, and A. Towse (2003) 'Differential Pricing for Pharmaceuticals: Reconciling Access, R&D and Patents,' *International Journal of Health Care Finance and Economics*, 3: 183–205.

J. DiMasi, R. Hansen, and H. Grabowski (2003) 'The Price of Innovation: New Estimates of Drug Development Costs,' *Journal of Health Economics*, 22: 151–185.

A. Farlow, D. Light, R. Mahoney, and R. Widdus (2005) 'Concerns Regarding the Center for Global Development Report 'Making Markets for Vaccines'. Submission to WHO Commission on Intellectual Property Rights, Innovation and Public Health,' (available at http://www.who.int/intellectualproperty/submissions/Vaccines.FarlowLight.pdf, accessed April 14th 2008).

A. Faunce, and H. Nasu (2008) 'Three Proposals for Rewarding Novel Health Technologies Benefiting People Living in Poverty: A Comparative Analysis of Prize Funds, Health Impact Funds and a Cost-Effectiveness/Competitive Tender Treaty,' *Public Health Ethics*, 1 (2):146–153.

Food and Drug Administration Amendments Act of 2007 (2007) Signed into law September 27, 2007 as Public Law 110–85, (available at http://frwebgate. access.gpo.gov/cgi-bin/getdoc.cgi?dbname=110 cong public laws&docid=f:publ085.110.pdf, accessed 14 April 2008).

A. Hollis (2005a) 'An Efficient Reward System for Pharmaceutical Innovation,' manuscript, (available at www.econ.ucalgary.ca/fac-files/ah/drugprizes.pdf, accessed April 14th 2008).

A. Hollis (2005b) 'Optional Rewards for New Drugs for Developing Countries,' manuscript, (available at www.who.int/entity/intellectualproperty/submissions/Submissions.AidanHollis.pdf, accessed April 14th 2008).

A. Hollis (2006) 'Neglected Disease Research: Health Needs and New Models for R&D,' in J. Cohen, P. Illingworth, and U. Schuklenk (eds.), *The Power of Pills* (London: Pluto Press) 125–133.

A. Hollis (2007a) 'Drugs for Neglected Diseases: New Incentives for Innovation,' in F. Sloan, and C. Hsieh (eds.), *Pharmaceutical Innovation: Incentives, Competition, and Cost-Benefit Analysis in International Perspective* (Cambridge: Cambridge University Press) 75–90.

A. Hollis (2007b) 'Incentive Mechanisms for Innovation,' manuscript, (available at www.iapr.ca/ iapr/files/iapr/iapr-tp-07005 0.pdf, accessed April 14th 2008).

A. Hollis (2007c) 'A Comprehensive Advance Market Commitment: A Useful Supplement to the Patent System?' manuscript, (available at http://www.patent2.org/files/PHE Hollis CAMC.pdf, accessed April 14th 2008).

A. Hollis (2008) 'The Health Impact Fund: A useful Supplement to the Patent System?' *Public Health Ethics*, 1: 124–133.

J. Johnston, and A. Wasunna (2007) 'Patents, Biomedical Research, and Treatments: Examining Concerns, Canvassing Solutions,' *Hastings Center Report*, 37: S1–S36.

M. Kremer, and R. Glennerster (2004) '*Strong Medicine: Creating Incentives for Pharmaceutical Research on Neglected Diseases,*' (Princeton, NJ: Princeton University Press).

T. Pogge (2008a) *World Poverty and Human Rights*, 2nd ed. (Cambridge: Polity).

T. Pogge (2008b). 'Medicines for the World: Boosting Innovation without Obstructing Free Access,' *Sur: Revista Internacional de Direitos Humanos*, 8, (available at http://www.patent2.org/files/sur6.pdf, accessed April 14th 2008).

D. Ridley, H. Grabowski, and J. Moe (2006) 'Developing Drugs for Developing Countries,' *Health Affairs*, 25: 313–324.

M. Rimmer (2008) 'Race Against Time: The Export of Essential Medicines to Rwanda,' *Public Health Ethics*, 2: 88–102.

E. Schwartz, and J. Hsu (2007) 'A Model of R&D Valuation and the Design of Research Incentives (with a Case Study on Vaccine Development),' (available at http://ssrn.com/abstract=1079609, accessed April 14th 2008).

M. Selgelid (2008) 'A Full-Pull Program for the Provision of Pharmaceuticals: Practical Issues,' *Public Health Ethics*, 2: 133–144.

S. Stolberg (2008) 'Global Battle on AIDS, Bush Creates Legacy,' New York Times, January 5th 2008, (available at http://www.nytimes.com/2008/01/05/washington/05aids.html?r=1&scp=1&sq=global+battle+aids+buh+legacy&oref=slogin, accessed April 14th 2008).

P. Trouiller, P. Olliaro, E. Torreele, J. Orbinski, R. Laing, and N. Ford (2002) 'Drug Development for Neglected Diseases: A Deficient Market and a Public-Health Policy Failure,' *The Lancet*, 359: 2188–2194.

US Food and Drug Administration (2005) 'CDER NDAs Approved in Calendar Years 1990–2004 by Therapeutic Potential and Chemical Types,' (available at http://www.fda.gov/cder/rdmt/pstable.htm, accessed April 14th 2008).

World Health Organization Commission on Intellectual Property Rights, Innovation, and Public Health. (2006a) *Public Health, Innovation and Intellectual Property Rights: Report of the Commission on Intellectual Property Rights, Innovation and Public Health*, (available at http://www.who.int/ intellectual-property/documents/thereport/ENPublic- HealthReport.pdf, accessed April 14th 2008).

World Health Organization (2006b) 'Evaluation of WHO's Contribution to '3 by 5',' (available at http:// www.who.int/hiv/topics/me/3by5%20Evaluation.pdf, accessed April 14th 2008).

World Health Organization Intergovernmental Working Group on Public Health, Innovation and Intellectual Property (2007) 'Draft Global Strategy and Plan of Action on Public Health, Innovation and Intellectual Property,' (available at http://www.who.int/gb/phi/pdf/igwg2/PHI_IGWG2_CP1Rev1-en.pdf, accessed April 14th 2008).

World Trade Organization (WTO) (2001) 'Declaration on the TRIPS Agreement and Public Health, Ministerial Conference,' (available at http://www.wto.org/English/thewto_e/minist_e/ min01_e/mindecl_trips_e.htm, accessed April 14th 2008).

World Trade Organization (2003) 'Implementation of Paragraph 6 of the Doha Declaration on the TRIPS Agreement and Public Health,' (available at http://www.wto.org/english/tratop_e/trips_e/implem para6_e.htm accessed April 14th 2008).

DISCUSSION QUESTIONS

1. Do you believe that citizens and/or governments of affluent nations have a responsibility help promote development of pharmaceutical technologies and dissemination mechanisms that would benefit the very poor in the world? Why or why not?

2. What do you think the most effective arguments are to convince the public that they should subsidize a system that would promote the development of drugs to alleviate the disease burden in other countries?

3. The author argues that push mechanisms are not an effective way of promoting innovation of technologies that would benefit the poor in developing nations. What reasons does he give for these conclusions? Do you think that he is correct? Why or why not?

4. Do you agree that the Health Impact Fund is the best approach to stimulating research on neglected diseases and making drugs more affordable for people in developing nations? Why or what not?

5. What barriers do you think there are to implementing a system like the Health Impact Fund? Do you think that pharmaceutical companies would have any reasons to resist it?

6. Do you think the case of pharmaceutical innovation and the poor in developing nations provides any general lessons on the relationship between technological innovation and market incentives? Can you think of other cases where free market mechanisms alone have been inadequate to deliver basic goods and services to people?

PART IV

HUMAN ENHANCEMENT TECHNOLOGIES

Human enhancement technologies are technologies that amplify human capacities well beyond what people are capable of when technologically unassisted or that provide people with completely novel capacities. Human enhancement technologies are nothing new. In fact, people have always technologically augmented their cognitive, physical, perceptual, and psychological capabilities through technology. Computational devices, optical lenses, pharmaceuticals, and communication systems are just a few examples of common enhancement technologies. They enable us to do and know things that we would not otherwise be capable of doing or knowing.

Enhancement technologies can be differentiated according to their magnitude (i.e. how large or novel the enhancement), reversibility (i.e. how readily the enhancement can be removed or turned off), and internality (i.e. the extent to which the technology integrates with or modifies a person's body). For example, personal computers significantly augment human cognitive and communication abilities – i.e. the magnitude of enhancement associated with them is large. However, they are highly reversible, since they are easily turned off; and they are external, in that they do not integrate with or modify the user's brain and body. In contrast, anabolic steroids provide a more modest enhancement than do computers, but they are also less reversible, since the technology persists some time after they intervention; and they are internal, since they work by modifying the users body. An enhancement is more robust, the greater its magnitude, irreversibility, and internality.

The chapters in this section consider the social and ethical issues associated with emerging and highly robust enhancement technologies, such has gene-based and intensive longitudinal pharmacological interventions. In *Beyond Therapy*, the United States President's Council on Bioethics raises several concerns regarding the use of biotechnology for enhancement purposes. These concerns include those that they call "familiar," such fairness, equality of access, coercion, and safety, as well as those that they call "essential," such as detrimental impacts on the parent–child relationships, human dignity, and individuality/identity.

In "Why I Want to be a Posthuman when I Grow Up," Nick Bostrom argues that it could be very good to be technologically enhanced to become a posthuman – i.e. to possess a "central capacity greatly exceeding the maximum attainable by any current human being" – with respect to healthspan, cognition, and emotion. Therefore, he argues, it is reasonable to want to become posthuman.

In "Moral Enhancement," Thomas Douglas focuses on a particular type of enhancement, the use of biotechnologies to alter emotions and affect motives in order to make people morally better. Douglas argues that in some cases people have good reasons to engage in moral enhancement and that it is permissible that they do so. In the course of developing his argument, Douglas considers and responds to several concerns raised in *Beyond Therapy*.

In "Enhancing Justice?" Tamara Garcia and I evaluate robust enhancement technologies from the perspective of social justice – i.e. whether they are likely to reduce, perpetuate, or increase existing unjust inequalities. We argue that, given the features of the contexts into which they are emerging, these enhancement technologies are likely to exacerbate existing injustices. For this reason, those supportive of the technologies must also advocate for addressing unjust inequalities in areas such as education and health care.

Taken together, the chapters in this section present and evaluate a wide range of concerns about and considerations in support of engineering ourselves in order to dramatically augment our capabilities. More detailed summaries, as well as a listing of related readings, are located at the start of the chapters.

"BEYOND THERAPY"[1]

United States President's Council on Bioethics

CHAPTER SUMMARY

In this chapter, the Presidential Commission on Bioethics identifies several possible problems with using biotechnologies for enhancement, rather than therapeutic, purposes. Several of the potential difficulties are familiar, in that they arise regarding health care and medical technologies generally – for example, safety, justice, and coercion. Several others are more particular to the power that novel biotechnologies provide and the motivation for and practice of trying to use that technology to make ourselves better than well. These include detrimental impacts on the parent–child relationship, undermining human dignity, compromising individuality and identity, and fostering an impoverished understanding of human flourishing.

RELATED READINGS

Introduction: Innovation Presumption (2.1); Power (2.4); Form of Life (2.5); EHS (2.6.1); Justice, Access, and Equality (2.6.2); Intrinsic Objections (2.7)

Other Chapters: S. Matthew Liao, *Selecting Children* (Ch. 8); Walter Glannon, *Psychopharmacology and Functional Neurosurgery* (Ch. 12); Thomas Douglas, *Moral Enhancement* (Ch. 16); Garcia and Sandler, *Enhancing Justice?* (Ch. 17)

1. INTRODUCTION

The four preceding chapters have examined how several prominent and (generally) salutary human pursuits may be aided or altered using a wide variety of biotechnologies that lend themselves to purposes "beyond therapy." In each case, we have discussed the character of the end, considered the novel means, and explored some possible implications, ethical and social. In surveying the pertinent technologies, we have taken a somewhat long-range view, looking at humanly significant technical possibilities that may soon – or not so soon – be available for general use, yet at the same time trying to separate fact from science fiction. In offering ethical analysis, we have tried to identify key issues pertinent to the case under discussion, asking questions about both ends and means, and looking

---

[1] This chapter is excerpted from President's Council on Bioethics (2003) *Beyond Therapy: Biotechnology and the Pursuit of Happiness* (Washington, DC).

always for the special significance of pursuing the old human ends by these new technological means. In this concluding chapter, we step back from the particular "case studies" to pull together some common threads and to offer some generalizations and conclusions to which the overall inquiry has led.

2. THE BIG PICTURE

The first generalization concerns the wide array of biotechnologies that are, or may conceivably be, useful in pursuing goals beyond therapy. Although not originally developed for such uses, the available and possible techniques we have considered – techniques for screening genes and testing embryos, choosing sex of children, modifying the behavior of children, augmenting muscle size and strength, enhancing athletic performance, slowing senescence, blunting painful memories, and brightening mood – do indeed promise us new powers that can serve age-old human desires. True, in some cases, the likelihood that the new technologies will be successfully applied to those purposes seems, at least for the foreseeable future, far-fetched: genetically engineered "designer babies" are not in the offing. In other cases, as with psychotropic drugs affecting memory, mood, and behavior, some uses beyond therapy are already with us. In still other cases, such as research aimed at retarding senescence, only time will tell what sort of powers may become available for increasing the maximum human lifespan, and by how much. Yet the array of biotechnologies potentially useful in these ventures should not be underestimated, especially when we consider how little we yet know about the human body and mind and how much our knowledge and technique will surely grow in the coming years. Once we acquire technical tools and the potential for their use based on fuller knowledge, we will likely be able to intervene much more knowingly, competently, and comprehensively.

Second, despite the heterogeneity of the techniques, the variety of purposes they may serve, and the different issues raised by pursuing these differing purposes by diverse means, we believe that all of these matters deserve to be considered together, just as we have done in this report. Notwithstanding the multiplicity of ends, means, and consequences that we have considered, this report offers less a list of many things to think about than a picture of *one big thing* to think about: the dawning age of biotechnology and the greatly augmented power it is providing us, not only for gaining better health but also for improving our natural capacities and pursuing our own happiness. The ambitious project for the mastery of nature, the project first envisioned by Francis Bacon and René Descartes in the early seventeenth century, is finally yielding its promised abilities to relieve man's estate – and then some. Though our society will, as a matter of public practice, be required to deal with each of these techniques and possibilities as they arrive, piecemeal and independently of one another, we should, as a matter of public understanding, try to see what they might all add up to, taken together. The Council's experience of considering these disparate subjects under this one big idea – "beyond therapy, for the pursuit of happiness" – and our discovery of overlapping ethical implications would seem to vindicate the starting assumption that led us to undertake this project in the first place: *biotechnology beyond therapy deserves to be examined not in fragments, but as a whole.*

Yet, third, the "whole" that offers us the most revealing insights into this subject is not itself technological. For the age of biotechnology is not so much about technology itself as it is about *human beings empowered by biotechnology*. Thus, to understand the human and social meaning of the new age, we must begin not from our tools and products but from where human beings begin, namely, with the very human desires that we have here identified in order to give shape to this report: desires for better children, superior performance, younger and more beautiful bodies, abler minds, happier souls. Looking at the big picture through this lens keeps one crucial fact always in focus: how people exploit the relatively unlimited uses of biotechnical power will be decisively determined by the perhaps still more unlimited desires of human beings, especially – and this is a vital point – as these desires themselves become transformed and inflated by the new technological powers they are all the while acquiring. Our desires to alter our consciousness or preserve our youthful strength, perhaps but modest to begin with, could swell considerably if and when we become more technically able to satisfy them. And as they grow, what would have been last year's satisfaction will only fuel this year's greater hunger for more.

Fourth, as the ubiquitous human desires are shaped and colored not only reactively by the tools that might serve them but also directly by surrounding cultural and social ideas and practices, the "one big picture" will be colored by the (albeit changeable) ruling opinions, mores, and institutions of the society in which we live and into which the technologies are being introduced. For example, the desire for performance-enhancing drugs will be affected by the social climate regarding competition; the eagerness to gain an edge for one's children will be affected by whether many other parents are doing so; and the willingness to use or forego medication for various sorts of psychic distress will be affected by the poverty or richness of private life, and the degree to which strong family or community support is (or is not) available for coping with that distress directly. Moreover, in a free and pluralistic society, we may expect a very diverse popular reaction to the invitation of the new technologies, ranging from exuberant enthusiasm to outright rejection, and the overall public response cannot be judged in advance. Yet because the choices made by some can, in their consequences, alter the shared life lived by all, it behooves all of us to consider the meaning of these developments, whether we are privately tempted by them or not. It is in part to contribute to a more thoughtful public appraisal of these possibilities that we have undertaken this report.

By beginning with the common human desires, we have sought to place what may be new and strange into a context provided by what is old and familiar. We recognize the temptation to add biotechnological means to our "tool kits" for pursuing happiness and self-improvement, and it is not difficult to appreciate, at least at first glance, the attractiveness of the goods being contemplated. We want to give our children the best start in life and every chance to succeed. We want to perform at our best, and better than we did before. We want to remain youthful and vigorous for as long as we can. We want to face life optimistically and with proper self-regard. And since we now avail ourselves of all sorts of means toward these ends, we will certainly not want to neglect the added advantages that biotechnologies may offer us, today and tomorrow.

At the same time, however, we have identified, in each of the previous four chapters, several reasonable sources of concern, ethical and social. And, in each case, we have called attention to some of the possible hidden costs of success, achieved by employing these means. The chapter on better children raised questions about the meaning and limits of parental control and about the character and rearing of children. The chapter on superior performance raised questions about the meaning of excellence and the "humanity" of human activity. The chapter on ageless bodies raised questions about the significance of the "natural" life cycle and lifespan, and their connection to the dynamic character of society and the prospects for its invigorating renewal. And the chapter on happy souls raised questions about the connections between experienced mood or self-esteem and the deeds or experiences that ordinarily are their foundation, as well as the connections between remembering truly and personal identity. Looking again at these subjects, now seen as part of "one big picture," we think it useful here to collect and organize the various issues into a semi-complete account, so that the reader may see in outline the most important and likely sources of concern.

Before proceeding, we wish to reiterate our intention in this inquiry, so as to avoid misunderstanding. In offering our synopsis of concerns, we are not making predictions; we are merely pointing to possible hazards, hazards that become visible only when one looks at "the big picture." More important, we are not condemning either biotechnological power or the pursuit of happiness, excellence, or self-perfection. Far from it. We eagerly embrace biotechnologies as aids for preventing or correcting bodily or mental ills and for restoring health and fitness. We even more eagerly embrace the pursuits of happiness, excellence, and self-improvement, for ourselves, our children, and our society. Desires for these goals are the source of much that is good in human life. Yet, as has long been known, these desires can be excessive. Worse, they can be badly educated regarding the nature of their object, sometimes with tragic result: we get what we ask for only to discover that it is very far from what we really wanted. Finally, they can be pursued in harmful ways and with improper means, often at the price of deforming the very goals being sought. To guard against such outcomes, we need to be alert in advance to the more likely risks and the more serious concerns. We begin with those that are more obvious and familiar.

3. FAMILIAR SOURCES OF CONCERN

The first concerns commonly expressed regarding any uses of biotechnology beyond therapy reflect, not surprisingly, the dominant values of modern America: health and safety, fairness and equality, and freedom. The following thumbnail sketches of the issues should suffice to open the questions – though of course not to settle them.

3.1 Health: issues of safety and bodily harm

In our health-conscious culture, the first reason people worry about any new biotechnical intervention, whatever its intended purpose, is safety. This will surely be true regarding "elective" uses of biotechnology that aim beyond therapy.

Athletes who take steroids to boost their strength may later suffer premature heart disease. College students who snort Ritalin to increase their concentration may become addicted. Melancholics taking mood-brighteners to change their outlook may experience impotence or apathy. To generalize: no biological agent used for purposes of self-perfection or self-satisfaction is likely to be entirely safe. This is good medical common sense: anything powerful enough to enhance system A is likely to be powerful enough to harm system B (or even system A itself), the body being a highly complex yet integrated whole in which one intervenes partially only at one's peril. And it surely makes sense, ethically speaking, that one should not risk basic health pursuing a condition of "better than well."

Yet some of the interventions that might aim beyond therapy – for example, genetic enhancement of muscle strength, retardation of aging, or pharmacologic blunting of horrible memories or increasing self-esteem – may, indirectly, lead also to improvements in general health. More important, many good things in life are filled with risks, and free people – even if properly informed about the magnitude of those risks – may choose to run them if they care enough about what they might gain thereby. If the interventions are shown to be *highly* dangerous, many people will (later if not sooner) avoid them, and the Food and Drug Administration or tort liability will constrain many a legitimate would-be producer. But if, on the other hand, the interventions work well and are indeed highly desired, people may freely accept, in trade-off, even considerable risk of later bodily harm for the sake of significant current benefits. Besides, the bigger ethical issues in this area have little to do with safety; the most basic questions concern not the hazards associated with the techniques but the benefits and harms of using the perfected powers, assuming that they may be safely used.

3.2 Unfairness

An obvious objection to the use of enhancement technologies, especially by participants in competitive activities, is that they give those who use them an unfair advantage: blood doping or steroids in athletes, stimulants in students taking the SATs, and so on. This issue, briefly discussed in Chapter Three, has been well aired by the International Olympic Committee and the many other athletic organizations who continue to try to formulate rules that can be enforced, even as the athletes and their pharmacists continue to devise ways to violate those rules and escape detection. Yet as we saw, the fairness question can be turned on its head, and some people see in biotechnical intervention a way to compensate for the "unfairness" of *natural* inequalities – say, in size, strength, drive, or native talent. Still, even if everyone had equal access to genetic improvement of muscle strength or mind-enhancing drugs, or even if these gifts of technology would be used only to rectify the inequalities produced by the unequal gifts of nature, an additional disquiet would still perhaps remain: The disquiet of using such new powers in the first place or at all, even were they fairly distributed. Besides, as we have emphasized, not all activities of life are competitive, and the uses of biotechnologies for purposes beyond therapy are more worrisome on other grounds.[2]

---

[2] This discussion depends heavily on Sandel (2002).

3.3 Equality of access

A related question concerns inequality of access to the benefits of biotechnology, a matter of great interest to many Members of this Council, though little discussed in the previous chapters. The issue of distributive justice is more important than the issue of unfairness in competitive activities, especially if there are systemic disparities between those who will and those who won't have access to the powers of biotechnical "improvement." Should these capabilities arrive, we may face severe aggravations of existing "unfairnesses" in the "game of life," especially if people who need certain agents to treat serious illness cannot get them while other people can enjoy them for less urgent or even dubious purposes. If, as is now often the case with expensive medical care, only the wealthy and privileged will be able to gain easy access to costly enhancing technologies, we might expect to see an ever-widening gap between "the best and the brightest" and the rest. The emergence of a biotechnologically improved "aristocracy" – augmenting the already cognitively stratified structure of American society – is indeed a worrisome possibility, and there is nothing in our current way of doing business that works against it. Indeed, unless something new intervenes, it would seem to be a natural outcome of mixing these elements of American society: our existing inequalities in wealth and status, the continued use of free markets to develop and obtain the new technologies, and our libertarian attitudes favoring unrestricted personal freedom for all choices in private life.

Yet the situation regarding rich and poor is more complex, especially if one considers actual benefits rather than equality or relative well-being. The advent of new technologies often brings great benefits to the less well off, if not at first, then after they come to be mass-produced and mass-marketed and the prices come down. (Consider, over the past half-century, the spread in the United States of refrigerators and radios, automobiles and washing machines, televisions and VCRs, cell phones and personal computers, and, in the domain of medicine, antibiotics, vaccines, and many expensive diagnostic and therapeutic procedures.) To be sure, the gap between the richest and the poorest may increase, but in absolute terms the poor may benefit more, when compared not to the rich but to where they were before. By many measures, the average American today enjoys a healthier, longer, safer, and more commodious life than did many a duke or prince but a few centuries back.

Nevertheless, worries about possible future bioenhanced stratification should not be ignored. And they become more poignant in the present, to the extent that one regards spending money and energy on goals beyond therapy as a misallocation of limited resources in a world in which the basic health needs of millions go unaddressed. Yet although the setting of priorities for research and development is an important matter for public policy, it is not unique to the domain of "beyond therapy." It cannot be addressed, much less solved, in this area alone. Moreover, and yet again, the inequality of access does not remove our uneasiness over the thing itself. It is, to say the least, paradoxical, in discussions of the dehumanizing dangers of, say, future eugenic selection of better children, that people vigorously complain that the poor will be denied equal access to the danger: "The food is contaminated, but why are my portions so small?" Huxley's *Brave New World* runs on a deplorable and impermeably rigid class system, but few people would want to live in that world even if offered the chance to enjoy it as an alpha (the privileged caste). Even an elite can be dehumanized, can dehumanize itself. The questions about access and distributive justice are, no doubt, socially important. Yet the more fundamental ethical questions about taking biotechnology "beyond therapy" concern not equality of access, but the goodness or badness of the things being offered and the wisdom of pursuing our purposes by such means.

3.4 Liberty: issues of freedom and coercion, overt and subtle

A concern for threats to freedom comes to the fore whenever biotechnical powers are exercised by some people upon other people. We encountered it in our discussion of "better children" (the choice of a child's sex or the drug-mediated alteration of his or her behavior; Chapter Two), as well as in the coerced use of anabolic steroids by the East German Olympic swimmers (Chapter Three). This problem will of course be worse in tyrannical regimes. But there are always dangers of despotism within families, as many parents already work their wills on their children with insufficient regard to a child's independence or long-term needs, jeopardizing even the "freedom to be a child." To the extent that even partial control over genotype – say, to take a relatively innocent example, musician parents selecting a child with genes for perfect pitch – would add to existing social instruments of parental control and its risks of despotic rule, this matter will need to be attended to.[3]

Leaving aside the special case of children, the risk of overt coercion does not loom large in a free society. On the contrary, many enthusiasts for using technology for personal enhancement are libertarian in outlook; they see here mainly the enlargement of human powers and possibilities and the multiplication of options for private choice, both of which they see as steps to greater human freedom. They look forward to growing opportunities for more people to earn more, learn more, see more, and do more, and to choose – perhaps several times in one lifetime – interesting new careers or avocations. And they look with suspicion at critics who they fear might want to limit their private freedom to develop and use new technologies for personal advancement or, indeed, for any purpose whatsoever. The coercion they fear comes not from advances in technology but from the state, acting to deny them their right to pursue happiness or self-improvement by the means they privately choose.

Yet no one can deny that people living in free societies, and even their most empowered citizens, already experience more subtle impingements on freedom and choice, operating, for example, through peer pressure. What is freely permitted and widely used may, under certain circumstances, become practically mandatory. If most children are receiving memory enhancement or stimulant drugs, failure to provide them for your child might be seen as a form of child neglect. If all the defensive linemen are on steroids, you risk mayhem if you go against them chemically pure. And, a point subtler still, some critics complain that, as with cosmetic surgery, Botox, and breast implants, many of the enhancement technologies of the

---

[3] The discussion that follows depends heavily on Kass (2003).

future will very likely be used in slavish adherence to certain socially defined and merely fashionable notions of "excellence" or improvement, very likely shallow and conformist. If these fears are realized, such exercises of individual freedom, suitably multiplied, might compromise the freedom to be an individual.[4]

This special kind of reduction of freedom – let's call it the problem of conformity or homogenization – is of more than individual concern. In an era of mass culture, itself the byproduct of previous advances in communication, manufacture, and marketing techniques, the exercise of un-coerced private choices may produce untoward consequences for society as a whole. Trends in popular culture lead some critics to worry that the self-selected nontherapeutic uses of the new biotechnical powers, should they become widespread, will be put in the service of the most common human desires, moving us toward still greater homogenization of human society – perhaps raising the floor but also lowering the ceiling of human possibility, and reducing the likelihood of genuine freedom, individuality, and greatness. (This is an extension of Tocqueville's concern about the leveling effects of democracy, now possibly augmented by the technological power to make those effects ingrained and perhaps irreversible.)

Indeed, such constriction of individual possibility could be the most important society-wide concern, if we consider the aggregated effects of the likely individual choices for biotechnical "self-improvement," each of which might be defended or at least not objected to on a case-by-case basis (the problem of what the economists call "negative externalities"). For example, it might be difficult to object to a personal choice for a life-extending technology that would extend the user's life by three healthy decades or a mood-brightened way of life that would make the individual more cheerful and untroubled by the world around him. Yet as we have suggested more than once, the aggregated social effects of such choices, widely made, could lead to a Tragedy of the Commons, where benefits gained by individuals are outweighed by the harms that return to them from the social costs of allowing everyone to share the goodies. And, as Huxley strongly suggests in *Brave New World*, when biotechnical powers are readily available to satisfy short-term desires or to produce easy contentment, the character of human striving changes profoundly and the desire for human excellence fades. Should this come to pass, the best thing to be hoped for might be the preservation of pockets of difference (as on the remote islands in *Brave New World*) where the desire for high achievement has not been entirely submerged or eroded.

4. ESSENTIAL SOURCES OF CONCERN

Our familiar worries about issues of safety, equality, and freedom, albeit very important, do not exhaust the sources of reasonable concern. When richly considered, they invite us to think about the deeper purposes for the sake of which we want to live safely, justly, and freely. And they enable us to recognize that even the safe, equally available, non-coerced and non-faddish uses of biomedical technologies to pursue happiness or self-improvement raise ethical and social questions,

---

[4]   See, for example, Jonas (1992); Kass (2002).

questions more directly connected with the essence of the activity itself: the use of technological means to intervene into the human body and mind, not to ameliorate their diseases but to change and improve their normal workings. Why, if at all, are we bothered by the voluntary *self-administration* of agents that would change our bodies or alter our minds? What is disquieting about our attempts to improve upon human nature, or even our own particular instance of it?

The subject being relatively novel, it is difficult to put this worry into words. We are in an area where initial revulsions are hard to translate into sound moral arguments. Many people are probably repelled by the idea of drugs that erase memories or that change personalities, or of interventions that enable seventy-year-olds to bear children or play professional sports, or, to engage in some wilder imaginings, of mechanical implants that would enable men to nurse infants or computer-brain hookups that would enable us to download the *Oxford English Dictionary*. But can our disquiet at such prospects withstand rational, anthropological, or ethical scrutiny? Taken one person at a time, with a properly prepared set of conditions and qualifications, it will be hard to say what is wrong with any biotechnical intervention that could improve our performances, give us (more) ageless bodies, or make it possible for us to have happier souls. Indeed, in many cases, we ought to be thankful for or pleased with the improvements our biotechnical ingenuity is making possible.

If there are essential reasons to be concerned about these activities and where they may lead us, we sense that it may have something to do with challenges to what is naturally human, what is humanly dignified, or to attitudes that show proper respect for what is naturally and dignifiedly human. As it happens, at least four such considerations have already been treated in one place or another in the previous chapters: appreciation of and respect for "the naturally given," threatened by hubris; the dignity of human activity, threatened by "unnatural" means; the preservation of identity, threatened by efforts at self-transformation; and full human flourishing, threatened by spurious or shallow substitutes.

4.1 Hubris or humility: respect for "the given"

A common, man-on-the-street reaction to the prospects of biotechnological engineering beyond therapy is the complaint of "man playing God." If properly unpacked, this worry is in fact shared by people holding various theological beliefs and by people holding none at all. Sometimes the charge means the sheer prideful presumption of trying to alter what God has ordained or nature has produced, or what should, for whatever reason, not be fiddled with. Sometimes the charge means not so much usurping Godlike powers, but doing so in the absence of God-like knowledge: the mere playing at being God, the hubris of acting with insufficient wisdom.

Over the past few decades, environmentalists, forcefully making the case for respecting Mother Nature, have urged upon us a "precautionary principle" regarding all our interventions into the natural world. Go slowly, they say, you can ruin everything. The point is certainly well taken in the present context. The human body and mind, highly complex and delicately balanced as a result of eons of gradual and exacting evolution, are almost certainly at risk from any ill considered attempt at "improvement." There is not only the matter of unintended consequences, a concern even with interventions aimed at therapy. There is also the matter of uncertain goals and absent natural standards, once one proceeds "beyond therapy." When a physician intervenes therapeutically to correct some deficiency or deviation from a patient's natural wholeness, he acts as a servant to the goal of health and as an assistant to nature's own powers of self-healing, themselves wondrous products of evolutionary selection. But when a bioengineer intervenes for nontherapeutic ends, he stands not as nature's servant but as her aspiring master, guided by nothing but his own will and serving ends of his own devising. It is far from clear that our delicately integrated natural bodily powers will take kindly to such impositions, however desirable the sought-for change may seem to the intervener. And there is the further question of the unqualified good-ness of the goals being sought, a matter to which we shall return.

One revealing way to formulate the problem of hubris is what one of our Council Members has called the temptation to "hyper-agency," a Promethean aspiration to remake nature, including human nature, to serve our purposes and to satisfy our desires. This attitude is to be faulted not only because it can lead to bad, unintended consequences; more fundamentally, it also represents a false understanding of, and an improper disposition toward, the naturally given world. The root of the difficulty seems to be both cognitive and moral: the failure prop-erly to appreciate and respect the "giftedness" of the world. Acknowledging the giftedness of life means recognizing that our talents and powers are not wholly our own doing, nor even fully ours, despite the efforts we expend to develop and to exercise them. It also means recognizing that not everything in the world is open to any use we may desire or devise. Such an appreciation of the giftedness of life would constrain the Promethean project and conduce to a much-needed humility. Although it is in part a religious sensibility, its resonance reaches beyond religion.

Human beings have long manifested both wondering appreciation for nature's beauty and grandeur and reverent awe before nature's sublime and mysterious power. From the elegance of an orchid to the splendor of the Grand Canyon, from the magnificence of embryological development to the miracle of sight or consciousness, the works of nature can still inspire in most human beings an attitude of respect, even in this age of technology. Nonetheless, the absence of a respectful attitude is today a problem in some – though by no means all – quarters of the biotechnical world. It is worrisome when people act toward, or even talk about, our bodies and minds – or human nature itself – as if they were mere raw material to be molded according to human will. It is worrisome when people speak as if they were wise enough to redesign human beings, improve the human brain, or reshape the human life cycle. In the face of such hubristic temptations, appreciating that the given world – including our natural powers to alter it – is not of our own making could induce a welcome attitude of modesty, restraint, and humility. Such a posture is surely recommended for anyone inclined to modify human beings or human nature for purposes beyond therapy.

Yet the respectful attitude toward the "given," while both necessary and desir-able as a restraint, is not by itself sufficient as a guide. The "giftedness of nature" also includes smallpox and malaria, cancer and Alzheimer disease, decline and decay. Moreover, nature is not equally generous with her gifts, even to man, the most gifted of her creatures. Modesty born of gratitude for the world's "givenness" may enable us to recognize that not everything in the world is open to any use we may desire or devise, but it will not *by itself* teach us *which* things can be tinkered with and which should be left inviolate. Respect for the "giftedness" of things cannot tell us which gifts are to be accepted as is, which are to be improved through use or training, which are to be housebroken through self-command or medication, and which opposed like the plague.

To guide the proper use of biotechnical power, we need something in addition to a generalized appreciation for nature's gifts. We would need also a particular regard and respect for the special gift that is our own given nature. For only if there is a *human* "givenness," or a given humanness, that is also good and worth respecting, either as we find it or as it could be perfected *without ceasing to be itself*, will the "given" serve as a *positive* guide for choosing what to alter and what to leave alone. Only if there is something precious in our given human nature – beyond the fact of its giftedness – can what is given guide us in resisting efforts that would degrade it. When it comes to human biotechnical engineering beyond therapy, only if there is something inherently good or dignified about, say, natural procreation, the human life cycle (with its rhythm of rise and fall), and human erotic longing and striving; only if there is something inherently good or dignified about the ways in which we engage the world as spectators and appreciators, as teachers and learners, leaders and followers, agents and makers, lovers and friends, parents and children, citizens and worshippers, and as seekers of our own special excellence and flourishing in whatever arena to which we are called – only then can we begin to see why those aspects of our nature need to be defended against our deliberate redesign.

We must move, therefore, from the danger of hubris in the powerful designer to the danger of degradation in the designed, considering how any proposed improvements might impinge upon the nature of the one being improved. With the question of human nature and human dignity in mind, we move to questions of means and ends.

4.2 "Unnatural" means: the dignity of human activity

Until only yesterday, teaching and learning or practice and training exhausted the alternatives for acquiring human excellence, perfecting our natural gifts through our own efforts. But perhaps no longer: biotechnology may be able to do nature one better, even to the point of requiring less teaching, training, or practice to permit an improved nature to shine forth. As we noted earlier, the insertion of the growth-factor gene into the muscles of rats and mice bulks them up and keeps them strong and sound without the need for nearly as much exertion. Drugs to improve alertness (today) or memory and amiability (tomorrow) could greatly relieve the need for exertion to acquire these powers, leaving time and effort for better things. What, if anything, is disquieting about such means of gaining improvement?

The problem cannot be that they are "artificial," in the sense of having man-made origins. Beginning with the needle and the fig leaf, man has from the start been the animal that uses art to improve his lot by altering or adding to what nature alone provides. Ordinary medicine makes extensive use of similar artificial means, from drugs to surgery to mechanical implants, in order to treat disease. If the use of artificial means is absolutely welcome in the activity of healing, it cannot be their unnaturalness alone that disquiets us when they are used to make people "better than well."

Still, in those areas of human life in which excellence has until now been achieved only by discipline and effort, the attainment of similar results by means of drugs, genetic engineering, or implanted devices looks to many people (including some Members of this Council) to be "cheating" or "cheap." Many people believe that each person should work hard for his achievements. Even if we prefer the grace of the natural athlete or the quickness of the natural mathematician – people whose performances deceptively appear to be effortless – we admire also those who overcome obstacles and struggle to try to achieve the excellence of the former. This matter of character – the merit of disciplined and dedicated striving – is surely pertinent. For character is not only the source of our deeds, but also their product. As we have already noted, healthy people whose disruptive behavior is "remedied" by pacifying drugs rather than by their own efforts are not learning self-control;† if anything, they may be learning to think it unnecessary. People who take pills to block out from memory the painful or hateful aspects of a new experience will not learn how to deal with suffering or sorrow. A drug that induces fearlessness does not produce courage.

Yet things are not so simple. Some biotechnical interventions may assist in the pursuit of excellence without in the least cheapening its attainment. And many of life's excellences have nothing to do with competition or overcoming adversity. Drugs to decrease drowsiness, increase alertness, sharpen memory, or reduce distraction may actually help people interested in their natural pursuits of learning or painting or performing their civic duty. Drugs to steady the hand of a neurosurgeon or to prevent sweaty palms in a concert pianist cannot be regarded as "cheating," for they are in no sense the source of the excellent activity or achievement. And, for people dealt a meager hand in the dispensing of nature's gifts, it should not be called cheating or cheap if biotechnology could assist them in becoming better equipped – whether in body or in mind.

Nevertheless, as we suggested at some length in Chapter Three, there remains a sense that the "naturalness" of means matters. It lies not in the fact that the assisting drugs and devices are artifacts, but in the danger of violating or deforming the nature of human agency and the dignity of the naturally human way of activity. In most of our ordinary efforts at self-improvement, whether by practice, training, or study, we sense the relation between our doings and the resulting improvement, between the means used and the end sought. There is an experiential and intelligible connection between means and ends; we can see how confronting fearful things might eventually enable us to cope with our fears. We can see how curbing our appetites produces self-command. Human education ordinarily proceeds by speech or symbolic deeds, whose meanings are at least in principle directly accessible to those upon whom they work.

In contrast, biotechnical interventions act directly on the human body and mind to bring about their effects on a passive subject, who plays little or no role at all. He can at best *feel* their effects *without understanding their meaning in human terms.* Thus, a drug that brightened our mood would alter us without our understanding how and why it did so – whereas a mood brightened as a fitting response to the arrival of a loved one or to an achievement in one's work is perfectly, because humanly, intelligible. And not only would this be true about our states of mind. All of our encounters with the world, both natural and interpersonal, would be mediated, filtered, and altered. Human experience under biological intervention becomes increasingly mediated by unintelligible forces and vehicles, separated from the human significance of the activities so altered. The relations between the knowing subject and his activities, and between his activities and their fulfillments and pleasures, are disrupted.

The importance of human effort in human achievement is here properly acknowledged: the point is less the exertions of good character against hardship, but the manifestation of an alert and self-experiencing agent making his deeds flow intentionally from his willing, knowing, and embodied soul. If human flourishing means not just the accumulation of external achievements and a full curriculum vitae but a lifelong being-at-work exercising one's *human* powers *well* and without great impediment, our genuine happiness requires that there be little gap, if any, between the dancer and the dance.

4.3 Identity and individuality

With biotechnical interventions that skip the realm of intelligible meaning, we cannot really own the transformations nor can we experience them as genuinely ours. And we will be at a loss to attest whether the resulting conditions and activities of our bodies and our minds are, in the fullest sense, our own as human. But our interest in identity is also more personal. For we do not live in a generic human way; we desire, act, flourish, and decline *as ourselves,* as individuals. To be human is to be someone, not anyone – with a given nature (male or female), given natural abilities (superior wit or musical talent), and – most important – a real history of attachments, memories, and experiences, acquired largely by living with others.

In myriad ways, new biotechnical powers promise (or threaten) to transform what it means to be an individual: giving increased control over our identity to others, as in the case of genetic screening or sex selection of offspring by parents; inducing psychic states divorced from real life and lived experience; blunting or numbing the memories we wish to escape; and achieving the results we could never achieve unaided, by acting as ourselves alone.

To be sure, in many cases, biomedical technology can restore or preserve a real identity that is slipping away: keeping our memory intact by holding off the scourge of Alzheimer disease; restoring our capacity to love and work by holding at bay the demons of self-destroying depression. In other cases, the effect of biotechnology on identity is much more ambiguous. By taking psychotropic drugs to reduce anxiety or overcome melancholy, we may become the person we always wished to be – more cheerful, ambitious, relaxed, content. But we also become a different person in the eyes of others, and in many cases we become dependent on the continued use of psychotropic drugs to remain the new person we now are.

As the power to transform our native powers increases, both in magnitude and refinement, so does the possibility for "self-alienation" – for losing, confounding, or abandoning our identity. I may get better, stronger, and happier – but I know not how. I am no longer the agent of self-transformation, but a passive patient of transforming powers. Indeed, to the extent that an achievement is the result of some extraneous intervention, it is detachable from the agent whose achievement it purports to be. "Personal achievements" impersonally achieved are not truly the achievements of persons. That I can use a calculator to do my arithmetic does not make *me* a knower of arithmetic; if computer chips in my brain were to "download" a textbook of physics, would that make *me* a knower of physics? Admittedly, the relation between biological boosters and personal identity is much less clear: if I make myself more alert through Ritalin, or if drugs can make up for lack of sleep, I may be able to learn more using my unimpeded native powers while it is still unquestionably *I* who am doing the learning. And yet, to find out that an athlete took steroids before the race or that a test-taker (without medical disability) took Ritalin before the test is to lessen our regard for the achievement of the doer. It is to see not just an acting self, but a dependent self, one who is less himself for becoming so dependent.

In the deepest sense, to have an identity is to have limits: my body, not someone else's – even when the pains of aging might tempt me to become young again; my memories, not someone else's – even when the traumas of the past might tempt me to have someone else's memories; my achievements and potential, not someone else's – even when the desire for excellence might tempt me to "trade myself in" for a "better model." We seek to be happy – to achieve, perform, take pleasure in our experiences, and catch the admiring eye of a beloved. But we do not, at least self-consciously, seek such happiness at the cost of losing our real identity.

4.4 Partial ends, full flourishing

Beyond the perils of achieving our desired goals in a "less-than-human way" or in ways "not fully our own," we must consider the meaning of the ends themselves: better children, superior performance, ageless bodies, and happy souls. Would their attainment in fact improve or perfect our lives as human beings? Are they – always or ever – reasonable and attainable goals?

Everything depends, as we have pointed out in each case, on how these goals are understood, on their specific and concrete content. Yet, that said, the first two human ends – better children and superior performance – do seem reasonable and attainable, sometimes if not always, to some degree if not totally. When asked what they wish for their children, most parents say: "We want them to be happy," or "We want them to live good lives" – in other words, to be better and to do better. The desire is a fitting one for any loving parent. The danger lies in misconceiving what "better children" really means, and thus coming to pursue this worthy goal in a misguided way, or with a false idea of what makes for a good or happy child.

Likewise, the goal of superior performance – the desire to be better or do better in all that we do – is good and noble, a fitting human aspiration. We admire excellence whenever we encounter it, and we properly seek to excel in those areas of life, large and small, where we ourselves are engaged and at-work. But the danger here is that we will become better in some area of life by diminishing ourselves in others, or that we will achieve superior results only by compromising our humanity, or by corrupting those activities that are not supposed to be "performances" measured in terms of external standards of "better and worse."

In many cases, biotechnologies can surely help us cultivate what is best in ourselves and in our children, providing new tools for realizing good ends, wisely pursued. But it is also possible that the new technological means may deform the ends themselves. In pursuit of better children, biotechnical powers risk making us "tyrants"; in pursuit of superior performance, they risk making us "artifacts." In both cases, the problem is not the ends themselves but our misguided idea of their attainment or our false way of seeking to attain them. And in both cases, there is the ubiquitous problem that "good" or "superior" will be reconceived to fit the sorts of goals that the technological interventions can help us attain. We may come to believe that genetic predisposition or brain chemistry holds the key to helping our children develop and improve, or that stimulant drugs or bulkier muscles hold the key to excellent human activity. If we are equipped with hammers, we will see only those things that can be improved by pounding.

The goals of ageless bodies and happy souls – and especially the ways biotechnology might shape our pursuit of these ends – are perhaps more complicated. The case for ageless bodies seems at first glance to look pretty good. The prevention of decay, decline, and disability, the avoidance of blindness, deafness, and debility, the elimination of feebleness, frailty, and fatigue, all seem to be conducive to living fully as a human being at the top of one's powers – of having, as they say, a "good quality of life" from beginning to end. We have come to expect organ transplantation for our worn-out parts. We will surely welcome stem-cell-based therapies for regenerative medicine, reversing by replacement the damaged tissues of Parkinson disease, spinal cord injury, and many other degenerative disorders. It is hard to see any objection to obtaining a genetic enhancement of our muscles in our youth that would not only prevent the muscular feebleness of old age but would empower us to do any physical task with greater strength and facility throughout our lives. And, should aging research deliver on its promise of adding not only extra life to years but also extra years to life, who would refuse it?

But as we suggested in Chapter Four, there may in fact be many human goods that are inseparable from our aging bodies, from our living in time, and especially from the natural human life cycle by which each generation gives way to the one that follows it. Because this argument is so counterintuitive, we need to begin not with the individual choice for an ageless body, but with what the individual's life might look like in a world in which everyone made the same choice. We need to make the choice universal, and see the meaning of that choice in the mirror of its becoming the norm.

What if everybody lived life to the hilt, even as they approached an ever-receding age of death in a body that looked and functioned – let's not be too greedy – like that of a thirty-year-old? Would it be good if each and all of us lived like light bulbs, burning as brightly from beginning to end, then popping off without warning, leaving those around us suddenly in the dark? Or is it perhaps better that there be a shape to life, everything in its due season, the shape also written, as it were, into the wrinkles of our bodies that live it – provided, of course, that we do not suffer years of painful or degraded old age and that we do not lose our wits? What would the relations between the generations be like if there never came a point at which a son surpassed his father in strength or vigor? What incentive would there be for the old to make way for the young, if the old slowed down little and had no reason to think of retiring – if Michael could play basketball until he were not forty but eighty? Might not even a moderate prolongation of lifespan with vigor lead to a prolongation in the young of functional immaturity – of the sort that has arguably already accompanied the great increase in average life expectancy experienced in the past century?

Going against both common intuition and native human desire, some commentators have argued that living with full awareness and acceptance of our finitude may be the condition of many of the best things in human life: engagement, seriousness, a taste for beauty, the possibility of virtue, the ties born of procreation, the quest for meaning. This might be true not just for immortality – an unlikely achievement, likely to produce only false expectations – but even for more modest prolongations of the maximum lifespan, especially in good health, that would permit us to live as if there were always tomorrow. The pursuit of perfect bodies and further life-extension might deflect us from realizing more fully the aspirations to which our lives naturally point, from living well rather than merely staying alive. A concern with one's own improving agelessness might finally be incompatible with accepting the need for procreation and human renewal. And far from bringing contentment, it might make us increasingly anxious over our health or dominated by the fear of death. Assume, merely for the sake of the argument, that even a few of these social consequences would follow from a world of much greater longevity and vigor: What would we then say about the simple goodness of seeking an ageless body?

What about the pursuit of happy souls, and especially of the sort that we might better attain with pharmacological assistance? Painful and shameful memories are disturbing; guilty consciences trouble sleep; low self-esteem, melancholy, and world-weariness besmirch the waking hours. Why not memory-blockers for the former, mood-brighteners for the latter, and a good euphoriant – without risks of hangovers or cirrhosis – when celebratory occasions fail to be jolly? For let us be clear: If it is imbalances of neurotransmitters that are largely responsible for our state of soul, would it not be sheer priggishness to refuse the help of pharmacology for our happiness, when we accept it guiltlessly to correct for an absence of insulin or thyroid hormone?

And yet, as we suggested in Chapter Five, there seems to be something misguided about the pursuit of utter and unbroken psychic tranquility or the attempt to eliminate all shame, guilt, and painful memories. Traumatic memories, shame, and guilt, are, it is true, psychic pains. In extreme doses, they can be crippling. Yet, short of the extreme, they can also be helpful and fitting. They are appropriate responses to horror, disgraceful conduct, injustice, and sin, and, as such, help teach us to avoid them or fight against them in the future. Witnessing a murder should be remembered as horrible; doing a beastly deed should trouble one's soul. Righteous indignation at injustice depends on being able to feel injustice's sting. And to deprive oneself of one's memory – including and especially its truthfulness of feeling – is to deprive oneself of one's own life and identity.

These feeling states of soul, though perhaps accompaniments of human flourishing, are not its essence. Ersatz pleasure or feelings of self-esteem are not the real McCoy. They are at most shadows divorced from the underlying human activities that are the essence of flourishing. Most people want both to feel good and to feel good about themselves, but only as a result of being good and doing good.

At the same time, there appears to be a connection between the possibility of feeling deep unhappiness and the prospects for achieving genuine happiness. If one cannot grieve, one has not truly loved. To be capable of aspiration, one must know and feel lack. As Wallace Stevens put it: Not to have is the beginning of desire. In short, if human fulfillment depends on our being creatures of need and finitude and therewith of longings and attachment, there may be a double-barreled error in the pursuit of ageless bodies and factitiously happy souls: far from bringing us what we really need, pursuing these partial goods could deprive us of the urge and energy to seek a richer and more genuine flourishing.

Looking into the future at goals pursuable with the aid of new biotechnologies enables us to turn a reflective glance at our own version of the human condition and the prospects now available to us (in principle) for a flourishing human life. For us today, assuming that we are blessed with good health and a sound mind, a flourishing human life is not a life lived with an ageless body or an untroubled soul, but rather a life lived in rhythmed time, mindful of time's limits, appreciative of each season and filled first of all with those intimate human relations that are ours only because we are born, age, replace ourselves, decline, and die – and know it. It is a life of aspiration, made possible by and born of experienced lack, of the disproportion between the transcendent longings of the soul and the limited capacities of our bodies and minds. It is a life that stretches towards some fulfillment to which our natural human soul has been oriented, and, unless we extirpate the source, will always be oriented. It is a life not of better genes and enhancing chemicals but of love and friendship, song and dance, speech and deed, working and learning, revering and worshipping.

If this is true, then the pursuit of an ageless body may prove finally to be a distraction and a deformation. And the pursuit of an untroubled and self-satisfied soul may prove to be deadly to desire, if finitude recognized spurs aspiration and fine aspiration acted upon *is itself* the core of happiness. Not the agelessness of the body, nor the contentment of the soul, nor even the list of external achievements and accomplishments of life, but the engaged and energetic being-at-work of what nature uniquely gave to us is what we need to treasure and defend. All other "perfections" may turn out to be at best but passing illusions, at worst a Faustian bargain that could cost us our full and flourishing humanity.

5. CONCLUSION

The concerns we have raised here emerge from a sense that tremendous new powers to serve certain familiar and often well-intentioned desires may blind us to the larger meaning of our ideals, and may narrow our sense of what it is to live, to be free, and to seek after happiness. If, by informing and moderating our desires and by grasping the limits of our new powers, we can keep in mind the meaning of our founding ideals, then we just might find the means to savor some fruits of the age of biotechnology, without succumbing to its most dangerous temptations.

To do so, we must first understand just what is at stake, and we must begin to imagine what the age of biotechnology might bring, and what human life in that age could look like. In these pages, we have sought to begin that vital project, in the hope that these first steps might spark and inform a public debate, so that however the nation proceeds, it will do so with its eyes wide open.

WORKS CITED

H. Jonas (1992) 'The Blessings and Burdens of Mortality", *Hastings Center Report*, January/February 1992.

L. Kass (2003) 'Beyond Therapy: Biotechnology and the Pursuit of Human Improvement' prepared for the President's Council on Bioethics, Washington, D.C.

L. Kass (2002) 'L'Chaim and Its Limits: Why Not Immortality', in *Life, Liberty, and the Defense of Dignity: The Challenge for Bio- ethics*, (San Francisco: Encounter Books).

M. J. Sandel (2002) 'What's Wrong with Enhancement', prepared for the President's Council on Bioethics, Washington, D.C.

DISCUSSION QUESTIONS

1. The Council suggests that we should think about the big picture when it comes to biotechnological enhancement, rather than considering each individual technology. What do you think they mean by "the big picture"? Do you agree that it is important to take up that perspective rather than focus on one technology at a time?

2. The Council argues that emerging biotechnologies deserve ethical attention because of the augmented powers that they provide us. What are the powers that they have in mind? Do you believe that the powers are novel or augmented enough such that they raise ethical concerns or issues that previous technologies have not?

3. One of the concerns that the Council raises has to do with the potential coercive effects of enhancement technologies. What do they mean by coercion? Do you agree that the technologies might be coercive in the ways they suggest? Why or why not?

4. Several of the Council's concerns regarding enhancement technologies involve how their existence will impact our drives, desires, and motivations. What precisely are their concerns in this regard, and do you believe that they are warranted?

5. By increasing the power of parents to choose traits for their children these technologies have the potential to restructure the parent–child relationship. How might they do this? Do you believe that they will do so in positive or negative ways, and why?

6. What is "Respect for the 'given'"? Do you agree that this respect is ethically important, and that the technological pursuit of enhancement is contrary to it or might undermine it?

7. What does the Council mean by the "dignity of human activity"? Can you think of any other conceptions of human dignity? Do you agree that these technologies might undermine an important sense of human dignity?

8. How might the pursuit of technological enhancement undermine identity and individuality? Do you think it is possible that it might actually promote these instead? Why or why not?

9. What is human flourishing? What are the components of it? How does the Council think it might be diminished by these technologies? Do you think it is possible that the technologies could expand conceptions or components of human flourishing instead of diminishing them?

15 WHY I WANT TO BE A POSTHUMAN WHEN I GROW UP[1]

Nick Bostrom

CHAPTER SUMMARY

In this chapter, Nick Bostrom discusses the possibility that extreme human enhancement could result in "posthuman" modes of being. After offering some definitions and conceptual clarifications, he argues for two theses. First, there are posthuman modes of being – including some related to healthspan, cognition, and emotion – that would be very worthwhile. Second, it could be very good for human beings to become posthuman in those ways. He then considers and responds to objections to his theses, including several raised by the President's Council on Bioethics in *Beyond Therapy* – for example, that personal identity could not be maintained through posthuman enhancement and that it constitutes a failure to be open to the gifted nature of life.

RELATED READINGS

Introduction: Form of Life (2.5); Autonomy, Authenticity, and Identity (2.6.4); Intrinsic Concerns (2.7)

Other Chapters: Walter Glannon, *Psychopharmacology and Functional Neurosurgery* (Ch. 12); President's Council on Bioethics, *Beyond Therapy* (Ch. 14); Thomas Douglas, *Moral Enhancement* (Ch. 16)

1. SETTING THE STAGE

The term "posthuman" has been used in very different senses by different authors.[2] I am sympathetic to the view that the word often causes more confusion than clarity, and that we might be better off replacing it with some alternative vocabulary. However, as the purpose of this paper is not to propose terminological reform but to argue for certain substantial normative theses (which one would

---

[1] This chapter is excerpted from Nick Bostrom (2008) 'Why I want to be a Posthuman When I Grow Up,' in *Medical Enhancement and Posthumanity*, eds. B. Gordijn and R. Chadwick (Springer). It appears here by permission of Springer and the author.

[2] The definition used here follows in the spirit of Bostrom (2003). A completely different concept of "posthuman" is used in e.g. Hayles (1999).

naturally search for in the literature under the label "posthuman"), I will instead attempt to achieve intelligibility by clarifying the meaning that I shall assign to the word. Such terminological clarification is surely a minimum precondition for having a meaningful discussion about whether it might be good for us to become posthuman.

I shall define a posthuman as a being that has at least one posthuman capacity. By a posthuman capacity, I mean a general central capacity greatly exceeding the maximum attainable by any current human being without recourse to new technological means. I will use general central capacity to refer to the following:

> *healthspan* – the capacity to remain fully healthy, active, and productive, both mentally and physically
>
> *cognition* – general intellectual capacities, such as memory, deductive and analogical reasoning, and attention, as well as special faculties such as the capacity to understand and appreciate music, humor, eroticism, narration, spirituality, mathematics, etc.
>
> *emotion* – the capacity to enjoy life and to respond with appropriate affect to life situations and other people

In limiting my list of general central capacities to these three, I do not mean to imply that no other capacity is of fundamental importance to human or posthuman beings. Nor do I claim that the three capacities in the list are sharply distinct or independent. Aspects of emotion and cognition, for instance, clearly overlap. But this short list may give at least a rough idea of what I mean when I speak of posthumans, adequate for present purposes.

In this paper, I will be advancing two main theses. The first is that some possible posthuman modes of being would be very good. I emphasize that the claim is not that *all* possible posthuman modes of being would be good. Just as some possible human modes of being are wretched and horrible, so too are some of the posthuman possibilities. Yet it would be of interest if we can show that there are some posthuman possibilities that would be very good. We might then, for example, specifically aim to realize those possibilities.

The second thesis is that it could be very good *for us* to become posthuman. It is possible to think that it could be good to be posthuman without it being good *for us* to become posthuman. This second thesis thus goes beyond the first. When I say "good for us", I do not mean to insist that for every single current human individual there is some posthuman mode of being such that it would be good for that individual to become posthuman in that way. I confine myself to making a weaker claim that allows for exceptions. The claim is that for *most* current human beings, there are possible posthuman modes of being such that it could be good for these humans to become posthuman in one of those ways.

I am setting aside issues of feasibility, costs, risks, side-effects, and social consequences. While those issues are obviously important when considering what we have most reason to do all things considered, they will not be addressed here.

In the next three sections we will look in a little more detail at each of the three general central capacities that I listed in the introduction section. I hope to show that the claim that it could be very good to be posthuman is not as radical as it might appear to some. In fact, we will find that individuals and society already in some ways seem to be implicitly placing a very high value on posthuman capacities – or at least, there are strong and widely accepted tendencies pointing that way. I therefore do not regard my claim as in any strong sense revisionary. On the contrary, I believe that the denial of my claim would be strongly revisionary in that it would force us to reject many commonly accepted ethical beliefs and approved behaviors. I see my position as a conservative extension of traditional ethics and values to accommodate the possibility of human enhancement through technological means.

2. HEALTHSPAN

It seems to me fairly obvious why one might have reason to desire to become a posthuman in the sense of having a greatly enhanced capacity to stay alive and stay healthy.[3] I suspect that the majority of humankind already has such a desire implicitly.

People seek to extend their healthspan, i.e. to remain healthy, active, and productive. This is one reason why we install air bags in cars. It may also explain why we go to the doctor when we are sick, why higher salaries need to be paid to get workers to do physically dangerous work, and why governments and charities give money to medical research.[4] Instances of individuals sacrificing their lives for the sake of some other goal, whether suicide bombers, martyrs, or drug addicts, attract our attention precisely because their behavior is unusual. Heroic rescue workers who endanger their lives on a dangerous mission are admired because we assume that they are putting at risk something that most people would be very reluctant to risk, their own survival.

For some three decades, economists have attempted to estimate individuals' preferences over mortality and morbidity risk in labor and product markets. While the tradeoff estimates vary considerably between studies, one recent meta-analysis puts the median value of the value of a statistical life for prime-aged workers to about $7 million in the United States (Viscusi and Aldy, 2003). A study by the EU's Environment Directorates-General recommends the use of a value in the interval €0.9 to €3.5 million (Johansson, 2002). Recent studies by health economists indicate that improvements in the health status of the U.S. population over the 20th century have made as large a contribution to raising the standards of living as all other forms of consumption growth combined (Murphy and Topel, 2003; Nordhaus, 2003). While the exact numbers are debatable, there is little doubt that most people place a very high value on their continued existence in a healthy state.

Admittedly, a desire to extend one's healthspan is not necessarily a desire to become posthuman. To become posthuman by virtue of healthspan extension, one would need to achieve the capacity for a healthspan that greatly exceeds

---

[3] Having such a capacity is compatible with also having the capacity to die at any desired age. One might thus desire a capacity for greatly extended healthspan even if one doubts that one would wish to live for more than, say, 80 years. A posthuman healthspan capacity would give one the option of much longer and healthier life, but one could at any point decide no longer to exercise the capacity.

[4] Although on the last item, see Hanson (2000) for an alternative view.

the maximum attainable by any current human being without recourse to new technological means. Since at least some human beings already manage to remain quite healthy, active, and productive until the age of 70, one would need to desire that one's healthspan were extended greatly beyond this age in order that it would count as having a desire to become posthuman.[5]

Many people will, if asked about how long they would wish their lives to be, name a figure between 85 and 90 years (Cohen and Langer, 2005). In many cases, no doubt, this is because they assume that a life significantly longer than that would be marred by deteriorating health – a factor from which we must abstract when considering the desirability of healthspan extension. People's stated willingness to pay to extend their life by a certain amount does in fact depend strongly on the health status and quality of that extra life (Johnson et al., 1998). Since life beyond 85 is very often beset by deteriorating health, it is possible that this figure substantially underestimates how long most people would wish to live if they could be guaranteed perfect health.

It is also possible that a stated preference for a certain lifespan is hypocritical. Estimates based on revealed preferences in actual market choices, such as fatality risk premiums in labor markets or willingness to pay for health care and other forms of fatality risk reduction might be more reliable. It would be interesting to know what fraction of those who claim to have no desire for healthspan extension would change their tune if they were ever actually handed a pill that would reliably achieve this effect. My conjecture would be that when presented with a real-world choice, most would choose the path of prolonged life, health, and youthful vigor over the default route of aging, disease, and death.

One survey asked: "Based on your own expectations of what old age is like, if it were up to you, how long would you personally like to live – to what age?" Only 27% of respondents said they would like to live to 100 or older (Cohen and Langer, 2005). A later question in the same survey asked: "Imagine you could live to 100 or older, but you'd have to be very careful about your diet, exercise regularly, not smoke, avoid alcohol, and avoid stress. Would it be worth it, or not?" To this, 64% answered in the affirmative! Why should *more* people want to live beyond 100 when restrictions on activity are imposed? Is it because it frames the question more as if it were a real practical choice rather than as an idle mind game? Perhaps when the question is framed as a mind game, respondents tend to answer in ways which they believe expresses culturally approved attitudes, or which they think signal socially desirable personal traits (such as having "come to terms" with one's own mortality), while this tendency is diminished when the framing suggests a practical choice with real consequences. We do not know for sure, but this kind of anomaly suggests that we should not take people's stated "preferences" about how long they would wish to live too seriously, and that revealed preferences might be a more reliable index of their guiding values.

---

[5] At least one human, Jeanne Calment, lived to 122. But although she remained in relatively fair health until close to her death, she clearly suffered substantial decline in her physical (and presumably mental) vigor compared to when she was in her twenties. She did not retain the capacity to be *fully* healthy, active, and productive for 122 years.

It is also worth noting that only a small fraction of us commit suicide, suggesting that our desire to live is almost always stronger than our desire to die.[6] Our desire to live, *conditional on our being able to enjoy full health*, is even stronger. This presumption in favor of life is in fact so strong that if somebody wishes to die soon, even though they are seemingly fully healthy, with a long remaining healthy life expectancy, and if their external circumstances in life are not catastrophically wretched, we would often tend suspect that they might be suffering from depression or other mental pathology. Suicidal ideation is listed as a diagnostic symptom of depression by the American Psychiatric Association.[7]

Even if a stated preference against healthspan extension were sincere, we would need to question how well-considered and informed it is. It is of relevance that those who know most about the situation and are most directly affected by the choice, namely the elderly, usually prefer life to death. They usually do so when their health is poor, and overwhelmingly choose life when their health is at least fair. Now one can argue that a mentally intact 90-year-old is in a better position to judge how their life would be affected by living for another year than she was when she was 20, or 40. If most healthy and mentally intact 90-year-olds prefer to live for another year (at least if they could be guaranteed that this extra year would be one of full health and vigor), this would be evidence against the claim that it would be better for these people that their lives end at 90.[8] Similarly, of course, for people of even older age.

One can compare this situation with the different case of somebody becoming paraplegic. Many able-bodied people believe that their lives would not be worth living if they became paraplegic. They claim that they would prefer to die rather than continuing life in a paraplegic state. Most people who have actually become paraplegic, however, find that their lives are worth living.[9] People who are paraplegic are typically better judges of whether paraplegic lives are worth continuing than are people who have never experienced what it is like to be paraplegic. Similarly, people who are 90 years old are in a better position to judge whether their lives are worth continuing than are younger people (including themselves at any earlier point in their lives).[10]

---

[6] For some, the reluctance to commit suicide might reflect a desire not to kill oneself rather than a desire not to die, or alternatively a fear of death rather than an authentic preference not to die.

[7] DSM-IV (American Psychiatric Association, 2000).

[8] This is a kind of Millian best-judge argument. However, if fear of death were irrational, one could argue that people who are closer to death are on average worse judges of the value for them of an extra year of life, because their judgments would tend to be more affected by irrational fear.

[9] This basic result is reflected in many chronic disease conditions (Ubel et al., 2003). The discrepancy of attitudes seems to be due to non-patient's failure to realize the extent to which patients psychologically adapt to their condition (Damschroder et al., 2005).

[10] The analogy with paraplegia is imperfect in at least one respect: when the issue is healthspan extension, we are considering whether it would be worth living an extended life in perfect health and vigor. If anything, this discrepancy strengthens the conclusion, since it is more worth continuing living in perfect health than in poor health, not less worth it.

One study assessed the will to live among 414 hospitalized patients aged 80 to 98 years, presumably representing the frailer end of the distribution of the "old old". 40.8% of respondents were unwilling to exchange any time in their current state of health for a shorter life in excellent health, and 27.8% were willing to give up at most 1 month of 12 in return for excellent health.[11] (Patients who were still alive one year later were even less inclined to give up life for better health, but with continued large individual variations in preferences.) The study also found that patients were willing to trade significantly less time for a healthy life than their surrogates assumed they would.

Research shows that life-satisfaction remains relatively stable into old age. One survey of 60,000 adults from 40 nations discovered a slight upward trend in life-satisfaction from the 20s to the 80s in age (Diener and Suh, 1998). Life satisfaction showed this upward trend even though there was some loss of positive affect. Perhaps life-satisfaction would be even higher if positive affect were improved (a possibility we shall discuss in a later section). Another study, using a cross-sectional sample (age range 70–103 years), found that controlling for functional health constraints reversed the direction of the relationship between age and positive affect and produced a negative association between age and negative affect (Kunzmann et al., 2000). These findings suggest that some dimensions of subjective well-being, such as life-satisfaction, do not decline with age but might actually increase somewhat, and that the decline in another dimension of subjective well-being (positive affect) is not due to aging per se but to health constraints.

Most people reveal through their behavior that they desire continued life and health,[12] and most of those who are in the best position to judge the value of continued healthy life, at any age, judge that it is worth having. This constitutes prima facie support for the claim that extended life is worth having even when it is not fully healthy. The fact that this holds true at all currently realized ages suggests that it is not a strongly revisionary view to hold that it could be good for many people to become posthuman through healthspan extension. Such a view might already be implicitly endorsed by many.

3. COGNITION

People also seem to be keen on improving cognition. Who wouldn't want to remember names and faces better, to be able more quickly to grasp difficult abstract ideas, and to be able to "see connections" better? Who would seriously object to being able to appreciate music at a deeper level? The value of optimal cognitive functioning is so obvious that to elaborate the point may be unnecessary.[13]

---

[11] Tsevat et al. (1998). See also McShine et al. (2000). For a methodological critique, see Arnesen and Norheim (2003).

[12] This is fully consistent with the fact that many people knowingly engage in risky behaviors such as smoking. This might simply mean that they are unable to quit smoking, or that they desire the pleasure of smoking more than they desire a longer healthier life. It does not imply that they do not desire longer healthier life.

This verdict is reflected in the vast resources that society allocates to education, which often explicitly aims not only to impart specific items of knowledge but also to improve general reasoning abilities, study skills, critical thinking, and problem solving capacity.[14] Many people are also keen to develop various particular talents that they may happen to have, for example musical or mathematical, or to develop other specific faculties such as aesthetic appreciation, narration, humor, eroticism, spirituality etc. We also reveal our desire for improving our cognitive functioning when take a cup of coffee to increase our alertness or when we regret our failure to obtain a full night's sleep because of the detrimental effects on our intellectual performance.

Again, the fact that there is a common desire for cognitive improvement does not imply that there is a common desire for becoming posthuman. To want to become posthuman through cognitive improvement, one would have to want a great deal of cognitive improvement. It is logically possible that each person would only want to become slightly more intelligent (or musical, or humorous) than he or she currently is and would not want any very large gain. I will offer two considerations regarding this possibility.

First, it seems to me (based on anecdotal evidence and personal observations) that people who are already endowed with above-average cognitive capacities are at least as eager, and, from what I can tell, actually *more* eager to obtain further improvements in these capacities than are people who are less talented in these regards. For instance, someone who is musically gifted is likely to spend more time and effort trying to further develop her musical capacities than is somebody who lacks a musical ear; and likewise for other kinds of cognitive gifts.

This phenomenon may in part reflect the external rewards that often accrue to those who excel in some particular domain. An extremely gifted musician might reap greater rewards in terms of money and esteem from a slight further improvement in her musicality than would somebody who is not musically gifted to begin with. That is, the difference in external rewards is sometimes greater for somebody who goes from very high capacity to outstandingly high capacity than it is for somebody who goes from average capacity to moderately high capacity. However, I would speculate that such differences in external rewards are only part of the explanation and that people who have high cognitive capacities are usually also more likely (or at least no less likely) to desire further increases in those capacities than are people of lower cognitive capacities even when only the intrinsic benefits of capacities are considered. Thus, if we imagine a group of people placed in solitary confinement for the remainder of their lives, but with access to books, musical instruments, paints and canvasses, and other prerequisites for the exercise of capacities, I would hypothesize that those with the highest pre-existing capacity in a given domain would be more likely (or at least not less likely) to work hard to further develop their capacities in that domain, for the sake of the intrinsic benefits that the possession and exercise of those capacities bestow, than would those with lower pre-existing capacities

---

[13] One might even argue that a desire for cognitive improvement is a constitutive element of human rationality, but I will not explore that hypothesis here.

[14] U.S. *public* expenditure on education in 2003 was 5.7% of its GDP (World Bank, 2003).

in the same domain.[15] While \$100 brings vastly less utility to a millionaire than to a pauper, the marginal utility of improved cognitive capacities does not seem to exhibit a similar decline.

These considerations suggest that there are continuing returns in the "intrinsic" (in the sense of non-instrumental, non-positional) utility of gains in cognitive capacities, at least within the range of capacity that we find instantiated within the current human population.[16] It would be implausible to suppose that the current range of human capacity, in all domains, is such that while increments of capacity within this range are intrinsically rewarding, yet any further increases outside the current human range would lack intrinsic value. Again, we have a prima facie reason for concluding that enhancement of cognitive capacity to the highest current human level, and probably beyond that, perhaps up to and including the posthuman level, would be intrinsically desirable for the enhanced individuals. We get this conclusion if we assume that those who have a certain high capacity are generally better judges of the value of having that capacity or of a further increment of that capacity than are those who do not possess the capacity in question to the same degree.

4. EMOTION

It is straightforward to determine what would count as an enhancement of healthspan. We have a clear enough idea of what it means to be healthy, active, and productive, and the difference between this state and that of being sick, incapacitated, or dead. An enhancement of healthspan is simply an intervention that prolongs the duration of the former state. It is more difficult to define precisely what would count as a cognitive enhancement because the measure of cognitive functioning is more multi-faceted, various cognitive capacities can interact in complex ways, and it is a more normatively complex problem to determine what combinations of particular cognitive competences are of value in different kinds of environments. For instance, it is not obvious what degree of tendency to forget certain kinds of facts and experiences is desirable. The answer might depend on a host of contextual factors. Nevertheless, we do have some general idea of how we might value various increments or decrements in many aspects of our cognitive functioning – a sufficiently clear idea, I suggest, to make it intelligible without much explanation what one might mean by phrases like "enhancing musical ability", "enhancing abstract reasoning ability," etc.

It is considerably more difficult to characterize what would count as emotional enhancement. Some instances are relatively straightforward. Most would readily

---

[15] Complication: if high capacity were solely a result from having spent a lot of effort in developing that capacity, then the people with high capacity in some domain might be precisely those that started out having an unusually strong desire for having a strong capacity in that domain. It would then not be surprising that those with high capacity would have the strongest desire for further increases in capacity. Their stronger desire for higher capacity might then not be the result of more information and better acquaintance with what is at stake, but might instead simply reflect a prior inclination.

[16] It would be more difficult to determine whether the marginal intrinsic utility of gains in capacity are constant, or diminishing, or increasing at higher levels of capacity, and if so by what amount.

agree that helping a person who suffers from persistent suicidal depression as the result of a simple neurochemical imbalance so that she once again becomes capable of enjoyment and of taking an interest in life would be to help her improve her emotional capacities. Yet beyond cases involving therapeutic interventions to cure evident psychopathology it is less clear what would count as an enhancement. One's assessment of such cases often depends on the exact nature of one's normative beliefs about different kinds of possible emotional constitutions and personalities.

It is correspondingly difficult to say what would constitute a "posthuman" level of emotional capacity. Nevertheless, people often do strive to improve their emotional capacities and functionings. We may seek to reduce feelings of hate, contempt, or aggression when we consciously recognize that these feelings are prejudiced or unconstructive. We may take up meditation or physical exercise to achieve greater calm and composure. We may train ourselves to respond more sensitively and empathetically to those we deem deserving of our trust and affection. We may try to overcome fears and phobias that we recognize as irrational, or we may wrestle with appetites that threaten to distract us from what we value more. Many of us expend life-long effort to educate and ennoble our sentiments, to build our character, and to try to become better people. Through these strivings, we seek to achieve goals involving modifying and improving our emotional capacities.

An appropriate conception of emotional capacity would be one that incorporates or reflects these kinds of goal, while allowing perhaps for there being a wide range of different ways of instantiating "high emotional capacity", that is to say, many different possible "characters" or combinations of propensities for feeling and reacting that could each count as excellent in its own way. If this is admitted, then we could make sense of emotional enhancement in a wide range of contexts, as being that which makes our emotional characters more excellent. A posthuman emotional capacity would be one which is much more excellent than that which any current human could achieve unaided by new technology.

One might perhaps question whether there are possible emotional capacities that would be *much* more excellent than those attainable now. Conceivably, there might be a maximum of possible excellence of emotional capacity, and those people who currently have the best emotional capacities might approach so closely to this ideal that there is not enough potential left for improvement to leave room for a posthuman realm of emotional capacity. I doubt this, because aside from the potential for fine-tuning and balancing the various emotional sensibilities we already have, I think there might also be entirely new psychological states and emotions that our species has not evolved the neurological machinery to experience, and some of these sensibilities might be ones we would recognize as extremely valuable if we became acquainted with them.

It is difficult intuitively to understand what such novel emotions and mental states might be like. This is unsurprising, since by assumption we currently lack the required neurological bases. It might help to consider a parallel case from within the normal range of human experience. The experience of romantic love is something that many of us place a high value on. Yet it is notoriously difficult for a child or a prepubescent teenager to comprehend the meaning of romantic love or why adults should make so much fuss about this experience. Perhaps we are all currently in the situation of children relative to the emotions, passions, and mental states that posthuman beings could experience. We may have no idea of what we are missing out on until we attain posthuman emotional capacities.

One dimension of emotional capacity that we can imagine enhanced is subjective well-being and its various flavors: joy, comfort, sensual pleasures, fun, positive interest and excitement. Hedonists claim that pleasure is the only intrinsic good, but one need not be a hedonist to appreciate pleasure as one important component of the good. The difference between a bleak, cold, horrid painful world and one that is teeming with fun and exciting opportunities, full of delightful quirks and lovely sensations, is often simply a difference in the hedonic tone of the observer. Much depends on that one parameter.

It is an interesting question how much subjective well-being could be enhanced without sacrificing other capacities that we may value. For human beings as we are currently constituted, there is perhaps an upper limit to the degree of subjective well-being that we can experience without succumbing to mania or some other mental unbalance that would prevent us from fully engaging with the world if the state were indefinitely prolonged. But it might be possible for differently constituted minds to have experiences more blissful than those that humans are capable of without thereby impairing their ability to respond adequately to their surroundings. Maybe for such beings, gradients of pleasure could play a role analogous to that which the scale ranging between pleasure and pain has for us (Pearce, 2004). When thinking the possibility of *posthumanly happy* beings, and their psychological properties, one must abstract from contingent features of the human psyche. An experience that would consume us might perhaps be merely "spicy" to a posthuman mind.

It is not necessary here to take a firm stand on whether posthuman levels of pleasure are possible, or even on whether posthuman emotional capacities more generally are possible. But we can be confident that, at least, there is vast scope for improvements for most of individuals in these dimensions because even within the range instantiated by currently exiting humans, there are levels of emotional capacities and degrees of subjective well-being that, for most of us, are practically unattainable to the point of exceeding our dreams. The fact that such improvements are eagerly sought by many suggests that if posthuman levels were possible, they too would be viewed as highly attractive.[17]

5. STRUCTURE OF THE ARGUMENT AND FURTHER SUPPORTING REASONS

It might be useful to pause briefly to reflect on the structure of the argument presented so far. I began by listing three general central capacities (healthspan, cognition, and emotion), and I defined a posthuman being as one who has at

---

[17] The quest for subjective well-being, in particular, seems to be a powerful motivator for billions of people even though arguably none of the various means that have been attempted in this quest has yet proved very efficacious in securing the goal (Brickman and Campbell, 1971).

least one of these capacities in a degree unattainable by any current human being unaided by new technology.

I offered some plausibility arguments suggesting that it could be highly desirable to have posthuman levels of these capacities. I did this partly by clarifying what having the capacities would encompass and by explaining how some possible objections would not apply because they rely on a misunderstanding of what is proposed. Furthermore, I tried to show that for each of the three capacities we find that many individuals actually desire to develop the capacities to higher levels and often undertake great effort and expense to achieve these aims. This desire is also reflected in social spending priorities, which devote significant resources to e.g. healthspan-extending medicine and cognition-improving education. Significantly, at least in the cases of healthspan extension and cognitive improvement, the persons best placed to judge the value and desirability of incremental improvements at the high end of the contemporary human capacity distribution seem to be especially likely to affirm the desirability of such additional improvements of capacity. For many cognitive faculties, it appears that the marginal utility of improvements *increases* with capacity levels. This suggests that improvements beyond the current human range would also viewed as desirable when evaluated by beings in a better position to judge than we currently are.

6. PERSONAL IDENTITY

Supposing the previous sections have succeeded in making it plausible that being a posthuman could be good, we can now turn to a further question: whether becoming posthuman could be good *for us*. It may be good to be Joseph Haydn. Let us suppose that Joseph Haydn had a better life than Joe Bloggs so that in some sense it is better to be Haydn and living the life that Haydn lived than to be Bloggs and living Bloggs' life. We may further suppose that this is so from Bloggs' evaluative standpoint. Bloggs might recognize that on all the objective criteria which he thinks makes for a better mode of being and a better life, Haydn's mode of being and life are better than his own. Yet it does not follow that it would be good for Bloggs to "become" Haydn (or to become some kind of future equivalent of Haydn) or to live Haydn's life (or a Haydn-like life). There are several possible reasons for this which we need to examine.

First, it might not be possible for Bloggs to become Haydn without ceasing to be Bloggs. While we can imagine a thought experiment in which Bloggs' body and mind are gradually transformed into those of Haydn (or of a Haydn-equivalent), it is not at all clear that personal identity could be preserved through such a transformation. If Bloggs' personal identity is essentially constituted by some core set of psychological features such as his memories and dispositions, then, since Haydn does not have these features, the person Bloggs could not become a Haydn-equivalent. Supposing that Bloggs has a life that is worth living, any transformation that causes the person Bloggs to cease to exist might be bad for Bloggs, including one that transforms him into Haydn.

Could a current human become posthuman while remaining the same person, or is the case like the one of Bloggs becoming Haydn, the person Bloggs necessarily ceasing to exist in the process? The case of becoming posthuman is different in an important respect. Bloggs would have to lose all the psychological characteristics that made him person Bloggs in order to become Haydn. In particular, he would have to lose all his memories, his goals, his unique skills, and his entire personality would be obliterated and replaced by that of Haydn. By contrast, a human being could retain her memories, her goals, her unique skills, and many important aspects of her personality even as she becomes posthuman. This could make it possible for personal identity to be preserved during the transformation into posthuman.[18]

It is obvious that personal identity could be preserved, at least in the short run, if posthuman status is achieved through radical healthspan enhancement. Suppose that I learnt that tonight after I go to bed, a scientist will perform some kind of molecular therapy on my cells while I'm sleeping to permanently disable the aging processes in my body. I might worry that I would not wake up tomorrow because the surgery might go wrong. I would not worry that I might not wake up tomorrow because the surgery succeeded. Healthspan enhancement would help preserve my personal identity.

The case that personal identify could be preserved is perhaps less clear-cut with regard to radical cognitive or emotional enhancement. Could a person become radically smarter, more musical, or come to possess much greater emotional capacities without ceasing to exist? Here the answer might depend more sensitively on precisely which changes we are envisaging, how those changes would be implemented, and on how the enhanced capacities would be used. The case for thinking that both personal identity and narrative identity would be preserved is arguably strongest if we posit that (a) the changes are in the form of addition of new capacities or enhancement of old ones, without sacrifice of preexisting capacities; and (b) the changes are implemented gradually over an extended period of time; (c) each step of the transformation process is freely and competently chosen by the subject; and (d) the new capacities do not prevent the preexisting capacities from being periodically exercised; (e) the subject retains her old memories and many of her basic desires and dispositions; (f) the subject retains many of her old personal relationships and social connections; and (g) the transformation fits into the life narrative and self-conception of the subject. Posthuman cognitive and emotional capacities could in principle be acquired in such a way that these conditions are satisfied.

Even if not all the conditions (a)–(g) were fully satisfied in some particular transformation process, the normatively relevant elements of a person's (numerical or narrative) identity could still be *sufficiently* preserved to avoid raising any fundamental identity-based objection to the prudentiality of undergoing such a transformation. We should not use a stricter standard for technological self-transformation than for other kinds of human transformation, such as migration, career change, or religious conversion.

---

[18] See, also, DeGrazia (2005). DeGrazia argues that identity-related challenges to human enhancement largely fails, both ones based on considerations of personal identity and ones based on narrative identity (authenticity), although he mainly discusses more moderate enhancements than those I focus on in this paper.

Consider again a familiar case of *radical* human transformation: maturation. You currently possess vastly greater cognitive capacities than you did as an infant. You have also lost some capacities, e.g. the ability to learn to speak a new language without an accent. Your emotional capacities have also changed and developed considerably since your babyhood. For each concept of identity which we might think has relevant normative significance – personal (numerical) identity, narrative identity, identity of personal character, or identity of core characteristics – we should ask whether identity in that sense has been preserved in this transformation.

The answer may depend on exactly how we understand these ideas of identity. For each of them, on a sufficiently generous conception of the identity criteria, identity was completely or in large part preserved through your maturation. But then we would expect that identity in that sense would also be preserved in many other transformations, including the ones that are *no more profound* as that of a child growing into an adult; and this would include transformations that would make you posthuman. Alternatively, we might adopt conceptions that impose more stringent criteria for the preservation of identity. On these conceptions, it might be impossible to become posthuman without wholly or in large part disrupting one form of identity or another. However, on such restrictive conceptions, identity would also be disrupted in the transformation of child into adult. Yet we do not think it is bad for a child to grow up. Disruptions of identity in those stringent senses form part of a normal life experience and they do not constitute a disaster, or a misfortune of any kind, for the individual concerned.

Why then should it bad for a person to continue to develop so that she one day matures into a being with posthuman capacities? Surely it is the other way around. If this had been our usual path of development, we would have easily recognized the failure to develop into a posthuman as a misfortune, just as we now see it as a misfortune for a child to fail to develop normal adult capacities.

Many people who hold religious beliefs are already accustomed to the prospect of an extremely radical transformation into a kind of posthuman being, which is expected to take place after the termination of their current physical incarnation. Most of those who hold such a view also hold that the transformation *could* be very good for the person who is transformed.

7. BRIEF SKETCHES OF SOME OBJECTIONS AND REPLIES

Objection: One might think that it would be bad for a person to be the only posthuman being since a solitary posthuman would not have any equals to interact with.

Reply: It is not necessary that there be only one posthuman.

Objection: The accumulated cultural treasures of humanity might lose their appeal to somebody whose capacities greatly exceeded those of the humans who produced them. More generally, challenges that seemed interesting to the person while she was still human might become trivial and therefore uninteresting to her when she acquires posthuman capacities. This could deprive posthumans of the good of meaningful achievements.

Reply: It is not clear why the ability to appreciate what is more complex or subtle should make it impossible to appreciate simpler things. Somebody who has learnt to appreciate Schoenberg may still delight in simple folk songs, even bird songs. A fan of Cézanne may still enjoy watching a sunrise.

Even if it were impossible for posthuman beings to appreciate some simple things, they could compensate by creating new cultural riches. If some challenges become too easy for posthumans, they could take on more difficult challenges. One might argue that an additional reason for developing posthuman cognitive capacities is that it would increase the range of interesting intellectual challenges open to us. At least within the human range of cognitive capacity, it seems that the greater one's capacity, the more numerous and meaningful the intellectual projects that one can embark on. When one's mind grows, not only does one get better at solving intellectual problems – entirely new possibilities of meaning and creative endeavor come into view.

Objection: A sense of vulnerability, dependence, and limitedness can sometimes add to the value of a life or help a human being grow as a person, especially along moral or spiritual dimensions.

Reply: A posthuman could be vulnerable, dependent, and limited. A posthuman could also be able to grow as a person in moral and spiritual dimensions without those extrinsic spurs that sometimes necessary to affect such growth in humans. The ability to spontaneously develop in these dimensions could be seen as an aspect of emotional capacity.

Objection: The very desire to overcome one's limits by the use of technological means rather than through one's own efforts and hard work could be seen as expressive of a failure to open oneself to the unbidden, gifted nature of life, or as a failure to accept oneself as one is, or as self-hate.

Reply: This paper makes no claims about the expressive significance of a desire to become posthuman, or about whether having such a desire marks one as a worse person, whether necessarily or statistically. The concern here rather is about whether being posthuman could be good, and whether it could be good for us to become posthuman.

Objection: A capacity obtained through a technological shortcut would not have the same value as one obtained through self-discipline and sacrifice.

Reply: I have argued that the possession of posthuman capacities could be extremely valuable even were the capacities effortlessly obtained. It is consistent with what I have said that achieving a capacity through a great expenditure of blood, sweat, and tears would further increase its value. I have not addressed what would be the *best* way of becoming posthuman. We may note, however, that is unlikely that we *could* in practice become posthuman other than via recourse to advanced technology.

Objection: The value of achieving a goal like winning a gold medal in the Olympics is reduced and perhaps annulled if the goal is achieved through inappropriate means (e.g. cheating). The value of possessing a capacity likewise depends on how the capacity was acquired. Even though having posthuman capacities might be extremely valuable if the capacities had been obtained by appropriate means, there are no humanly possible means that are appropriate. Any means by which humans could obtain posthuman capacities would negate the value of having such capacities.

Reply: The analogy with winning an Olympic medal is misleading. It is in the nature of sports competitions that the value of achievement is intimately connected with the process by which it was achieved. We may say that what is at stake in the analogy is not really the value of a medal, nor even the value of winning a medal, but rather (something like) winning the medal by certain specified means in a fair competition, in a non-fluke-like way, etc. Many other goods are not like this. When we visit the doctor in the hope of getting well, we do not usually think that the value of getting well is strongly dependent on the process by which health is achieved; health and the enjoyment of health are valuable in their own right, independently of how these states come about. Of course, we are concerned with the value of the means to getting well – the means themselves can have negative value (involving perhaps pain and inconvenience), and in evaluating the value of the consequences of an action, we take the value of the means into account as well as the value of the goal that they achieve. But usually, the fact that some means have negative value does not reduce the value of obtaining the goal state. The values that I have alleged could be derived from posthuman capacities are not like the value of an Olympic gold medal, but rather like the value of health… I am aware of no logical, metaphysical, or "in principle" reason why humans could not obtain post-human capacities in ways that would avoid recourse to immoral means of the sort that would "taint" the outcome.

8. CONCLUSION

I have argued, first, that some posthuman modes of being would be extremely worthwhile; and, second, that it could be good for most human beings to become posthuman.

I have discussed three general central capacities – healthspan, cognition, and emotion – separately for most of this paper. However, some of my arguments are strengthened if one considers the possibility of combining these enhancements. A longer healthspan is more valuable when one has the cognitive capacity to find virtually inexhaustible sources of meaning in creative endeavors and intellectual growth. Both healthspan and cognition are more valuable when one has the emotional capacity to relish being alive and to take pleasure in mental activity.

It follows trivially from the definition of "posthuman" given in this paper that we are not posthuman at the time of writing. It does not follow, at least not in any obvious way, that a posthuman could not also remain a human being. Whether or not this is so depends on what meaning we assign to the word "human". One might well take an expansive view of what it means to be human, in which case "posthuman" is to be understood as denoting a certain possible type of human mode of being – if I am right, an exceedingly worthwhile type.

WORKS CITED

American Psychiatric Association. (2000). *Diagnostic criteria from DSM-IV-TR.* (Washington, D.C.: American Psychiatric Association).

T. M. Arnesen, and O. F. Norheim (2003) 'Quantifying quality of life for economic analysis: time out for time trade off', *Medical Humanities* 29 (2): 81–86.

T. L. Beauchamp, and J. F. Childress (2001) *Principles of biomedical ethics* (New York, N.Y.: Oxford University Press).

N. Bostrom (2003). "The Transhumanist FAQ: v 2.1." World Transhumanist Association. Available from http://transhumanism.org/index.php/WTA/faq/.

N. Bostrom (2004). Transhumanist Values, in *Ethical Issues for the 21st Century*. F. Adams, ed., Philosophical Documentation Center Press.

N. Bostrom (2005). "The Fable of the Dragon-Tyrant." *Journal of Medical Ethics* 31(5): 273–277.

P. Brickman. and D. T. Campbell (1971). Hedonic relativism and planning the good society, in *Adaptation-level theory: A symposium*. M. H. Apley, ed. New York, Academic Press: 287–301.

J. Cohen and G. Langer (2005). "Most Wish for a Longer Life – Despite Broad Aging Concerns." ABC News/USA Today Poll. Available from http://abcnews.go.com/images/Politics/995a1Longevity.pdf.

L. J. Damschroder, B. J. Zikmund-Fisher, et al. (2005). "The impact of considering adaptation in health state valuation." *Social Science & Medicine* 61(2): 267–277.

D. DeGrazia (2005). "Enhancement Technologies and Human Identity." *Journal of Medicine and Philosophy* 30: 261283.

E. Diener and E. M. Suh (1998). "Subjective well-being and age: An international analysis." *Annual Review of Gerontology and Geriatrics* 17: 304–324.

W. Glannon (2002). "Identity, prudential concern, and extended lives." *Bioethics* 16(3): 266–83.

J. B. S. Haldane (1924). *Daedalus; or, Science and the future*. London, K. Paul, Trench, Trubner & co., ltd.

R. Hanson (2000). "Showing That You Care: The Evolution of Health Altruism." Available from http://hanson.gmu.edu/showcare.pdf.

N. K. Hayles (1999). *How we became posthuman: virtual bodies in cybernetics, literature, and informatics*. Chicago, Ill., University of Chicago Press.

P. O. Johansson (2002). "On the definition and age-dependency of the value of a statistical life." *Journal of Risk and Uncertainty* 25(3): 251–263.

F. R. Johnson, W. H. Desvousges, et al. (1998). "Eliciting stated health preferences: An application to willingness to pay for longevity." *Medical Decision Making* 18(2): S57-S67.

U. Kunzmann, T. Little, et al. (2000). "Is age-related stability of subjective well-being a paradox? Cross-sectional and longitudinal evidence from the Berlin Aging Study." *Psychology and Aging* 15(3): 511–426.

D. Lewis (1989). "Dispositional Theories of Value." *Proceedings of the Aristotelian Society, supp.* 63: 113–137.

R. McShine, G. T. Lesser, et al. (2000). "Older Americans hold on to life dearly." *British Medical Journal* 320(7243): 1206–1207.

G. E. Moore (1903). *Principia Ethica*. Cambridge, Cambridge University Press.

K. Murphy and R. Topel (2003). 'The Economic Value of Medical Research,' *Measuring the gains from medical research: an economic approach.* Chicago, University of Chicago Press.

W. Nordhaus (2003). The Health of Nations: The Contribution of Improved Health to Living Standards. *Measuring the gains from medical research: an economic approach,* in K. M. Murphy and R. H. Topel, eds. Chicago, University of Chicago Press: vi, 263 p.

D. Pearce (2004). "The Hedonistic Imperative." Available from http://www.hedweb.com/hedab.htm.

M. Sandel (2004). "The Case Against Perfection." *The Atlantic Monthly* 293(3).

J. Tsevat, N. V. Dawson, et al. (1998). "Health values of hospitalized patients 80 years or older." *Jama-Journal of the American Medical Association* 279(5): 371–375.

P. A. Ubel, G. Loewenstein, et al. (2003). "Whose Quality of Life? A Commentary exploring discrepancies between health state evaluations of patients and the general public." *Quality of Life Research* 12(6): 599–607.

W. K. Viscusi and J. E. Aldy (2003). "The value of a statistical life: A critical review of market estimates throughout the world." *Journal of Risk and Uncertainty* 27(1): 5–76.

World Bank. (2003). "EdStats – the World Bank's comprehensive Database of education statistics." 2006. Available from http://www1.worldbank.org/education/edstats/index.html.

DISCUSSION QUESTIONS

1. Bostrom argues that it could be good for a person to become posthuman with respect to cognitive ability, healthspan, and emotional capacity. Do you agree? Would you want to become posthuman in any of these respects (assuming the process of doing so is safe)? Why or why not?

2. Bostrom argues that personal identity could be preserved through a transition to being posthuman. What do you think constitutes personal identity? Given your understanding of personal identity, would an individual be the same person when they became posthuman?

3. Did you find Bostrom's response to the possible objections to his view persuasive? Why or why not?

4. Can you think of any objections to becoming posthuman that Bostrom does not consider?

5. If it is good for a person to become cognitively posthuman, as Bostrom argues, do you think that parents ought to technologically enhance their children to be cognitively posthuman? Would there be any ethical difference between what parents do now to help their children increase their cognitive capacities and the use of genetic and pharmacological technologies to help them do so?

6. This chapter focuses primarily on whether it would be good to become posthuman for the people who undergo the enhancement. Do you think it would be socially good – i.e. good for society as a whole – if there were widespread implementation of posthuman technological enhancement? Why or why not?

# 16

MORAL ENHANCEMENT[1]

Thomas Douglas

CHAPTER SUMMARY

Opponents of biomedical enhancement often claim that, even if such enhancement would benefit the enhanced, it would harm others. Thomas Douglas argues that this objection looks unpersuasive when the enhancement in question is a moral enhancement – i.e. a modification that will expectably leave the enhanced person with morally better motives than she had previously. In this chapter he: (1) describes one type of psychological alteration that would plausibly qualify as a moral enhancement; (2) argues that we will, in the medium-term future, probably be able to induce such alterations via biomedical intervention; and (3) defends future engagement in such moral enhancements against possible objections. His aim is to present this kind of moral enhancement as a counter-example to the view that biomedical enhancement is always morally impermissible.

RELATED READINGS

Introduction: Extrinsic Concerns (2.6); Respecting Nature (2.7.3)

Related Readings: Walter Glannon, *Psychopharmacology and Functional Neurosurgery* (Ch. 12); President's Council on Bioethics, *Beyond Therapy* (Ch. 14); Nick Bostrom, *Why I Want to be a Posthuman When I Grow Up* (Ch. 15)

1. INTRODUCTION

Biomedical technologies are routinely employed in attempts to maintain or restore health. But many can also be used to alter the characteristics of already healthy persons. Without thereby attributing any value to these latter alterations, I will refer to them as *biomedical enhancements*.

Biomedical enhancement is perhaps most apparent in sport, where drugs have long been used to improve performance (Verroken, 2005), but it is also

---

[1] This chapter is excerpted from Thomas Douglas (2008) 'Moral Enhancement,' *Journal of Applied Philosophy* (25): 228–245. It appears here by permission of John Wiley and Sons, Inc.

widespread in other spheres. Some musicians take beta-blockers to calm their nerves before performances (Tindall, 2004), a significant proportion of American college students report taking methylphenidate (Ritalin) while studying in order to improve performance in examinations (Johnston et al., 2003; Teter et al., 2006), and then, of course, there is cosmetic surgery. Research on drugs that may enhance memory (Lynch, 2002; Scott et al., 2002; Tully et al., 2003), the retention of complex skills (Yesavage et al., 2001), and alertness (Caldwell et al., 2000; Turner et al., 2003) suggests that the possibilities for biomedical enhancement are likely to grow rapidly in coming years. However, the morality of using biomedical technologies to enhance remains a matter of controversy. Some argue that it would be better if people were more intelligent, longer-lived, and physically stronger, and that there is no objection to using biomedical technologies to achieve these goals. But others hold that biomedical enhancement ought to be avoided.

2. THE BIOCONSERVATIVE THESIS

The opponents of enhancement do not all set out to defend a common and clearly specified thesis. However, several would either assent or be attracted to the following claim (henceforth, the Bioconservative Thesis):

> Even if it were technically possible and legally permissible for people to engage in biomedical enhancement, it would not be *morally permissible* for them to do so.[2]

The scope of this thesis needs to be clarified. By 'people', I mean here to include all currently existing people, as well as those people that may exist in the medium term future – say, the next one hundred years – but not people who may exist in the more distant future. Similarly, I mean to include under 'biomedical enhancement' only those enhancement practices that may plausibly become technically feasible in the medium term future. The opponents of enhancement may justifiably have little to say about enhancements that would take place in the distant future, or would require far-fetched technologies.

In what follows, I argue that the Bioconservative Thesis, thus qualified, is false.

3. A POSSIBLE COUNTER-EXAMPLE TO THE BIOCONSERVATIVE THESIS

The Bioconservative Thesis may be defended in various ways. But many of the most prevalent arguments for it are based on social considerations: though enhancement might be good for the enhanced individuals, it could well be bad

---

[2] Some writers may be opposed only to certain kinds of enhancement, but others appear to find *all* enhancement problematic, and perhaps impermissible, preferring that biomedical technology is used only maintain and restore health. The most prominent recent exponent of this view is Michael Sandel. See Sandel (2004, 2007 at pp. 12, 47–49).

for others.[3] Thus, regarding intelligence enhancement it could be argued that if one person makes herself more intelligent she will disadvantage the unenhanced by, for example, out-competing them for jobs, or by discriminating against them on the basis of their lower intelligence.[4]

These arguments may be persuasive when directed against the most commonly discussed biomedical enhancements – physical ability enhancements, intelligence and memory enhancements, and natural lifespan enhancements. But there are other types of biomedical enhancement against which they appear much less persuasive. In this paper I will focus on one possibility: that future people might use biomedical technology to *morally* enhance themselves.

There are various ways in which we could understand the suggestion that we morally enhance ourselves. To name a few, we could take it as a suggestion that we make ourselves more virtuous, more praiseworthy, more capable of moral responsibility, or that we make ourselves act or behave more morally. But I will understand it in none of these ways. Rather, I will take it as a suggestion that we cause ourselves to have morally better motives (henceforth often omitting the 'morally'). I understand motives to be the psychological – mental or neural – states or processes that will, given the absence of opposing motives, cause a person to act.[5]

Since I focus only on motives, I will not claim that the morally enhanced person *is* more moral, has a more moral character, or will necessarily act more morally than her earlier, unenhanced self. I will also try to avoid committing myself to any particular view about what determines the moral goodness of a motive. For example, I will, insofar as possible, remain neutral between the views that the moral goodness of a motive is determined by the sort of acts it motivates, the character traits it partially constitutes, the consequences of its existence, or its intrinsic properties.

With these qualifications in hand, I now set out my formula for moral enhancement:

> A person morally enhances herself if she alters herself in a way that may reasonably be expected to result in her having morally better future motives, taken in sum, than she would otherwise have had.

[...] I will argue that, when performed under certain conditions, there would be no good objection – social or other – to biomedical moral enhancement. I will suggest that it would, contrary to the Bioconservative Thesis, be morally

---

[3] See, for example, Annas (2002); Fukuyama (2002, p. 97); McKibben (2005).

[4] For competitiveness-based objections to enhancement, see Buchanan et al., (2000, pp. 188–191); McKibben (2003); Farah et al. (2004, p. 423); Sandel (2007, pp. 8–12). For discrimination based objections, see, for example, Sandel (2007, p. 15).

[5] I focus on the morality of motives because I take this to be common ground. Some Kantians might deny that acts or behaviour are the proper objects of moral appraisal, and some of those who regard acts as the most basic units of moral appraisal might shy away from making judgments of moral character. But I think that all, or nearly all, would accept that motives come in varying degrees of morality, even if their morality derives ultimately from the behaviour that they motivate or the virtues they derive from or constitute.

permissible for people to undergo such enhancements. Before proceeding to my argument, however, it is necessary to say something more about how moral enhancement might work.

4. THE NATURE OF MORAL ENHANCEMENT

There is clearly scope for most people to morally enhance themselves. According to every plausible moral theory, people often have bad or suboptimally good motives. And according to many plausible theories, some of the world's most important problems – such as developing world poverty, climate change and war – can be attributed to these moral deficits....

I think it would be possible to identify several kinds of psychological change that would, for some people under some circumstances, uncontroversially qualify as moral enhancements. I will focus solely on one possibility here. My thought is that there are some emotions – henceforth, the counter-moral emotions – whose attenuation would sometimes count as a moral enhancement regardless of which plausible moral and psychological theories one accepted. I have in mind those emotions which may interfere with all of the putative good motives (moral emotions, reasoning processes, and combinations thereof) and/or which are themselves uncontroversially *bad* motives. Attenuating such emotions would plausibly leave a person with better future motives, taken in sum.

One example of a counter-moral emotion might be a strong aversion to certain racial groups. Such an aversion would, I think, be an uncontroversial example of a bad motive. It might also *interfere with* what would otherwise be good motives. It might, for example, lead to a kind of subconscious bias in a person who is attempting to weigh up the claims of competing individuals as part of some reasoning process. Alternatively, it might limit the extent to which a person is able to feel sympathy for a member of the racial group in question.

A second example would be the impulse towards violent aggression. This impulse may occasionally count as a good motive. If I am present when one person attacks another on the street, impulsive aggression may be exactly what is required of me. But, on many occasions, impulsive aggression seems to be a morally bad motive to have – for example, when one has just been mildly provoked. Moreover, as with racial aversion, it could also interfere with good motives. It might, for example, cloud a person's mind in such a way that reasoning becomes difficult and the moral emotions are unlikely to be experienced.

I suspect, then, that for many people the mitigation of an aversion to certain racial groups or a reduction in impulsive violent aggression would qualify as a moral enhancement – that is, it would lead those people to expectably have better motives, taken in sum, than they would otherwise have had. However, I do not want, or need, to commit myself to this claim here. Rather, I will stake myself to the following weaker claim: there are some emotions such that a reduction in the degree to which an agent experiences those emotions would, under some circumstances, constitute a moral enhancement....

5. THE POSSIBILITY OF BIOMEDICAL MORAL ENHANCEMENT

I will tentatively argue that it would sometimes be morally permissible for people to biomedically mitigate their counter-moral emotions. But first I want to briefly consider what might appear to be a prior question. Will this sort of biomedical moral enhancement be possible within the medium-term time span that we are considering?

There are two obvious reasons for doubting that biomedical moral enhancement will, in the medium term, become possible. The first is that there are, on some views about the relationship between mind and brain, some aspects of our moral psychology that cannot in principle be altered through biological intervention.[6] This is not the place to explore this claim. I hope it suffices merely to note that it is not a mainstream philosophical position. The second ground for doubt is that our moral psychology is presumably highly complex – arguably, so complex that we will not, within the medium term future, gain sufficient understanding of its neuroscientific basis to allow the informed development of appropriate biomedical interventions.

Consider the two emotions that I mentioned earlier – aversion to certain racial groups, and impulses towards violent aggression. Work in behavioural genetics and neuroscience has lead to an early but growing understanding of the biological underpinnings of both. There has long been evidence from adoption and twin studies of a genetic contribution to aggression (Crowe, 1974; Cadoret, 1978; Grove et al., 1990), and there is now growing evidence implicating a polymorphism in the Monoamine Oxidase A gene (Brunner et al., 1993a, Brunner et al., 1993b), and, at the neuro-physiological level, derangements in the serotonergic neurotransmitter system (Caspi and McClay, 2002; De Almeida et al., 2005). Race aversion has been less well studied. However, a series of recent functional magnetic resonance imaging studies suggest that the amygdala – part of the brain already implicated in the regulation of emotions – plays an important role (Hart et al., 2000; Phelps et al., 2000; Cunningham et al., 2004). Given this progress in neuroscience, it does not seem unreasonable to suppose that moral enhancement technologies which operate on relatively simple emotional drives could be developed in the medium term.

6. THE SCENARIO

I am now in a position to set out the conditions under which it would, I will argue, be morally permissible for people to morally enhance themselves. These conditions are captured in a scenario consisting of five assumptions.[7]

---

[6] Most obviously, this would be held by mind-body parallelists who believe that mind and brain are causally insulated from one another. The most famous exponent of this view is Leibniz (1973).

[7] I also assume, as a background to the listed assumptions, that Smith is a normal person living in a world similar to our own – that is, a world governed by the scientific and social scientific principles that we take to govern our own world.

The first assumption simply specifies that we are dealing with an enhancement that satisfies my formula for moral enhancement:

> **Assumption 1**. Through undergoing some biomedical intervention (for example, taking a pill) at time T, an agent Smith can bring it about that he will expectably have better post-T motives than he would otherwise have had.

In order to focus on the situation where the case for moral enhancement is, I think, strongest, I introduce a second assumption as follows:

> **Assumption 2**. If Smith does not undergo the intervention, he will expectably have at least some bad (rather than merely suboptimally good) motives.

A third assumption captures my earlier claim about how, as a matter of psychology, moral enhancement might work:

> **Assumption 3**. The biomedical intervention will work by attenuating some emotion(s) of Smith's.

And finally, the fourth and fifth assumptions rule out what I take to be uninteresting objections to moral enhancement: that it might have adverse side effects, and that it might be done coercively or for other unnecessarily bad reasons:

> **Assumption 4**. The only effects of Smith's intervention will be (a) to alter Smith's psychology in those (and only those) ways necessary to bring it about that he expectably has better post-T motives, and (b) consequences of these psychological changes.
>
> **Assumption 5**. Smith can, at T, freely choose whether or not to morally enhance himself, and if he chooses to do so, he will make this choice for the best possible reasons (whatever they might be).[8]

Would it be morally permissible for Smith to morally enhance himself in these circumstances? I will argue that, probably, it would.

7. REASONS TO ENHANCE

Smith clearly has some moral reason to morally enhance himself: if he does, he will expectably have a better set of motives than he would otherwise have had, and I take it to be uncontroversial that he has some moral reason to bring this result about. (I henceforth omit the 'moral' of 'moral reason'.)

Precisely why he has such reason is open to question. One explanation would run as follows. If Smith brings it about that he expectably has better motives, he expectably brings at least one good consequence about: namely, his having better

---

[8]   I take it that Assumption 5 entails at least that there is no physical or legal constraint on Smith's morally enhancing himself.

motives.[9] And plausibly, we all have at least some moral reason to expectably bring about any good consequence.

This explanation is weakly consequentialist in that it relies on the premise that we have good reasons to expectably bring about any good consequence. But thoroughgoing nonconsequentialists could offer an alternative explanation. They could, for example, maintain that Smith's *act* of moral enhancement has some intrinsic property – such as the property of being an act of self-improvement – that gives him reason to perform it.

But regardless of *why* Smith has reason to morally enhance himself in our scenario, I take it to be intuitively clear that he has such reason. This intuition can, moreover, be buttressed by intuitions about closely related cases. Suppose that some agent Jones is in precisely the same position as Smith, except that in her case, the moral enhancement can be attained not through biomedical means but through some form of self- education – for example, by reflecting on and striving to attenuate her counter-moral emotions. Intuitively, Jones has some reason to morally enhance herself – or so it seems to me. And if pressed on why she has such reason, it seems natural to point to features of her situation that are shared with Smith's – for example, that her morally enhancing herself would have expectably good consequences, or that it may express a concern for the interests of others.[10]

8. REASONS NOT TO ENHANCE

Smith may also, of course, have reasons not to morally enhance himself, and I now turn to consider what these reasons might be.[11]

8.1 Objectionable motives

One possibility is that Smith has reason not to enhance himself because he could only do so from some bad motive. I assumed, in setting up the Smith scenario, that if he enhances himself, he will do so from the best possible motive. But the best possible motive may not be good enough.

There are various motives that Smith *could* have for morally enhancing himself. And some of these seem quite unobjectionable: he may believe that he ought to morally enhance himself, he may have a desire to act morally in the future, or he may be moved simply by a concern for the public good. However, we should consider, at this point, an objection due to Michael Sandel. Sandel argues that engaging in enhancement expresses an excessive desire to change oneself, or insufficient acceptance of 'the given'. And since we have reasons to avoid such motives,

---

[9] Smith might also bring many other expectably good consequences about – for example, those that follow from his expectably having good motives.

[10] I do not claim that Smith's reason to engage in moral enhancement is as strong as Jones's.

[11] The reasons considered in this section are based on a range of different substantive moral views. I do not claim that there is any one moral viewpoint which could accommodate all of the putative reasons discussed.

we have, he thinks, reasons to refrain from enhancing ourselves (Sandel, 2004, pp 50–65; Sandel, 2007).

It would be difficult to deny that Smith's moral enhancement would, like any voluntary instance of enhancement, be driven to some extent by an unwillingness to accept the given (though this need not be his conscious motive). Here, we must agree with Sandel. But what is less clear is that this gives Smith any reason to refrain from enhancement. Leaving aside any general problems with Sandel's suggestion, it faces a specific problem when applied to the case of Smith. Applied to that case, Sandel's claim would be that Smith has reason to accept his bad motives, as well as that which interferes with his good motives. But this is implausible. Surely, if there are any features of himself that he should not accept, his bad motives and impediments to his good motives are among them. The appropriate attitude to take towards such properties is precisely one of *non-acceptance* and a *desire for self-change*.

8.2 Objectionable means

A second reason that Smith might have not to morally enhance himself is that the biomedical means by which he would do so are objectionable.

We can distinguish between a weak and a strong version of the view that Smith's proposed means are objectionable. On the weak version, his means are objectionable in the sense that it would be *better* if he morally enhanced himself via non-biomedical means. There is certainly some intuitive appeal to this view. It might seem preferable for Smith to enhance himself through some sort of moral training or self-education.

When compared with self-education, taking a pill might seem 'all too easy' or too disconnected from ordinary human understanding (Kass, 2003, pp. 21–24; President's Council on Bioethics, 2003, pp. 290–293). Arguably, given the choice between biomedical moral enhancement and moral enhancement via self-education, Smith would have strong reasons to opt for the latter.

Note, however, that Smith's choice is not between alternative means of enhancement, but simply between engaging in biomedical moral enhancement or not. Reasons that Smith has to engage in moral enhancement via other means will be relevant to Smith's choice only to the extent that whether he engages in biomedical moral enhancement will influence the extent to which he seeks moral enhancement through those other means. If Smith's morally enhancing himself through biomedical means would lead him to engage in *less* moral enhancement through some superior means (say, via self- education), then Smith may have some reason not to engage in biomedical moral enhancement. But it is difficult to see why Smith would regard biomedical enhancement and self-education as substitutes in this way. It seems at least as likely that he would regard them as complementary; having morally enhanced himself in one way, he may feel more inclined to morally enhance himself in the other (say, because he enjoys the experience of acting on good motives).

One might, at this point, turn to a stronger version of the 'objectionable means' claim, arguing that to adopt biomedical means to moral enhancement is objectionable not just relative to other alternative means, but in an *absolute* sense. Indeed, it is so absolutely objectionable that any moral benefits of Smith's morally enhancing himself would be outweighed or trumped by the moral costs of using biomedical intervention as a means.

Any claim that biomedical means to moral enhancement are absolutely objectionable is likely to be based on a claim that they are unnatural. Certainly, this is a common means-based criticism levelled at biomedical enhancement.[12] But the problem is to come up with some account of naturalness (or unnaturalness) such that it is true both that:

[1] using biomedical means to morally enhance oneself is unnatural, and that:
[2] this unnaturalness gives a person reason not to engage in such enhancement.

Can any such account be found? David Hume distinguished between three different concepts of nature; one which may be opposed to 'miracles', one to 'the rare and unusual', and one to 'artifice' (Hume, 1888, pp. 473–475). This taxonomy suggests a similar approach to the concept of unnaturalness. We might equate unnaturalness with miraculousness (or supernaturalness), with rarity or unusualness, or with artificiality. In what follows I will consider whether any of these concepts of naturalness succeeds in rendering both [1] and [2] plausible.

8.3 Unnaturalness as supernaturalness

Consider first the concept of unnaturalness as supernaturalness. On one popular account of this concept, something like the following is true: something is unnatural if, or to the extent that, it lies outside the world that can be studied by the sciences.[13] It seems clear, on this view, that biomedical interventions are not at all unnatural, for such interventions are precisely the sort of thing that *could* be studied by the sciences. The concept of unnaturalness as supernaturalness thus renders [1] clearly false.

8.4 Unnaturalness as unusualness

The second concept of unnaturalness suggested by Hume's analysis is that which can be equated with unusualness or unfamiliarity. Leon Kass's idea of unnaturalness as disconnectedness from everyday human understanding may be a variant of this concept.

Unusualness and unfamiliarity are relative concepts in the following way: something has to be unusual or unfamiliar *for* or *to* someone. Thus, whether Smith's biomedical intervention would qualify as unnatural may depend on whom we relativise unusualness and unfamiliarity to. For us inhabitants of the present day, the use of biomedical technology for the purposes of moral enhancement

---

[12] See, for example, Kass (2003, pp. 17, 20–24); President's Council on Bioethics (2003, pp. 290–293).
[13] See, for example, Moore (1903, p. 92).

certainly does qualify as unusual and unfamiliar, and thus, perhaps, as unnatural. But for some future persons, it might not. Absent any specification of how to relativise unusualness or unfamiliarity, it is indeterminate whether [1] is true.

We need not pursue these complications, however, since regardless of whether [1] comes out as true on the current concept of unnaturalness, [2] appears to come out false. It is doubtful whether we have any reason to avoid adopting means merely because they are unusual or unfamiliar, or, for that matter, disconnected from everyday human understanding. We may often prefer familiar means to unfamiliar ones on the grounds that predictions about their effects will generally be better informed by evidence, and therefore more certain. Thus, if I am offered the choice between two different drugs for some medical condition, where both are thought to be equally safe and effective, I may choose the more familiar one on the grounds that it will probably have been better studied and thus have more certain effects. But the concern here is not ultimately with the unnaturalness – or any other objectionable feature – of the means, but rather with the effects of adopting it. I will return to the possible adverse effects of Smith's enhancement below. The position I am interested in here is whether the unfamiliarity of some means gives us reasons not to use it *regardless* of its effects. To affirm that it does seems to me to involve taking a stance that is inexplicably averse to novelty.

8.5 Unnaturalness as artificiality

Consider finally the concept of unnaturalness as artificiality. This is arguably the most prevalent concept of naturalness to be found in modern philosophy. It may be roughly characterised as follows: something is unnatural if it involves human action, or certain types of human action (such as intentional action).

Claim [1] is quite plausible on this concept of unnaturalness. Biomedical interventions clearly involve human action – and almost always intentional action. However, [2] now looks rather implausible. *Whenever* we intentionally adopt some means to some end, that means involves intentional human action. But it does not follow that we have reason not to adopt that means. If it did, we would have reason not to intentionally adopt any means to any end. And this surely cannot be right.

The implausibility of [2] on the current concept of unnaturalness can also be brought out by returning to the case where moral enhancement is achieved through self-education rather than biomedical intervention. Such enhancement seems unproblematic, yet it clearly involves unnatural means if unnaturalness is analysed as involving or being the product of (intentional) human action.

We should consider, at this point, a more restrictive account of unnaturalness as artificiality: one which holds that, in order to qualify as unnatural, something must not only involve (intentional) human action, it must also involve *technology* – the products of highly complex and sophisticated social practices such as science and industry. Moving to this account perhaps avoids the need to classify practices such as training and education as unnatural. But it still renders unnatural many practices which, intuitively, we may have no means-based reasons to avoid. Consider, for example, the treatment of disease. This frequently involves biomedical technology, yet it is not clear that we have any means-based reasons not to engage in it. To avoid this problem, the concept of unnaturalness as artificiality would have to be limited still further, such that technology-involving means count as unnatural only if they are not aimed at the treatment of disease. On this view, Smith's means are not unnatural in themselves. Rather the unnaturalness arises from the combination of his means with certain intentions or aims. Perhaps by restricting the concept of unnaturalness in this way, we avoid classifying as unnatural practices (such as self-education, or the medical treatment of diseases) that seem clearly unobjectionable. However, it remains unclear *why*, on this account of the unnatural, we should have reasons to avoid unnatural practices. In attempting to show that Smith has reason not to engage in biomedical moral enhancement, it is not enough to simply stipulate some concept of unnaturalness according to which his engaging in moral enhancement comes out as unnatural while seemingly less problematic practices come out as natural. It must be shown that a practice's being unnatural *makes* it problematic, or at least provides evidence for its being problematic. Without such a demonstration, the allegation of unnaturalness does no philosophical work, but merely serves as a way of asserting that we have reasons to refrain from biomedical moral enhancement.

8.6 Objectionable means?

I have argued that none of the three concepts of unnaturalness suggested by Hume's analysis renders both [1] and [2] plausible. If my conclusions are correct, it follows that none of these concepts of unnaturalness point to any means-based reason for Smith to refrain from moral enhancement. There may be some further concept of unnaturalness on the basis of which one could argue more convincingly for [1] and [2]. Or there may be some way of showing that biomedical moral enhancement involves means that are objectionable for reasons other than their unnaturalness. But I am not sure what the content of these concepts and arguments would be.

8.7 Objectionable consequences

Would the consequences of Smith's enhancement provide him with reasons to refrain from engaging in that enhancement? Two points about this possibility need to be noted up front. First, since we are assuming that Smith's moral enhancement will have no side-effects (Assumption 4), the only consequences that his action will have are:

(a) That he will expectably have better post-T motives than he would otherwise have had
(b) Those, and only those, psychological changes necessary to bring about (a)
(c) Consequences that follow from (a) and (b)

Thus, if Smith has consequence-based reasons to avoid moral enhancement, those reasons must be grounded on the features – presumably the intrinsic badness – of (a), (b) or (c).

Second, there are some moral theories which constrain whether, or to what extent, consequences (a) and (c) could be bad. Consider theories according to which only hedonic states (such as states of pleasure or pain) can be intrinsically good or bad. On these theories, (a) could not be intrinsically bad since motives are not hedonic states. Consider, alternatively, a consequentialist moral theory according to which the moral goodness of a motive is determined by the goodness of the consequences of a person's having it. On this theory, if Smith indeed has better post-T motives, then the consequences of his having those motives – these fall under (c) – must be better than the corresponding consequences that would have come about had he had worse motives. Smith's having better motives is guaranteed to have better consequences than his having worse motives because having good consequences is what makes a motive good. In what follows, I will assume, for the sake of argument, that moral theories which limit the possible badness of (a) and (c) in these ways are false.

8.8 Identity change

One bad effect of Smith's morally enhancing himself might be that he loses his identity. Worries about identity loss have been raised as general objections to enhancement, and there is no obvious reason why they should not apply to cases of moral enhancement.[14] Clearly, moral enhancement of the sort we are considering need not be identity-altering in the strong sense that Smith will, post-enhancement, be a different person than he was before. Our moral psychologies change all the time, and sometimes they change dramatically, for example, following particularly traumatic experiences. When these changes occur, we do not think that one person has literally been replaced by another. However, perhaps Smith's moral enhancement would be identity-altering in the weaker sense that it would change some of his most fundamental psychological characteristics – characteristics that are, for example, central to how he views himself and his relationships with others, or that pervade his personality.[15] Suppose we concede that Smith's moral enhancement would be identity-altering in this weaker sense. This may not give Smith any reason to refrain from undergoing the change. Plausibly, we have reasons to preserve our fundamental psychological characteristics only where those characteristics have some positive value. But though Smith's counter-moral emotions *may* have some value (Smith may, for example, find their experience pleasurable), they need not.

8.9 Restricted freedom

By morally enhancing himself Smith will bring it about that he has better post-T motives, taken in sum, than he would otherwise have had. However, it might

---

[14] See, for example, Wolpe (2003, pp. 393–394); President's Council on Bioethics (2003, p. 294).

[15] See, for a discussion of this weaker sense of 'identity', Schechtman (1996, esp. at pp. 74 –76).

be thought that this result will come at a cost to his freedom: namely, he will, after *T*, lack the freedom to have and to act upon certain bad motives. And even though having and acting upon bad motives may itself have little value, it might be thought that the *freedom* to hold and act upon them is valuable. Indeed, this freedom might seem to be a central element of human rational agency. Arguably, Smith has reasons not to place restrictions on this freedom. The objection that I am considering here can be captured in the following two claims:

[3] Smith's morally enhancing himself will result in his having less freedom to have and to act upon bad motives.
[4] Smith has reason not to restrict his freedom to have and act upon bad motives.

Claim [4] is, I think, problematic. It is not obvious that the freedom referred to therein has any value. Moreover, even if this freedom does have value, there may be no problem with restricting it provided that the restriction is itself self-chosen, as in Smith's case it is. However, I will focus here on [3]. The proponent of [3] is committed to a certain understanding of freedom. She would have to maintain that freedom consists not merely in the absence of external constraints, but also in the absence of internal psychological constraints, for it is only Smith's internal characteristics that would be altered by his moral enhancement. This view could be sustained by regarding the self as being divided into two parts – the true or authentic self, and a brute self that is external to this true self. One could then regard any aspect of the brute self which constrains the true self as a constraint on freedom.[16] And one could defend [3] on the ground that Smith's enhancement will alter his brute self in such a way that it will constrain his autonomous self.

There would be some justification for thinking that Smith's moral enhancement would alter his brute self rather than his true self. We are assuming that Smith's enhancement will attenuate certain emotions, so it will presumably work by altering the brain's emotion-generating mechanisms, and these mechanisms are arguably best thought of as part of the brute self. Certainly, it would be strange to think of the predominantly subconscious mechanisms which typically call forth racial aversion or impulsive aggression as part of the true autonomous self.

However, the view that moral enhancement would alter Smith's brute self in a way that would *interfere with* his autonomous self seems to be at odds with my assumption (Assumption 3) about the mechanism of that enhancement. Since Smith's enhancement is assumed to attenuate certain emotions, it presumably works by *suppressing* those brute mechanisms that generate the relevant emotions. The enhancement seems to work by *reducing* the influence of Smith's brute self and thus allowing his true self *greater* freedom. It would be more accurate to say that the enhancement increases Smith's freedom to have and to act upon good motives than to say that it diminishes his freedom to have and to act upon bad ones…

---

[16] See, for an example of this approach, Taylor (1979, pp. 175–193).

9. IMPLICATIONS

I have argued that Smith has some reason to morally enhance himself via biomedical means. I have also rejected several arguments for the existence of good countervailing reasons. Thus, I hope that I have offered some support for the claim that it would be morally permissible for Smith to engage in biomedical moral enhancement. But if it would be permissible for Smith to morally enhance himself, then the Bioconservative Thesis is almost certainly false. For as I claimed earlier, it is plausible that biomedical moral enhancement technologies will become technically feasible in the medium term future. And it is almost certain that, if they do become feasible, some – probably many – actual future people will find themselves in scenarios sufficiently like Smith's that our conclusions about Smith will apply to them also: contrary to the Bioconservative Thesis, there will be people for whom it would be morally permissible to engage in biomedical enhancement.

I should end, however, by noting that the Bioconservative Thesis is not the only claim advanced by the opponents of enhancement. As well as claiming that it would not be morally permissible for people to enhance themselves, many bioconservatives would assert that it would not be permissible for us to *develop* technologies for enhancement purposes, nor for us to *permit* enhancement. For all that I have said, these claims may well be true. It would not follow straightforwardly from the fact that it would be permissible for some future people to morally enhance themselves – given the presence of the necessary technology and the absence of legal barriers – that they could permissibly be allowed to do so, or that we could permissibly develop the technologies whose availability we are taking as given. Other factors would need to be considered here. It may be, for example, that if we were to develop moral enhancement technologies, we would be unable to prevent their being used in undesirable ways – for example, to enhance self-interestedness or *im*morality. Whether we could permissibly develop or permit the use of moral enhancement technologies might thus depend on a weighing of the possible good uses of those technologies against the possible bad ones.

ACKNOWLEDGEMENTS

I would like to thank, for their comments on earlier versions of this paper, Julian Savulescu, David Wasserman, S. Matthew Liao, Ingmar Persson, Allen Buchanan, Rebecca Roache, Roger Crisp, two anonymous reviewers for the *Journal of Applied Philosophy*, and audiences at Otago, Oxford and Hong Kong Baptist Universities.

WORKS CITED

R. M. M. de Almeida, P. F. Ferari, S. Parmigiani, *et al.* (2005) 'Escalated aggressive behavior: Dopamine, serotonin and GABA,' *European Journal of Pharmacology*, 526: 51–64.

G. A. Annas (2002) 'Cell division,' *Boston Globe*, 21 April.

H. G. Brunner, M. R. Nelen, X. O. Breakefield, *et al.* (1993) 'Abnormal behaviour associated with a point mutation in the structural gene for Monoamine Oxidase A,' *Science* 262, 5133: 578–580.

H. G. Brunner, M. R. Nelen, P. van Zandvoort, *et al.* (1993) 'X-linked borderline mental retardation with prominent behavioural disturbance: phenotype, genetic localization, and evidence for disturbed monoamine metabolism,' *American Journal of Human Genetics*, 52, 6: 1032–1039.

A. Buchanan, D. Brock, N. Daniels, and D. Wikler (2000) *From Chance to Choice: Genetics and Justice* (Cambridge: Cambridge University Press).

R. J. Cadoret (1978) 'Psychopathology in adopted-away offspring of biologic parents with antisocial behavior,' *Archives of General Psychiatry* 35: 176–184.

A. Caldwell, J. L. Caldwell & N. K. Smythe, 'A double-blind, placebo-controlled investigation of the efficacy of modafinil for sustaining the alertness and performance of aviators: a helicopter simulator study,' *Psychopharmacology* 150 (2000): 272–282.

A. Caspi & J. McClay, 'Evidence that the cycle of violence in maltreated children depends on genotype', *Science* 297 (2002): 851–854.

R. R. Crowe (1974) 'An adoption study of antisocial personality,' *Archives of General Psychiatry* 31: 785–791.

W. A. Cunningham, M. K. Johnson, C. L. Raye, *et al.* (2004) 'Separable neural components in the processing of black and white faces,' *Psychological Science* 15: 806–813.

M. J. Farah, J. Illes, R. Cook-Deegan, *et al.* (2004) 'Neurocognitive enhancement: what can we do and what should we do?' *Nature Reviews Neuroscience* 5: 421–425.

F. Fukuyama (2002) *Our Posthuman Future: Consequences of the Biotechnology Revolution* (New York: Farrar, Straus, and Giroux).

W. M. Grove, E. D. Eckert, L. Heston, *et al.* (1990) 'Heritability of substance abuse and antisocial behavior: a study of monozygotic twins reared apart,' *Biological Psychiatry* 27: 1293–1304.

A. J. Hart, P. J. Whalen, L. M. Shin *et al.*, 'Differential response in the human amygdala to racial outgroup vs. ingroup face stimuli', *Neuroreport: For Rapid Communication of Neuroscience Research* 11 (2000): 2351–2355.

D. Heyd, 'Human nature: an oxymoron?' *Journal of Medicine and Philosophy* 28, 2 (2003): 151–169.

D. Hume, *A Treatise of Human Nature*, L. A. Selby-Bigge, ed. (Oxford: Clarendon, 1888).

D. Johnston, P. M. O'Malley & J. G. Bachman, *Monitoring the Future National Survey Results on Drug Use, 1975 –2002: II. College Students and Adults Ages 19 – 40* (Washington DC: US Department of Health and Human Services, 2003).

L. R. Kass, 'Ageless bodies, happy souls: biotechnology and the pursuit of perfection,' *The New Atlantis* 1 (2003): 9–28.

G. W. Leibniz. 'New system, and explanation of the new system' in his *Philosophical Writings*, G. H. R. Parkinson, ed., M. Morris, trans. (London: Dent, 1973).

Lynch, 'Memory enhancement: the search for mechanism-based drugs,' *Nature Neuroscience* 5 (2002): 1035–1038.

B. McKibben, 'Designer genes,' *Orion*, 30 April 2003.

M. J. Mehlman, 'Genetic enhancement: plan now to act later', *Kennedy Institute of Ethics Journal* 15, 1 (2005): 77–82.

G. E. Moore, *Principia Ethica* (Cambridge: Cambridge University Press, 1903), p. 92.

E. A. Phelps, K. J. O'Connor, W. A. Cunningham *et al.*, 'Performance on indirect measures of race evaluation predicts amygdala activation,' *Journal of Cognitive Neuroscience* 12 (2000): 729–738.

President's Council on Bioethics, *Beyond Therapy: Biotechnology and the Pursuit of Happiness* (Washington DC: President's Council on Bioethics, 2003) at pp. 290–293.

M. J. Sandel (2004) 'The case against perfection,' *The Atlantic Monthly* 293, 3: 50–65.

M. J. Sandel (2007) *The Case Against Perfection: Ethics in the Age of Genetic Engineering* (Cambridge, MA: Harvard University Press).

M. Schechtman (1996) *The Constitution of Selves* (Ithaca, NY: Cornell University Press).

R. Scott, R. Bourtchouladze, S. Gossweiler, *et al.* (2002) 'CREB and the discovery of cognitive enhancers,' *Journal of Molecular Neuroscience* 19: 171–177.

C. Taylor (1979) 'What's wrong with negative liberty' in A. Ryan (ed.) *The Idea of Freedom* (London: Oxford University Press), pp. 175–193.

C. J. Teter, S. E. McCabe, K. LaGrange, *et al.* (2006) 'Illicit use of specific stimulants among college students: prevalence, motives, and routes of administration,' *Pharmacotherapy*, 26, 10:1501–1510.

B. Tindall (2004) 'Better playing through chemistry,' *New York Times*, 17 Oct.

T. Tully, R. Bourtchouladze, R. Scott *et al.* (2003) 'Targeting the CREB pathway for memory enhancers,' *Nature Reviews Drug Discovery* 2: 267–277.

D. C. Turner, T. W. Robbins, L. Clark, A. R. Aron, J. Dowson, and B. J. Sahakian (2003) 'Cognitive enhancing effects of modafinil in healthy volunteers,' *Psychopharmacology* 165: 260–269.

M. Verroken (2005) 'Drug use and abuse in sport,' in D. R. Mottram (ed.) *Drugs in Sport* (London: Routledge).

P. R. Wolpe (2002) 'Treatment, enhancement, and the ethics of neurotherapeutics,' *Brain and Cognition* 50: 387–395.

Yesavage, M. Mumenthaler, J. Taylor, *et al.* (2001) 'Donezepil and flight simulator performance: effects on retention of complex skills,' *Neurology* 59: 123–125.

DISCUSSION QUESTIONS

1. Would you be more hesitant to technologically enhance or modify psychological aspects of yourself than you would be to enhance you cognitive capacities? If so, why would you be more reluctant?

2. Do you agree that, when it comes to improving oneself morally, the most important thing is that you improve yourself, rather than the means by which you do so? Do you think the means is ethically significant at all?

3. In what senses is technological enhancement "unnatural"? Do any of those senses of "unnatural" imply wrongness?

4. The author suggests that moral enhancement could actually increase a person's autonomy by reducing the influence of brute self on her behavior,

thereby increasing the influence of her autonomous self. What is the difference between a person's brute self and her autonomous self? Do you agree that the autonomous self is more authentic, and that by increasing its influence on behavior moral enhancement could increase autonomy?

5. Do you think that moral enhancement has the potential to alter personal identity and responsibility? Why or why not?

6. Can you think of any considerations, in addition to the ones discussed in this chapter, that might be offered in support of the bioconservative thesis?

ENHANCING JUSTICE?[1]

Tamara Garcia and Ronald Sandler

CHAPTER SUMMARY

In this chapter, Tamara Garcia and Ronald L. Sandler focus on the following question: Are human enhancement technologies likely to be justice impairing or justice promoting? After defining human enhancement and defending a basic principle of justice, they present the core argument for the conclusion that they will be justice impairing. They then consider several responses that have been offered to this argument, and argue that each one of the responses fails. They conclude that human enhancement technologies may not be inherently just or unjust, but when situated within obtaining social contexts they are likely to exacerbate rather than alleviate social injustices. They do not conclude from this that enhancement technologies should not be developed, but that proponents of human enhancement technologies should be concerned as much about education and health care reform as they are about technology development.

RELATED READINGS

Introduction: Justice, Access, and Equality (2.6.2)

Related Readings: President's Council on Bioethics, *Beyond Therapy* (Ch. 14); Thomas Douglas, *Moral Enhancement* (Ch. 16)

1. INTRODUCTION

Emerging technologies situated at the intersections of nanotechnology, biotechnology, information technology, cognitive science, computer science, and robotics have the potential to significantly increase or augment human cognitive, psychological, and physical capabilities, e.g., learning speed, information retention, perception, endurance, strength, longevity, and emotional regulation. This article focuses on the following question: Are such human enhancement technologies likely to be justice impairing or justice promoting? We begin by defining the conception of human enhancement and the principle of distributive justice that are operative in this article. We then review the standard argument that development

---

[1] This chapter originally appeared as Tamara Garcia and Ronald Sandler (2008) 'Enhancing Justice?' *Nanoethics* (2): 277–287. It appears here by permission of Springer and the authors.

and dissemination of robust human enhancement technologies is likely to be social justice impairing, before presenting and critically evaluating several arguments that it is likely to be justice promoting (or, at least, not justice impairing). We conclude that these technologies may not be inherently just or unjust, but when situated within obtaining social contexts, they are likely to exacerbate rather than alleviate social injustices. Moreover, responding to the social justice challenges associated with human enhancement technologies cannot be accomplished by technological design and innovation alone. It requires addressing problematic features of social, political, and economic practices, policies, and institutions.

2. DEFINING HUMAN ENHANCEMENT

It is difficult to provide a precise, exceptionless definition of human enhancement, just as it is difficult to do so for technology, natural, and therapeutic. However, a functional account, one adequate for present purposes, is possible.

Human enhancement technologies are technologies that improve or augment some core cognitive, physical, perceptual, or psychological human capacity, or enable some novel capacity not standardly among human capacities. Human enhancement through technology is ubiquitous. Education technologies, computational devices, nutritional supplements, steroids, pharmaceuticals, communication systems, and optical lenses are each a type of human enhancement technology. Several distinctions are needed to distinguish robust human enhancement from mundane human enhancement.

Some enhancement technologies are *episodic enhancements* – the enhancement persists only so long as the technological intervention is enabled. Computers are this sort of enhancement. When a person is interfacing with her computer, several of her cognitive and communication capacities are significantly augmented, as well as novel capacities enabled. But these enhancement effects do not persist when the computer is shut down. Other enhancement technologies are *sustained enhancements* – the enhancement persists some duration after the technological intervention is complete. The non-therapeutic use of anabolic steroids can be this sort of enhancement. Their enhancement effects – increased muscle mass – can last some time after they are used.

Another distinction among enhancements is between external and internal enhancements. *External enhancements* augment existing or enable novel human capacities without modifying core biological, psychological, or cognitive systems and without introducing some novel system. Again, computers are this sort of enhancement. They enable users to do many things that they would not otherwise be able to accomplish, but they do not do so by altering, for example, how their cognitive or perceptual systems function. *Internal enhancements* augment existing or enable novel human capacities by altering a particular aspect of some core biological, psychological, or cognitive system, or by introducing some novel system. Increasing longevity by altering the mechanism of cell division so that it occurs with less telomerase shortening would be an example of an internal enhancement.

The type of human enhancement under consideration in this article is internal, sustained enhancement, i.e., the technological alteration of some system/process or the introduction of some novel system/process that augments some core biological capability beyond the range of capacity attainable by technologically-unassisted human beings or introduces a capacity not had by technologically-unassisted human beings. Hereafter, "human enhancement" refers to this type of enhancement. This account of human enhancement is not without limitations. Some of the terms employed are vague, e.g., 'core biological capability.' It also leaves out several important distinctions within human enhancement, e.g., between permanent and reversible enhancement, germ line and somatic enhancement, and different methods and mechanisms of enhancement. Nevertheless, it provides a clear sense of what is distinctive of the type of enhancement at issue: they alter *us* in a way that gets at *the kind of creature that we are.*[2]

3. A PRINCIPLE OF JUSTICE

'Justice' is a complex and contested concept. There are multiple domains of justice that differ with respect to the activities over which principles are operative, e.g., economic justice, political justice, and criminal justice. There are multiple theories of justice that are distinguished by contrary conceptual frameworks, e.g., communitarian, utilitarian, liberal, and libertarian. And there is substantial diversity within each theory of justice with respect to, for example, who are appropriate subjects of justice (e.g., whether it is limited only to human beings), what are the appropriate objects of justice (e.g., distribution of goods, recognition, participation, or promotion of welfare), and the relationship between justice and other ethical concepts (e.g., whether it is one ethical concept among many, a central ethical concept, or the superordinate concept).

Rather than defend a particular instantiation of a particular theory of justice, we will proceed, following Peter Wenz, by endorsing a particular principle of distributive justice that most instantiations of most theories of justice would accept:

> Justice increases when the benefits and burdens of social cooperation are born more equally, except when moral considerations or other values justify greater inequality. This principle is uncontroversial because it basically restates the principle of the equal consideration of interests … which rests on the uncontroversial claim that all human beings are of equal moral considerability. Unequal treatment of human beings (some reaping extra benefits or bearing extra burdens related to social cooperation) must therefore be justified, and such justification requires recourse to moral considerations or other values (2007: 58).

Theories and conceptions of justice differ substantially in their accounts of what justifies inequality, e.g., that it increases well-being overall or is the result of free and fair exchange of goods earned by non-coercive means (under appropriate conditions). However, they often converge on what does not justify inequality or unequal treatment. In fact, any theory of justice that would endorse

---

[2] This account of human enhancement is neutral with regard to evaluative judgments. It leaves as an open question whether enhancement of a particular capacity is beneficial/valuable or not. It is, however, worth exploring whether this should be the case, or whether enhancement should be conceptualized the same way as adaptation, i.e., as something that has a positive valence, indexed to particular environments/circumstances.

considerations such as race, ethnicity, or class as a basis for unequal treatment would be unacceptable for that reason.

4. FORMULATING THE ISSUE

Given the principle of justice endorsed above, if the implementation and dissemination of human enhancement technologies increases inequality or unequal treatment as a result of pre-existing, unjustified economic or social inequalities, those technologies would be justice impairing. If those inequalities are reduced, they would be justice promoting. The remainder of this article focuses on the following question: Are human enhancement technologies likely to be justice impairing or justice promoting, in this respect?[3]

5. ARGUMENT THAT HUMAN ENHANCEMENT TECHNOLOGIES WILL BE JUSTICE IMPAIRING

The standard argument that human enhancement technologies are likely to be justice impairing is as follows. Once a human enhancement technology is developed and made available to the public, initial access to it will be limited due to its high cost. Those who can afford the technology when it first becomes available will enjoy a compounding benefit: the increased capabilities that the technology provides (whether cognitive, psychological, and/or physical) will further advantage the individual (who is already advantaged in virtue of their position and resources, which provided them access to the technologies) in pursuit of competitive and positional goods that are relevant to one's quality of life (Tamburrini 2006). Those without resources to gain access to the technology will have a corresponding compounding disadvantage: the absence of enhanced capabilities is added to the initial resource shortage (material and/or social) that prevented them from acquiring the technology in the first place. In this way, initial social inequalities – which in most societies are not fully justified – are increased rather than diminished by human enhancement technologies.

Several variations of this general argument have been offered. For example, Maxwell Mehlman argues that due to the probable high cost associated with human enhancement technologies, particularly initially, only those with sufficient economic resources or whose health insurance covers them are likely to have access to them (Mehlman 2004). Because enhancement technologies are, by

---

[3] The "in this respect" qualification is necessary because there are other domains or principles of justice that also are relevant to whether human enhancement technologies are likely to be justice promoting or justice impairing. For example, widespread dissemination of human enhancement technologies are likely to be socially transformative, so decisions regarding them (which to implement and how to do it) have a procedural justice dimension: To what extent should decision be made by "elite" actors in science and industry communities? To what extent (and how) should they be made more "democratic"? What is the appropriate role of the marketplace and of governmental bodies? Each of these has justice components that are distinct from the largely distributive justice issues that are the focus of this article.

definition, non-therapeutic, they are not likely to be covered by health insurance companies, which tend to cover only therapeutic (or, in some cases, preventative) care. As Mehlman points out, some health insurance companies already are refusing to cover minor enhancement drugs such as Viagra. Moreover, in cases such as Doe vs. Mutual of Omaha (1999), the courts have interpreted the law in ways that do not obligate health insurance companies to cover all disabilities under the Americans with Disabilities Act. If insurance companies are not even required to cover certain therapeutic procedures, it would be wildly optimistic to assume that they will cover elective, non-therapeutic technologies. Even if higher-priced health insurances choose to cover non-therapeutic treatments, they will remain inaccessible to middle-income and low-income individuals who cannot afford such plans.

In addition, the number of uninsured U.S. residents grew by 2.2 million in 2006, raising the number to 47 million, up from 44.8 million in 2005. 73.2% of the uninsured were U.S. citizens; 11.7% of all children lacked health insurance (19.3% of children in families with annual incomes below the federal poverty level lacked health insurance); and the percentages of uninsured individuals differed by race (34.1% of Hispanics were uninsured in 2006; 20.5% of blacks; 10.8% of whites; 15.5% of Asian-Americans) (DeNavas-Walt et al. 2007). So, even assuming that health insurers will cover non-therapeutic procedures (an unlikely assumption), there will still be over 40 million individuals in the United States without access to human enhancement technologies, as well as significant racial disparities with respect to access to them.

Access to health care is not the only social context factor relevant to whether human enhancement technologies will promote or impair social justice. As Wenz (2005) argues, per capita, more resources are spent educating children from wealthy families than children from low-income families, when arguably the latter needs more resources due to socio-economic disadvantages. An unjust distribution of educational resources leads to (further) unjust difference in socio-economic outcomes, which results in unjust disparities in technology access. Wenz (2005) also argues that the U.S taxation system is increasingly unjust. According to Walden Bellow (1999, cited in Wenz 2005), tax "reform" in the 1980s increased the tax share for those in the bottom 10 percent of income earners in the country by 28 percent, while reducing the share of those in the top 1 percent by 14 percent. This shifting of the relative tax burden from those who are wealthy to those who are low income has continued in this century with the 2001 U.S. federal tax cuts.

In a just society, human enhancement technologies may well not be justice impairing. However, according to Wenz and Mehlman, the United States is not a just society with respect to factors that are crucial to access to emerging technologies. As a result, human enhancement technologies are likely to be justice impairing.

6. ARGUMENTS THAT HUMAN ENHANCEMENT TECHNOLOGIES WILL NOT BE JUSTICE IMPAIRING

An adequate response to the argument that human enhancement technologies are likely to be justice impairing would need to show one of the following: (1)

pre-existing social and economic inequalities are not unjust; (2) pre-existing social and economic inequalities, while unjust, are not likely to result in disparities in access to human enhancement technologies; or (3) although pre-existing social and economic inequalities are unjust and likely to result in disparities in access to human enhancement technologies, the disparities themselves and their effects are not likely to be unjust. Most respondents to the argument that human enhancement technologies are likely to be justice impairing concede that there are significant unjust inequalities both within societies and between societies. Therefore, their responses are of types (2) or (3). In what follows we review several of these, some of which have been put forward by prominent proponents of human enhancement. Each of which, we argue, has significant shortcomings.

6.1 Argument from the law of accelerating returns

Ray Kurzweil, technologist and author of *The Singularity is Near*, defends what he calls the law of accelerating returns. The following quotation illustrates his view:

> The ongoing acceleration of technology is the implication and inevitable result of what I call the law of accelerating returns, which describes the acceleration of the pace of and the exponential growth of the products of an evolutionary process. These products include, in particular, information bearing technologies such as computation, and their acceleration extends substantially beyond the predictions made by what has become known as Moore's Law (2005, 35–36).

Kurzweil believes that, as a consequence of the law of accelerating returns, enhancement technologies are likely to become very inexpensive – indeed, almost free – very quickly. Initially, information technologies are extremely expensive, do not work particularly well, and are available only to the economically well-off. However, as technologies improve, the price comes down and the result is technology that works well and is widely available.

> Another concern expressed … is the "terrifying" possibility that through these technologies the rich may gain certain advantages and opportunities to which the rest of humankind does not have access. Such inequality, of course, would be nothing new, but with regard to this issue the law of accelerating returns has an important and beneficial impact. Because of the ongoing exponential growth of price-performance, all of these technologies quickly become so inexpensive as to become almost free (2005, 469).

Kurzweil claims that the amount of time that it currently takes for a technology to go from expensive to inexpensive is approximately ten years, but as the rate approximately doubles each decade (expressing the law of accelerating returns), the lag would be reduced to five years in one decade and only three years in two decades. The horizon for realizing human enhancement technologies appears to be at least a few decades away (even on Kurzweil's reckoning). By the time the technologies are in place, the duration between first appearance and widespread availability will be quite short. Therefore, they are not likely to be justice impairing.

6.2 Response to the argument from the law of accelerating returns

Let us assume that the law of accelerating returns is true. Even still, there will be a lag between the release of the initial technology and its subsequent widespread availability. The three year lag that he predicts in two decades is still three years – *at a time of rapid technological advancement*. In the time this technology becomes inexpensive enough for those of limited resources to acquire it, there will (according to the law of accelerating returns) be exponential advancement with the latest technologies available to those who can continually afford (or otherwise have access to) them. If the law holds, this will be true in each case, no matter how short the time lag. What matters is not the absolute duration of the time lag, but the absolute duration relative to the pace of technological change. If the lag decreases exponentially while the rate of change increases exponentially, those who are economically disadvantaged will remain a generation of technology (or more) behind.

Moreover, even given the cost reductions that would occur (if the law holds) over the next several decades, human enhancement technologies are likely to remain out of reach for many people. Certainly, computers and iPods that went on the market a few years ago cost less today, but this has not made them universally accessible. 2.7 billion people in the world live on less than $2 ppp/day (World Bank 2008), with 1.1 billion on less than $1 ppp/day (United Nations 2006). These individuals struggle to obtain even basic necessities of living. They have no marginal resources to spend on last decade's technologies. Even in the United States, whose poorest are better off than those in the least developed countries, many of those living below the poverty line – 36.5 million people in 2006 (US Census Bureau 2007) – struggle to afford technology that has already been reduced in price. Even assuming the law of accelerating returns, human enhancement technologies are likely to remain out of reach for many people long after (relative to the rate of technological change) those with access to them have accelerated off.

6.3 Argument from marginal enhancement

Like Kurzweil, Ramez Naam believes that, over time, the cost of enhancement technologies will significantly drop as newer, more efficient technologies are introduced. He recognizes that, at least at first, the latest technologies will only be affordable to a small minority, but he does not find this problematic because he believes the enhancement effect (per dollar) will diminish with each subsequent version of an enhancement technology. On his view, the basic enhancement technology will provide the most significant enhancement effect, so that when those who are economically well-off purchase the latest version of an enhancement technology, they will be paying exorbitantly for only minor refinements (see, also, Bostrom & Sandberg forthcoming). Thus, rather than accelerating off, they will be puttering along only marginally ahead of those who cannot afford (or lack access to) the latest version of the technology.

Consider automobiles. In most places in the United States, a few thousand dollars will purchase a fairly safe, comfortable vehicle for getting from point A to point B.

> Alternately, many tens of thousands of dollars will buy you a new luxury vehicle. A luxury car can cost ten times what the most affordable new economy car would cost, yet the luxury car isn't ten times as fast, or ten times as fuel-efficient, or ten times as safe. The extra money spent on it buys only incremental advantages. In terms of basic mobility, the inexpensive step between no car and a cheap car is larger than the very expensive step from a cheap car to a BMW (Naam 2005, 66–67).

Although low-income consumers may never afford the latest refinements in human enhancement technologies, this is, according to Naam, far less problematic than it seems, since the bulk of an enhancement effect will be available in low-cost versions.

6.4 Response to the argument from marginal enhancement

Even if, as Naam supposes, the enhancement differential between a basic enhancement technology and the latest (luxury) version of the technology is much less than the enhancement effect associated with the basic enhancement, there will remain a difference in (enhanced) capabilities between those who have access only to the basic technology and those who have access to the latest version. The enhancement differential, even if small, will not be nil. The issue, then, is whether even reduced differentials in capabilities resulting from unequal access to enhancement technologies are likely to exacerbate or reduce unjustified inequalities.

There are reasons to believe that it likely would exacerbate them, or, at least, perpetuate them. First, minor differences in attributes – even when they are functionally irrelevant – often matter in the distribution of competitive and positional goods (Tamburrini 2006). Research from the University of Florida and the University of North Carolina has shown that tall people earn more than their shorter counterparts, at a rate of approximately $789 a year per inch (Judge 2004). A survey of 11,000 thirty-three year olds done at London Guildhall University found that those deemed less attractive earn 10 to 15 percent less than their counterparts (Harper 2000). These results suggest that even minor enhancement differentials, on seemingly insignificant traits that appear unrelated to functionality, can increase or perpetuate inequality.

Moreover, most human enhancements are not targeted at functionally insignificant traits (relative to competitive and positional goods) but are relevant capabilities such as memory and problem solving (which, of course, have associated non-competitive, non-positional goods as well). In the case of human enhancement, then, differences among people will be biologically real (since that is the nature of the intervention at issue), and the enhanced trait will have a positive evaluative valence, i.e., it will have been intentionally sought due to its perceived desirability. Concerns have been raised that advances in genomics that enable identifying genetic differences among populations (for therapeutic purposes) will invite discrimination, i.e., that some populations will be viewed as genetically inferior. If there is any merit to this concern at all, there is more so in the case of human enhancement, where the biological difference is not merely descriptive and unintended (as they are in the unenhanced genomics case), but desired and intended.

Returning to Naam's case of the economy automobile versus the luxury automobile, while it is true that both provide mobility, the luxury car exceeds the other in many other respects, both performance and aesthetic. This is precisely why most people, if given the option, would prefer the luxury automobile, all other things being equal. The fact that both are, at basic, transportation, and that is the most important thing about them, does not imply that their other differences are not going to be the basis for discrimination (justified or unjustified). The same is true of the enhancement differential between basic and luxury versions of an enhancement technology.

6.5 Argument from cognitive and/or virtue enhancement

Another possible response to the social justice challenge is that those who are cognitively enhanced will be more capable of identifying and developing effective responses to the causes of unjust inequalities than those who are not enhanced, and those who are psychologically enhanced will be more disposed to addressing them than those who are not enhanced. As a result, differential access to (at least some) enhancement technologies will ultimately be justice promoting, rather than justice impairing. Although this argument has not been made explicitly (that we are aware of), it is suggested by the claim, frequently made by proponents of human enhancement, that cognitive enhancement would have net good consequences for society overall (Bostrom 2006). The rationale for this is that cognitive abilities are relevant to identifying and analyzing complex problems, as well as devising and executing solutions to them – in general, increased cognitive functioning is associated with increased problem solving capacities.

One version of the argument from cognitive enhancement merely applies this rationale to the social justice problem: increased cognitive capacity should be associated with increased ability to understand and devise effective solutions for the causes of unjust social inequalities, including unjust disparities in access to human enhancement technologies. A second version of the argument appeals to the benefits that would accrue to those who are socially and economically worst off – through, for example, advances in agricultural, energy, communication, and environmental remediation technologies that would be made by cognitively enhanced individuals – even if social inequalities persist. So while pre-existing and access inequalities might remain or even increase, they would be justified by their benefiting even the economically worst off, and therefore would not remain unjust inequalities.

It also is often claimed that technological interventions to improve people's character, i.e., virtue enhancement, is desirable and may be possible. Here, for example, is Julian Savulescu:

> Buchanan and colleagues [Buchanan et al. 2000] have discussed the value of "all purpose goods." These are traits which are valuable regardless of which kind of life a person chooses to live. They give us greater all around capacities to live a vast array of lives. Examples include intelligence, memory, self-discipline, patience, empathy, a sense of humour, optimism and just having a sunny temperament. All of these characteristics –

sometimes may include virtues – may have some biological and psychological basis capable of manipulation with technology.

Technology might even be used to improve our *moral character*. We certainly seek through good instruction and example, discipline and other methods to make better children. It may be possible to alter biology to make people predisposed to be more moral by promoting empathy, imagination, sympathy, fairness, honesty, etc.

In so far as these characteristics have some genetic basis, genetic manipulation could benefit us. There is reason to believe that complex virtues like fair-mindedness may have a biological basis. In one famous experiment, a monkey was trained to perform a task and rewarded either a grape or piece of cucumber. He preferred the grape. On one occasion, he performed the task successfully and was given a piece of cucumber. He watched as another monkey who had not performed the task was given a grape. He became very angry. This shows that even monkeys have a sense of fairness and desert – or at least self-interest! (Savulescu, 2007, 7)

The argument from virtue enhancement merely applies this view to the social justice problem, i.e., that it would be desirable and may be possible to develop technologies and interventions that increase a person's sense of and commitment to social justice.

Taken together, the arguments from cognitive enhancement and virtue enhancement imply that human enhancement technologies might result in individuals that are more capable of and committed to reducing social injustice than would exist if the technologies were not developed and disseminated. So even if there were a disparity in access to those technologies, the ultimate result would be a reduction in social injustice.

6.6 Response to the argument from cognitive and/or virtue enhancement

The argument from cognitive enhancement (in each of its formulations), when considered independently from the argument from virtue enhancement, has an obvious limitation: there is little reason to believe that those with greater cognitive capacities will, in virtue of those capacities, have increased concern for and commitment to social justice. If a person who enjoys unjust advantages is inclined to protect and promote those advantages, when she gains access to cognitive enhancement technologies she will be more likely to use her additional capabilities to increase those advantages rather than to reduce them. As a result, disparities in access to cognitive enhancement technologies (of the sort under consideration) are as likely to perpetuate unjust social inequalities as to result in their reduction.

This would not be the case if the primary barrier to addressing the relevant social injustices were cognitive, i.e., that people did not know about the inequalities, did not recognize them as unjust, or there were no good strategies developed for addressing them. However, it is not so – substantial behavioral, institutional, and salience barriers exist as well. Often, people lack resources and commitment, have other priorities (often worthwhile ones), or exhibit moral weakness, i.e., they are aware of what they ought to do but are not particularly motivated to do it. Moral shortcomings are not normally the result of not knowing what one ought to do or how to go about doing it (particularly in a way likely to be remedied by cognitive enhancement). It is, of course, an empirical question whether there is a causal relationship between cognitive capacity and concern about and motivation to address social injustice. But in the absence of such empirical evidence, which would suggest that cognitive enhancement might enhance social justice commitment as well (and thereby belie the considerations raised above), the argument from cognitive enhancement is, at best, unjustified.

Kean Birch (2005) has suggested that cognitive enhancement not only will not increase concerns about social justice, but may in fact have a diminishing effect. He points out (as does Wenz 2005) that current educational resource distributions and institutions favor a system that advantages children from middle and upper income families. There is no reason to believe cognitive enhancement (in itself) will change this. Indeed, when these families allocate resources for their children to receive some cognitive enhancement, they likely will expect certain outcomes, and may provide additional resources to promote them further. Birch worries that allocations of resources for enhancements, as well as allocations to ensure positive outcomes from them, could be accompanied by a concomitant decrease in resources for, and concerns about, the outcomes of those who are not enhanced. Moreover, even if such a decrease in concern and resources for those who are not enhanced does not occur, the resource gap between families that have access to the technologies and those that do not is widened and a capabilities gap between students who are enhanced and those that are not is initiated in virtue of the enhancement.

Another potential (though speculative) difficulty with the argument from cognitive enhancement suggested by Birch is that cognitively-enhanced individuals, rather than having an increased commitment to justice, may develop a sense of inherent (nor merely cognitive) superiority.

> The preoccupations at the beginning of the twentieth century established certain traits as the preserve of specific individuals, dependent upon their class and race, which affect their role in society. Thus it would not be difficult to foresee a time when the intelligently 'enhanced' could assert a claim to superiority based upon their inherent sense of justness (2005, 25).

Birch's concern here builds off of one raised earlier – that a potential result of differential access to enhancement technologies will be intentional and value-laden biological (or bio-machine) differences among people. This might result in further discrimination and biases, beyond the already widespread discrimination and biases that have been predicated on unintended, undesigned biological differences.

The argument from virtue enhancement, either on its own or in combination with the argument from cognitive enhancement, is a better response to the social justice problem than is the argument from cognitive enhancement alone, in this sense: if virtue enhancement were realized, then it would result in increased concern about, and propensity to act to reduce, social injustice, and so it might result in reductions in social injustice. There are other difficulties with the argument, however.

First, it appears unlikely that virtue enhancement will be at the forefront of human enhancement. Technologies that increase healthful longevity and cognitive capacities may significantly precede it, for both technical and social reasons. The relevant underlying biological mechanisms and genetic components of physical health and cognitive functioning are better understood and are receiving more research attention than those associated with a sense of and commitment to fairness or justice. So even if, as Savulescu hypothesizes, there is a genetic or otherwise biological component, and even if effective technological intervention is possible, virtue enhancement technologies likely will not be developed as quickly as enhancement technologies associated with other capacities, leaving the social justice problem (at least temporarily) unaddressed. There also may be greater reluctance among people to engage in deep psychological interventions – as an intervention associated with a sense of justice would be – in comparison to cognitive and longevity enhancements. This is, of course, speculative, but the former might be implicated in people's self-conception and self-identity more strongly than the latter, resulting in more reluctance to engage in virtue enhancement, again resulting (at least temporarily) in the social justice problem being unaddressed.

Second, the virtue enhancement approach to addressing the social justice challenge has obvious deficiencies when compared to addressing directly the social context factors in place prior to the widespread implementation of human enhancement technologies. Addressing the social context directly is more immediate – there would not be a lag between adopting human enhancement technologies and addressing the social justice problem. Addressing the social context directly also is more likely to be successful. As suggested above, it is far from certain that virtue enhancement will be effectively developed and widely implemented. Moreover, directly addressing the social context would address injustices that already exist, i.e., prior to the implementation of human enhancement technologies, and therefore merit concern and attention, even independent from their relationship to human enhancement technologies.[4]

So while virtue enhancement, either alone or in combination with cognitive enhancement, could promote rather than impair social injustice, its realization and implementation is by no means assured and there is a readily available and preferable alternative (although not a contrary one).

---

[4] A similar response is appropriate to the second formulation of the argument from cognitive enhancement, i.e., that even those who do not have access to the enhancements (including the world's socially and economically worst off) will be better off if the enhancements are developed and disseminated, since they will benefit from what those who are enhanced are able to accomplish (for example, improvements in agricultural, energy, and environmental remediation technologies). Human enhancement technologies are not necessary to develop these technologies. The lack of pro-poor technology research and dissemination is not due to a lack of technological capabilities, but to social and political factors (e.g., incentive structures, funding priorities, and governance). Moreover, addressing those factors is more immediate, likely to succeed, and addresses an already urgent social problem (independent of human enhancement technologies), and is therefore preferable to an approach mediated through the implementation of human enhancement technologies.

The arguments against there being a social justice problem are not compelling. Bostrom, Naam, and Kurzweil are likely to be mistaken about how inexpensive and readily available the technologies will become, how much of an advantage even minor enhancement differentials are likely to provide, and how significant the time lag between initial adoption and widespread availability will be from a competitive and positional goods perspective. Moreover, there is no evidence (or other justification) to think that those who are cognitively enhanced will be ethically enhanced, i.e., more motivated to address social injustices, and virtue enhancement is likely to be more difficult, more controversial, and, therefore, later realized than cognitive enhancement. Finally, developing human enhancement technologies is neither necessary nor the most efficient route available for addressing social injustice and the needs of the socially and economically worst off.

7. CONCLUSION

Are human enhancement technologies likely to be social justice promoting or social justice impairing? Several conclusions and comments relevant to the issue are warranted:

1. Given the obtaining features of the social contexts into which they would emerge, human enhancement technologies are likely to be justice impairing. There are pre-existing unjust inequalities that are likely to result in unjust disparities in access to human enhancement technologies. Human enhancement technologies are likely to perpetuate or increase these inequalities, given the advantages they would provide with respect to competitive and positional goods. The reasons proffered for why the resultant inequalities would not be justice impairing are not plausible.

2. Human enhancement technologies may not be inherently unjust. Human enhancement technologies are not responsible for the obtaining unjust social inequalities. Moreover, if they were introduced into a comprehensively just society there is no reason to believe (on the basis of arguments discussed in this paper) that they would result in unjust inequalities with respect to either access or outcomes (Buchanan et al. 2000). However, what matters to the social justice issue are the possible technologies that are likely to be created and distributed in actual societies, with their particular social arrangements and institutions. So even if human enhancement technologies are not inherently unjust, this does not alleviate the social justice challenges associated with them.

3. This social justice problem associated with human enhancement cannot be effectively addressed through technology development and design. The social justice challenge for human enhancement technologies could be reduced by addressing the problematic social, economic, and institutional features that contribute to them, e.g., health care, taxation, and education policies and institutions. For reasons discussed above, i.e., immediacy, plausibility, and effectiveness, addressing the problematic features of the social context is far preferable to 'virtue engineering' approaches, which are the proffered techno-fix. We must fix social injustice; the technologies will not do it for us.

4. This social justice issue is only one ethical consideration relevant to the development and dissemination of human enhancement technologies. The

conclusion that human enhancement technologies are likely to be justice impairing unless the relevant features of the relevant social contexts are addressed is not an all-things-considered ethical evaluation of either human enhancement technologies as such or particular enhancements by particular means. (Different human enhancement technologies will have different ethical profiles.) As discussed earlier, there are other justice considerations that are relevant, e.g., participatory justice and autonomy. In addition, there are ethical considerations that fall outside the domain of justice (as it is often conceived), e.g., environmental values, individual flourishing, and aggregate welfare, that are crucial to any all-things-considered ethical assessment of any particular form or method of human enhancement.

There is a social justice problem associated with emerging human enhancement technologies. This does not (alone) imply that there should be a moratorium for research on technologies that have human enhancement potentials or even that public funding of it should be eliminated, for example. It implies that if the goal for the development and dissemination of human enhancement technologies is to promote flourishing in sustainable and socially just ways, there is considerable and difficult social and political work to be done to accomplish the justice component. Proponents of ethical development of human enhancement technologies should be concerned as much about education and health care reform as they are about public funding for research on regenerative medicine or regulation of synthetic biology.

WORKS CITED

W. Bello, S. Cunningham, and B. Rau (1999) *'Dark victory: the United States and global poverty,'* (Oakland, CA: Food First).

K. Birch (2005) 'Beneficence, determinism, and justice: an engagement with the argument for the genetic selection of intelligence,' *Bioethics* 19 (1): 12–28.

N. Bostrom, and T. Ord (2006) 'The reversal test: eliminating status quo bias in applied ethics,' *Ethics* 116 (4): 656–679.

N. Bostrom, and A. Sandberg (Forthcoming) 'Cognitive enhancement: methods, ethics, regulatory challenges,' *Science and Engineering Ethics*, Pre-publication version *available at* http://www.nickbostrom.com/cognitive.pdf *(accessed January 17, 2008)*.

A. Buchanan, D. W. Brock, and N. Daniels (2000) *'From chance to choice: genetics and justice,'* (New York: Cambridge University Press).

C. DeNavas-Walt, B. D. Proctor, and C. H. Lee (2005) 'Income, poverty, and health insurance coverage in the United States: 2007,' Current Population Reports (Washington, DC: US Census Bureau, US Government Printing Office).

Doe vs. Omaha. (1999). 179 F.3d 557 (7th Cir.).

B. Harper (2000) 'Beauty, stature and the labour market: a British cohort study,' *Oxford Bulletin of Economics and Statistics* 62 (1): 771–800.

T. Judge (2004) 'The effect of physical height on workplace success and income: preliminary test of a theoretical model,' *Journal of Applied Psychology* 89 (3): 428–441.

R. Kurzweil (2005) *'The singularity is near: when humans transcend biology,'* (New York: Viking Penguin).

M. Mehlman (2004) 'Cognition-enhancing drugs,' *Milbank Quarterly* 82: 483–506.

Millennium Development Goals Report 2006 (2006) United Nations. Available at http://mdgs.un.org/unsd/mdg/Resources/Static/Products/Progress2006/MDGReport2006.pdf (accessed January 8, 2008).

R. Naam (2005) *More than human: embracing the promise of biological enhancement* (New York: Broadway Books).

J. Savulescu (2007) 'Genetic interventions and the ethics of advancements on human beings' In B. Steinbock (ed.), *The Oxford Handbook of Bioethics*, 516–535. (Oxford: Oxford University Press). Also available electronically at http://www.abc.net.au/rn/backgroundbriefing/stories/2008/2122476.htm (accessed January 17, 2008).

C. Tamburrini (2006) 'Enhanced bodies: distributive aspects of genetic enhancement in elite sports and society,' Unpublished paper presented at ENHANCE Conference, Beijing, China in August 2006.

U.S. Census Bureau (2007) 'Current population survey, 2007 annual social and economic (ASEC) supplement,' Available at http://www.census.gov/apsd/techdoc/cps/cpsmar07.pdf (accessed January 8, 2008).

P. Wenz (2007) 'Does environmentalism promote injustice for the poor?' In R. Sandler and P. C. Pezzullo (eds.), *Environmental justice and environmentalism: the social justice challenge to the environmental movement* (Cambridge, MA: MIT Press).

Wenz, P. (2005). "Engineering genetic injustice." *Bioethics* 19.1: 1–11.

World Bank (2008) 'Poverty: Overview,' Available at http://go.worldbank.org/RQBDCTUXW0 (accessed January 8, 2008).

DISCUSSION QUESTIONS

1. Garcia and Sandler argue that the principle of justice they use to evaluate enhancement technologies is uncontroversial in that any reasonable theory of justice must accept it. Do you agree with the principle? Can you think of any legitimate grounds on which the principle might be rejected?

2. What is the core argument that human enhancement technologies are likely to be justice impairing? Do you find this argument compelling – why or why not? Can you think of any other reasons that human enhancement technologies might be justice impairing?

3. Do you think there is merit to any of the arguments that human enhancement technologies might actually promote justice? Can you think of any other reasons, besides those discussed by the authors, for why enhancement technologies might be socially good or promote justice?

4. Do you agree with the authors that even if enhancement technologies are likely to be justice impairing that it could still be permissible to develop them? How can it be that something could promote injustice and that it still be permissible to do?

5. Do you agree that human enhancement technologies are not inherently unjust? What would it mean for a technology to be inherently unjust? Are any technologies inherently unjust?

PART V

INFORMATION TECHNOLOGIES

Information technologies are technologies that generate, transmit, receive, store, process, access, or analyze data. Information technologies always have had large impacts on society – consider, for example, the significance of the invention of writing, the printing press, and the telegraph. In recent decades, of course, it has been computer and digital based information technologies that have been transforming social, political, economic, and individual activities. The internet, personal computing, and smart phones, for example, have transformed how people work (and what work they do), how they interact with each other, and how they take their recreation. They have facilitated globalization, created and destroyed entire industries, and required development of an enormous physical infrastructure to support them. Moreover, they are powerful enabling technologies. Increases in the capacity to collect, share, store, and analyze date have accelerate research and development in every science and engineering field, from conservation biology to robotics, and in every area of application, from medical to military. For those that have access to them, information and computer technologies (ICTs) have restructured the conditions of human existence more than anything else over the last few decades. (Specific instances of this – for example, the impact of ICTs on universities – are discussed in the Introduction [1.1].)

Given the tremendous social impacts of information technologies, it is no surprise that they raise important, complex, urgent, and, in some cases, novel ethical issues. The first four chapters in this section discuss ethical issues associated with particular forms or applications of ICTs. In "Bigger Monster, Weaker Chains," Jan Stanley and Barry Steinhardt discuss increases in surveillance activities resulting from the development of more powerful information technologies and changes to privacy protection laws. Their focus is the United States, but the issues they raise apply to any society in which there is growth in surveillance capabilities and a decrease in surveillance restrictions and oversight.

In "Nanotechnology and Privacy: The Instructive Case of RFID," Jeroen Van den Hoven reflects upon the introduction and use of radio frequency identification tags (RFID) in order to help anticipate privacy concerns that might be associated with nanoscale information technologies. The chapter includes an extended discussion of the value of privacy and what makes it worth protecting.

In "Intellectual Property: Legal and Moral Challenges of Online File Sharing," Richard Spinello discusses whether the use of peer-to-peer (P2P) networks to share movies and music constitutes an ethically problematic form of theft. He also considers whether those who develop and disseminate P2P software are ethically and legally culpable in virtue of knowingly and intentionally enabling file sharing.

In "Virtual Reality and Computer Simulation," Philip Brey surveys ethical issues associated with virtual reality and video games. The issues he discusses includes those that arise within virtual reality and gaming, as well as those associated with the tremendous growth in the amount of time people spend engaged in the activities. The chapter also includes an extended discussion of the differences between virtual and non-virtual (or physical) reality.

In the fifth and final chapter of this section, "The Digital Divide," Kenneth Himma and Maria Bottis discuss the responsibility of those with access to ICTs to support extending access to those who currently do not have it. The first half of their chapter focuses on whether people in affluent nations have an obligation to assist those who are very poor, while the second half of the chapter focuses on whether closing the digital divide is an effective means to doing so.

A theme that runs throughout several of the chapters in this section is the importance of value sensitivity in the development and adoption of ICTs. ICTs alter power relationships, organize our social worlds, influence our perspectives (for example, by the information and representations we are exposed to online), and structure our daily activities (for example, many of us spend much of our days punching keys with our fingers). It is therefore crucial that we are attentive to the values that we embed in them, the ways that they frame our choices, the judgments that they favor, and the behaviors that they encourage; and this attentiveness needs to inform ICT design and dissemination.

More detailed summaries, as well as a listing of related readings, are located at the start of the chapters.

BIGGER MONSTER, WEAKER CHAINS: THE GROWTH OF AN AMERICAN SURVEILLANCE SOCIETY[1]

Jan Stanley and Barry Steinhardt

CHAPTER SUMMARY

In this chapter, Jan Stanley and Barry Steinhardt argue that privacy and liberty in the United States are at risk. They believe that a combination of lightning-fast technological innovation and the erosion of privacy protections threaten to transform the United States into a Surveillance Society. They begin by discussing the explosion of technologies – e.g. computers, cameras, sensors, wireless communication, GPS, and biometrics – that have dramatically increased surveillance power. They then turn to discussing the weakening of legal restraints on the use of that power since September 11th, 2001. Although their focus is on the United States, the same factors – increases surveillance capabilities and decreased restrictions and oversight of their use – apply in many other countries as well.

RELATED READINGS

Introduction: Form of Life (2.5); Rights and Liberties (2.6.3); Dual Use (2.6.5)

Other Chapters: Jeroen van den Hoven, *Nanotechnology and Privacy* (Ch. 19); Fritz Allhoff, *The Coming Era of Nanomedicine* (Ch. 11)

1. THE GROWING SURVEILLANCE MONSTER

In the film *Minority Report*, which takes place in the United States in the year 2050, people called "Pre-cogs" can supposedly predict future crimes, and the nation has become a perfect surveillance society. The frightening thing is that except for the psychic Pre-cogs, the technologies of surveillance portrayed in the film already exist or are in the pipeline. Replace the Pre-cogs with "brain fingerprinting" – the

---

[1] This chapter is excerpted from Jan Stanley and Barry Steinhardt (2003) *Bigger Monster Weaker Chains: The Growth of an American Surveillance Society* (American Civil Liberties Union). It appears here by permission of the American Civil Liberties Union and the authors.

supposed ability to ferret out dangerous tendencies by reading brain waves – and the film's entire vision no longer lies far in the future. Other new privacy invasions are coming at us from all directions, from video and data surveillance to DNA scanning to new data-gathering gadgets. In this chapter we discuss these new technologies and their collective significance for the future of privacy and civil liberties. The focus is on the United States, but similar technologies and techniques are being deployed in countries around the world.

1.1 Video surveillance

Surveillance video cameras are rapidly spreading throughout the public arena. A survey of surveillance cameras in Manhattan, for example, found that it is impossible to walk around the city without being recorded nearly every step of the way. And since September 11 the pace has quickened, with new cameras being placed not only in some of our most sacred public spaces, such as the National Mall in Washington and the Statue of Liberty in New York harbor, but on ordinary public streets all over America.

As common as video cameras have become, there are strong signs that, without public action, video surveillance may be on the verge of a revolutionary expansion in American life. There are three factors propelling this revolution.

Improved technology: Advances such as the digitization of video mean cheaper cameras, cheaper transmission of far-flung video feeds, and cheaper storage and retrieval of images.

Centralized surveillance: A new centralized surveillance center in Washington, DC is an early indicator of what technology may bring. It allows officers to view images from video cameras across the city – public buildings and streets, neighborhoods, Metro stations, and even schools. With the flip of a switch, officers can zoom in on people from cameras a half-mile away (Bravin, 2003).

Unexamined assumptions that cameras provide security: In the wake of the September 11 attacks, many embraced surveillance as the way to prevent future attacks and prevent crime. But it is far from clear how cameras will increase security. U.S. government experts on security technology, noting that "monitoring video screens is both boring and mesmerizing," have found in experiments that after only 20 minutes of watching video monitors, "the attention of most individuals has degenerated to well below acceptable levels."[2] In addition, studies of cameras' effect on crime in Britain, where they have been extensively deployed, have found no conclusive evidence that they have reduced crime.[3]

These developments are creating powerful momentum toward pervasive video surveillance of our public spaces. If centralized video facilities are permitted in Washington and around the nation, it is inevitable that they will be expanded – not only in the number of cameras but also in their power and ability. It is easy to foresee inexpensive, one-dollar cameras being distributed throughout our cities and tied via wireless technology into a centralized police facility where the life

---

[2] See http://www.ncjrs.org/school/ch2a_5.html.
[3] See http://www.scotcrim.u-net.com/researchc2.htm.

of the city can be monitored. Those video signals could be stored indefinitely in digital form in giant but inexpensive databases, and called up with the click of a mouse at any time. With face recognition, the video records could even be indexed and searched based on who the systems identify – correctly, or all too often, incorrectly.

Several airports around the nation, a handful of cities, and even the National Park Service at the Statue of Liberty have installed face recognition technology. While not nearly reliable enough to be effective as a security application,[4] such a system could still violate the privacy of a significant percentage of the citizens who appeared before it (as well as the privacy of those who do not appear before it but are falsely identified as having done so). Unlike, say, an iris scan, face recognition doesn't require the knowledge, consent, or participation of the subject; modern cameras can easily view faces from over 100 yards away.

Further possibilities for the expansion of video surveillance lie with unmanned aircraft, or drones, which have been used by the military and the CIA overseas for reconnaissance, surveillance, and targeting. Controlled from the ground, they can stay airborne for days at a time. Now there is talk of deploying them domestically. Senate Armed Services Committee Chairman John Warner (R, VA) said in December 2002 that he wants to explore their use in Homeland Security, and a number of domestic government agencies have expressed interest in deploying them. Drones are likely to be just one of many ways in which improving robotics technology will be applied to surveillance (Sia, 2002). The bottom line is that surveillance systems, once installed, rarely remain confined to their original purpose. Once the nation decides to go down the path of seeking security through video surveillance, the imperative to make it work will become overwhelming, and the monitoring of citizens in public places will quickly become pervasive.

1.2 Data surveillance

An insidious new type of surveillance is becoming possible that is just as intrusive as video surveillance – what we might call "data surveillance." Data surveillance is *the collection of information about an identifiable individual, often from multiple sources, that can be assembled into a portrait of that person's activities.* Most computers are programmed to automatically store and track usage data, and the spread of computer chips in our daily lives means that more and more of our activities leave behind "data trails." It will soon be possible to combine information from different sources to recreate an individual's activities with such detail that it becomes no different from being followed around all day by a detective with a video camera.

Some think comprehensive public tracking will make no difference, since life in public places is not "private" in the same way as life inside the home. This

---

[4] The success rate of face recognition technology has been dismal. The many independent findings to that effect include a trial conducted by the U.S. military in 2002, which found that with a reasonably low false-positive rate, the technology had less than a 20% chance of successfully identifying a person in its database who appeared before the camera. See http://www.aclu.org/issues/privacy/FINAL_1_Final_Steve_King.pdf, 17th slide.

is wrong; such tracking would represent a radical change in American life. A woman who leaves her house, drives to a store, meets a friend for coffee, visits a museum, and then returns home may be in public all day, but her life is still private in that she is the only one who has an overall view of how she spent her day. In America, she does not expect that her activities are being watched or tracked in any systematic way – she expects to be left alone. But if current trends continue, it will be impossible to have any contact with the outside world that is not watched and recorded.

1.3 The commodification of information

A major factor driving the trend toward data surveillance forward is the commodification of personal information by corporations. As computer technology exploded in recent decades, making it much easier to collect information about what Americans buy and do, companies came to realize that such data is often very valuable. The expense of marketing efforts gives businesses a strong incentive to know as much about consumers as possible so they can focus on the most likely new customers. Surveys, sweepstakes questionnaires, loyalty programs and detailed product registration forms have proliferated in American life – all aimed at gathering information about consumers. Today, any consumer activity that is *not* being tracked and recorded is increasingly being viewed by businesses as money left on the table.

On the Internet, where every mouse click can be recorded, the tracking and profiling of consumers is even more prevalent. Web sites can not only track what consumers buy, but what they *look at* – and for how long, and in what order. With the end of the Dot Com era, personal information has become an even more precious source of hard cash for those Internet ventures that survive. And of course Americans use the Internet not just as a shopping mall, but to research topics of interest, debate political issues, seek support for personal problems, and many other purposes that can generate deeply private information about their thoughts, interests, lifestyles, habits, and activities.

1.4 Genetic privacy

The relentless commercialization of information has also led to the breakdown of some longstanding traditions, such as doctor–patient confidentiality. Citizens share some of their most intimate and embarrassing secrets with their doctors on the old-fashioned assumption that their conversations are confidential. Yet those details are routinely shared with insurance companies, researchers, marketers, and employers. An insurance trade organization called the Medical Information Bureau even keeps a centralized medical database with records on millions of patients. Weak, new medical privacy rules will do little to stop this behavior.

An even greater threat to medical privacy is looming: genetic information. The increase in DNA analysis for medical testing, research, and other purposes will accelerate sharply in coming years, and will increasingly be incorporated into routine health care.

Unlike other medical information, genetic data provides data that is both difficult to keep confidential and extremely revealing about us. DNA is very easy to acquire because we constantly slough off hair, saliva, skin cells and other samples of our DNA (household dust, for example, is made up primarily of dead human skin cells). That means that no matter how hard we strive to keep our genetic code private, we are always vulnerable to other parties' secretly testing samples of our DNA. The issue will be intensified by the development of cheap and efficient DNA chips capable of reading parts of our genetic sequences.

Already, it is possible to send away a DNA sample for analysis. A testing company called Genelex reports that it has amassed 50,000 DNA samples, many gathered surreptitiously for paternity testing. "You'd be amazed," the company's CEO told *U.S. News & World Report*. "Siblings have sent in mom's discarded Kleenex and wax from her hearing aid to resolve the family rumors" (Hawkins, 2002).

Not only is DNA easier to acquire than other medical information, revealing it can also have more profound consequences. Genetic markers are rapidly being identified for all sorts of genetic diseases, risk factors, and other characteristics. None of us knows what time bombs are lurking in our genomes.

The consequences of increased genetic transparency will likely include the following:

Discrimination by insurers: Health and life insurance companies could collect DNA for use in deciding who to insure and what to charge them, with the result that a certain proportion of the population could become uninsurable. The insurance industry has already vigorously opposed efforts in Congress to pass meaningful genetic privacy and discrimination bills.

Employment discrimination: Genetic workplace testing is already on the rise, and the courts have heard many cases. Employers desiring healthy, capable workers will always have an incentive to discriminate based on DNA – an incentive that will be even stronger as long as health insurance is provided through the workplace.

Genetic spying: Cheap technology could allow everyone from schoolchildren to dating couples to nosy neighbors to routinely check out each other's genetic codes. A likely, high-profile example: online posting of the genetic profiles of celebrities or politicians.

1.5 New data-gathering technologies

The discovery by businesses of the monetary value of personal information and the vast new project of tracking the habits of consumers has been made possible by advances in computers, databases and the Internet. In the near future, other new technologies will continue to fill out the mosaic of information it is possible to collect on every individual. Examples include:

Cell phone location data: The government has mandated that manufacturers make cell phones capable of automatically reporting their location when an owner dials 911. Of course, those phones are capable of tracking their location at other times as well. And in applying the rules that protect the privacy of telephone records to this location data, the government is weakening those rules in a way that allows phone companies to collect and share data about the location and movements of their customers.

Biometrics: Technologies that identify us by unique bodily attributes such as our fingerprints, faces, iris patterns, or DNA are already being proposed for inclusion on national ID cards and to identify airline passengers. Face recognition is spreading. Fingerprint scanners have been introduced as security or payment mechanisms in office buildings, college campuses, grocery stores and even fast-food restaurants. And several companies are working on DNA chips that will be able to instantly identify individuals by the DNA we leave behind everywhere we go.

Black boxes: All cars built today contain computers, and some of those computers are being programmed in ways that are not necessarily in the interest of owners. An increasing number of cars contain devices akin to the "black boxes" on aircraft that record details about a vehicle's operation and movement. Those devices can "tattle" on car owners to the police or insurance investigators. Already, one car rental agency tried to charge a customer for speeding after a GPS device in the car reported the transgression back to the company. And cars are just one example of how products and possessions can be programmed to spy and inform on their owners.

RFID chips: RFID chips, which are already used in such applications as toll-booth speed passes, emit a short-range radio signal containing a unique code that identifies each chip. Once the cost of these chips falls to a few pennies each, plans are underway to affix them to products in stores, down to every can of soup and tube of toothpaste. They will allow everyday objects to "talk" to each other – or to anyone else who is listening. For example, they could let market researchers scan the contents of your purse or car from five feet away, or let police officers scan your identification when they pass you on the street.

Implantable GPS chips: Computer chips that can record and broadcast their location have also been developed. In addition to practical uses such as building them into shipping containers, they can also serve as location "bugs" when, for example, hidden by a suspicious husband in a wife's purse. And they can be implanted under the skin (as can RFID chips).

If we do not act to reverse the current trend, data surveillance – like video surveillance – will allow corporations or the government to constantly monitor what individual Americans do every day. Data surveillance would cover *everyone*, with records of every transaction and activity squirreled away until they are sucked up by powerful search engines whether as part of routine security checks, a general sweep for suspects in an unsolved crime, or a program of harassment against some future Martin Luther King.

1.6 Government surveillance

Data surveillance is made possible by the growing ocean of privately collected personal data. But who would conduct that surveillance? There are certainly business incentives for doing so; companies called data aggregators (such as Acxiom and ChoicePoint) are in the business of compiling detailed databases on individuals and then selling that information to others. Although these companies are invisible to the average person, data aggregation is an enormous, multi-billion-dollar industry. Some databases are even "co-ops" where participants agree to contribute data about their customers in return for the ability to pull out cross-merchant profiles of customers' activities.

The biggest threat to privacy, however, comes from the government. Many Americans are naturally concerned about corporate surveillance, but only the government has the power to legally take away liberty – as has been demonstrated starkly by the post-September 11 detention of suspects without trial as "enemy combatants."

In addition, the government has unmatched power to centralize all the private sector data that is being generated. In fact, the distinction between government and private-sector privacy invasions is fading quickly. The Justice Department, for example, reportedly has an $8 million contract with data aggregator ChoicePoint that allows government agents to tap into the company's vast database of personal information on individuals (Simpson, 2001). Although the Privacy Act of 1974 banned the government from maintaining information on citizens who are not the targets of investigations, the FBI can now evade that requirement by simply purchasing information that has been collected by the private sector. Other proposals – such as the Pentagon's "Total Information Awareness" project and airline passenger profiling programs – would institutionalize government access to consumer data in even more far-reaching ways (see below).

1.7 Government databases

The government's access to personal information begins with the thousands of databases it maintains on the lives of Americans and others. For instance:

- The FBI maintains a giant database that contains millions of records covering everything from criminal records to stolen boats and databases with millions of computerized fingerprints and DNA records.
- The Treasury Department runs a database that collects financial information reported to the government by thousands of banks and other financial institutions.
- A "new hires" database maintained by the Department of Health and Human Services, which contains the name, address, social security number, and quarterly wages of every working person in the U.S.
- The federal Department of Education maintains an enormous information bank holding years worth of educational records on individuals stretching from their primary school years through higher education. After September 11th, Congress gave the FBI permission to access the database without probable cause.
- State departments of motor vehicles possess millions of up-to-date files containing a variety of personal data, including photographs of most adults living in the United States.

1.8 Communications surveillance

The government also performs an increasing amount of eavesdropping on electronic communications. While technologies like telephone wiretapping have been around for decades, today's technologies cast a far broader net. The FBI's controversial "Carnivore" program, for example, is supposed to be used to tap into the e-mail traffic of a particular individual. Unlike a telephone wiretap, however, it doesn't cover just one device but (because of how the Internet is built) filters through *all* the traffic on the Internet Service Provider to which it has been attached. The only thing keeping the government from trolling through all this traffic is software instructions that are written by the government itself. (Despite that clear conflict of interest, the FBI has refused to allow independent inspection and oversight of the device's operation.)

Another example is the international eavesdropping program codenamed Echelon. Operated by a partnership consisting of the United States, Britain, Canada, Australia, and New Zealand, Echelon reportedly grabs e-mail, phone calls, and other electronic communications from its far-flung listening posts across most of the earth. (U.S. eavesdroppers are not supposed to listen in on the conversations of Americans, but the question about Echelon has always been whether the intelligence agencies of participating nations can set up reciprocal, back-scratching arrangements to spy on each others' citizens.) Like Carnivore, Echelon may be used against particular targets, but to do so its operators must sort through massive amounts of information about potentially millions of people. That is worlds away from the popular conception of the old wiretap where an FBI agent listens to one line. Not only the volume of intercepts but the potential for abuse is now exponentially higher.

1.9 The Patriot Act

The potential for the abuse of surveillance powers in the United States has risen sharply due to a dramatic post-9/11 erosion of legal protections against government surveillance of citizens. Just six weeks after the September 11 attacks, a panicked Congress passed the USA PATRIOT Act, an overnight revision of the nation's surveillance laws that vastly expanded the government's authority to collect data on and monitor its own citizens as well as reduced checks and balances on those powers, such as judicial oversight. The government never demonstrated that restraints on surveillance had contributed to the attack, and indeed much of the new legislation had nothing to do with fighting terrorism. Rather, the bill represented a successful use of the terrorist attacks by the FBI to roll back unwanted checks on its power. The most powerful provisions of the law allow for:

Easy access to records: Under the PATRIOT Act, the FBI can force anyone to turn over records on their customers or clients, giving the government unchecked power to rifle through individuals' financial records, medical histories, Internet usage, travel patterns, or any other records. Some of the most invasive and disturbing uses permitted by the Act involve government access to citizens' reading habits from libraries and bookstores. The FBI does not have to show suspicion of a crime, can gag the recipient of a search order from disclosing the search to anyone, and is subject to no meaningful judicial oversight.

Expansion of the "pen register" exception in wiretap law: The PATRIOT Act expands exceptions to the normal requirement for probable cause in wiretap law. As with its new power to search records, the FBI need not show probable cause or even reasonable suspicion of criminal activity, and judicial oversight is essentially nil.

Expansion of the intelligence exception in wiretap law: The PATRIOT Act also loosens the evidence needed by the government to justify an intelligence wiretap or physical search. Previously the law allowed exceptions to the Fourth Amendment for these kinds of searches only if "the purpose" of the search was to gather foreign intelligence. But the Act changes "the purpose" to "a significant purpose," which lets the government circumvent the Constitution's probable cause requirement even when its main goal is ordinary law enforcement.

More secret searches: Except in rare cases, the law has always required that the subject of a search be notified that a search is taking place. Such notice is a crucial check on the government's power because it forces the authorities to operate in the open and allows the subject of searches to challenge their validity in court. But, the PATRIOT Act allows the government to conduct searches without notifying the subjects until long after the search has been executed.

Under these changes and other authorities asserted by the Bush Administration, U.S. intelligence agents could conduct a secret search of an American citizen's home, use evidence found there to declare him an "enemy combatant," and imprison him without trial. The courts would have no chance to review these decisions – indeed, they might never even find out about them.[5]

1.10 Loosened domestic spying regulations

In May 2002, Attorney General John Ashcroft issued new guidelines on domestic spying that significantly increase the freedom of federal agents to conduct surveillance on American individuals and organizations. Under the new guidelines, FBI agents can infiltrate "any event that is open to the public," from public meetings and demonstrations to political conventions to church services to 12-step programs. This was the same basis upon which abuses were carried out by the FBI in the 1950s and 1960s, including surveillance of political groups that disagreed with the government, anonymous letters sent to the spouses of targets to try to ruin their marriages, and the infamous campaign against Martin Luther King who was investigated and harassed for decades. The new guidelines are purely for spying on Americans; there is a separate set of Foreign Guidelines that cover investigations inside the U.S. of foreign powers and terrorist organizations such as Al Qaeda.

---

[5] See, New York Times (2002); Washington Post (2002); National Journal (2002).

Like the TIPS program, Ashcroft's guidelines sow suspicion among citizens and extend the government's surveillance power into the capillaries of American life. It is not just the reality of government surveillance that chills free expression and the freedom that Americans enjoy. The same negative effects come when we are constantly forced to wonder whether we *might* be under observation – whether the person sitting next to us is secretly informing the government that we are "suspicious."

2. THE SYNERGIES OF SURVEILLANCE

Multiple surveillance techniques added together are greater than the sum of their parts. One example is face recognition which combines the power of computerized software analysis, cameras, and databases to seek matches between facial images. But the real synergies of surveillance come into play with data collection.

The growing piles of data being collected on Americans represent an enormous invasion of privacy, but our privacy has actually been protected by the fact that all this information still remains scattered across many different databases. As a result, there exists a pent-up capacity for surveillance in American life today – a capacity that will be fully realized if the government, landlords, employers, or other powerful forces gain the ability to *draw together* all this information. A particular piece of data about you – such as the fact that you entered your office at 10:29 AM on July 5, 2001 – is normally innocuous. But when enough pieces of that kind of data are assembled together, they add up to an extremely detailed and intrusive picture of an individual's life and habits.

2.1 Data profiling and "total information awareness"

Just how real this scenario is has been demonstrated by another ominous surveillance plan to emerge from the effort against terrorism: the Pentagon's "Total Information Awareness" (TIA) program. The aim of this program is to give officials easy, unified access to every possible government and commercial database in the world.[6] According to program director John Poindexter, the program's goal is to develop "ultra-large-scale" database technologies with the goal of "treating the world-wide, distributed, legacy databases as if they were one centralized database." The program envisions a "full-coverage database containing all information relevant to identifying" potential terrorists and their supporters. As we have seen, the amount of available information is mushrooming by the day, and will soon be rich enough to reveal much of our lives.

The TIA program, which is run by the Defense Advanced Research Projects Agency (DARPA), not only seeks to bring together the oceans of data that are already being collected on people, but would be designed to afford what DARPA calls "easy future scaling" to embrace new sources of data as they become available. It would also incorporate other work being done by the military, such as their

---

[6] See New York Times (Nov. 9, 2002); Washington Post (2002); National Journal (2002).

"Human Identification at a Distance" program, which seeks to allow identification and tracking of people from a distance, and therefore without their permission or knowledge.[7]

In short, the government is working furiously to bring disparate sources of information about us together into one view, just as privacy advocates have been warning about for years. That would represent a radical branching off from the centuries-old, Anglo-American tradition that the police conduct surveillance only where there is evidence of involvement in wrongdoing. It would seek to protect us by monitoring *everyone* for signs of wrongdoing – in short, by instituting a giant dragnet capable of sifting through the personal lives of Americans in search of "suspicious" patterns. The potential for abuse of such a system is staggering.

The massive defense research capabilities of the United States have always involved the search for ways of outwardly defending our nation. Programs like TIA[8] involve turning those capabilities inward and applying them to the American people – something that should be done, if at all, only with extreme caution and plenty of public input, political debate, checks and balances, and Congressional oversight. So far, none of those things have been present with TIA.

2.2 National ID cards

If Americans allow it, another convergence of surveillance technologies will probably center around a national ID card. A national ID would immediately combine new technologies such as biometrics and RFID chips along with an enormously powerful database (possibly distributed among the 50 states). Before long, it would become an overarching means of facilitating surveillance by allowing far-flung pools of information to be pulled together into a single, incredibly rich dossier or profile of our lives. Before long, office buildings, doctors' offices, gas stations, highway tolls, subways and buses would incorporate the ID card into their security or payment systems for greater efficiency, and data that is currently scattered and disconnected will get organized around the ID and lead to the creation of what amounts to a national database of sensitive information about American citizens.

History has shown that databases created for one purpose are almost inevitably expanded to other uses; Social Security, which was prohibited by federal law from being used as an identifier when it was first created, is a prime example. Over time, a national ID database would inevitably contain a wider and wider range of information and become accessible to more and more people for more and more purposes that are further and further removed from its original justification.

---

[7] Quotes are from the TIA homepage at http://www.darpa.mil/iao/ and from public 8/2/02 and remarks by Poindexter, online at http://www.fas.org/irp/agency/dod/poindexter.html.

[8] The TIA is just one part of a larger post-9/11 expansion of federal research and development efforts. The budget for military R&D spending alone has been increased by 18% in the current fiscal year to a record $58.8 billion. See David (2002).

The most likely route to a national ID is through our driver's licenses. Since September 11, the American Association of Motor Vehicle Administrators has been forcefully lobbying Congress for funds to establish nationwide uniformity in the design and content of driver's licenses – and more importantly, for tightly interconnecting the databases that lie behind the physical licenses themselves.

An attempt to retrofit driver's licenses into national ID cards will launch a predictable series of events bringing us toward a surveillance society.

Proponents will promise that the IDs will be implemented in limited ways that won't devastate privacy and other liberties.

Once a limited version of the proposals is put in place, its limits as an anti-terrorism measure will quickly become apparent. Like a dam built halfway across a river, the IDs cannot possibly be effective unless their coverage is total.

The scheme's ineffectiveness – starkly demonstrated, perhaps, by a new terrorist attack – will create an overwhelming imperative to "fix" and "complete" it, which will turn it into the totalitarian tool that proponents promised it would never become.

Once in place, it is easy to imagine how national IDs could be combined with an RFID chip to allow for convenient, at-a-distance verification of ID. The IDs could then be tied to access control points around our public places, so that the unauthorized could be kept out of office buildings, apartments, public transit, and secure public buildings. Citizens with criminal records, or low incomes could be barred from accessing airports, sports arenas, stores, or other facilities. Retailers might add RFID readers to find out exactly who is browsing their aisles, gawking at their window displays from the sidewalk or passing by without looking. A network of automated RFID listening posts on the sidewalks and roads could even reveal the location of all citizens at all times. Pocket ID readers could be used by FBI agents to sweep up the identities of everyone at a political meeting, protest march, or Islamic prayer service.

3. CONCLUSION

If we do not take steps to control and regulate surveillance to bring it into conformity with our values, we will find ourselves being tracked, analyzed, profiled, and flagged in our daily lives to a degree we can scarcely imagine today. We will be forced into an impossible struggle to conform to the letter of every rule, law, and guideline, lest we create ammunition for enemies in the government or elsewhere. Our transgressions will become permanent Scarlet Letters that follow us throughout our lives, visible to all and used by the government, landlords, employers, insurance companies and other powerful parties to increase their leverage over average people. Americans will not be able to engage in political protest or go about their daily lives without the constant awareness that we are – or could be – under surveillance. We will be forced to constantly ask of even the smallest action taken in public, "Will this make me look suspicious? Will this hurt my chances for future employment? Will this reduce my ability to get insurance?" The exercise of free speech will be chilled as Americans become conscious that their every word may be reported to the government by FBI infiltrators, suspicious fellow citizens or an Internet Service Provider.

Many well-known commentators like Sun Microsystems CEO Scott McNealy have already pronounced privacy dead. The truth is that a surveillance society does loom over us, and privacy, while not yet dead, is on life support. Heroic measures are required to save it.

Four main goals need to be attained to prevent this dark potential from being realized: a change in the terms of the debate, passage of comprehensive privacy laws, passage of new laws to regulate the powerful and invasive new technologies that have and will continue to appear, and a revival of the Fourth Amendment to the U.S. Constitution.

3.1 Changing the terms of the debate

In the public debates over every new surveillance technology, the forest too often gets lost for the trees, and we lose sight of the larger trend: the seemingly inexorable movement toward a surveillance society. It will always be important to understand and publicly debate every new technology and every new technique for spying on people. But unless each new development is also understood as just one piece of the larger surveillance mosaic that is rapidly being constructed around us, Americans are not likely to get excited about a given incremental loss of privacy like the tracking of cars through toll booths or the growing practice of tracking consumers' supermarket purchases.

We are being confronted with fundamental choices about what sort of society we want to live in. But unless the terms of the debate are changed to focus on the forest instead of individual trees, too many Americans will never even recognize the choice we face, and a decision against preserving privacy will be made by default.

3.2 Comprehensive privacy laws

Although broad-based protections against government surveillance, such as the wiretap laws, are being weakened, at least they exist. But surveillance is increasingly being carried out by the private sector – frequently at the behest of government – and the laws protecting Americans against non-governmental privacy invasions are pitifully weak.

In contrast to the rest of the developed world, the U.S. has no strong, comprehensive law protecting privacy – only a patchwork of largely inadequate protections. For example, as a result of many legislators' discomfort over the disclosure of Judge Robert Bork's video rental choices during his Supreme Court confirmation battle, video records are now protected by a strong privacy law. Medical records are governed by a separate, far weaker law that allows for wide-spread access to extremely personal information. Financial data is governed by yet another "privacy" law – Gramm-Leach – which really amounts to a license to share financial information. Another law protects only the privacy of children under age 13 on the Internet. And layered on top of this sectoral approach to privacy by the federal government is a geographical patchwork of constitutional and statutory privacy protections in the states.

The patchwork approach to privacy is grossly inadequate. As invasive practices grow, Americans will face constant uncertainty about when and how these complex laws protect them, contributing to a pervasive sense of insecurity. We need to develop a baseline of simple and clear privacy protections that crosses all sectors of our lives and give it the force of law.

3.3 New technologies and new laws

The technologies of surveillance are developing at the speed of light, but the body of law that protects us is stuck back in the Stone Age. In the past, new technologies that threatened our privacy, such as telephone wiretapping, were assimilated over time into our society. The legal system had time to adapt and reinterpret existing laws, the political system had time to consider and enact new laws or regulations, and the culture had time to absorb the implications of the new technology for daily life. Today, however, change is happening so fast that none of this adaptation has time to take place – a problem that is being intensified by the scramble to enact unexamined anti-terrorism measures. The result is a significant danger that surveillance practices will become entrenched in American life that would never be accepted if we had more time to digest them.

Since a comprehensive privacy law may never be passed in the U.S. – and certainly not in the near future – law and legal principles must be developed or adapted to rein in particular new technologies such as surveillance cameras, location-tracking devices, and biometrics. Surveillance cameras, for example, must be subject to force-of-law rules covering important details like when they will be used, how long images will be stored, and when and with whom they will be shared.

3.4 Reviving the fourth amendment

> The right of the people to be secure in their persons, houses, papers, and effects, against unreasonable searches and seizures, shall not be violated, and no warrants shall issue, but upon probable cause, supported by oath or affirmation, and particularly describing the place to be searched, and the persons or things to be seized.
>
> – Fourth Amendment to the U.S. Constitution

The Fourth Amendment, the primary Constitutional bulwark against Government invasion of our privacy, was a direct response to the British authorities' use of "general warrants" to conduct broad searches of the rebellious colonists.

Historically, the courts have been slow to adapt the Fourth Amendment to the realities of developing technologies. It took almost 40 years for the U.S. Supreme Court to recognize that the Constitution applies to the wiretapping of telephone conversations.[9] In recent years – in no small part as the result of the failed "war on drugs" – Fourth Amendment principles have been steadily eroding. The circumstances under which police and other government officials may conduct warrantless searches has been rapidly expanding. The courts have

---

[9] In 1967 the Supreme Court finally recognized the right to privacy in telephone conversations in the case *Katz v. U.S.* (389 US 347), reversing the 1928 opinion *Olmstead v. U.S.* (277 US 438).

allowed for increased surveillance and searches on the nation's highways and at our "borders" (the legal definition of which actually extends hundreds of miles inland from the actual border). And despite the Constitution's plain language covering "persons" and "effects," the courts have increasingly allowed for warrantless searches when we are outside of our homes and "in public." Here the courts have increasingly found we have no "reasonable expectation" of privacy and that therefore the Fourth Amendment does not apply.

But like other Constitutional provisions, the Fourth Amendment needs to be understood in contemporary terms. New technologies are endowing the government with the 21st Century equivalent of Superman's X-ray vision. Using everything from powerful video technologies that can literally see in the dark, to biometric identification techniques like face recognition, to "brain fingerprinting" that can purportedly read our thoughts, the government is now capable of conducting broad searches of our "persons and effects" while we are going about our daily lives – even while we are in "public."

The Fourth Amendment is in desperate need of a revival. The reasonable expectation of privacy cannot be defined by the power that technology affords the government to spy on us. Since that power is increasingly limitless, the "reasonable expectation" standard will leave our privacy dead indeed.

But all is not yet lost. There is some reason for hope. In an important pre-9/11 case, *Kyllo vs. U.S.*,[10] the Supreme Court held that the reasonable expectation of privacy could not be determined by the power of new technologies. In a remarkable opinion written by conservative Justice Antonin Scalia, the Court held that without a warrant the police could not use a new thermal imaging device that searches for heat sources to conduct what was the functional equivalent of a warrantless search for marijuana cultivation in Danny Kyllo's home.

The Court specifically declined to leave Kyllo "at the mercy of advancing technology." While Kyllo involved a search of a home, it enunciates an important principle: the Fourth Amendment must adapt to new technologies. That principle can and should be expanded to general use. The Framers never expected the Constitution to be read exclusively in terms of the circumstances of 1791.

WORKS CITED

J. Bravin, (2002) 'Washington Police to Play `I Spy' With Cameras, Raising Concerns,' *Wall Street Journal*, Feb. 13, 2002.

D. Hawkins, 'As DNA Banks Quietly Multiply, Who is Guarding the Safe?' *U.S. News & World Report*, Dec. 2, 2002.

B. Davis, 'Massive Federal R&D Initiative To Fight Terror Is Under Way,' Wall Street Journal, November 25, 2002.

R. H.P. Sia, (2002) 'Pilotless Aircraft Makers Seek Role For Domestic Uses,' CongressDaily, December 17, 2002.

G. R. Simpson, 'Big Brother-in-Law: If the FBI Hopes to Get The Goods on You, It May Ask ChoicePoint,' *Wall Street Journal*, April 13, 2001.

---

[10]   190 F.3d 1041, 2001.

'Pentagon Plans a Computer System That Would Peek at Personal Data of Americans,' *New York Times*, Nov. 9, 2002.

'US Hopes to Check Computers Globally,' *Washington Post*, Nov. 12, 2002

'The Poindexter Plan,' *National Journal*, Sept. 7, 2002.

DISCUSSION QUESTIONS

1. What do you think grounds the right to privacy? Why do you have a right not to be tracked?
2. Do the potential security benefits of a surveillance society outweigh concerns about privacy? Why or why not?
3. Do you think that the right to privacy precludes collection of data about you without cause or consent, even when you are in public places?
4. Should we be concerned about the capture or collection of information about us? Or do problems only arise when the information is misused? For example, is the presence of surveillance cameras throughout a city problematic in itself, or is it only problematic if those with access to the data they collect misuse them?
5. What benefits are worth your giving up a significant amount of your privacy? Do any of those benefits justify requiring others to give up significant amounts of their privacy?
6. How might surveillance technologies alter human behavior for the better? For the worse?
7. How do the fact that people we are now constantly leaving data trails – for example on surveillance cameras, by using store cards, and by searching the web – change the nature of our actions or restructure our activities, practices, and institutions?

NANOTECHNOLOGY AND PRIVACY: THE INSTRUCTIVE CASE OF RFID[1]

Jeroen van den Hoven

CHAPTER SUMMARY

The development of ever smaller integrated circuits at the sub-micron and nanoscale drives the production of very small tags, smart cards, smart labels and sensors. Nanoelectronics and submicron technology supports surveillance technology that is practically invisible. In this chapter, Jeroen van den Hoven argues that, as a result, one of the most urgent and immediate concerns associated with nanotechnology is privacy. Computing in the twenty-first century will not only be pervasive and ubiquitous, but also inconspicuous. If these features are not counteracted in design, they will facilitate ubiquitous surveillance practices that are widely available, cheap, and intrusive. Van den Hoven looks to RFID (radio frequency identification tag) technology as an instructive example of what nanotechnology has in store for privacy.

RELATED READINGS

Introduction: Justice, Access, and Equality (2.6.2); Rights and Liberties (2.6.3); Responsible Development (2.8)

Other Chapters: Langdon Winner, *Technologies as Forms of Life* (Ch. 4); Jan Stanley and Barry Steinhardt, *Bigger Monster, Weaker Chains* (Ch. 18).

1. INTRODUCTION

One of the problems with nanoethics is that it addresses problems of applications of nanoscience which are yet to come. In the first decade of the twenty-first century, we still have very few examples of widely used nanotechnology. It is advisable to start thinking about ethical implications of new technology at the early stages of its development, but it does not make reflection and analysis particularly easy. This predicament is a version of the Collingridge dilemma: in the first stages of the

---

[1] This chapter is excerpted from Jeroen van den Hoven (2006) 'Nanotechnology and Privacy: The Instructive Case of RFID,' *International Journal of Applied Philosophy* (20): 215–228. It appears here by permission of the Philosophy Documentation Center and the author.

development of a new technology, it is still possible to influence the development of the technology, although there is little information about its effects. When the technology is entrenched and widely used, there is information about its effects, but there is little room to change the course of the development of the technology.

One of the areas where we have already a relatively clear picture of the impact of nanotechnology at this stage is the area of the privacy implications of submicron and nano-electronics. This is an area where we can move beyond mere antici-pation, speculation, and science fiction. Practically invisible badges, integrated circuits, tags, minute sensors, or "smart dust" and wearable electronics are gradu-ally finding their way to the world of retail, supply chains, logistics, shops and warehouses, workplace, criminal justice, and homeland security.[2] New sensor and surveillance technology is the result of the rapid development of submicron and nanotechnology-in accordance with Moore's law, which states that the number of transistors on a chip doubles every eighteen months. When combined with middle-ware and back-end data bases, as well as a range of wireless and mobile communication modalities, such as Wi-Fi, Ultra Wide Band, and Bluetooth, and connections to computer networks and the Internet, the technology will give rise to a panoply of privacy issues.

People will knowingly or unknowingly carry around tagged items ranging from clothing to watches, mobile phones, chip cards, identity documents, bank notes, or jewelry. These can all be read and uniquely identified from a distance (ranging from centimeters to hundreds of meters) by readers which may be hidden or not in the line of sight. This will make objects, and the people carrying or accompa-nying them, traceable wherever they go. They may be followed from shelf to shelf or from shop to shop and identified as the buyer, carrier, or user of an item, which can lead to further identifications and knowledge discovery in the associated data bases.

Kris Pister, one of the leading researchers in the field, sketches the following picture on the basis of current research:

> In 2010 your house and office will be aware of your presence, and even orientation, in a given room. In 2010 everything you own that is worth more than a few dollars will know that it's yours, and you'll be able to find it whenever you want it. Stealing cars, furniture, stereos, or other valuables will be unusual, because any of your valu-ables that leave your house will check in on their way out the door, and scream like a troll's magic purse if removed without permission (they may scream at 2.4 GHz rather than in audio).... In 2010 a speck of dust on each of your fingernails will continuously transmit fingertip motion to your computer. Your computer will understand when you type, point, click, gesture, sculpt, or play air guitar. (http://robotics.eecs.berkeley. edu/~pister/SmartDust/in2010)

---

[2] See EU paper (EU, 1995). For an overview of the legal issues, see Kardasiadou and Talidou (2006).

2. RFID

The core technology of this type of tracking and tracing is the widely used Radio Frequency Identity Chip (RFID). An RFID chip or tag consists of a small integrated circuit attached to a tiny radio antenna, which can receive and transmit a radio signal. The storage capacity of the chip can be up to 128 bits. The chip can either supply its own energy (active tag) from a battery or get its energy input from a radio signal from the antenna of the reader (passive tag). Like in the case of bar codes, there is an international number organization which provides and registers the unique ID numbers of RFID chips. RFID is ideally suited for the tracking and tracing of objects such as boxes, containers and vehicles in logistic chains. RFID tags are now also being used to trace and track consumer products and everyday objects on the item level as a replacement of barcodes. Governments and the global business world are preparing for a large-scale implementation of RFID technology in the first decades of the twenty-first century for these purposes.

Apart from a race to the bottom and the aim of making RFIDs smaller, one of the research challenges is to make the chips self-sufficient and energy saving, or even energy "scavenging," in which case they will get energy from their environment in the form of heat, light, or movement. The other challenge is to make them cheaper. One way to lower the unit cost of RFID chips is to find mass applications such as chipping bank notes, which the European Union (EU) is considering (Yoshida, 2005).

With RFID each object has its own unique identifier, and individuals will – apart from being walking repositories of biometric data – also show up in data bases as clouds of tagged objects and become entangled in an "internet of things."[3] RFID foreshadows what nano-electronics has in store for our privacy: invisible surveillance.

RFID chips are also referred to as "contactless technology," "contactless chips," or "proximity chips." Many authors on RFID have argued that there are privacy threats associated with the introduction of millions or even billions of smart tags and labels and RFIDs in health-care, retail, travel, and law enforcement. As a result of opposition and critique of consumer organizations such as NOTAGS and CASPIAN, RFID has received serious negative moral connotations. Benetton planned to put RFIDs in all of their clothing with the help of Philips. This gave rise to vehement consumer protest and tainted the reputation of both Benetton and Philips. Tesco in the UK experimented with a photo camera in the store which was activated when consumers took a packet of Gillette razors off the shelf. The picture taken was then added to a consumer data base. This also gave rise to intense public debate, for understandable reasons. Terms with a more neutral meaning are therefore welcomed by the industry and governments, since there are many advantages to be had from RFID technology in health care, safety, security, and industry that may go unnoticed because of bad publicity and bad reputation of a few relatively frivolous first applications.

---

[3] This is the Title of a study of the International Telecom Union; www.itu.int/internetofthings

The following examples give further evidence of a development toward tracking and tracing, monitoring, and surveillance. Precise real-time location systems using radio frequency identification tags have gone commercial.[4] They use Ultra Wide Band communication to help locate tagged persons and objects in buildings with a precision of thirty cm. Several hospitals around the world monitor the location and movement of equipment and persons in the hospital with the help of RFID. The US Department of Agriculture conducts experiments with smart dust and nano-sensors which register properties of the environment and may help to detect the use of forbidden chemicals. The project is called "little brother." CLENS is a strategic US defense initiative and stands for *Camouflaged Long Endurance Nano Sensors*, which allow precise location and tracking of soldiers during missions. Kris Pisters's group in California has many fascinating MEMS (Micro Electronic Mechanical Systems) applications on display, microphones of 500 micrometer diameter and research on cameras of one cubic millimeter. Extreme miniaturization in sensor technology and location based services is clearly well underway.[5]

Not only objects and artifacts may be tagged on the item level, also living creatures-animals and human beings-may be tagged. The US Food and Drug Administration has decided to make chip implants in humans legal in the case of medical records. The company Applied Digital Solutions introduced, as mentioned above, the Verichip for subcutaneous implantation in humans, and there are more examples:[6]

(1) In Japan, school children are chipped subcutaneously and are traced by a computer at school on their way to and from school
(2) The Baya Beach club in Rotterdam and Barcelona offers people the possibility of having a chip for payments in the club to be placed under their skin by a doctor who is present in the club
(3) 160 people at the Ministry of Justice in Mexico received a chip under their skin to make it easier to trace them in case of kidnapping
(4) Millions of pets in the USA have implanted chips to make it easier to find them when they run away.

The US federal government has recently experimented with RFID cards in immigration documents for foreign visitors in the context of the US Visit program, but the CEO of the company Digital Applications has taken this idea one step further: the CEO stated on national television that the chip could be used to tag immigrants and monitor their movements.[7]

---

[4] See www.ubisense.com.
[5] See the project Smart Dust, autonomous sensing and communication in a cubic millimeter, http://robotics.eecs.berkeley.edu/~pister/SmartDust/.
[6] See for these examples Weinberg (2005).
[7] http://biz.yahoo.com/bw/060515/20060515005981.html?.v=1.

3. PRIVACY

Privacy is one of the major moral issues that is discussed in connection with the development and applications of nanotechnology (Guttierrez, 2004; Mehta, 2003).

In this section, I present a framework for structuring debates on privacy in the context of nanotechnology. This framework provides a taxonomy of moral reasons for data-protection. It also provides us with suggestions for the Value Sensitive Design of RFID and nano-surveillance technology.

3.1 The importance of privacy

Laws and regulations to protect the personal sphere and the privacy of persons have been formulated and implemented in the last 100 years around the world, but not without debate and controversy. A good deal of practical and legal consensus has emerged. Data protection laws and regulations define constraints on the processing of personal information, which function as *de facto* norms. These norms were already articulated in the OECD principles for data protection of 1980.[8] The main idea here – familiar in medicine and medical ethics – is informed consent, which forms the moral core of the European data protection laws (1995) and has started to influence thinking about privacy in the rest of the world. It states that before personal data can be processed, informed consent of the data subject is required, the person has to be notified, s/he must be offered the opportunity to correct the data if they are wrong, the use is limited to the purpose for which the data were collected; those who process data must guarantee accuracy, integrity, and security and are accountable for doing so and are accountable for acting in compliance with the requirements of data protection laws.

The requirements of security and accuracy are problematic in the case of RFID, since radio signals can in principle be sent and read by anyone. Even if individuals are aware of the tracking and tracing of objects and people as described above, there still would be problems with the technology from a data protection point of view: there could be cases of "sniffing, skimming, and spoofing" when unauthorized readers are trying to get hold of the information stored on RFIDs in one's possession. A group at Johns Hopkins University demonstrated that the minimal crypto on RFID chips can be cracked. A low-cost spoofing and cloning attack has been demonstrated on some RFID tags used for transport road tolling and the purchase of fuel at petrol stations. The researchers created a cheap code-cracking device for a brute force attack on the forty bit cryptographic key space on the tag. A group from the free university of Amsterdam, led by Andy Tanenbaum, has shown that RFIDs can be infected by viruses that can spread via middle ware into data bases and propagate (Tanenbaum et al).[9]

---

[8] The OECD identified eight basic principles: collection limitation principle, data quality principle, purpose specification principle, the use limitation principle, security safeguards principle, openness principle, individual participation principle, accountability principle.

[9] See their "Is Your Cat Infected with a Computer Virus?" (www.rfidvirus.org/papers/percom.06.pdf).

Why should we have such a stringent regime for data protection at the level of the principles of the OECD and the EU Directive of 1995? Why make the requirement of informed consent by individuals a necessary condition for the processing of their information? This is often not spelled out in full detail, but if privacy and data protection are important, it is important to know exactly why they are important.

Privacy has been the subject of much philosophical discussion (Nissenbaum, 2004; Roessler 2005; Decew, 1997, Van den Hoven 2007) and different authors have presented different accounts of privacy. Although there are many different accounts, I think the following taxonomy of moral reasons is useful for justifying the protection of personal information and constraints on the design and use of a new generation of nano-surveillance devices. The taxonomy has the advantage of turning the privacy discussion into a more or less tractable problem where moral reasons for data protection in a specific area or in a particular case can be spelled out and confronted with moral reasons of the same type which would seem to support arguments against data protection.

The following moral reasons can account for the importance given to individual control over personal data: 1) prevention of information-based harm; 2) prevention of informational inequalities; 3) prevention of informational injustice and discrimination; and 4) respect for moral autonomy and identity. In claiming privacy, we do not simply and nondescriptly want to be "left alone" or to be "private," but more concretely we want to prevent others from harming us, wronging us by making use of knowledge about us, or we want fair treatment and equality of opportunity, and do not want to be discriminated against.

3.2 Information-based harm

The first type of moral reason for data-protection is concerned with the prevention of harm, more specifically harm that is done to persons by making use of personal information about them. Criminals are known to have used data bases and the Internet to get information on their victims in order to prepare and stage their crimes. The most important moral problem with "identity theft," for example, is the risk of financial and physical damages. One's bank account may get plundered and one's credit reports may be irreversible tainted so as to exclude one from future financial benefits and services. Stalkers and rapists have used the Internet and online data bases to track down their victims. They could not have done what they did without tapping into these resources. In an information society there is a new vulnerability to information-based harm. The prevention of information-based harm provides government with the strongest possible justification for limiting the freedom of individual citizens and to constrain access to personal data.

RFID information could be sniffed, people could be monitored, accurate pictures could be made of what they carry with them, and their identity could be stolen. We would also like to prevent that people deceive others by stealing someone else's identity in order to manipulate the information about the nature of objects and present goods as new, when they are old, as edible when they are poisonous, as legitimate when they are stolen, as having cleared customs when they were in fact smuggled.

No other moral principle than John Stuart Mill's Harm Principle is needed to justify limitations of the freedom of persons who cause, threaten to cause, or are likely to cause, information-based harms to people. Protecting personal information, instead of leaving it in the open, diminishes the likelihood that people will come to harm, analogous to the way in which restricting the access to firearms diminishes the likelihood that people will get shot in the street. We know that if we do not establish a legal regime that somehow constrains citizens' access to weapons, the likelihood that innocent people will get shot increases.

3.3 Informational equality

The second type of moral reason to justify data-protection is concerned with equality and fairness. More and more people are keenly aware of the benefits that a market for personal data can provide. If a consumer buys coffee at the shopping mall, information about that transaction can be generated and stored. Many consumers have come to realize that every time they come to the counter to buy something, they can also sell something, namely, information about their purchase or transaction (transactional data). Likewise, sharing information about ourselves-on the Internet with web sites, or through sensor technology may pay off in terms of more and more adequate information (or discounts and convenience) later. Many privacy concerns have been and will be resolved in *quid pro quo* practices and private contracts about the use and secondary use of personal data.

RFID sensor, tracking, and tracing would turn our environment into a transaction space, where information is generated constantly and systematically. But although a market mechanism for trading personal data seems to be kicking in on a global scale, not all individual consumers are aware of this economic opportunity, and if they are, they are not always trading their data in a transparent and fair market environment.

Moreover they do not always know what the implications are of what they are consenting to when they sign a contract or agree to be monitored. We simply cannot assume that the conditions of the developing market for personal data guarantee fair transactions by independent standards. Data-protection laws can help to guarantee equality and a fair market for personal data. Data-protection laws in these types of cases protect individual citizens by requiring openness, transparency, participation, and notification on the part of business firms and direct marketers to secure fair contracts. For example, Amazon.com has already been accused of price targeting. In general, if a retailer knows that I like product X, bought lots of it, irrespective of its price, then they may charge me more for X than someone who does not know the product and needs to be enticed by means of low prices and discounts.

3.4 Informational injustice

A third and important moral reason to justify the protection of personal data is concerned with justice in a sense which is associated with the work of the political philosopher Michael Walzer (Walzer, 1983). Walzer has objected to the simplicity of John Rawls's conception of primary goods and universal rules of distributive justice by pointing out that "there is no set of basic goods across all moral and material worlds, or they would have to be so abstract

that they would be of little use in thinking about particular distributions"
(Walzer, 1983, p. 8). Goods have no natural meaning; their meaning is the
result of socio-cultural construction and interpretation. In order to determine
what a just distribution of the good is, we have to determine what it means
to those for whom it is a good. In the medical, the political, the commercial
sphere, there are different goods (medical treatment, political office, money)
which are allocated by means of different allocation or distributive practices:
medical treatment on the basis of need, political office on the basis of desert,
and money on the basis of free exchange. What ought to be prevented, and
often is prevented as a matter of fact, is dominance of particular goods. Walzer
calls a good *dominant* if the individuals that have it, because they have it, can
command a wide range of other goods. A monopoly is a way of controlling
certain social goods in order to exploit their dominance. In that case advantages in one sphere can be converted as a matter of course to advantages in
other spheres. This happens when money (commercial sphere) could buy
you a vote (political sphere) and would give you preferential treatment in
healthcare (medical), would get you a university degree (educational), etc.
We resist the dominance of money – and other social goods for that matter
(property, physical strength) – and think that political arrangements allowing
for it are unjust. No social good X should be distributed to men and women
who possess some other good Y merely because they possess Y and without
regard to the meaning of X.

What is especially offensive to our sense of justice, Walzer argues, is the
allocation of goods internal to sphere A on the basis of the distributive logic
or the allocation scheme associated with sphere B, second, the transfer of
goods across the boundaries of separate spheres and third, the dominance
and tyranny of some goods over others. In order to prevent this, the 'art of
separation' of spheres has to be practiced and 'blocked exchanges' between
them have to be put in place. If the art of separation is effectively practiced and
the autonomy of the spheres of justice is guaranteed, then 'complex equality'
is established. One's status in terms of the holdings and properties in one
sphere are irrelevant, *ceteris paribus*, to the distribution of the goods internal
to another sphere.

Walzer's analysis also applies to information (Van den Hoven, 1999). The
meaning and value of information is local, and allocation schemes and local
practices that distribute access to information should accommodate local
meaning and should therefore be associated with specific spheres. Many people
do not object to the use of their personal medical data for *medical* purposes,
whether these are directly related to their own personal health affairs, to those
of their family, perhaps even to their community or the world population at
large, as long as they can be absolutely certain that the only use that is made of
it is to cure people of disease. They do object, however, to their medical data
being used to disadvantage them socio-economically, to discriminate against
them in the workplace, to refuse them commercial services, to deny them
social benefits, or to turn them down for mortgages or political office on the
basis of their medical records. They do not mind if their library search data are
used to provide them or others with better *library* services, but they do mind

if these data are used to criticize their tastes and character.[10] They would also object to these informational cross-contaminations when they would benefit from them, as when the librarian would advise them a book on low-fat meals on the basis of knowledge of their medical record and cholesterol values, or a doctor poses questions, on the basis of the information that one has borrowed a book from the public library about AIDS.

We may thus distinguish another form of informational wrongdoing: "informational injustice," i.e., disrespect for the boundaries of what we may refer to as "spheres of access." I think that what is often seen as a violation of privacy is often more adequately construed as the morally inappropriate transfer of data across the boundaries of what we intuitively think of as separate "spheres of access."

RFIDs allow for a wide range of cross-domain profiling and information processing practices, which do not respect the boundaries of these spheres of access, unless they are explicitly designed to do so.

3.5 Respect for moral autonomy and identity

Some philosophical theories of privacy account for its importance in terms of *moral* autonomy (i.e., the capacity to shape our own moral biographies, to reflect on our moral careers) in order to evaluate and identify with our own moral choices, without the critical gaze and interference of others and a pressure to conform to the "normal" or socially desired identities (Van den Hoven, 1998; Van den Hoven, 2005; Van den Hoven, 2007). Privacy, conceived along these lines, would only provide protection to the individual in his quality of a *moral* person engaged in self-definition, self-presentation, and self-improvement against the normative pressures which public opinions and moral judgements exert on the person to conform to a socially desired identity. Information about some individual, whether fully accurate or not, facilitates the formation of beliefs and judgements about that individual. Judgements and beliefs of others about that individual, when s/he learns about them, or suspects that they are made, fears that they are made, may bring about a change in his view of himself, may induce him or her to behave or think differently than s/he would have otherwise done. They pre-empt acts and choices of self-determination and compromise one's status as a self-presenter. When individuals fail in this respect, that is a source of shame; it reveals that one cannot manage the way one presents oneself.

To modern individuals, who have cast aside the ideas of historical and religious necessity, living in a highly volatile socio-economic environment, and a great diversity of audiences and settings before which they make their appearance, the *fixation* of one's moral identity by means of the judgements of others is felt as an obstacle to 'experiments in living' as Mill called them. The modern liberal individual wants to be able to determine himself morally or to undo his previous determinations, on the basis of more profuse experiences in life, or additional factual information. Data-protection laws provide the individual with the leeway to do just that.

---

[10] The world of books and libraries is one of the most likely candidates for complete item level tagging RFID.

This conception of the person as being morally autonomous-as being the author and experimenter of his or her own moral career-provides a justification for protecting his personal data. Data-protection laws thus provide protection against the freezing of one's moral identity by those other than one's self and convey to citizens that they are morally autonomous.

A further explanation for the importance of respect for moral autonomy may be provided along the following lines. Factual knowledge of another person is always knowledge by description. The person himself however, does not only know the facts of his biography, but is the only person who is *acquainted* with the associated thoughts, desires, and aspirations. However detailed and elaborate our files and profiles on a person may be, we are never able to refer to the data subject as he himself is able to do. We may only approximate his knowledge and self-understanding. Bernard Williams (1985) has pointed out that respecting a person involves 'identification' in a very special sense, which could be referred to as 'moral identification' (quoted in Van den Hoven, 1999):

> in professional relations and the world of work, a man operates, and his activities come up for criticism, under a variety of professional or technical titles, such as 'miner' or 'agricultural labourer' or 'junior executive.' The technical or professional attitude is that which regards the man solely under that title, the human approach that which regards him as a man who has that title (among others), willingly, unwillingly, through lack of alternatives, with pride, etc.... each man is owed an effort at identification: that he should not be regarded as the surface to which a certain label can be applied, but one should try to see the world (including the label) from his point of view.

Moral identification thus presupposes knowledge of the point of view of the data-subject and a concern with what it is for a person to live that life. Persons have aspirations, higher-order evaluations and attitudes and they see the things they do in a certain light. Representation of this aspect of persons seems exactly what is missing when personal data are piled up in our data-bases and persons are represented in administrative procedures. The identifications made on the basis of our data fall short of respecting the individual person, because they will never match the identity as it is experienced by the data-subject. It fails because it does not conceive of the other on her terms. Respect for privacy of persons can thus be seen to have a distinctly epistemic dimension. It represents an acknowledgement of the fact that it is impossible to really know other persons as they know and experience themselves.

Ubiquitous and pervasive computing with surveillance and monitoring as a permanent, but invisible feature may change our conception of ourselves as self-presenters. Under such a technological regime the notion of 'self-presentation' and the associated forms of autonomy, may disappear and become obsolete. The dominant view which is associated with the use of profiles and data bases fails to *morally identify* individuals in Williams's sense. Only if citizens can have a warranted belief that those who process their data adopt a moral stance towards them and are genuinely concerned with moral identification next to other forms of identification, can a universal surveillance and an entanglement of individuals in 'an internet of things' be construed as morally acceptable.

4. THE CHALLENGE OF INVISIBILITY

Privacy was construed above in terms of moral reasons for protecting personal information (i.e., moral reasons for putting constraints on the acquisition, processing, and dissemination of personal information). The central constraint was *informed consent*; personal information can only be processed if the data-subject has provided informed consent. Four moral reasons for making informed consent a necessary condition were discussed above. This indicates that the core problem concerning privacy with nanotechnology is epistemic in nature: it is the fact that we do not know that we are monitored, tracked, and traced. Stanley Benn (1984, p. 230) already clearly stated what the problem with this epistemic condition is. We need to distinguish between two cases. First, if the information processing is covert, then it is clear that this interferes with our autonomy, because our thinking and choices are tainted by our false assumption (i.e., that we assume we are unobserved). Many of our assumptions and reasoning can be defeated just by adding the information that we are observed. Second, if the information processing is overt, we can adjust to being observed, but we no longer have the prior choice to be unobserved. In both ways our autonomy is compromised.

A related but slightly different aspect of invisibility and lack of relevant knowledge was articulated by Jeffrey Reiman in his essay on Automated Vehicle Registration Systems (Reiman, 1997). If unbeknownst to me, my passage from A to B is registered, something strange happens. If asked what I did, I will respond that I drove from A to B. But this is only part of the story. My action could be more adequately described as "I drove from A to B and thereby created a record in the data base of the system."

In the same way people will have to become aware of the fact that when they buy clothing they could be buying invisible transponders and memory sticks as well. It changes the conditions under which people consent to actions and intend things. Actions like 'trying on a coat,' 'carrying a gift out of a shop,' or 'drive from A to B,' are no longer what they appear to be to the agent. What actually happens is that one buys a gift *and* lets the store know which route one follows through the shop.

A socio-technological system which obfuscates these mechanisms robs individuals of chances to describe their actions more adequately. Moreover it seems to violate a requirement of publicity or transparency, as articulated by Rawls and Williams among others. The functioning of social institutions should not depend on a wrong understanding of how they work by those who are subject to them (Fontana, 1985, p. 101). Suppose that ubiquitous and covert surveillance arrangements work well and to the satisfaction of a majority, then they seem to work because those affected by them have a false understanding of why and how they work. This seems to violate a reasonable requirement of transparency.

A further fundamental problem needs to be discussed which is relevant to nanotechnology and ubiquitous surveillance by means of RFID and functionally equivalent technology. Is the information concerned *personal information* and do the data protection laws apply by implication?

The answer is affirmative. Although an RFID tag does not necessarily contain personal information, but if that information can be linked without too much trouble and cost in a back-end data base to a file which does contain data about a natural person, it counts as personal data.

5. VALUE-SENSITIVE DESIGN

Ed Felten, a Professor of Security and Computer Science at Princeton, has observed that:

> It seems that the decision to use contactless technology was made without fully understanding its consequences, relying on technical assurances from people who had products to sell. Now that the problems with that decision have become obvious, it's late in the process and would be expensive and embarrassing to back out. In short, this looks like another flawed technology procurement program. (http://michaelzimmer. blogspot.com/2005/04/rfid-passports-need-for-valuesin.html)

Value-sensitive Design is a way of doing ethics that aims at making moral values part of technological design, research, and development. It works with the assumption that human values and norms, our ethics for short, can be imparted to the things we make and use. It construes technology as a formidable force which can be used to make the world a better place, especially when we take the trouble of reflecting on its ethical aspects in advance.

Information technology has become a constitutive technology; it partly constitutes the things to which it is applied. It shapes our practices and institutions in important ways. What health care, public administration, politics, education, science, transport, and logistics will be within twenty years from now will be, in important ways, determined by the ICT applications we decide to use in these domains.

– If our moral talk about the user's autonomy, talk of patient centredness and citizen centredness, his privacy, her security is to be more than an empty promise, these values will have to be expressed in the (chip-) design, systems-architecture, and standards and specifications.
– If our laws, politics, and public policy about corporate governance, accountability, and transparency are to be more than just cheap talk we have to make sure that they are incorporated in the systems we need to support the relevant policies in a global business environment.
– If we want our nanotechnology-and the use that is made of it-to be just, fair, safe, environmentally friendly, transparent, we must see to it that our technology inherits our good intentions. Moreover, we want them to be seen to have those properties, and we want to be able to demonstrate that they possess these morally desirable features. We want to be able to compare different architectures from these value perspectives and motivate political choices and justify investments from this perspective.

Nanotechnology will take privacy discussions to the level of the design of materials, surfaces, properties of artifacts, and fabrics. This will require an adjustment in our thinking about legal and moral constraints on their development and use in addition to thinking about constraints on the use of the personal information they help to generate, store and distribute. Gildas Avoine and Philippe Oechslin have already argued that it is not sufficient to discuss data protection at the level of the application or at the level of the communication. They argue that also the physical level needs to be looked at. (Avoine, et al., 2004).

There are various ways in which we could start to incorporate our values in our designs. IBM work on antennae of RFIDs shows, e.g., that they can be easily torn off a label on a product, to limit the range in which they can be read. The tag has a couple of indentations, the more you tear off, the more you limit the range in which the tag can be read. There are several ways in which the tags could be protected by means of encryption, made visible, comparable to the way we notify people that they are on CCTV camera and the way we warn people that there are additives in food. In the same way we could notify people, empower them and give them means of controlling the flow of their information.[11] The ideal of restoring a power balance regarding the use of one's personal data is sometimes referred to as *sousveillance*. We can think about simple measures which create a Faraday cage (wrapping your passport in aluminum foil for example) by introducing ways in which sensors can be 'killed,' or put in 'privacy mode,' or signals are jammed, or tags get blocked by blocker tags.

Customers have good moral reasons to want to keep control over how they are perceived in stores, in hospitals, and in the street, for moral reasons outlined above. They may fear harm, unfair treatment, discrimination, or they will start not to feel free to be the person they want to be.

6. CONCLUSION

Typically privacy is about information and in a normative sense it refers to a nonabsolute moral right of persons to have direct or indirect control over access to 1) information about oneself, 2) situations where others could acquire information about oneself, 3) technology and/or artifacts that can be used to support the processing of personal data. Not only data base specialists and ICT professionals, security and cryptographers should think about privacy in the future, also nanotechnologists and designers of material, fabrics, sensors and sensor networks, and supply chain managers and retail people will have to think in terms of privacy designs. They will have to worry about how existing nanotechnology can be made visible, can be detected and neutralized or be read, and how the information it helps to generate and store can be protected by design.

WORKS CITED

G. Avoine, and P. Oechslin (2004) 'RFID Traceability: A Multilayer Problem' (available at lasecwww.epfl.ch/pub/lasec/doc/AO05b.pdf)

---

[11] See, e.g., the work of Juels, et al. (2005, 2006), Garfinkel, et al. (2005), Gao, et al. (2004), and others who have studied cryptography for RFID.

S. Benn (1984) 'Privacy, freedom, and respect for persons,' F. Schoeman (ed.) *Philosophical Dimensions of Privacy: An Anthology* (Cambridge: Cambridge University Press).

J. DeCew (1997) *In Pursuit of Privacy: Law, Ethics, and the Rise of Technology* (Ithaca: Cornell University Press).

EU Data Protection Laws (1995) EU Directive 95 (available at http://europa. eu.int/comm/internal_market/ privacy/index_en.htm).

E. Felten, http://michaelzimmer.blogspot.com/2005/04/rfid-passports-need-for-valuesin.html.

X. Gao, et al. (2004) 'An Approach to Security and Privacy of RFID System for Supply Chain,' in *E-Commerce Technology for Dynamic E-Business, 2004. IEEE International Conference* (available at http://pmlab.iecs.fcu.edu.tw/PP/Papers/RF/GXWS04.pdf).

S. L. Garfinkel, A. Juels, and R. Pappu (2005) 'RFID Privacy: An Overview of Problems and Proposed Solutions,' *IEEE Security and Privacy*, 34–43.

E. Gutierrez (2004) 'Privacy Implications of Nanotechnology,' *EPIC* (available at www.epic.org/ privacy/nano Last updated April 26 2004 (accessed June 28, 2004).

A. Juels, P. Syverson, and D. Bailey (2005) 'High-Power Proxies for Enhancing RFID Privacy and Utility,' (available at www.rsasecurity.com/rsalabs/node. asp?id=2948).

A. Juels, and S. A. Weis (2006) 'Defining Strong Privacy for RFID,' (available at ieeexplore.ieee. org/iel5/49/ 33490/01589116.pdf?tp=a&arnumber=1589116).

Z. Kardasiadou, and Z. Talidou (2006) *Legal Issues of RFID Technology* (available at http://rfidconsultation.eu/docs.ficheiros/legal_issues_of_RFID_technology_LEGAL_IST.pdf).

M. D. Michael (2003) 'On Nano-Panopticism: A Sociological Perspective,' (available at http://chem4823.usask.ca/~cassidyr/OnNano-Panopticism-ASociologicalPerspective.htm).

D. Molnar, and D. Wagner (2004) 'Privacy and Security in Library RFID. Issues, Practices, and Architectures,' (available at www.cs.berkeley.edu/~dmolnar/library.pdf).

H. Nissenbaum (2004) 'Privacy as Contextual Integrity,' *Washington Law Review*, 79: 101–39.

Reiman, "Driving to the Panopticon," in *Critical Moral Liberalism* (Lanham, MD: Rowman and Littlefield): 169–188.

B. Roessler (2005) *The Value of Privacy* (Oxford: Polity Press)

J. Van den Hoven (1999) "Privacy and the Varieties of Informational Wrongdoing," *Australian Journal of Professional and Applied Ethics*, vol. 1, no. 1: 30–44 .

J. Van den Hoven (2005) 'Privacy,' in *Encyclopedia for Ethics and Technology*, (ed.) C. Mitcham and D. Johnson.

J. Van den Hoven (2007) "Privacy and Data protection," in *Information Technology and Moral Philosophy* (ed.) J. Weckert, J. Van den Hoven (Cambridge: Cambridge University Press).

M. Walzer (1983) *Spheres of Justice* (Basic Books, New York).

S. D. Wanczyk (2004) 'The Nano-threats to Privacy: Sci-fi or Sci-fact?' *Culture, Communication* & *Technology Program*, Georgetown University, 3,

Spring 2004 (available at www.gnovis.georgetown.edu/includes/ac.cfm?documentNum=31).

J. Weinberg (2005) 'RFID and Privacy,' (available at www.law.wayne.edu/weinberg/rfid.paper.new.pdf).

B. Williams (1985) *Ethics and the Limits of Philosophy* (London: Fontana).

DISCUSSION QUESTIONS

1. How would you define privacy? Does your definition differ from the conception of privacy that the author uses in this chapter?

2. To what extent do you think people should have an expectation of privacy? How does the expectation of privacy differ in public and private spaces?

3. The author suggests several ways in which privacy violations can lead to harm. What are these ways? Can you think of any other ways that an inability to control or restrict information about yourself (or others) can lead to harm?

4. Is privacy important even if it is not related to harm? Suppose a person enjoys looking through the windows of people's homes without their knowledge or consent. Would there be anything wrong with his doing so if it did not harm anyone or violate any property rights?

5. What is value sensitive design? How might value sensitive design be used to mitigate the privacy concerns regarding invisible and powerful information technologies?

6. Do you think we need to revise our conceptions of privacy in light of the proliferation of powerful information technologies, such as the internet, smart phones, and RFID tags?

INTELLECTUAL PROPERTY: LEGAL AND MORAL CHALLENGES OF ONLINE FILE SHARING[1]

Richard A. Spinello

CHAPTER SUMMARY

In this chapter, Richard Spinello discusses the ethics of the use of peer-to-peer (P2P) networks, such as Napster and Grokster, for online file sharing. The first issue he addresses is whether using P2P networks to share music and movie files, for example, constitutes a sort of theft from those who own the copyrights to them. After presenting and assessing arguments on both sides of the issue, he concludes that such sharing is ethically problematic because it involves unfair and unauthorized use of another person's property. The second issue that he addresses is whether those that enable P2P sharing by writing and publishing the software that makes it possible are culpable and should be held responsible. He argues that their behavior is ethically problematic, since it induces and encourages others to act unethically. Therefore, it is appropriate that they be held responsible.

RELATED READINGS

Introduction: Power (2.4); Rights and Liberties (2.6.3); Responsible Development (2.8)

1. INTRODUCTION

The recording industry in the United States has filed multiple lawsuits against purveyors of file sharing software. It has even initiated lawsuits against individuals who make substantial use of this technology (*RIAA v. Verizon*, 2003). The industry contends that the unauthorized "sharing" of copyrighted files is actually tantamount to the theft of intellectual property. The industry also contends that those companies that provide this software, such as Grokster and StreamCast, are culpable of secondary or indirect liability for such theft. This assertion is contentious, but recent court cases have tended to support the recording industry's claims about secondary liability, especially when there is evidence of inducement.

[1] This chapter is excerpted from Richard Spinello (2008) 'Intellectual Property: Legal and Moral Challenges of Online File Sharing,' *The Handbook on Information and Computer Ethics*, eds. Himma and Tavani (Wiley). It appears here by permission of Wiley and Sons, Inc.

Lost in the thicket of lawsuits and policy challenges are the ethical issues associated with the use and distribution of file sharing software. Is the downloading or "sharing" of copyrighted music morally reprehensible? Quite simply, are we talking about sharing or pilfering? Is social welfare enhanced by a legal regime of indirect liability? And should we hold companies like Napster, Grokster, or BitTorrent morally accountable for the direct infringement of their users, particularly if they intentionally design the code to enable the avoidance of copyright liability? Or does such accountability stretch the apposite moral notion of cooperation too far? In this overview, we will present the conflicting arguments on both sides of this provocative debate. Although our primary focus will be on the ethical dimension of this controversy, we cannot neglect the complex and intertwined legal issues. We will take as a main axis of discussion the recent *MGM v. Grokster* (2005) case, in which all of these issues have surfaced. We begin, however, with a brief summary of this technology's functionality.

2. PEER-TO-PEER NETWORKS

The technology at the center of these copyright disputes is a software that enables computer users to share digital files over a peer-to-peer (P2P) network. Although the P2P architecture is evolving, a genuine P2P network is still defined as the one in which "two or more computers share [files] without requiring a separate server computer or server software" (Cope, 2002). Unlike the traditional client/server model, where data are only available from a single server (or group of servers), data can be accessed and distributed from any node in a P2P network. Each computer in the network can function as a server when it is serving or distributing information to others. Or it can assume the role of a client when it is accessing information from another system. As a result of this decentralization, the P2P network has the potential to be a more reliable information distribution system. For example, in the client/server model, if the server that hosts the data crashes, no data are available. But if a computer in a P2P network crashes, data are still available from other nodes in the network. Files can also be shared more expeditiously and more widely than ever before.

Thus, the most distinctive feature of this architecture is that each node in the system is a "peer" or an equal. There is no need for a central authority to mediate and control the exchange of information. The "purest" P2P architecture is flat and nonhierarchical. However, the diminished control associated with such a completely decentralized network leads to obvious scalability problems. As Wu (2003) observes, as the network grows "the loss of control makes it difficult to ensure performance on a mass scale, to establish network trust, and even to perform simple tasks like keeping statistics."

P2P software programs are usually free and easy to install. Once they are installed, a user can prompt his or her personal computer to ask other PCs in a peer-to-peer network if they have a certain digital file. That request is passed along from computer to computer within the network until the file is located and a copy is sent along to the requester's system. Each time a P2P user makes a copy of a digital file, by default that copy becomes available on the user's computer so that it can be copied by other P2P users. This process, which is known as "uploading," results in "an exponentially multiplying redistribution of perfect digital copies" (Petition for Writ of Certiorari, 2004).

Peer-to-peer networks require some method of indexing the information about the digital files available across the network so that user queries can be handled efficiently. There are three different methods of indexing: a centralized index system, in which the index is located on a central server; a decentralized indexing system; and a supernode system, in which a special group of computers act as indexing servers. The first method, which was adopted by Napster, relies on central servers to maintain an index of all the files available on the network; users search that index, and they are then referred to peers with a copy of the desired file. This method was abandoned by Napster's successors after Napster lost the court case defending its technology (*A&M Records, Inc. v. Napster, 2001*). The supernode system, developed as part of KaZaA, BV's FastTrack technology, relies on selected computers within the network with large memory capacity; these index servers perform the searches and return the search results to the user. The supernodes change periodically, and a given computer may never realize that it is serving in this capacity. The supernode has been the preferred solution in recent years, since it combines the advantages of the first two methods. While Grokster has depended on the supernode approach, some versions of the Gnutella protocol have relied on the decentralized method with no supernodes.

The P2P architecture represents a powerful communications technology with obvious social benefits. These include properties such as anonymity and resistance to censorship. The problem with P2P software, however, is that it has facilitated the unauthorized reproduction and distribution of copyrighted works in violation of the Copyright Act. Approximately 2.6 billion copyrighted music files are downloaded each month (Grossman, 2003), and about 500,000 infringing movie files are downloaded each day (MPAA Press Release, 2004). Companies supplying this software are obviously aware that their users are downloading copyrighted files, but they do not know which specific files are being copied or when this copying is occurring.

3. SHARING OR THEFT

The Web has emboldened free riders and engendered a new ethic on copying proprietary material based on the belief that cultural products such as movies and music should be freely available online to anyone who wants them. Whatever enters the realm of cyberspace as a digital file is fair game. According to this ethic, there is nothing wrong with the use of P2P networks for sharing copyrighted material. Freenet's project leader, for example, has described copyright as "economic censorship," since it retards the free flow of information for purely economic reasons (Roblimo, 2000). He and others support an anticopyright model, which calls for the repudiation of exclusive property rights in cyberspace. Echoes of this viewpoint can also be found in the writings of other information libertarians such as Barlow (1994):

> ... all the goods of the Information Age – all of the expressions once contained in books or film strips or newsletters – will exist as thought or something very much

like thought: voltage conditions darting around the Net at the speed of light, in conditions that one might behold in effect, as glowing pixels or transmitted sounds, but never touch or claim to "own" in the old sense of the word.

One could infer that Barlow would not be troubled by the "darting around" of copyrighted works on P2P networks, since these digital networks help the Net to realize its true potential and thereby accelerate the abandonment of archaic notions of intellectual property "ownership."

Nonetheless, not everyone agrees with Barlow's radical vision or the anticopyright approach. For those who recognize the value of P2P networks for sharing digital content and also respect the beneficial dynamic effects of intellectual property rights, some important questions need consideration. Are those who copy copyrighted files by means of a P2P system legally responsible for breaking the law? Does their action constitute direct infringement of a copyright? According to the U.S. Copyright Act (2004), an infringer is "anyone who violates any of the exclusive rights of the copyright owner ...," including the right to make copies.

Defenders of the unfettered use of P2P networks come in many different stripes. But they typically concur that the personal copying of protected files on P2P networks is legally acceptable. Some legal scholars in this camp maintain that users who download files are not making copies of those files but simply "sharing" digital information over a conduit.[2] They also argue that even if sharing digital copies over a network is equivalent to making an unauthorized reproduction of a copyrighted work, that action comes under the fair use exception. Litman (2002), for example, contends that it is far from evident under current law whether individual users are liable for copyright infringement if they engage in "personal copying." On the contrary, Zittrain admits that "it is generally an infringement to download large amounts of copyrighted material without permission; even if you already own the corresponding CD, the case could be made that a network-derived copy is infringing" (Gantz and Rochester, 2005).

What about the moral propriety of sharing copyrighted files without permission? While David Lange (2003) does not argue that such file sharing is morally acceptable, he claims that there is considerable "softness" on the side of those who make the opposite claim. He maintains that those who argue that file sharing is morally wrong do so along the following lines: "Taking property without permission is wrong. Recorded music is property. Taking recorded music without permission is therefore wrong as well." But the problem in this line of reasoning lies in the minor premise. Many do not accept that music is property and, in Lange's view, there is some merit to this claim. Therefore, the issue of a legitimate property right in such intellectual objects "is still very much unsettled ... [and] it may yet be that the idea of property and exclusivity will prove unable to withstand the popular will" (Lange, 2003).

---

[2] Peter Mennell et al. point out that "file sharing" is a misnomer: "a more accurate characterization of what such technology accomplishes is file search, reproduction, and distribution.... following a peer-to-peer transaction, one copy of the file remains on the host computer and another identical copy resides on the recipient's computer." See Amici Curiae Brief of Law Professors (2005).

Lange seems to assume an asymmetry between physical, tangible property, and intellectual property. He does not question a right to physical property (ownership of a house or an automobile), but intellectual property rights are more ambiguous, given the peculiar characteristics of that property. Unlike its physical counterpart, an intellectual object is nonrivalrous and nonexcludable: its consumption doesn't reduce the supply available to others, and it's difficult to "exclude" or fence out those who haven't paid.[3]

We cannot resolve this complex issue here, but let it suffice to say that a potent case can be made for a right to intellectual property based on the classic labor-desert argument first proposed by Locke. I have argued elsewhere that the application of Locke's theory to intellectual property seems plausible enough (Spinello, 2003). Locke's core argument is that each person has a property right in herself and in the labor she performs. Therefore, it follows that a property right should also extend to the final product of that labor. Surely one is entitled to that right by virtue of creating value that would not otherwise exist except for one's creative efforts. Who else would have a valid claim to the fruits of this person's efforts? This property right must satisfy one condition summarized in the Lockean proviso: a person has such a property right "at least where there is enough, and as good left in common for others" (Locke, 1952). The Lockean-inspired argument for an intellectual property right is that one's intellectual labor, which builds on the ideas, algorithms, generic plots, and other material in the intellectual commons, should entitle one to have a natural property right in the finished product of that work such as a novel or a musical composition. The ideas remain in the commons and only the final expression is protected, so the common domain is undiminished and the proviso satisfied.

This labor-based approach gives intellectual property rights a stronger normative foundation than consequentialist arguments, which, in my opinion, are ultimately indeterminate (Spinello, 2003). A creator or author should have a basic right to exclude others from her work because she created that work through her own labor.[4] Even the U.S. Supreme Court has recognized the general suitability of this argument: "sacrificial days devoted to… creative activities deserve rewards commensurate with the services rendered" (*Mazer v. Stein*, 1954). We should not focus on the nature and qualities of the product (tangible or intangible, excludable or nonexcludable), but on the value inherent in that product that is the result of labor and initiative. Also, contrary to Lange's comments, given the normative underpinnings of intellectual property rights, they should not be contingent on the support of the majority. What's of primary importance is the creator's interests – she has expended time and energy in the creative process. At the same time, while the consumers' interests cannot be completely discounted, their desire for "content" should not give rise to some sort of morally or legally protected interest.[5]

---

[3] It should be pointed out that even if intellectual objects are not strictly speaking property, this does not negate the creator's right to limited control over his or her created products. See Himma (2007a) for more elaboration on this issue.

[4] The recognition of this right is only a starting point for policy makers who may choose to qualify it in certain ways for the sake of justice and the common good.

[5] Himma (2007b) develops this line of reasoning quite cogently. He argues that "the interests of content creators in controlling the disposition of the content they create outweigh[s] the interests of other persons in using that content in most, but not all, cases."

If we grant the premise that "recorded music is property," then it seems clearer that there might be something wrong with copying this music without permission. But in order to classify copyright infringement as a form of theft we need to understand more precisely what is entailed by a property right, particularly when that right is viewed from a distinctly moral perspective. The essence of such a right is the liberty to use a physical or intellectual object at one's discretion, that is, the right to determine what is to be done with this created object. It includes a right to exclude others from appropriating or using that object without my permission. According to Nozick (1974):

> The central core of the notion of a property right in X, relative to which other parts of the notion are to be explained, is the right to determine what shall be done with X; the right to choose which of the constrained set of options concerning X shall be realized or attempted.

Theft, therefore, should be understood as a misuse, an "unfair taking," or a misappropriation of another's property contrary to the owner's reasonable will.[6] In the case of intellectual property (such as digital movies and music), unless the copyright holder's consent can be reasonably presumed, using that copyright holder's creative expression in a way that exceeds her permission is using it contrary to her will. We can be quite certain that downloading a copyrighted Disney movie and then uploading it for millions of other users to copy is an action that is contrary to the will of Disney. When one uses something contrary to the creator's (or owner's) will, this kind of act is equivalent to *taking something* from that owner without her volition and consent. The use of a piece of intellectual property without the copyright holder's permission (and therefore against his will) is unfair to that copyright holder since it violates his right to determine what is to be done with that property. And this unfair use of another's intellectual property constitutes a form of theft (Grisez, 1997).

This presumes that the copyright holder's will is reasonable and that his or her consent would not be given. In the case of digital music files, one can also safely presume that the copyright holder will not want his or her music uploaded on peer-to-peer networks for everyone else to copy, since this action will erode legitimate sales of this product. On the contrary, one might presume that making another copy of a purchased online song or MP3 file for one's own personal use would be reasonable and acceptable. An additional CD burn of music I already own, so I can have an extra copy to keep in my office, seems perfectly legitimate (unless the owner indicates otherwise).

When seen in this light, the moral argument against downloading and sharing music files with other P2P users seems more persuasive. But what about the objection that "sharing" within the online community is a noble act and that the sharing of digital information serves the common good through a de facto expansion of

---

[6] It should be presumed that the owner's will is reasonable unless a strong case can be established to the contrary. This qualification leaves room for exigent circumstances in which an owner's failure to share or license his property may yield dire consequences, and hence the appropriation contrary to his will is arguably not unfair.

the public domain? After all, we are taught at an early age that sharing with others is a good thing to do. The Internet and P2P software facilitate sharing, so why should there be constraints that hold back the full potential of this technology?

Grodzinsky and Tavani (2005) make some noteworthy arguments along these lines as they underscore the fundamental importance of sharing. In their view, it is important to "defend the principle 'information wants to be shared,' which presumes against the fencing off or enclosing of information in favor of a view of information that should be communicated and shared." Of course, information has no "wants," so the argument being proposed here is the normative claim that the distribution of information as widely as possible should be promoted and encouraged.[7]

We must be careful, however, not to overestimate the value of sharing. Sharing is not a core value or a basic human good, even though it often contributes to the harmony of community and the furtherance of knowledge, which are basic human goods. The basic human goods (including life and health, knowledge, the harmony of friendship and community, and so forth) are basic not because we need them to survive but because we cannot flourish as human beings without them. These goods, which are intrinsic aspects of human well-being and fulfillment, are the primary ends or reasons for action and the ultimate source of normativity. Also, goods intrinsic to the human person are greater than instrumental goods. For example, life is a higher good than property. Hence, the basic human goods are of primary significance for justifying ethical policies.

But information sharing does not fall in this category. The obvious problem is that such sharing does not always promote the basic human goods. Sharing is an instrumental good, a means to an end, but that end may not always be morally acceptable. For example, the act of "sharing" pornographic material with young children is certainly immoral, since it contributes to the corruption of those children by hindering the proper integration of sexuality into their lives. Exposure of immature children to pornography puts their relationships at risk and often yields social disharmony. Similarly, child pornography is being "shared" with increasing frequency over P2P networks, but no reasonable person would be in favor of this form of sharing (Amici Curiae Brief of Kids First Coalition, 2005). Therefore, we cannot assume that sharing information, either as a means or as an end, always contributes to the common good. We must consider the quality and the nature of the information to be shared.

Information may "want to be shared," but for the sake of the common good we must sometimes put restrictions on the sharing of information. Some information (such as child pornography) shouldn't be shared at all, and other types of information should be shared according to the wishes of its creator, assuming a valid property right has been established. Since an intellectual property right or copyright embodies the labor-desert principle, it is difficult to dispute this right from a moral perspective. As noted earlier, "sharing" is a misnomer since what

---

[7] There are some scholars such as John Perry Barlow who do maintain that information has the quality of being a life-form. See Himma (2005) for a helpful discussion of this topic.

is really going on is the search for a digital file followed by its reproduction and distribution. The real question is whether or not an intellectual work, such as a movie created by Disney at considerable expense, should be shared with impunity against the will of the creator and rightful owner of this intellectual property.

But what about compulsory sharing, that is, a scheme whereby noncommercial file sharing would become lawful, and copyright owners would be compensated through a tax or levy on Internet services and equipment? Grodzinsky and Tavani (2005) favor this approach because they recognize that creators should be compensated for their efforts. This model, which has been carefully developed by Fisher (2004), would essentially displace exclusive property rights with mandatory compensation. According to Lessig (2001), it is "compensation without control." In legal terms, liability rules would take the place of property rules.

Compulsory licensing certainly has potential, and its benefits should not be discounted by policy makers. At the same time, advocates of compulsory licensing often gloss over the practical implementation problems. How and by whom will a fair compensation plan be determined? What about the high costs of administering such a cumbersome regulatory system, which will undoubtedly be subject to the vagaries of the political process? The potential for economic waste cannot be casually dismissed. Finally, what will be the effects of compulsory licensing for digital networks on the whole copyright regime – will there be a perception that copyrighted works in any format or venue are "up for grabs?" Above all, we must not naively assume that nonvoluntary licensing is a panacea. Epstein (2004) clearly summarizes the Hobson's choice confronted by policy makers: "any system of private property imposes heavy costs of exclusion; however, these costs can only be eliminated by adopting some system of collective ownership that for its part imposes heavy costs of governance – the only choice that we have is to pick the lesser of two evils."

4. SECONDARY LIABILITY

Now that we have considered the issue of direct copyright infringement, we can focus attention on the question of contributory or secondary infringement. If we accept, however cautiously, the reasonable assumption that downloading and uploading copyrighted files is direct infringement, what about the liability of those companies providing the software for this purpose? Do they "cooperate" in the wrongdoing in any morally significant way?

Purveyors of P2P systems fall in the category of "technological gatekeepers." Kraakman (1986) elaborated upon the limits of "primary enforcement" of certain laws and the need for strict "gatekeeper liability." According to Zittrain (2006), such secondary liability "asks intermediaries who provide some form of support to wrongdoing to withhold it, and penalizes them if they do not." To some extent, the emergence of the decentralized P2P architecture, which eliminates intermediaries, represents the undermining of the gatekeeper regime. According to Wu (2003), "the closer a network comes to a pure P2P design, the more disparate the targets for copyright infringement …." For this reason, Gnutella has described itself as a protocol instead of an application. The problem is that efficiency requires some centralization. As we saw above, the FastTrack system used by Grokster relies on supernodes with generous bandwidth for the storage of an index, and it uses a central server to maintain user registrations and to help locate other peers in the network. Thus, there is some conflict between developing a P2P system that will optimize avoidance of legal liability and the technical goals of efficiency and scalability.

The debate over imposing secondary liability has been intense in recent years. Proponents argue that it's a major deterrent of infringement. According to the Supreme Court, "when a widely shared service or product is used to commit infringement, it may be impossible to enforce rights in the protected work effectively against all direct infringers, the only practical alternative being to go against the distributor of the copying device for secondary liability …"(MGM v. Grokster, 2005). However, as Lemley and Reese (2004) point out, "going after makers of technology for the uses to which their technologies may be put threatens to stifle innovation." Hence the tension in enforcing secondary liability or pressing these cases too vigorously: how to enforce the rights of copyright holders in a cost-effective way without stifling innovation. The issue has received extraordinary attention in the legal literature (see Dogan, 2001; Fagin et al., 2002; Kraakman, 1986; Lichtman and Landes, 2003), but it has not yet generated much interest among ethicists….

5. MORAL CONSIDERATIONS

As we have noted, ethicists have not subjected the issue of secondary liability, especially as it pertains to P2P networks, to much moral scrutiny. But we can identify two salient moral issues, one at the "macro" level and the other at a more "micro" level of the individual moral agent. First, can secondary liability law itself be normatively justified in social welfare terms? Second, how can we understand indirect copyright liability from a strictly moral viewpoint?

The utilitarian arguments for maintaining a strong legal tradition of secondary liability seem especially persuasive. Given the enormous difficulties of enforcing copyright protection by pursuing direct infringers and the threats posed to content providers by dynamic technologies such as P2P software, the need for this liability seems indisputable. As Zimmerman (2006) indicates, bringing infringement cases against "private copyists" is difficult, since "private copying often takes place out of public view." Pursuing these private copyists would also be expensive since it would require frequent and repeated litigation. Moreover, it stands to reason that copyright holders will not pay to enforce their rights where the costs of doing so exceed the expected returns.

On the contrary, intermediaries are "highly visible," and a single lawsuit can deter the actions of many egregious infringers, provided that the intermediary has contributed "in some palpable way to the creation of unlicensed private copies" (Zimmerman, 2006). Thus, bringing suits against these intermediaries overcomes the disutility of pursuing private copyists, and so a compelling case can be advanced that indirect liability is a much more efficient mechanism for achieving justice. As Mennell and his coauthors point out in their Amici Curiae

Brief (2005), "The social and systemic benefits of being able to protect copyrights at the indirect infringement level rather than at the end user level, are substantial. Suing thousands of end users who waste both private and public resources is not nearly as effective as confronting enterprises whose business model is based on distributing software that is used predominantly for infringing uses." In addition, third parties such as software providers often have a reasonable opportunity to deter copyright infringement by means of monitoring user activities or designing code in a way that impedes infringement.

On the contrary, there must be reasonable restrictions on the scope of secondary liability claims so that they do not stifle innovation – dual-use technologies, capable of substantial noninfringing use, should be immune from secondary liability so long as there is no evidence of inducement. But the bottom line is that indirect liability, carefully implemented, promotes efficiency in the enforcement of copyright law, which is necessary to maximize the production of creative expression in the first place. Arguably, although we must bear in mind the difficulty of measuring welfare effects, this policy enhances social welfare, since it appears to be so strongly justified in utilitarian terms.

The second question is how we assess the moral propriety of actions that appear to facilitate the wrongdoing of others. The moral case for indirect liability centers on the question of cooperation. Under normal circumstances, cooperation associated with communal activities certainly poses no moral problems. But what about the "community" of online file sharers, which is made possible by software providers such as Grokster? Is there something problematic about the online file-sharing community, given that the primary function of the software is for sharing copyrighted music and movie files?

The basic moral imperative at stake here can be stated succinctly: a moral agent should not cooperate in or contribute to the wrongdoing of another. This simple principle seems self-evident and axiomatic. If someone intentionally helps another individual carry out an objectively wrong choice, that person shares in the wrong intention and bad will of the person who is executing such a choice. In the domain of criminal law, if person X helps his friend commit or conceal a crime, person X can be charged as an accessory. Hence, it is common moral sense that a person who willingly helps or cooperates with a wrongdoer deserves part of the blame for the evil that is perpetrated. But this principle needs some qualification since under certain circumstances cooperation at some level is unavoidable and not morally reprehensible.

First, a distinction must be made between formal and material cooperation. Second, while all forms of formal cooperation are considered immoral, we must differentiate between material cooperation that is justifiable and material cooperation that is morally unacceptable. The most concise articulation of the first distinction is found in the writings of the eighteenth century moral theologian and philosopher St. Alphonsus Liguori (1905):

> But a better formulation can be expressed that [cooperation] is formal which concurs in the bad will of the other, and it cannot be without fault; that cooperation is

material which concurs only in the bad action of the other, apart from the cooperator's intention.[8]

The question of material cooperation is quite complex, so we will confine our analysis to formal cooperation, which is more straightforward and apposite in this context. Is there a case to be made that Grokster and StreamCast are culpable of such formal cooperation which, in Liguori's words, cannot be "without fault" (*sine peccato*)? Formal cooperation means that one intentionally shares in another person's or group's wrongdoing. In other words, what one chooses to do coincides with or includes what is objectively wrong in the other's choice. For example, a scientist provides his laboratory and thereby willingly assists in harmful medical experiments conducted on human beings by a group of medical doctors because he is interested in the results for his research. This scientist shares in the wrongful intentions and actions of the doctors.

Is there evidence of such "formal cooperation" among P2P software companies? Has code been designed and implemented by commercial enterprises like StreamCast and Grokster in order to avoid copyright law? Are these companies deliberately seeking to help users to evade the law, perhaps for their own material gain? If so, one can justifiably press the claim that there is formal cooperation and hence moral irresponsibility.

Although it is often difficult to assess intentionality, the assertions of both companies in the Grokster case seemed to betray their true motives. Both companies "aggressively sought to attract Napster's infringing users," referring to themselves as "the next Napster" (Plaintiffs. Joint Excerpts of Record, 2003). StreamCast's CEO boldly proclaimed that "we are the logical choice to pick up the bulk of the 74 million users that are about to 'turn Napster off'" (Plaintiffs. Joint Excerpts of Record, 2003). Moreover, StreamCast executives monitored the number of songs by famous commercial artists available on their network because "they aimed to have a larger number "of copyrighted songs available on their networks than other file-sharing networks"(*MGM v. Grokster*, 2005). Mindful of Napster's legal problems, they were careful to avoid the centralized index design employed by Napster in order to circumvent legal liability. Some products were even designed without key functional advantages (such as specialized servers to monitor performance) in order to evade legal liability. Given this evidence, even the District Court conceded that the "defendants may have intentionally structured their businesses to avoid secondary liability for copyright infringement, while benefiting financially from the illicit draw of their wares" (*MGM v. Grokster*, 2003). At the same time, both companies derived substantial advertising revenues from their users who downloaded a massive volume of copyrighted music and movie files.

It surely appears that these companies deliberately designed their software to help users share copyrighted files and evade the law, and to profit from this collusive activity. Even if the *Sony* standard is blind to design issues (since it only

---

[8] "Sed melius cum aliis dicendum, illam esse formalem, quae concurrit ad malam voluntatem alterius, et nequit esse sine peccato; materialem vero illam, quae concurrit tantum ad malam actionem alterius, praeter intentionem cooperantis."

requires that technologies be capable of substantial noninfringing use), we cannot ignore the moral problem of designing code as a mechanism for avoiding the law. On the contrary, there is a moral requirement to design products that will support and respect valid laws. In this case, the code should have included filtering and monitoring tools that would have minimized the software's potential for misuse, unless it would have been "disproportionately costly" to do so (*In re Aimster Copyright Litigation*, 2003).

Therefore, the evidence is strong that Grokster and StreamCast have acted in "bad faith." Given the rhetoric and behavior of both companies, this case is a classic example of formal cooperation where a moral agent's will coincides with the moral wrongdoing (evasion of copyright law) of another, and the moral agent helps to bring about that wrongdoing. In this context, the illicit cooperation takes the form of providing the mechanism to download and upload copyrighted files. These companies deliberately blinded themselves to this type of content being distributed over their network, and failed to take any sincere affirmative steps to deter the exchange of these protected files. On the contrary, these companies took positive steps to facilitate an illegal and immoral activity and to materially benefit from that activity. They made no secret of their true purpose by repeatedly stating their desire to be the heirs of Napster's notoriety.

In general, companies cannot develop code with the intention of minimizing the burden of just law, including copyright law. This type of "antiregulatory code" (Wu, 2003) that has been deliberately designed to undermine the legal system cannot be morally justified. But if code, including permutations of the P2P architecture, has been designed for the purpose of a legitimate functionality (sharing of information), and its misuse as a mechanism of legal evasion is accepted as an unwanted side effect, the code designer cannot be held morally accountable. Moral prudence also dictates that reasonable efforts must be made to anticipate such misuse. One such effort would be the inclusion of tools such as filters that might mitigate or curtail the misuse of one's product. Under these circumstances, it is safe to claim that there would be no formal cooperation. What is morally decisive, therefore, is the intentions and purpose of the code designer. When code is designed as a deliberate mechanism for the evasion of a legitimate law, there is complicity and moral liability. But software developers cannot be held morally accountable when prudently designed code, created for a valid purpose, is exploited by some users for copyright infringement or some other mischief. In the case of Grokster, intention and purpose seem pretty clear; but in other situations involving potential gatekeepers, the question of moral liability will be much more ambiguous.

6. CONCLUSIONS

In the first part of this essay we explained the likelihood that the unauthorized downloading of copyrighted files constituted direct infringement. We also delineated the limits and problems associated with primary enforcement. This explains why secondary liability has become such a salient issue.

The Internet has many "gatekeepers," from Internet Service Providers (ISPs) and search engines to purveyors of certain types of network software. These gatekeepers are sometimes in a position to impede or curtail various online harms such as defamation or copyright infringement. Thanks to the Digital Millennium Copyright Act (1998), ISPs have been immunized from strict gatekeeper liability since they are rightly regarded as passive conduits for the free flow of information. But other companies such as Napster and Grokster have had a more difficult time navigating toward a safe harbor. We have concentrated on these popular gatekeepers, purveyors of P2P network software such as Grokster, and we have discussed the scope of their legal and moral liability.

The Supreme Court has introduced an inducement standard while upholding the basic doctrine of *Sony*. "Purposeful, culpable expression and conduct" must be evident in order to impose legal liability under this sensible standard (*MGM v. Grokster*, 2005). In this way, the Court has judiciously sought to balance the protection of copyright and the need to protect manufacturers. If P2P developers succeed in developing a pure decentralized system that does not sacrifice efficiencies they may succeed in undermining the gateway regime, which has been so vital for preserving the rights of copyright holders.

Of course, as we have intimated, inducement is also problematic from a moral point of view. It is not morally permissible to encourage or facilitate the immoral acts of others, especially when one profits by doing so through advertising revenues. If we assume that direct infringement is morally wrong, inducement and the correlation of profits to the volume of infringement represents formal cooperation in another individual's wrongdoing.

Finally, as Lessig (1999) has reminded us, "code is law," and given the great power of software code as a logical constraint, software providers have a moral obligation to eschew the temptations of writing antiregulatory code. This type of code includes some P2P programs that facilitate and promote copyright infringement. Instead, developers must design their code responsibly and embed within that code ethical values in the form of tools that will discourage and minimize misuse. This assumes, of course, that such modifications would be feasible and cost effective.

WORKS CITED

A&M Records, Inc. v. Napster (2001) 239 F 3d 1004 [9th Cir.].

Amici Curiae Brief of Kids First Coalition et al. (2005) 'On Petition for Writ of Certiorari, to the Supreme Court of the United States, in review of Metro-Goldwyn-Mayer Studios v. Grokster, Ltd' (2004). 380 F.3d 1154 [9th Cir.].

Amici Curiae Brief of Law Professors (Mennell, Nimmer, Merges, and Hughes) (2005). On Petition for Writ of Certiorari, to the Supreme Court of the United States, in review *of Metro- Goldwyn-Mayer Studios v. Grokster, Ltd* (2004). 380 F.3d 1154 [9th Cir.].

J. P. Barlow (1994) 'The Economy of Ideas,' *Wired*, 2 (3): 84–88.

J. Cope (2002) 'Peer-to-Peer Network,' (available at: http://www.computerworld.com/networkingtopics/ networking/story).

Digital Millennium Copyright Act (1998) Section 512 (c).

S. Dogan (2001) 'Is Napster a VCR? The implications of Sony for Napster and other Internet- Technologies,' *Hastings Law Journal* 52: 939.

R. A. Epstein (2004) 'Liberty vs. Property? Cracks in the Foundation of Copyright Law,' *University of Chicago Law & Economics, Olin Working Paper No. 204.*

M. Fagin, F. Pasquale, and K. Weatherall (2002) 'Beyond Napster: using anti-trust law to advance and enhance online music distribution,' *Boston University Journal of Science and Technology Law*, 8, 451.

W. Fisher (2004) *Promises to Keep* (Cambridge: Harvard University Press).

J. Gantz, and J. B. Rochester (2005) *Pirates of the Digital Millennium* (Upper Saddle River, NJ: Financial Times Prentice-Hall).

Gershwin Publishing Corp. v. Columbia Artists Management, Inc. (1971). 443 F.2d 1159 [2nd Cir.].

G. Grisez (1997) *Difficult Moral Questions* (Quincy, IL: Franciscan Herald Press) 589–598.

F. Grodzinsky, and H. Tavani (2005) 'P2P Networks and the *Verizon v. RIAA* case: implications for personal privacy and intellectual property,' *Ethics and Information Technology*, 7 (4): 243–250.

L. Grossman (2003) 'It's All Free,' *Time* 166: 88.

In re Aimster Copyright Litigation (2003). 334 F. 3d 643 [7th Cir.], cert. denied sub nom., 124 S. Ct. 1069 (2004).

K. E. Himma (2007a) 'The justification of intellectual property rights: contemporary philosophical disputes,' *Journal of the American Society for Information Science and Technology*, 58: 2–5.

K. E. Himma (2007b) 'Justifying property protection: why the interests of content creators usually win over everyone else's,' in E. Rooksby, and J. W (eds.) *Information Technology and Social Justice* (Hershey, PA: Idea Group) 47–68.

K. E. Himma (2005) 'Information and intellectual property protection: evaluating the claim that information should be free,' in R. Spinello (ed.) *Newsletter of Law and Philosophy*, The American Philosophical Association, 2–9.

R. Kraakman (1986) 'Gatekeepers: the anatomy of a third party enforcement strategy,' *Journal of Law, Economics, and Organizations*, 2: 53.

D. Lange (2003) 'Students, music and the net: a comment on peer-to-peer file sharing,' *Duke Law and Technology Review*, 23: 21.

M. Lemley, and R. Reese (2004) 'Reducing copyright infringement without restricting innovation,' *Stanford Law Review* 56: 1345.

L. Lessig (2001) *The Future of Ideas: The Fate of the Commons in a Connected World* (New York: Random House).

L. Lessig (1999) *Code and Other Laws of Cyberspace* (New York: Basic Books)

D. Lichtman, and W. Landes (2003) 'Indirect liability for copyright infringement: an economic perspective,' *Harvard Journal of Law & Technology*, 16: 395.

St. A. Liguori (1905–1912), in L. Gaude (ed.), *Theologia Moralis*. 4 vol. (Vaticana, Rome: Ex Typographia) 357.

J. Litman (2002) 'War stories,' *Cardozo Arts & Entertainment Law Journal*, 20: 337.

J. Locke (1952) *The Second Treatise of Government* (Indianapolis, IN: Bobbs-Merrill).

Mazer v. Stein (1954) 347 U.S. 201.

MPAA Press Release (2004) MPAA Launches New Phase of Aggressive Education Campaign against Movie Piracy (available at http://mpaa.org/MPAAPress/).

Metro-Goldwyn-Mayer Studios v. Grokster, Ltd. (2005) 125 U.S. 2764.

Metro-Goldwyn-Mayer Studios v. Grokster, Ltd. (2004) 380 F.3d 1154 [9th Cir.].

Metro-Goldwyn-Mayer Studios v. Grokster, Ltd. (2003) 259 F. Supp. 2d 1029 [C.D. Cal.].

R. Nozick (1974) Anarchy, State, and Utopia (Oxford: Basil Blackwell).

Oak Industries, Inc. v. Zenith Electronics Corp. (1988). 697 F. Supp. 988 [ND Ill.].

Petition for Writ of Certiorari (2004). To the Supreme Court of the United States, in review of *Metro-Goldwyn-Mayer Studios v. Grokster, Ltd.* 380 F.3d 1154 [9th Cir.].

Plaintiffs. Joint Excerpts of Record (2003). *Metro-Goldwyn-Mayer Studios, Inc. v. Grokster, Ltd.* 259 F. Supp. 2d [C.D. Cal].

RIAA v. Verizon (2003). No. 03–7015[D.C. Cir].

L. Roblimo (2000) Posting to Slashdot (available at http://slashdot.org/article.pl).

P. Samuelson (2005) 'Did MGM really win the Grokster case?' *Communications of the ACM*, 19–24.

Sony Corp of America v. Universal City Studios, Inc. (1984) 464 U.S. 417.

R. A. Spinello (2003) 'The future of intellectual property,' *Ethics and Information Technology*, 5 (1): 1–16.

U.S. Copyright Act, 17 U.S.C. (2004). Section 501(a).

T. Wu (2003) 'When code isn't law,' *Virginia Law Review*, 89: 679.

D. Zimmerman (2006) 'Daddy are we there yet? Lost in Grokster-Land,' *New York University Journal of Legislation and Public Policy*, 9, 75.

J. Zittrain (2006) 'A history of online gatekeeping,' *Harvard Journal of Law & Technology*, 19: 253.

DISCUSSION QUESTIONS

1. Spinello argues that those whose labor goes into creating intellectual property have rights over it, and that it is theft when someone uses their intellectual property without authorization. Do you agree with this account of the basis for intellectual property rights and the definition of what constitutes theft?

2. Do you think that the fact that intellectual property often is nonrivalrous and nonexcludable is a good reason for considering intellectual property rights to be weaker than property rights over physical objects?

3. Do you agree that online file sharing is ethically problematic? Why or why not?

4. What are the distinctive features of digital media in comparison to prior forms of media? Who is empowered by these features and who is disempowered?

5. How is value sensitive design being used to try to address the issue of unauthorized P2P file sharing? Do you think it can be successful? Why or why not?

6. What is the author's argument that those who write the software that makes P2P sharing possible are acting unethically? Do you agree with this argument?

7. What is the phrase "code is law" intended to convey, and how does it relate to the issue of P2P file sharing?

21 VIRTUAL REALITY AND COMPUTER SIMULATION[1]

Philip Brey

CHAPTER SUMMARY

Virtual reality and computer simulation are becoming increasingly immersive and interactive, and people are spending more and more time and money in virtual and simulated environments. In this chapter, Philip Brey begins with an overview of what virtual reality is – including the senses in which it is "virtual" and the senses in which it is "real." He then discusses a set of issues that are connected to the representational nature of virtuality – i.e. the possibility of misrepresentation, biased representation, and indecent representation – before examining behavior in virtual environments. Brey argues that behavior in virtual environments can be evaluated according to how it affects users of the environments – for example, property can be stolen in virtual environments and too much time and emotion spent in virtual environments can be detrimental to living well. Finally, Brey discusses several issues associated with video games, including their impact on children and gender representation and bias in them. A theme that runs throughout the chapter is the way in which the designs of virtual environments are value laden – they express values, structure choices, and encourage evaluative attitudes.

RELATED READINGS

Introduction: Form of Life (2.5); Justice, Access, and Equality (2.6.2); Responsible Development (2.8)

Other Chapters: Langdon Winner, *Technologies as Forms of Life* (Ch. 4)

1. INTRODUCTION

Virtual reality and computer simulation have not received much attention from ethicists. It is argued in this essay that this relative neglect is unjustified, and that there are important ethical questions that can be raised in relation to these technologies. First of all, these technologies raise important ethical questions about the way in which they represent reality and the misrepresentations, biased representations, and offensive representations that they may contain. In addition, actions in virtual environments can be harmful to others and raise moral issues within all major traditions

---

[1] This Chapter is excerpted from Philip Brey (2008) 'Virtual Reality and Computer Simulation,' *The Handbook on Information and Computer Ethics*, eds. Himma and Tavani (Wiley). It appears here by permission of Wiley and Sons, Inc.

in ethics, including consequentialism, deontology, and virtue ethics. Although immersive virtual reality systems are not yet used on a large scale, nonimmersive virtual reality is regularly experienced by hundreds of millions of users, in the form of computer games and virtual environments for exploration and social networking. These forms of virtual reality also raise ethical questions regarding their benefits and harms to users and society, and the values and biases contained in them.

This paper has the following structure. The next section will describe what virtual reality is. This is followed by a section that analyzes the relation between virtuality and reality. Three subsequent sections discuss ethical aspects of representation in virtual reality, the ethics of behavior in virtual reality, and the ethics of computer games.

2. BACKGROUND: THE TECHNOLOGY AND ITS APPLICATIONS

Virtual reality (VR) technology emerged in the 1980s, with the development and marketing of systems consisting of a head-mounted display (HMD) and datasuit or dataglove attached to a computer. These technologies simulated three-dimensional (3D) environments displayed in surround stereoscopic vision on the head-mounted display. The user could navigate and interact with simulated environments through the datasuit and dataglove, items that tracked the positions and motions of body parts and allowed the computer to modify its output depending on the recorded positions. This original technology has helped define what is often meant by "virtual reality": an immersive, interactive three-dimensional computer-generated environment in which interaction takes place over multiple sensory channels and includes tactile and positioning feedback.

According to Sherman and Craig (2003), there are four essential elements in virtual reality: a virtual world, immersion, sensory feedback, and interactivity. A *virtual world* is a description of a collection of objects in a space and rules and relationships governing these objects. In virtual reality systems, such virtual worlds are generated by a computer. *Immersion* is the sensation of being present in an environment, rather than just observing an environment from the outside. *Sensory feedback* is the selective provision of sensory data about the environment based on user input. The actions and position of the user provide a perspective on reality and determine what sensory feedback is given. *Interactivity*, finally, is the responsiveness of the virtual world to user actions. Interactivity includes the ability to navigate virtual worlds and to interact with objects, characters, and places.

These four elements can be realized to a greater or lesser degree with a computer, and that is why there are both broad and narrow definitions of virtual reality. A narrow definition would only define fully immersive and fully interactive virtual environments as VR. However, there are many virtual environments that do not meet all these criteria to the fullest extent possible, but can still be categorized as VR. Computer games played on a desktop with a keyboard and mouse, like *Doom* and *Half-Life*, are not fully immersive, and sensory feedback and interactivity in them are more limited than in immersive VR systems that include a head-mounted display and datasuit. Yet they do present virtual worlds that are immersive to an extent, and that are interactive and involve visual and auditory feedback. Brey (1999) therefore proposed a broader definition of virtual reality as *a three-dimensional interactive computer-generated environment that incorporates a first-person perspective.* This definition includes both immersive and nonimmersive (screen-based) forms of VR.

The notion of a virtual world, or *virtual environment*, as defined by Sherman and Craig, is broader than that of virtual reality. A virtual world can be defined so as to provide sensory feedback of objects, in which case it yields virtual reality, but it can also be defined without such feedback. Classical text-based adventure games like *Zork*, for example, play in interactive virtual worlds, but users are informed about the state of this world through text. They provide textual inputs, and the game responds with textual information rather than sensory feedback about changes in the world. A virtual world is hence an interactive computer-generated environment, and virtual reality is a special type of virtual world that involves location- and movement-relative sensory feedback.

Next to the term "virtual reality," there is the term "virtuality" and its derivative adjective "virtual." This term has a much broader meaning than the term "virtual reality" or even "virtual environment." As explained more extensively in the following section, the term "virtual" refers to anything that is created or carried by a computer and that mimics a "real," physically localized entity, as in "virtual memory" and "virtual organization." In this essay, the focus will be on virtual reality and virtual environments, but occasionally, especially in the following section, the broader phenomenon of virtuality will be discussed as well....

3. VIRTUALITY AND REALITY

In the computer era, the term "virtual" is often contrasted with "real." Virtual things, it is often believed, are things that only have a simulated existence on a computer and are therefore not real, like physical things. Take, for example, rocks and trees in a virtual reality environment. They may look like real rocks and trees, but we know that they have no mass, no weight, and no identifiable location in the physical world and are just illusions generated through electrical processes in microprocessors and the resulting projection of images on a computer screen. "Virtual" hence means "imaginary," "make believe," "fake," and contrasts with "real," "actual," and "physical". A virtual reality is therefore always only a make-believe reality and can as such be used for entertainment or training, but it would be a big mistake, in this view, to call anything in virtual reality real and to start treating it as such.

This popular conception of the contrast between virtuality and reality can, however, be demonstrated to be incorrect. "Virtual" is not the perfect opposite of "real," and some things can be virtual and real at the same time. To see how this is so, let us start by considering the semantics of "virtual." The word "virtual" has two traditional, precomputer meanings. On the first, most salient meaning, it refers to things almost having certain qualities, or having certain qualities in essence or in effect, but not in name. For instance, if a floor only has a few spots, one can say that the floor is virtually spotless, spotless for all practical purposes, even though it is not formally or actually spotless. Second, virtual can also mean imaginary, and therefore not real, as in optics, where reference is made to virtual foci and images. Note that only on the second, less salient meaning does "virtual" contrast with "real." On the more salient meaning, it does not mean "unreal" but rather "practically but not formally real."

In the computer era, the word "virtual" came to refer to things simulated by a computer, like virtual memory, which is memory that is not actually built into a processor but nevertheless functions as such. Later, the scope of the term "virtual" expanded to include anything that is created or carried by a computer and that mimics a "real" equivalent, like a virtual library and a virtual group meeting. The computer-based meaning of "virtual" conforms more with the traditional meaning of "virtual" as "practically but not formally real" than with "unreal." Virtual memory, for example, is not unreal memory, but rather a simulation of physical memory that can effectively function as real memory.

Under the above definition of "virtual" as "created or carried by a computer and mimicking a 'real' equivalent," virtual things and processes are simulations of real things, but this need not preclude them from also being real themselves. A virtual game of chess, for example, is also a real game of chess. It is just not played with a physically realized board and pieces. I have argued (Brey, 2003) that a distinction can be made between two types of virtual entities: simulations and ontological reproductions.

Simulations are virtual versions of real-world entities that have a perceptual or functional similarity to them but do not have the pragmatic value or actual consequences of the corresponding real-world equivalent. *Ontological reproductions* are computer imitations of real-world entities that have (nearly) the same value or pragmatic effects as their real-world counterparts. They hence have a real-world significance that extends beyond the domain of the virtual environment and is roughly equal to that of their physical counterpart.

To appreciate this contrast, consider the difference between a virtual chess game and a virtual beer. A virtual beer is necessarily a mere simulation of a real beer: it may look much like a real one and may be lifted and consumed in a virtual sense, but it does not provide the taste and nourishment of a real beer and will never get one drunk. A virtual chess game, in contrast, may lack the physical sensation of moving real chess pieces on a board, but this sensation is considered peripheral to the game, and in relevant other respects playing virtual chess is equivalent to playing chess with physical pieces. This is not to say that the distinction between simulations and ontological reproductions is unproblematic; a virtual entity will be classified as one or the other depending on whether it is judged to share enough of the essential features of its physical counterpart, and pragmatic considerations may come into play in deciding when enough features are present....

It can be concluded that many virtual entities can be just as real as their physical counterparts. Virtuality and reality are therefore not each other's opposites. Nevertheless, a large part of ordinary reality, which includes most physical objects and processes, cannot be ontologically reproduced in virtual form. In addition, institutional virtual entities can both possess and lack real-world implications. Sometimes virtual money can also be used as real money, whereas at other times it is only a simulation of real money. People can also disagree on the status of virtual money, with some accepting it as legal tender and others distrusting it. The ontological distinction between reality and virtuality is for these reasons confusing, and the ontological status of encountered virtual objects will often not be immediately clear....

4. REPRESENTATION AND SIMULATION: MISREPRESENTATIONS, BIASED REPRESENTATIONS AND INDECENT REPRESENTATIONS

VR and computer simulations are representational media: they represent real or fictional objects and events. They do so by means of different types of representations: pictorial images, sounds, words, and symbols. In this section, ethical aspects of such representations will be investigated. It will be investigated whether representations are morally neutral and whether their manufacture and use in VR and computer simulations involves ethical choices.

I will argue that representations in VR or computer simulations can become morally problematic for any of three reasons. First, they may cause harm by failing to uphold *standards of accuracy*. That is, they may *misrepresent* reality. Such representations will be called *misrepresentations*. Second, they may fail to uphold *standards of fairness*, thereby unfairly disadvantaging certain individuals or groups. Such representations will be called *biased representations*. Third, they may violate standards of decency and public morality. I will call such representations *indecent representations*.

Misrepresentation in VR and computer simulation occurs when it is part of the aim of a simulation to realistically depict aspects of the real world, yet the simulation fails to accurately depict these features (Brey, 1999). Many simulations aim to faithfully depict existing structures, persons, state of affairs, processes, or events. For example, VR applications have been developed that simulate in great detail the visual features of existing buildings such as the Louvre or Taj Mahal or the behavior of existing automobiles or airplanes. Other simulations do not aim to represent particular existing structures, but nevertheless aim to be realistic in their portrayal of people, things, and events. For example, a VR simulation of military combat will often be intended to contain realistic portrayals of people, weaponry, and landscapes without intending to represent particular individuals or a particular landscape.

When simulations aim to be realistic, they are subject to certain *standards of accuracy*. These are standards that define the degree of freedom that exists in the depiction of a phenomenon and that specify what kinds of features must be included in a representation for it to be accurate, what level of detail is required, and what kinds of idealizations are permitted. Standards of accuracy are fixed in part by the aim of a simulation. For example, a simulation of surgery room procedures should be highly accurate if it is used for medical training, should be somewhat accurate when sold as edutainment, and need not be accurate at all when part of a casual game. Standards of accuracy can also be fixed by promises or claims made by manufacturers. For example, if a game promises that surgery room procedures in it are completely realistic, the standards of accuracy for the simulation of these procedures will be high. People may also disagree about the standards of accuracy that are appropriate for a particular simulation. For example, a VR simulation of military combat that does not represent killings in graphic detail may be discounted as inaccurate and misleading by antiwar activists, but may be judged to be sufficiently realistic for the military for training purposes.

Misrepresentations of reality in VR and computer simulations are morally problematic to the extent that they can result in harm. The greater these harms are, and the greater the chance that they occur, the greater the moral responsibility of designers and manufacturers to ensure accuracy of representations. Obviously, inaccuracies in VR simulations of surgical procedures for medical training or computer simulations to test the bearing power of bridges can lead to grave consequences. A misrepresentation of the workings of an engine in educational software causes a lesser or less straightforward harm: it causes students to have false beliefs, some of which could cause harms at a later point in time.

Biased representations constitute a second category of morally problematic representations in VR modeling and computer simulation (Brey, 1999). A biased representation is a representation that unfairly disadvantages certain individuals or groups or that unjustifiably promotes certain values or interests over others. A representation can be biased in the way it idealizes or selectively represents phenomena. For example, a simulation of global warming may be accurate overall but unjustifiably ignore the contribution to global warming made by certain types of industries or countries. Representations can also be biased by stereotyping people, things, and events. For example, a computer game may contain racial or gender stereotypes in its depiction of people and their behaviors. Representations can moreover be biased by containing implicit assumptions about the user, as in a computer game that plays out male heterosexual fantasies, thereby assuming that players will generally be male and heterosexual. They can also be biased by representing affordances and interactive properties in objects that make them supportive of certain values and uses but not of others. For example, a gun in a game may be programmed so that it can be used to kill enemies but not to knock them unconscious.

Indecent representations constitute a third and final category of morally problematic representations. Indecent representations are representations that are considered shocking or offensive or that are held to break established rules of good behavior or morality and that are somehow shocking to the senses or moral sensibilities.

Decency standards vary widely across different individuals and cultures, however, and what is shocking or immoral to some will not be so to others. Some will find any depiction of nudity, violence, or physical deformities indecent, whereas others will find any such depiction acceptable. The depiction of particular acts, persons, or objects may be considered blasphemous in certain religions but not outside these religions. For this reason, the notion of an indecent representation is a relative notion, barring the existence of universally indecent acts or objects, and there will usually be disagreement about what representations count as indecent. In addition, the context in which a representation takes place may also influence whether it is considered decent. For example, the representation of open heart surgery, with some patients surviving the procedure but others dying on the operation table, may be inoffensive in the context of a medical simulator but offensive in the context of a game that makes light of such a procedure....

5. BEHAVIOR IN VIRTUAL ENVIRONMENTS: ETHICAL ISSUES

This section focuses on ethical issues in the use of VR and interactive computer simulations. Specifically, the focus will be on the question of whether actions within the worlds generated by these technologies can be unethical. This issue will be analyzed for both single-user and multiuser systems. Before it is taken up, I will first consider how actions in virtual environments take place and what is the relation between users and the characters as that they appear in virtual environments.

5.1 Avatars, agency and identity

In virtual environments, users assume control over a graphically realized character called an *avatar*. Avatars can be built after the likeness of the user, but more often they are generic persons or fantasy characters. Avatars can be controlled from a first-person perspective, in which the user sees the world through the avatar's eyes, or from a third-person perspective. In multiuser virtual environments, there will be multiple avatars corresponding to different users. Virtual environments also frequently contain *bots*, which are programmed or scripted characters that behave autonomously and are controlled by no one.

The identity that users assume in a virtual environment is a combination of the features of the avatar they choose, the behaviors that they choose to display with it, and the way others respond to the avatar and its behaviors. Avatars can function as a manifestation of the user, who behaves and acts like himself, and to whom others respond as if it is the user himself, or as a character that has no direct relation to the user and that merely plays out a role. The actions performed by avatars can therefore range from authentic expressions of the personality and identity of the user to experimentation with identities that are the opposite of who the user normally is, whether in appearance, character, status, or other personal characteristics.

Whether or not the actions of an avatar correspond with how a user would respond in real life, there is no question that the user is causally and morally responsible for actions performed by his or her avatar. This is because users normally have full control over the behavior of their avatars through one or more input devices. There are occasional exceptions to this rule, because avatars are sometimes taken over by the computer and then behave as bots. The responsibility for the behavior of bots could be assigned to either their programmer or to whomever introduced them into a particular environment, or even to the programmer of the environment for not disallowing harmful actions by bots (Ford, 2001).

5.2 Behavior in single-user VR

Single-user VR offers much fewer possibilities for unethical behavior than multiuser VR because there are no other human beings that could be directly affected by the behavior of a user. The question is whether there are any behaviors in single-user VR that could qualify as unethical. In Brey (1999), I considered the possibility that certain actions that are unethical when performed in real life could also be unethical when performed in single-user VR. My focus was particularly on violent and degrading behavior toward virtual human characters, such as murder, torture, and rape. I considered two arguments for this position, the argument from moral development and the argument from psychological harm.

According to the argument from moral development, it is wrong to treat virtual humans cruelly because doing so will make it more likely that we will treat real humans cruelly. The reason for this is that the emotions appealed to in the treatment of virtual humans are the same emotions that are appealed to in the treatment of real humans because these actions resemble each other so closely. This argument has recently gained empirical support (Slater et al., 2006). The argument from psychological harm is that third parties may be harmed by the knowledge or observation that people engage in violent, degrading, or offensive behavior in single-user VR and that therefore this behavior is immoral. This argument is similar to the argument attributed to Sandin in my earlier discussion of indecent representations. I claimed in Brey (1999) that although harm may be caused by particular actions in single-user VR because people may be offended by them, it does not necessarily follow that the actions are immoral, but only that they cause indirect harm to some people. One would have to balance such harms against any benefits, such as pleasurable experiences to the user.

McCormick (2001) has offered yet another argument according to which violent and degrading behavior in single-user VR can be construed as unethical. He argues that repeated engagement in such behavior erodes one's character and reinforces "virtueless" habits. He follows Aristotelian virtue ethics in arguing that this is bad because it makes it difficult for us to lead fulfilling lives, because as Aristotle has argued, a fulfilling life can only be lived by those who are of virtuous character. More generally, the argument can be made that the excessive use of single-user VR keeps one from leading a good life, even if one's actions in it are virtuous, because one invests into fictional worlds and fictional experiences that seem to fulfill one's desires but do not actually do so (Brey, 2007).

5.3 Behavior in multiuser VR

Many unethical behaviors between persons in the real world can also occur in multiuser virtual environments. As discussed earlier in the section on reality and virtuality, there are ... real-world phenomena that can also exist in virtual form: institutional entities that derive their status from collective agreements, like money, marriage, and conversations, and certain physical and formal entities, like images and musical pieces, which computers are capable of physically realizing. Consequently, unethical behaviors involving such entities can also occur in VR, and it is possible for there to be real thefts, insults, deceptions, invasions of privacy, breaches of contract, or damage to property in virtual environments.

Immoral behaviors that cannot really happen in virtual environments are those that are necessarily defined over physically realized entities. For example, there can be real insults in virtual environments, but not real murders, because real murders are defined over persons in the physical world, and the medium of VR does not equip users with the power to kill persons in the physical world. It may, of course, be possible to kill avatars in VR, but these are, of course, not killings of real persons. It may also be possible to plan a real murder in VR, for example by using VR to meet up with a hitman, but this cannot then be followed up by the *execution* of a real murder in VR.

Even though virtual environments can be the site of real events with real consequences, they are often recognized as fictional worlds in which characters merely play out roles. In such cases, even an insult may not be a real insult, in the sense of an insult made by a real person to another real person, because it may only have the status of an insult between two virtual characters. The insult is then only real in the context of the virtual world, but is not real in the real world. Ambiguities arise, however, because it will not always be clear when actions and events in virtual environments should be seen as fictional or real (Turkle, 1995). Users may assign different statuses to objects and events, and some users may identify closely with their avatar, so that anything that happens to their avatar also happens to them, whereas others may see their avatar as an object detached from themselves with which they do not identify closely. For this reason, some users may feel insulted when their avatar is insulted, whereas others will not feel insulted at all.

This ambiguity in the status of many actions and events in virtual worlds can lead to moral confusion as to when an act that takes place in VR is genuinely unethical and when it merely resembles a certain unethical act. The most famous case of this is the case of the "rape in cyberspace" reported by Dibbell (1993). Dibbell reported an instance of a "cyberrape" in *LambdaMOO*, a text-only virtual environment in which users interact with user-programmable avatars. One user used a subprogram that took control of avatars and made them perform sex acts on each other. Users felt their characters were raped, and some felt that they themselves were indirectly raped or violated as well. But is it ever possible for someone to be sexually assaulted through a sexual assault on her avatar, or does sexual assault require a direct violation of someone's body? Similar ambiguities exist for many other immoral practices in virtual environments, like adultery and theft. If it would constitute adultery when two persons were to have sex with each other, does it also constitute adultery when their avatars have sex? When a user steals virtual money or property from other users, should he be considered a thief in real life?

5.4 Virtual property and virtual economies

For any object or structure found in a virtual world, one may ask the question: who owns it? This question is already ambiguous, however, because there may be both virtual and real-life owners of virtual entities. For example, a user may be considered to be the owner of an island in a virtual world by fellow users, but the whole world, including the island, may be owned by the company that has created it and permits users to act out roles in it. Users may also become creators of virtual objects, structures, and scripted events, and some put hundreds of hours of work into their creations. May they therefore also assert intellectual property rights to their creations? Or can the company that owns the world in which the objects are found and the software with which they were created assert ownership? What kind of framework of rights and duties should be applied to virtual property (Burk, 2005)?

The question of property rights in virtual worlds is further complicated by the emergence of so-called virtual economies. Virtual economies are economies that exist within the context of a persistent multiuser virtual world. Such economies have emerged in virtual worlds like *Second Life* and *The Sims Online*, and in massively multiplayer online role-playing games (MMORPGs) like *Entropia Universe*, *World of Warcraft*, *Everquest*, and *EVE Online*. Many of these worlds have millions of users. Economies can emerge in virtual worlds if there are scarce goods and services in them for which users are willing to spend time, effort, or money, if users can also develop specialized skills to produce such goods and services, if users are able to assert property rights on goods and resources, and if they can transfer goods and services between them.

Some economies in these worlds are primitive barter economies, whereas other make use of recognized currencies. *Second Life*, for example, makes use of the Linden Dollar (L$) and *Entropia Universe* has the Project Entropia Dollar (PED), both of which have an exchange rate against real U.S. dollars. Users of these worlds can hence choose to acquire such virtual money by doing work in the virtual world (e.g., by selling services or opening a virtual shop) or by making money in the real world and exchanging it for virtual money. Virtual objects are now frequently traded for real money outside the virtual worlds that contain them, on online trading and auction sites like eBay. Some worlds also allow for the trade of land. In December 2006, the average price of a square meter of land in *Second Life* was L$ 9.68 or U.S. $0.014 (up from L$6.67 in November), and over 36,000,000 square meters were sold. Users have been known to pay thousands of dollars for cherished virtual objects, and over $100,000 for real estate.

The emergence of virtual economies in virtual environments raises the stakes for their users, and increases the likelihood that moral controversies ensue. People will naturally be more likely to act immorally if money is to be made or if valuable property is to be had. In one incident that took place in China, a man lent a precious sword to another man in the online game *Legend of Mir 3*, who then sold it to a third party. When the lender found out about this, he visited the borrower at his home and killed him. Cases have also been reported of Chinese sweatshop laborers who work day and night in conditions of practical slavery to collect resources in games like *World of Warcraft* and *Lineage*, which are then sold for real money.

There have also been reported cases of virtual prostitution, for instance on *Second Life*, where users are paid to (use their avatar to) perform sex acts or to serve as escorts. There have also been controversies over property rights. On *Second Life*, for example, controversy ensued when someone introduced a program called CopyBot that could copy any item in the world. This program wreaked havoc on the economy, undermining the livelihood of thousands of business owners *in Second Life*, and was eventually banned after mass protests. Clearly, then, the emergence of virtual economies and serious investments in virtual property generates many new ethical issues in virtual worlds. The more time, money, and social capital people invest in virtual worlds, the more such ethical issues will come to the front.

6. THE ETHICS OF COMPUTER GAMES

Contemporary computer and video games often play out in virtual environments or include computer simulations, as defined earlier. Computer games are nowadays mass media. A recent study shows that the average American 8-to 18-year old spends almost 6 hours per week playing computer games, and that 83% have access to a video game console at home (Rideout et al., 2005). Adults are also players, with four in ten playing computer games on a regular basis. In 2005, the revenue in the U.S. generated by the computer and game industry was over U.S. $7 billion, far surpassing the film industry's annual box office results (Entertainment Software Association, 2006). Computer games have had a vast impact on youth culture, but also significantly influence the lives of adults. For these reasons alone, an evaluation of their social and ethical aspects is needed.

Some important issues bearing on the ethics of computer games have already been discussed in previous sections, and therefore will be covered less extensively here. These include, among others, ethical issues regarding biased and indecent representations; issues of responsibility and identity in the relation between avatars, users, and bots; the ethics of behavior in virtual environments; and moral issues regarding virtual property and virtual economies. These issues and the conclusions reached regarding them all fully apply to computer games. The focus in this section will be on three important ethical questions that apply to computer games specifically: Do computer games contribute to individual well-being and the social good? What values should govern the design and use of computer games? Do computer games contribute to gender inequality?

6.1 The goods and ills of computer games

Are computer games a benefit to society? Many parents do not think so. They worry about the extraordinary amount of time their children spend playing computer games, and about the excessive violence that takes place in many games. They worry about negative effects on family life, schoolwork, and the social and moral development of their kids. In the media, there has been much negative reporting about computer games. There have been stories about computer game addiction and about players dying from exhaustion and starvation after playing video games for days on end. There have also been stories about ultraviolent and otherwise controversial video games, and the ease by which children can gain access to them. The Columbine High School massacre, in 1999, in which two teenage students went out on a shooting rampage, was reported in the media to have been inspired by the video game *Doom*, and since then, other mass shootings have also been claimed to have been inspired by video games. Many have become doubtful, therefore, as to whether computer games are indeed a benefit to society rather than a social ill.

The case against computer games tends to center on three perceived negative consequences: addiction, aggression, and maladjustment. The perceived problem of addiction is that many gamers get so caught up in playing that their health, work or study, family life, and social relations suffer. How large this problem really is has not yet been adequately documented (but see Chiu et al., 2004). There is clearly a widespread problem, as there has been a worldwide emergence of clinics for video addicts in recent years. Not all hard-core gamers will be genuine addicts in the psychiatric sense, but many do engage in overconsumption, resulting in the neglect described above. The partners of adults who engage in such over-consumption are sometimes called gamer widows, analogous to soccer widows, denoting that they have a relationship with a gamer who pays more attention to the game than to them.

Whereas there is no doubt that addiction to video games is a real social phenom-enon, there is somewhat less certainty that playing video games can be correlated with increased aggression, as some have claimed. A large percentage of contempo-rary video games involve violence. The preponderance of the evidence seems to indicate that the playing of violent video games can be correlated with increases in aggression, including increases in aggressive thoughts, aggressive feeling, aggressive behaviors, a desensitization to real-life violence, and a decrease in helpful behaviors (Bartholow, 2005; Carnagey et al., 2007). However, some studies have found no such correlations, and present findings remain controver-sial. Whatever the precise relation between violent video games and aggression turns out to be, it is clear now that there is a huge difference between the way that children are taught to behave toward others by their parents and how they learn to behave in violent video games. This at least raises the question of how their understanding of and attitude towards violence and aggression is influenced by violent video games.

A third hypothesized ill of video games is that they cause individuals to be socially and cognitively slighted and maladjusted. This maladjustment is attrib-uted in part to the neglect of studies and social relations due to an overindulgence in video games and to increased aggression levels from playing violent games. But it is also held to be due to the specific skills and understandings that users gain from video games. Children who play video games are exposed to conceptions of human relations and the workings of the world that have been designed into them by game developers. These conceptions have not been designed to be realistic or pedagogical, and often rely on stereotypes and simplistic modes of interaction and solutions to problems. It is therefore conceivable that children develop ideas and behavioral routines while playing computer games that leave much to be desired.

The case in favor of computer games begins with the observation that they are a new and powerful medium that brings users pleasure and excitement, and that seems to allow for new forms of creative expression and new ways of acting out fantasies. Moreover, although playing computer games may contribute to social isolation, it can also stimulate social interaction. Playing multiplayer games is a social activity that involves interactions with other players, and that can even help solitary individuals find new friends. Computer games may moreover induce social learning and train social skills. This is especially true for role-playing games and games that involve verbal interactions with other characters. Such games let players experiment with social behavior in different social settings, and role-playing games can also make users intimately familiar with the point of view and experiences of persons other than themselves. Computer games have moreover been claimed to improve perceptual, cognitive, and motor skills, for example by improving hand-eye coordination and improving visual recognition skills (Green and Bavelier, 2003; Johnson, 2005).

6.2 Computer games and values

It has long been argued in computer ethics that computer systems and software are not value-neutral but are instead value-laden (Brey, 2000; Nissenbaum, 1998). Computer games are no exception. Computer games may suggest, stimulate, promote, or reward certain values while shunning or discouraging others. Computer games are value-laden, first of all, in the way they represent the world. As discussed earlier, such representations may contain a variety of biases. They may, for example, promote racial and gender stereotypes (Chan, 2005; Ray, 2003), and they may contain implicit, biased assumptions about the abilities, interests, or gender of the player. Simulation games like *SimCity* may suggest all kinds of unproven causal relations, for example, between poverty and crime, which may help shape attitudes and feed prejudices. Computer games may also be value-laden in the interactions that they make possible. They, for example, may be designed to make violent action the only solution to problems faced by a player. Computer games can also be value-laden in the storylines they suggest for players and in the feedback and rewards that are given. Some first-person shooters awards extra points, for example, for not killing innocent bystanders, whereas others instead award extra points for killing as many as possible.

A popular game like *The Sims* can serve to illustrate how values are embedded in games. *The Sims* is a game that simulates the everyday lives and social relationships of ordinary persons. The goal of characters in the game is happiness, which is attained through the satisfaction of needs like hunger, comfort, hygiene, and fun. These needs can be satisfied through success in one's career, and through consumption and social interaction. As Sicart (2003) has argued, *The Sims* thus presents an idealized version of a progressive liberal consumer society in which the goal in life is happiness, gained by being a good worker and consumer. The team-based first-person shooter *America's Army* presents another example. This game is offered as a free download by the U.S. government, who uses it to stimulate U.S. Army recruitment. The game is designed to give a positive impression of the U.S. Army. Players play as servicemen who obey orders and work together to combat terrorists. The game claims to be highly realistic, yet it has been criticized for not showing certain realistic aspects of military life, such as collateral damage, harassment, and gore. It may hence prioritize certain values and interests over others by presenting an idealized version of military life that serves the interests of recruiters but not necessarily those of the recruit or of other categories of people depicted in the game.

The question is how much influence computer games actually have on the values of players. The amount of psychological research done of this topic is still limited. However, psychological research on the effects of other media, such as television, has shown that it is very influential in affecting the values of media users, especially children. Since many children are avid consumers of computer games, there are reasons to be concerned about the values projected on them by such games. Children are still involved in a process of social, moral, and cognitive development, and computer games seem to have an increasing role in this developmental process. Concern about the values embedded in video games therefore seems warranted. On the contrary, computer games are *games*, and therefore should allow for experimentation, fantasy, and going beyond socially accepted boundaries. The question is how games can support such social and moral freedom without also supporting the development of skewed values in younger players.

Players do not just develop values on the basis of the structure of the game itself, they also develop them by interacting with other players. Players communicate messages to each other about game rules and acceptable in-game behavior. They can respond positively or negatively to certain behaviors, and may praise or berate other players. In this way, social interactions in games may become part of the socialization of individuals and influence their values and social beliefs. Some of these values and norms may remain limited to the game itself, for example, norms governing the permissibility of cheating (Kimppa and Bissett, 2005). In some games, however, like massively multiplayer online role-playing games (MMORPGs), socialization processes are so complex as to resemble real life (Warner and Raiter, 2005), and values learned in such games may be applied to real life as well.

6.3 Computer games and gender

Games magazines and game advertisements foster the impression that computer games are a medium for boys and men. Most pictured gamers are male, and many recurring elements in images, such as scantily clad, big-breasted women, big guns, and fast cars, seem to be geared toward men. The impression that computer games are mainly a medium for men is further supported by usage statistics. Research has consistently shown that fewer girls and women play computer games than boys and men, and those that do spend less time playing than men. According to research performed by Electronic Arts, a game developer, among teenagers only 40% of girls play computer games, compared to 90% of boys. Moreover, when they reach high school, most girls lose interest, whereas most boys keep playing (BBC, 2006). A study by the UK games trade body, the Entertainment and Leisure Publishers Association, found that in Europe, women gamers make up only a quarter of the gaming population (ELSPA, 2004).

The question of whether there is a gender bias in computer games is morally significant because it is a question about gender equality. If it is the case that computer games tend to be designed and marketed for men, then women are at an unfair disadvantage, as they consequently have less opportunity to enjoy computer games and their possible benefits. Among such benefits may be greater computer literacy, an important quality in today's marketplace. But is the gender gap between usage of computer games really the result of gender bias in the gaming industry, or could it be the case that women are simply less interested in computer games than men, regardless of how games are designed and marketed?

Most analysts hold that the gaming industry is largely to blame. They point to the fact that almost all game developers are male, and that there have been few efforts to develop games suitable for women. To appeal to women, it has been suggested, computer games should be less aggressive, because women have been socialized to be non-aggressive (Norris, 2004). It has also been suggested that women have a greater interest in multiplayer games, games with complex characters, games that contain puzzles, and games that are about human relationships. Games should also avoid assumptions that the player is male and avoid stereotypical representations of women. Few existing games contain good role models for women. Studies have found that most female characters in games have unrealistic body images and display stereotypical female behaviors, and that a disproportionate number of them are prostitutes and strippers (Children Now, 2001)....

In Brey (1999), I have argued that designers of simulations and virtual environments have a responsibility to incorporate proper values into their creations. It has been argued earlier that representations and interfaces are not value-free but may contain values and biases. Designers have a responsibility to reflect on the values and biases contained in their creations and to ensure that they do not violate important ethical principles. The responsibility to do this follows from the ethical codes that are in use in different branches of engineering and computer science, especially the principle that professional expertise should be used for the enhancement of human welfare. If technology is to promote human welfare, it should not contain biases and should regard the values and interests of stakeholders or society at large. Taking into account such values and avoiding biases in design cannot be done without a proper methodology. Fortunately, a detailed proposal for such a methodology has recently been made by Batya Friedman and her associates, and has been termed *value-sensitive design* (Friedman et al., 2006)....

Virtual reality and computer simulation will continue to present new challenges for ethics, because new and more advanced applications are still being developed and their use is more and more widespread. Moreover, as has been argued, virtual environments can mimic many of the properties of real life and, therefore, contain many of the ethical dilemmas found in real life. It is for this reason that they will continue to present new ethical challenges not only for professional developers and users but also for society at large.

WORKS CITED

B. Bartholow (2005) 'Correlates and Consequences of Exposure to Video Game Violence: Hostile Personality, Empathy, and Aggressive Behavior,' *Personality and Social Psychology Bulletin*, 31(11): 1573–1586.

J. Baudrillard (1995) *Simulacra and Simulation* Trans. S. Fraser. (Ann Arbor, MI: University of Michigan Press).

BBC News (2006) 'Games Industry is 'Failing Women,'' Available at: http://news.bbc.co.uk/2/hi/technology/5271852.stm

A. Borgmann (1999) *Holding On to Reality: The Nature of Information at the Turn of the Millennium* (Chicago, Il: University Of Chicago Press).

P. Brey (1998) 'New Media and the Quality of Life,' *Techné: Journal of the Society for Philosophy and Technology*, 3(1): 1–23.

P. Brey (1999) 'The Ethics of Representation and Action in Virtual Reality,' *Ethics and Information Technology*, 1(1): 5–14.

P. Brey (2000) 'Disclosive Computer Ethics,' *Computers and Society*, 30(4): 10–16.

P. Brey (2003) 'The Social Ontology of Virtual Environments,' *American Journal of Economics and Sociology*, 62(1): 269–282.

P. Brey (2007) 'Theorizing the Cultural Quality of New Media,' *Techné: Research in Philosophy and Technology*.

G. Burdea and P. Coiffet (2003) *Virtual Reality Technology*, 2nd ed. (Hoboken, NJ: John Wiley & Sons).

D. Burk (2005) 'Electronic Gaming and the Ethics of Information Ownership,' *International Review of Information Ethics*, 4: 39–45.

N. Carnagey, C. Anderson, and B. Bushman (Forthcoming) 'The effect of video game violence on physiological desensitization to real-life violence,' *Journal of Experimental Social Psychology*.

D. Chan (2005) 'Playing with Race: The Ethics of Racialized Representations in E-Games,' *International Review of Information Ethics*, 4: 24–30.

Children Now (2001) '*Fair Play: Violence, Gender and Race in Video Games,*' Available at: http://publications.childrennow.org/

S. Chiu, J. Lee, and D. Huang (2004) 'Video Game Addiction in Children and Teenagers in Taiwan,' *Cyberpsychology & Behavior*, 7(5): 571–581.

J. Dibbell (1993) 'A Rape in Cyberspace,' *The Village Voice*, December 21, 36–42. Reprinted in *Reading Digital Culture*, ed. D. Trend, (MA, and Oxford: Blackwell), 199–213.

H. Dreyfus (2001) *On the Internet* (New York, NY: Routledge).

ELSPA (2004) '*Chicks and Joysticks, An Exploration of Women and Gaming,*' Available at: www.elspa.com/assets/files/c/chicksandjoysticksanexploration ofwomenandgaming_176.pdf

Entertainment Software Association (2006) '*Essential Facts about the Computer and Video Game Industry,*' Available at: http://www.theesa.com/archives/files/ Essential%20Facts%202006.pdf

P. Ford (2001) 'A further analysis of the ethics of representation in virtual reality: Multi-user Environments,' *Ethics and Information Technology*, 3: 113–121.

B. Friedman, P. H. Kahn, P. H., Jr., & Borning, A. 2006. Value Sensitive Design and Information Systems, in P. Zhang & D. Galletta (Eds.), *Human-Computer Interaction in Management Information Systems: Foundations* (M.E. Sharpe, Inc: NY).

C. S. Green, and D. Bavelier (2003) 'Action video game modifies visual selective attention,' *Nature*, 423, 29 May 2003: 534–537.

M. Heim (1994) *The Metaphysics of Virtual Reality* (New York: Oxford University Press).

S. Johnson (2005) *Everything Bad Is Good for You: How Today's Popular Culture Is Actually Making Us Smarter* (New York: Riverhead Books).

K. Kimppa, and A. K. Bissett (2005) 'The Ethical Significance of Cheating in Online Computer Games,' *International Review of Information Ethics*, 4: 31–37.

N. Levy (2002) 'Virtual child pornography: The Eroticization of Inequality,' *Ethics and Information Technology*, 4: 319–323.

M. McCormick (2001) 'Is it wrong to play violent video games?' *Ethics and Information Technology*, 3(4): 277–287.

J. McLeod (1983) 'Professional Ethics and Simulation,' *Proceedings of the 1983 Winter Simulation Conference*. Ed. S. Roberts, J. Banks, B. Schmeiser (La Jolla, CA: Society for Computer Simulation), 371–3.

H. Nissenbaum (1998) 'Values in the Design of Computer Systems,' *Computers and Society*, March 1998, 38–39.

K. Norris (2004) 'Gender Stereotypes, Aggression, and Computer Games: An Online Survey of Women,' *Cyberpsychology & Behavior*, 7(6): 714–27.

T. I. Ören, M. S. Elzas, I Smit, and L.G. Birta (2002) 'A Code of Professional Ethics for Simulationists,' *Proceedings of the 2002 Summer Computer Simulation Conference*, San Diego, CA, 434–435.

W. Prosser (1960) 'Privacy,' *California Law Review*, 48(4): 383–423.

S. Ray (2003) *Gender Inclusive Game Design: Expanding The Market* (Hingham, MA: Charles River Media).

V. Rideout, D. Roberts, and U. Foehr (2005) '*Generation M: Media in the Lives of 8–18 Year-olds*,' AKaiser Family Foundation Study, March 2005. http://www.kff.org/entmedia/entmedia030905pkg.cfm

P. Sandin (2004) 'Virtual Child Pornography and Utilitarianism,' *Journal of Information, Communication & Ethics in Society*, 2(4): 217–223.

J. Searle (1995) *The Construction of Social Reality* (Cambridge, MA: MIT Press).

W. Sherman, and W. Craig (2003) *Understanding Virtual Reality. Interface, Application and Design* (San Francisco, CA : Morgan Kaufmann Publishers).

M. Sicart (2003) 'Family Values: Ideology, Computer Games & Sims,' *Level Up Conference Proceedings*, Utrecht: University of Utrecht (CD-ROM).

M. Slater, M. Antley, A. Davison, et al. (2006) 'A Virtual Reprise of the Stanley Milgram Obedience Experiments,' *PLoS ONE*, 1(1), e39 (Open access).

J. Tabach-Bank (2004) 'Missing the right of publicity boat: How Tyne v. Time Warner Entertainment Co. threatens to "Sink" the First Amendment,' *Loyola of Los Angeles Entertainment Law Review*, 24(2): 247–88.

S. Turkle (1995) *Life on the Screen. Identity in the Age of the Internet* (New York, NY: Simon & Schuster).

D. Warner, and M. Raiter (2005) 'Social Context in Massively-Multiplayer Online Games (MMOGs): Ethical Questions in Shared Space,' *International Review of Information Ethics*, 4: 47–52.

B. Wiederhold, and M. Wiederhold (2004) *Virtual Reality Therapy for Anxiety Disorders: Advances in Evaluation and Treatment* (Washington, D.C.: American Psychological Association).

P. Zhai (1998) *Get Real: A Philosophical Adventure in Virtual Reality* (Oxford: Rowman & Littlefield Publishers).

P. Zhang & D. Galletta (eds.), *Human-computer interaction in management information systems: Foundations* (pp. 348–372) (Armonk, New York; London, England: M.E. Sharpe).

DISCUSSION QUESTIONS

1. Is virtual reality any less "real" than physical reality? If so, in what sense is it less real and what makes it less real?

2. Do you agree with the author that mere representations can be ethically problematic – for example, in virtue of the values they express or attitudes they encourage? Can they be problematic even if they do not cause harm to anyone?

3. Do you agree with the author that a person can act unethically in a single-user virtual reality, one in which there are no other players or users? Why or why not?

4. What grounds ethics in multiuser virtual environments? Do the ethics from physical reality carry over to virtual reality? Or is the basis for ethical values and norms internal to the environments themselves – e.g. in their structure or the wants and desires of users or creators?

5. The author discusses the idea that virtual "rape" and "adultery" can occur and be ethically problematic. What are these and do you agree that they are ethically problematic? Why or why not?

6. Imagine a multiuser virtual reality that is comprehensive and completely immersive and interactive, such that it is indistinguishable from physical reality. Would what constitutes ethical behavior in that environment be any different from ethical behavior in physical reality? If so, why would it be different? If not, why would it be the same?

7. In what respects do you think the design of virtual realities and video games should be value sensitive? That is to ask, in what ways, if any, are they value laden?

THE DIGITAL DIVIDE: INFORMATION TECHNOLOGIES AND THE OBLIGATION TO ALLEVIATE POVERTY

Kenneth Himma and Maria Bottis

CHAPTER SUMMARY

In this chapter, Kenneth Himma and Maria Bottis discuss the obligation of affluent people in wealthy nations to help to close the digital divide by promoting dissemination of information and computer technologies (ICTs) to those in poverty who currently do not have access to them. They begin with a discussion of global economic inequality and the relationship between poverty and ICT access. They then ague that according to almost any western ethical and theological tradition the affluent have an obligation to assist the very poor. Given this, the question is whether expanding access to ICTs is an effective and efficient way of doing so. After reviewing several studies relevant to this question they conclude that promoting access to ICTs will not by itself greatly reduce poverty, but that it can be a crucial component of an integrated approach to doing so.

RELATED READINGS

Introduction: Justice, Access, and Equity (2.6.2); Types of Theories (3.2)

Other Chapters: Michael Ravvin, *Incentivizing Access and Innovation for Essential Medicines* (Ch. 13)

1. INTRODUCTION

The term 'digital divide' roughly refers to a series of gaps between information and computer technology (ICT) haves and have-nots. These gaps include differences in access to information, access to appropriate ICT hardware and software, literacy rates, and ICT skill-sets. Moreover, these gaps include those between developed and developing nations and those between rich and poor in developed nations.

The digital divide bears on the world's inequitable distribution of resources and contributes to the perpetuation of absolute and relative poverty. For this reason, these inequalities raise important moral issues in distributive justice and individual fairness. This essay provides an introduction to these issues and argues that the affluent have an obligation to help the poor, which includes, but is not limited to, an obligation to address the problems associated with the digital divides.

There are two moral issues here – one concerning the state and the other concerning individuals. States are political entities subject to different moral principles than individuals. For example, legitimate states are presumed to be morally justified in enforcing their rules for how citizens are required to behave by coercive measures like incarceration; no individual is permitted to do this in his capacity as an individual.

Our focus in this chapter will largely, albeit not exclusively, be limited to the moral obligations of individuals. Claims about the state's obligations are somewhat more tentative, since fully developing them involves more political theory than can be incorporated here, but they are included to give the reader a sense of the two different dimensions of the problem.

2. LOCAL AND GLOBAL INEQUALITIES IN THE DISTRIBUTION OF WEALTH

Some 1.2 billion of the world's seven billion people live on less than $1 per day, while nearly 3 billion live on less than $2 per day (United Nations 2012). This is the kind of poverty that is *absolute* in the sense that people do not have sufficient resources to meet their basic needs on a consistent basis.

Many people in the developing world characteristically live in conditions of absolute poverty, lacking consistent access to adequate nutrition, clean water, health care, and education, as well as facing a significant probability of death from a variety of causes easily prevented or treated in affluent nations: 925 million people lack sufficient nutrition; 1 billion lack access to clean water; 1 billion are illiterate; 10.9 million children under 5 die every year of malnutrition in a world where the food disposed of as *garbage* by affluent persons is enough to save most, if not all, of these lives (United Nations 2012). 98% of the world's hungry live in the developing world (United Nations 2012).

Poverty in the developed or *affluent* world is generally "relative" in the sense that someone who is "poor" has enough to meet her needs but has significantly less than what others around him have. Although relative poverty is thus not life threatening, it remains a problem because moral worth – and thus one's sense of self-worth – is frequently associated with economic worth and because relative poverty correlates highly with crime.

There is some absolute poverty in the US, which can be seen in the form of homelessness and in the use of food banks and other government services intended to reduce poverty. For example, estimates of the number of people experiencing a period of homeless in 2009 range from between 1.6 million and 3.5 million (National Coalition for the Homeless 2009) – i.e. between 0.5% and 1% of the total population. The total poverty rate, as of 2010, in the US is 15.1% (approximately 45,000,000 people), which is the *highest* rate since 1993 (National Poverty Center 2012). The percentage of poor that is absolutely poor ranges from 4% to 8%. Similarly, 5.6 million *households* utilized a food bank more than once in 2010 (Feeding America 2012). But, as can be seen from these statistics, relative poverty, as is likely true of other affluent nations, is the vast majority of poverty in the US.

3. THE MEANING OF 'DIGITAL DIVIDE'

The term 'digital divide' is of comparatively recent vintage. At the most abstract level, the term refers to a gap between the information haves of the world and the information have-nots of the world. But this fundamental problem can be identified from a number of different conceptual angles. It can be seen, for example, from a global context (digital information inequalities between nations); from a national context (digital information inequalities within a state); or even from more local contexts (such inequalities between tribes, etc.). The list of markers to measure the digital divide is long and includes many known factors indicating other forms of discrimination: age, gender, economic and social status, education, ethnicity, type of household (urban/rural), and so on.

The fundamental problem can be seen, more specifically, in terms of an inequality with respect to a gap in *meaningful* access to information and communication technologies (ICTs), which requires not only the availability of the technology but also the ability to use it to economic and cultural advantage. On this conception, someone with the relevant ICTs who can do no more with them than download music has access to ICTs. But she does not have *meaningful* access to those ICTs because she does not have the ability, opportunity, or disposition to use them in a way that promotes her cultural knowledge or economic well-being. In addition, someone who can find information that can ground economically productive activity, but lacks the ability (perhaps because of underdeveloped analytic skills) or opportunity to put it to use would suffer from an information gap relative to someone who is succeeding in the "information" or "knowledge" society. Although there are many complex distinctive problems, they all fall under the rubric of the digital divide.

4. BI-DIRECTIONAL RELATIONSHIP BETWEEN POVERTY AND THE DIGITAL AND INFORMATION DIVIDES

There are gaps in access to ICTs within nations and between nations. Within the US, for example, there are gaps between rich and poor citizens, as well as whites and blacks. In 2010, 57% of individuals earning less than $30,000, 80% of individuals earning $30,000–$49,999, 86% of individuals earning $50,000–$74,999, and 95% of individuals earning $75,000 and more used the internet (Pew 2010). In addition, 68% of whites and 49% of blacks have broadband internet access at home (CNN 2010).

Similar gaps exist between the affluent developed world and the impoverished developing world. Although Internet access is increasing across the world, it is still the case that a comparatively small percentage of the developing world's poor has internet access. Studies indicate that about 33% of the world's population has internet access but people in the affluent world have disproportionate access; 78% of people in North America and 61% in Europe have internet access but only 13% of Africans do (Internet Usage Statistics 2011).

One would think there is a causal relation between the digital divide and the gap between rich and poor. Obviously, people who are too poor to fully meet their immediate survival needs cannot afford either ICT access or the training that prepares one to take advantage of such access. But not being able to afford such training and access is likely to perpetuate poverty in a global economy increasingly requiring the ability to access, process, and evaluate information. Lack of access owing to poverty is a vicious circle that helps to ensure continuing poverty.

5. THE MORAL DIMENSIONS OF THE DIGITAL DIVIDES

Most theorists focus on the benefits of bridging the digital gap. Having meaningful access to ICTs, which includes the skills to be able to process information in a way that creates marketable value, results in benefits that are economic and non-economic in character.

But there are potential downsides to ameliorating the digital divide. The worldwide availability of mass media featuring content from all over the world can have the effect of reducing cultural diversity that, as a moral matter, should be preserved. Many persons share the intuition that the progressive Americanization of western and eastern cultures (in the form of a proliferation of American corporate franchises, like McDonald's and Starbucks, in an increasing number of international cities) has a clear moral downside. Likewise, since 75% of the Web's content is in English and less than 50% of the world are native English speakers, the continuing increase in the percentage of Web content that is in English will force people to become more fluent in English, ignoring their own native languages. Resolving the fundamental problem of the digital divide might threaten as many as 6,000 languages currently being spoken, the majority of which are in Africa.

But it is as important to avoid a cultural paternalism that attempts to insulate existing indigenous cultures from outside influences as it is to avoid the sort of cultural imperialization of which the U.S. is often accused – especially in places where life-threatening poverty is endemic. There are not always easy choices here with respect to the kind of gaps with which we are concerned. But it is fair to assume that, while the value of preserving culture is an important moral value, the values associated with making possible a more economically affluent life for the 1 billion people who live on less than $1 a day and the nearly 3 billion who live on less than $2 a day outweigh the admittedly important moral value of preserving diversity. A full stomach is more important than cultural integrity.

6. DO THE AFFLUENT HAVE A MORAL OBLIGATION TO HELP OVERCOME POVERTY AND THE DIGITAL DIVIDES?

It is largely uncontroversial that it is *morally good* for affluent persons or nations to help impoverished persons or nations, but there is disagreement about whether affluent persons and nations are *morally obligated* to alleviate the effects of absolute poverty. Many persons believe that the only moral obligations we have are *negative* in the sense that they require people only to abstain from certain acts; we are obligated, for example, to refrain from killing, stealing, lying, and so on. On this view, we have no moral obligations that are *positive* in the sense that they require us to improve people's lives or the state of the world in some way. It follows, on this view, that we have no moral obligation to help the poor; helping the poor is good, but beyond the demands of obligation.

We argue below that this view is inconsistent with the ethics of every classically theistic Western religion, ordinary intuitions about cases, and each of the two main approaches to normative ethical theory, consequentialism and deontological ethical theory. Taken together, these arguments make a compelling case for thinking the affluent are morally obligated to help alleviate absolute poverty wherever it is found.

6.1 Theological considerations

It is clear that Christian ethics entail a robust moral obligation to help the poor. There are 3000 references in the Bible to alleviating poverty. Jesus frequently speaks of helping the poor as a *constituent* of authentic religious faith in God. In Matthew 25:31–49 ("Parable of Sheep and Goats"), Jesus explains what qualities separate those who are saved from those who are not, and what matters is how one acts and not what one believes. Jesus identifies his interests, in this parable, with those of all people in need; as he explains, the ultimate fate of saved and condemned are determined by how they treat those in need: "And the king will answer them, 'Truly I tell you, just as you did it to one of the least of these who are members of my family, you did it to me'" (Matthew 25:40).

The implicit conception of authentic faith here is that it is not just about believing certain propositions; it is also about doing things – and one of those things is to help the poor. Not helping others in need is tantamount to rejecting Jesus. Since (1) this is justifiably punished, and (2) punishment is justified only for failures to do what is obligatory, it follows that helping others is morally obligatory.

In Judaism, *Tzedakah* includes an obligation to help the poor. Leviticus 19:18 states the law: "You shall not take vengeance or bear a grudge against any of your people, but you shall love your neighbor as yourself: I am the Lord." Leviticus 23:22 puts the point in terms of agricultural products: "And when you reap the harvest of your land, you shall not reap all the way to the edges of your field, or gather the gleanings of your harvest; you shall leave them for the poor and the stranger." As Rabbi Maurice Lamm sums up the Jewish view: "Support for the disadvantaged in Judaism is not altruism." It is "justice." And to do justice, of course, is obligatory; in the case of Judaism, it is necessary to save the Jew from a "meaningless death." As such, it is a commandment and an obligation.

Finally, Islam regards the obligation to help the poor (*Zakat*) as one of the five basic obligations (or "pillars," as these obligations are commonly called) of its faith. These pillars obligate Muslims (1) to declare that there is no God but Allah and Muhammad is the Messenger of God (*Shahada*); (2) to worship in prayer five times daily while facing Mecca (*Salat*); (3) to fast from sunrise to sunset during the holy month of Ramadan (*Sawm*); (4) to make a pilgrimage to Mecca (*Hajj*); and (5) to give to the poor and needy (*Zakat*). Once a tax collected by the government, satisfaction of the obligation to help the poor is now often left to the conscience of the believer.

6.2 Intuitions: Peter Singer's drowning infant case

Peter Singer asks us to consider the following situation. An adult notices an infant face down at the edge of a nearby pond in some very shallow water and can see the infant is flailing. Instead of simply bending over and removing the infant from the water, a gesture that would cost no more than a few seconds and some wet hands, he walks by without doing anything and allows the infant to drown. People almost universally react to this case with a judgment that the adult has done something grievously wrong, which is inconsistent, of course, with the view that the only moral obligations we have are negative.

This case suggests we all hold the view that *individuals* have a moral obligation to save the life of an innocent person if we can do so without incurring a significant cost to ourselves. This is strong enough to entail a robust obligation on the part of the affluent to alleviate the life-threatening conditions of absolute poverty. Sacrificing a $30 shirt one does not need in order to save the life of a desperately malnourished child for one month is a trivial cost for someone who makes $50,000 per year, about the average income in the U.S. (US Census 2012).

Is the situation any different for *states* than for individuals? As noted above, this is a complicated issue but this much seems true: we appear, as a society, to accept that this principle applies at least in some instances to the US government. Most people believe, for example, that the US is sometimes obligated to intervene to prevent genocide – which comes at the expense of the lives of US soldiers and the costs of using our military. This surely harmonizes with (although it does not logically imply) the principle extracted from the Singer case: an affluent state has a moral obligation to save innocent people if it can do so without incurring significant costs (either to itself or to taxpayers).

It is not clear that US foreign aid expenditures satisfy even this modest principle. A national commitment of even $100 billion per year to foreign aid is insignificant in an economy worth $12 trillion dollars. Indeed, $100 billion is about 2.6% of the $3.8 trillion federal budget approved in 2011 (US Census Bureau 2012). However, in 2011, the U.S. spent about $28 billion in foreign aid for humanitarian purposes (US Census Bureau 2012), about 0.74% of the total budget and 0.19% of a GDP of $14.5 trillion.

This foreign humanitarian expenditure figure seems to be a morally insignificant percentage of the GDP and federal budget but it also seems to be a morally insignificant figure relative to population – at least for reasonably affluent people. $28 billion spread over 314 million people is approximately $89 per person. For someone earning an average income of $50,000, $89 is 0.18% of her income. Federal expenditures reducing national poverty are significant on any reasonable definitions of 'significant,' but those reducing foreign poverty are not.

6.3 Alleviating life-threatening effects of bad luck

It might be tempting to think that merit largely determines how material resources are distributed in the world. We are affluent and they are not, on this line of thinking, because we have earned it and they have not. While poverty is always regrettable, it does not necessarily involve justice: as long as people have gotten everything they deserve, there is no injustice in their having less than they need. We are our own keepers, and our respective merits determine what distributions are just. In other words, we have what we have because we have earned and hence deserve it.

While desert plays a role in explaining why people have what they have, luck also is a large factor. Had, for example, Bill Gates's parents lived in conditions of absolute poverty in a developing nation instead of an affluent suburb of Seattle, he would not be living anything like the kind of life he now lives. He would surely not be one of the world's richest men or the former head of Microsoft, because he would not have had access to the resources available in an affluent nation like the US, including an education that made it possible for him to achieve the level of digital and business sophistication needed to start a successful corporation. Indeed, the probability that Gates would not also be mired in conditions of absolute poverty is extremely low.

Further, Gates also benefitted greatly from having been born (1) with certain native intellectual abilities (2) into a world that had reached a level of technological advancement that made Microsoft a technological possibility (3) to loving skilled parents who knew how to raise him well. Had Gates been born to parents who were neglectful and abusive, or had he been born with average native abilities, he would not have enjoyed the success he has enjoyed. Obviously, Gates did not have a choice in any of these matters; as far as these factors were concerned, Gates got very lucky where others did not.

The same is true of almost anyone who lives in the affluent developed world. Most of us who enjoy affluence in these nations have done something to deserve it, but we also owe what we have to not having had the misfortune of being born to parents living in conditions of life-threatening poverty who lack access to the basic resources affluent persons take for granted: adequate nutrition, water and shelter, as well as 12 years of free education and government funding available for a university education.

There is, of course, nothing morally wrong with being lucky *per se*. What we luck into is, by definition, beyond our control and hence not subject to moral evaluation. Nor is it necessarily wrong to keep what you have lucked into. If, for example, my neighbors and I contribute a modest amount to fund a lottery game we will all play, it seems reasonable to think that, other things being equal, there is no injustice with the winner keeping the prize – even though the result of the game is determined by luck and no one can antecedently claim to deserve the winnings.

But when a person cannot opt out of a game of chance and the results of that game largely determine whether she will have *much* more than she needs to survive or whether she will instead struggle (and sometimes fail) to survive, those who have the good fortune to draw birth in the affluent world owe an obligation of justice to those who have the misfortune to draw birth in conditions of absolute and hence life-threatening poverty. That is, we have an obligation to share the bounty of our good luck with those whose luck was bad. Here, again, it seems reasonable to think that the obligation applies to both individuals and states (but it is important to keep in mind the distinction between the two).

6.4 Consequentialism and deontological moral theories

There are two main species of normative ethical theory that evaluate acts rather than character: consequentialism and deontology. Consequentialism is the view that the moral value of any action is entirely determined by its consequences; for example, act utilitarianism holds that the first principle of ethics is the obligation to maximize "utility," which may be defined in terms of pleasure, well-being, happiness, or satisfied preferences (Mill 1879). Strictly speaking, deontology can be accurately described as the negation of consequentialism: the moral value of at least one act is partly determined by features intrinsic to the act, rather than the consequences of the act. For example, an act utilitarian would have to explain the wrongness of lying in terms of features extrinsic to the lie (namely, the effects of the lie on total utility), whereas a deontologist can explain the wrongness in lying in terms of its inherent features (namely, its deceptive character).

While there are different consequentialist theories and different deontological theories, a brief consideration of two of the most historically influential will suffice to show these different theories generally converge on the view that we have a moral obligation to help the poor. Consider act utilitarianism's claim that our sole obligation is to maximize utility. Here it is important to note that material resources have diminishing marginal utility once basic needs are satisfied. Once basic needs are met, each successive increment of disposable income has less value to us than the last increment of the same amount. For example, a person making $45,000 per year, other things being equal, will derive less utility from a $5,000 raise than someone making $30,000 per year. If this is correct, then utility will generally be maximized by moving it from people who have more than they need to people who have less than they need.

Some act-utilitarian theorists argue that we are obligated to distribute material resources so everyone has an equal share; if you have $50,000 and I have $40,000, the utility of an additional $5,000 to me exceeds the utility of the $5,000 you have over $50,000. Accordingly, to satisfy your obligation to maximize utility, you should give me $5,000, which would equalize our shares of the resources.

Not all act utilitarians are egalitarians. Many would argue that, notwithstanding the diminishing marginal utility of non-necessities, an equal distribution of income would ultimately reduce utility by eliminating the incentive for work that increases the community's material resources. But, one way or another, the diminishing marginal utility of non-basic material resources pretty clearly implies, on an act-utilitarian view, an obligation to move disposable income to persons who lack the basic necessities.

Deontological theories almost universally hold that we have an obligation to help the poor. Consider Immanuel Kant's view that the first principle of ethics, the Categorical Imperative – to act on only those principles that we can consistently universalize as a law governing everyone's behavior – entails an obligation to help the poor:

> A fourth, who is in prosperity, while he sees that others have to contend with great wretchedness and that he could help them, thinks: "What concern is it of mine? Let everyone be as happy as Heaven pleases, or as he can make himself; I will take nothing from him nor even envy him, only I do not wish to contribute anything to his welfare or to his assistance in distress!" Now no doubt if such a mode of thinking were a universal law, the human race might very well subsist …. But although it is possible that a universal law of nature might exist in accordance with that maxim, it is impossible to will that such a principle should have the universal validity of a law of nature. For a will which resolved this would contradict itself, inasmuch as many cases might occur in which one would have need of the love and sympathy of others, and in which, by such a law of nature, sprung from his own will, he would deprive himself of all hope of the aid he desires (Kant 1765).

Another influential deontological theorist, W.D. Ross, took the position that we have a number of *prima facie* duties that, taken together, determine what we are obligated to do on any given occasion. A duty is *prima facie* in the sense that it is presumptive and can be overridden by a stronger *prima facie* duty; what we are ultimately obligated to do is determined by the strongest of these presumptive (*prima facie*) duties.

Ross gives what he takes to be a complete list of *prima facie* duties: (1) Those resting on a promise; (2) on a previous wrongful act; (3) on previous acts of other persons (e.g., services that may give rise to a duty of gratitude); (4) on a distribution of pleasure or happiness; (5) on persons whose conditions we can make better (duties of beneficence); (6) on the ability to improve our own conditions (duties of self-improvement); (7) on the harmfulness of certain behaviors on others (duties not to harm others) (Ross 1930). Proposition (5), of course, describes a *prima facie* obligation to help the poor. Consequentialist and deontological theories typically (though not necessarily) agree that the affluent have a moral obligation to help the absolutely poor.

Secular and theological ethical traditions converge on the view that there is an obligation for affluent people to help those in absolute poverty, particularly when the cost of their doing so is very low.

7. EMPIRICAL SKEPTICISM ABOUT THE RELATIONSHIP BETWEEN DIGITAL DIVIDES AND ABSOLUTE POVERTY

It has been almost universally assumed that addressing the digital divide would effectively reduce absolute poverty, producing benefits that far outweigh the costs – especially in the form of eliminating poverty. The network had to expand by all means and at any cost: access to unlimited digital information meant unlimited financial opportunities, entertainment, personal growth, unlocked working potential, even spirituality, easing the path towards democracy and freedom of speech for so many countries.

These expectations were so high that it became inevitable that people would become skeptical about the prospects of solving the problem of absolute poverty by bridging the digital divide. To adjudicate disagreements about the potential for closing the digital divide to reduce absolute poverty we need evidence from studies examining attempts to do so.

Here are some results from US studies. Singleton and Ross (2000) deny that new information technologies in the classroom necessarily reduce the number of students who lack basic skills when they leave school, which is one of the most reliable determinants of individual income and wealth levels. First, students can (and have for many years) achieved a high level of academic achievement and later on financial achievement without the use of digital technologies. Second, exposure to unlimited technological resources might not lead to any significant skill development at all:

> One prime example is the Kansas City School System. Under court order, the Kansas City School Board was told to design a "money is no object" program to integrate the school system and to raise the test scores of African-American students. The board added, among other new facilities, computers everywhere, television sets and compact-disc players, television studios, and a robotics lab, and it boasted a student-teacher ratio of 12 or 13 to 1. But the test scores of the minority students did not rise. The board ultimately concluded that paying more attention to hiring good teachers and firing bad ones would have made a greater difference.

Another study is indicative of failures in the attempts to bridge the digital divide. In 2000, LaGrange, Georgia became the first city in the world to implement 'The Free Internet Initiative', a municipal program offering broadband access to the internet for every citizen (Keil 2003). Internet access was provided through a digital cable set-top box that was distributed free of charge and every citizen could also received free training. The project did little to better the economic prospects of people in lower income groups.

Although these studies might cast doubt on the relationship between poverty and technology, they show only that digital and information inequalities cannot be resolved *overnight*; indeed, a study like this can tell us little, if anything, that would challenge the connection between information inequalities and economic inequalities. The problems that cause absolute poverty in the global south and relative poverty in affluent nations are much too complex to be solved by a one-time, short-term investment of information capital. Attitudes may have to be changed, while educational systems will have to be improved. But, even in the long-term, there are many contingent cultural difficulties – at least in the US and presumably in other countries with a history of institutionalized systemic racism – that will have to be overcome. Solving the problems associated with the digital divide is, not surprisingly, a long-term commitment.

Still, it is clear that not all poverty, relative or otherwise, can be explained by lack of access to the relevant technologies. One study dealing with information poverty and homeless people in the US concluded that homeless people may lack needed financial resources, but this did not translate to a lack of access to their more frequently articulated information needs (mainly: how to find permanent housing; how to help children; how to find a job; how to deal with finances; how to cope with substance abuse and domestic violence). Homeless people found information mostly by person-to-person contact.

This is of somewhat more relevance in assessing the relationship between bridging the digital divide and reducing absolute poverty. However, the context is too specific to tell us much about the more general connections between the two. There are many reasons for this. One is that many US homeless people suffer from serious mental health disorders and substance abuse problems. Another, it is impossible to lawfully get a job without an address in the US. But, again, this tells us no more than that the problems comprising the digital divide and its relation to poverty are enormously complex and their solution requires a long-term, multi-faceted approach.

There are studies calling attention to different obstacles faced in bridging the global digital divide as a means of addressing poverty. In Costa Rica, for example, the Little Intelligence Communities project (LINCOS), founded by MIT, Microsoft, Alcatel and the Costa Rican government, aimed at helping poor rural Costa Rican communities through telecenters failed. It was not the poor, but the rich coffee farmers who tried to take advantage of the project; local residents either did not care at all, or were mostly interested in using the technology in ways that did not material improve their situation – e.g. viewing virtual pornography. This study shows that efforts to bridge the digital divide will not succeed unless people are properly educated about what these technologies can accomplish economically – and they must also want to produce those results.

Some problems are simply technological in character and require more time to resolve. In Greece, a very expensive software program funded by the European Community, a telemedicine program aimed at connecting sick people and their primary care doctors with the most specialized physicians in the biggest Athens hospital for trauma (KAT), failed in its entirety when the physicians realized that all their orders, based upon digitally sent exam results, scans, etc. from the remote islands, would be stored and that the question of medical liability was not resolved. The attempt to use telemedicine to close the digital gap between people in remote islands (who do not have access to digital medical diagnostic technology) and people in the center (who do), in this case, was a total failure. It seems that the money would have been better spent by funding the salaries of specialized physicians who would work at these remote islands and by financing some medical equipment there.

These studies, even taken together, are inconclusive. They do not justify a robust digital skepticism. All technologies that resolve morally important problems take time to develop. While it is widely believed that gene therapy will make possible cures for diseases that are currently incurable, the research has progressed very slowly. The same should be expected of ICTs and related measures intended to alleviate the conditions of poverty. There is simply nothing one can do to make the economic injustices of the world disappear tomorrow.

8. CONCLUSION: TOWARDS SOLVING THE PROBLEMS OF THE DIGITAL DIVIDE AND ABSOLUTE POVERTY

A few closing observations about solving poverty and the digital divide are worth making. First, and most obviously, you cannot eat ICTs, Internet access, or information; if we are dealing with countries with life-threatening poverty, then the very first step in providing meaningful access to ICTs is to ensure that people's more basic needs are met. Someone who is malnourished and sick will not be in a position to take advantage of ICTs no matter what else is done. So part of the program must include provision of foodstuffs, clean water, and healthcare.

Second, other kinds of physical infrastructure are needed in developing nations to ensure that people have access to the opportunity to participate in the online economy. The affluent have no problem ordering goods from Amazon. com because they have homes with road access making it possible for UPS or Fed-Ex to deliver those goods. In many places in Africa, for example, people live away from roadways and must walk long distances to school and work, and they cannot get UPS and Fed-Ex service there. Further, impoverished people in the developing world do not have the credit cards needed to use such services.

Third, and most importantly, people must not only have the relevant ICTs, but also the ability to utilize these ICTs to produce output that is ultimately marketable in a global economy. People once thought that having access to radio technology would improve the economic lot of poor persons in the developing world; evidence now suggests there are more radios in South Africa than mattresses, but unemployment is at more than 30%. People must be taught to extract marketable knowledge from relevant information. Internet access does no good in alleviating poverty if all that is done with it is to download films or music.

What is needed is a particular type of skill – the type that enables a person to use ICTs and information to produce output that is in demand, or to be able to identify, find and use information that is beneficial to them. Only where impoverished persons are in a position to produce something other people want to buy can they raise their standards of living. Obviously, these skills include programming, designing websites, and so on; less obviously, they require at this point in time training in English, which is increasingly becoming the world's language of commerce – although it would clearly be ideal to make efforts to ensure the easy availability of devices that accurately translate the contents of a website in one language into any other of the world's written languages.

But training means nothing if the resulting skills do not elicit a fair wage. To improve the lot of poor countries, affluent countries must provide *fair*, competitive opportunities for a person to take advantage of her skills. While more and more people are getting some opportunities through corporate outsourcing, they do not receive a fair wage – though what they receive is more than what they could otherwise earn.

What the foregoing discussion shows is that addressing the digital divide will not itself eliminate, or even very greatly reduce, global poverty. However, it needs to be a component of any comprehensive effort to address it. As the global information economy continues to develop, meaningful access to ICTs is necessary to enable people, communities and nations to achieve significant economic progress.

WORKS CITED

A. Amighetti, and N. Reader (2003) 'Internet Project for Poor Attracts Rich,' *The Christian Science Monitor*, (available at http://www.csmonitor.com/2003/0724/p16s01-stin.html, accessed September 12, 2012).

Feeding America (2012) 'Hunger Statistics: Hunger and Poverty Facts,' (available at: http://feedingamerica.org/hunger-in-america/hunger-facts/hunger-and-poverty-statistics.aspx, accessed September 12, 2012).

J. Hersberger (2003) 'Are the Economically Poor Information Poor? Does the Digital Divide Affect the Homeless and Access to Information?' *Canadian Journal of Information and Library Science*, 27 (3): 45–63.

J. Jansen (2010) 'Use of the internet by higher-income households,' Pew Internet (available at http://www.pewinternet.org/Reports/2010/Better-off-households/Overview.aspx, accessed September 12, 2012).

I. Kant (1785) *Groundwork for the Metaphysics of Morals* (available at http://evans-experientialism.freewebspace.com/kant_groundwork_metaphysics_morals01.htm, accessed September 12, 2012).

M. Keil, G. Meader, and L. Kvasny (2003) 'Bridging the Digital Divide: The Story of the Free Internet Initiative,' 36th-Hawaii-International Conference on Systems Sciences (available at csdl.computer.org/comp/proceedings/hicss/2003/1874/05/187450140b.pdf, accessed September 12, 2012).

J. S. Mill (1879) *Utilitarianism* (available at http://scholar.google.com/scholar_url?hl=en&q=http://www.olimon.org/uan/stuart-mill_Utilitarianism.pdf&sa=X&scisig=AAGBfm3tg94V1KRW1iV7pFRsRwzaD35_Jg&oi=scholarr, accessed September 12, 2012).

National Coalition for the Homeless (2009) 'How Many People Experience Homelessness?' (available at http://www.nationalhomeless.org/factsheets/How_Many.html, accessed September 12, 2012).

National Poverty Center (2012) 'Poverty in the US: Frequently Asked Questions,' (available at http://www.npc.umich.edu/poverty/, accessed September 12, 2012).

W. D. Ross (1930) *The Right and the Good* (Oxford: Oxford University Press).

S. Singleton and L. Mast (2000) 'How Does the Empty Glass Fill? A Modern Philosophy of the Digital Divide,' *EDUCASE-Review*, 35(6): 30–6.

J. Sutter (2010) 'Racial inequalities persist online,' *CNN Report* (available at http://edition.cnn.com/2010/TECH/web/11/08/broadband.digital.divide/index.html, accessed September 12, 2012).

United Nations (2012) 'Child Hunger,' *UN Resources for Speakers on Global Issues* (available at http://www.un.org/en/globalissues/briefingpapers/food/childhunger.shtml, accessed September 12, 2012).

US Census Bureau (2012) 'National Data Book, Statistic Abstract' (available at htt p://www.census.gov/compendia/statab/cats/federal_govt_finances_employment/federal_budget– receipts_outlays_and_debt.html, accessed September 12, 2012).

DISCUSSION QUESTIONS

1. What do the authors mean when they say that the relationship between poverty and ICT access is "bi-directional"? Why is poverty and ICT access being linked in this way?

2. The authors argue that nearly every western ethical tradition maintains that the affluent have an obligation to assist those in poverty? Were their arguments for this compelling? Can you think of any ethical perspective from which this might not be the case?

3. If there is an obligation to assist the very poor, how far does this obligation extend? That is to ask, how much of one's own resources is one obligated to give to those in need? Only resources you are not otherwise using? Only resources that are going toward satisfying your mere wants (as opposed to basic or significant needs)? Until you are giving up something very important to your well-being? Until everyone is equally well off? Or some other standard?

4. Do you agree that assisting those in need is an obligation, rather than a charitable act – i.e. one that it is good to do but that one is not obligated to do? Why or why not?

5. Do you agree with the authors' conclusion that closing the digital divide is not by itself likely to lift people out of poverty, but that ICT access is a crucial component to doing so? Why or why not?

PART VI

ROBOTICS AND ARTIFICIAL INTELLIGENCE

The reality of automated and intelligent machines has lagged behind early expectations for them. In the 1950s and 1960s many technologists and futurists expected that by now we would live in a world in which fully autonomous robots and conscious, self-aware artificial intelligences (AIs) were seamlessly integrated into every aspect of our lives. While we are not quite there yet, robots and intelligent machines have become commonplace in our homes, factories, and offices. They do everything from helping to build (and drive) cars, to conducting wars, to making stock market trading decisions, and predicting what movies we will like. Moreover, we may finally be on the cusp of creating machine intelligences that are comparable to or exceeds human intelligence and perhaps that even involve consciousness.

Concerns about the development and widespread dissemination of robots and AI have tended to focus on the risks involved – for example, that they will malfunction or turn against us – and on the ways that they could change our way of life for the worse – for example, by our become too dependent on them or their rendering us (particularly our work and labor) superfluous. In "Ethics, War, and Robots," Patrick Lin, Keith Abney, and George Bekey survey the legal and ethical issues raised by the military use of robots. Several of the issues they discuss are associated with how the rapid proliferation of military robots is changing the nature of warfare, impacting everything from decisions about going to war to the roles of soldiers in military operations. Other issues they discuss concern the risks associated with the use of robots in warfare – for example, that the robots could get out of control, could be hacked by the enemy, or could have difficulty distinguishing combatants from noncombatants.

In "Ethics, Law and Governance in the Development of Robots," Wendell Wallach discusses two strands of robot ethics. The first, roboethics, concerns ethical issues associated with the dissemination of robots throughout society, for example as companions or caregivers. The second, machine ethics, concerns the prospects of and approaches to developing machines that are moral agents or that make moral decisions.

While most of the ethical discourse around robots and AI continues to focus on what their impacts and risk could be for us, as well as on how they can be designed and programmed so as to avoid detrimental outcomes, it is also important to consider what we might owe them. In "What to do About Artificial Consciousness," John Basl considers what the moral status of artificial consciousnesses would be and whether we could have duties or obligations to them once they are created.

In the final chapter in the section, Ray Kurzweil argues that that the rate of technological innovation increases exponentially, and that as a result we will soon enter a period of such rapid technological change that all aspects of our form of life will be radically transformed. The key to this transformation is the merging of human and machine intelligences, as well as of virtual and nonvirtual reality. Kurzweil predicts that in a matter of decades their will be little or no distinction between human and artificial intelligence, and that the vast majority of our intelligence will be nonbiological.

More detailed summaries, as well as a listing of related readings, are located at the start of the chapters.

ETHICS, WAR, AND ROBOTS[1]

Patrick Lin, Keith Abney, and George Bekey

CHAPTER SUMMARY

There has been a rapid proliferation of the use of robots in warfare in recent years, and the United States military (which accounts for ~40% of global military spending) has made further roboticization a priority within its research and development program. In this chapter, Patrick Lin, Keith Abney, and George Bekey review the state of robots within warfare and survey the wide array of legal and ethical issues raised by them. These include issues related to conducting a just war and determining responsibility for events in war, as well as issues related to the effects of the widespread use of robots on those involved in war and on society more generally. The authors argue that these technologies warrant a great deal more attention than they currently receive, given both the nature of them and the pace at which they are emerging.

RELATED READINGS

Introduction: Form of Life (2.5); Extrinsic Concerns (2.6); Responsible Development (2.8)

Other Chapters: Wendell Wallach, *Ethics, Law, and Governance in the Development of Robots* (Ch. 24)

1. INTRODUCTION

"No catalogue of horrors ever kept men from war. Before the war you always think that it's not you that dies. But you will die, brother, if you go to it long enough." – Ernest Hemingway (1935, p.156)

Imagine the face of warfare with advanced robotics. Instead of soldiers returning home in flag-draped caskets to heartbroken parents, autonomous robots – mobile machines that can make decisions, such as whether to fire upon a target, without human intervention – will increasingly carry out

---

[1] This chapter is adapted from a report (Lin et al. 2008) sponsored by the US Department of the Navy, Office of Naval Research, under Award # N00014–07–1-1152. We also thank the College of Liberal Arts and the College of Engineering at Cal Poly for their support, as well as our colleagues for their discussions and contributions to our investigation here, particularly: Colin Allen, Ron Arkin, Peter Asaro, and Wendell Wallach.

dangerous missions: tunneling through dark caves in search of terrorists, securing urban streets rife with sniper fire, patrolling the skies and waterways where there is little cover from attacks, clearing roads and seas of improvised explosive devices (IEDs), surveying damage from biochemical weapons, guarding borders and buildings, controlling potentially hostile crowds, and serving as frontline infantry.

Technology, however, is a double-edge sword that has both benefits and risks, critics and advocates. Advanced, autonomous military robotics is no exception, no matter how compelling the scenario imagined above. In this chapter, we discuss a range of ethical and policy issues arising from current and future military robotics. Though we speak from a US perspective and refer to reports originating from US defense organizations, our discussion may also be applied to advanced military robotics developed or used by other nations.

2. BACKGROUND

Robotics is a game-changer in national security (Ramo 2011; US Dept. of Defense 2011). We now find military robots in just about every environment: land, sea, air, and even outer space. They have a wide range of forms, from tiny robots that look like insects to aerial drones with wingspans greater than a Boeing 737 airliner. Some are fixed onto battleships, while others patrol borders in Israel and South Korea. Some even have fully-auto modes and can make their own targeting and attack decisions, such as the US Navy's Phalanx Close-In Weapons System (CIWS). As you might expect, military robots have fierce names: TALON SWORDS, Crusher, BEAR, Big Dog, Predator, Reaper, Harpy, Raven, Global Hawk, Vulture, and Switchblade. But not all are weapons – for instance, BEAR is designed to retrieve wounded soldiers on an active battlefield. Research is underway with micro robots, swarm robots, humanoids, chemical robots, and biological-machine integrations, so we can expect their applications to expand, from unmanned extraditions to perhaps robotic interrogation and torture (Lin 2011).

The usual justification for using robots for national security and intelligence is that they can do jobs known as the 3 "D"s: *Dull* jobs, such as extended reconnaissance or patrol beyond limits of human endurance, and standing guard over perimeters; *dirty* jobs, such as work with hazardous materials and after nuclear or biochemical attacks, as well as in environments unsuitable for humans, such as underwater and outer space; and *dangerous* jobs, such as tunneling in terrorist caves, controlling hostile crowds, or clearing IEDs.

Future military robots would be "smart" enough to make decisions that are currently reserved for humans. Moreover, as conflicts increase in tempo and require much quicker information processing and responses, robots have a distinct advantage over the limited and fallible cognitive capabilities of *Homo sapiens*. Not only would robots expand the battlespace over difficult, larger areas of terrain, they also represent a significant force-multiplier: each effectively doing the work of many human soldiers, while immune to sleep deprivation, fatigue, low morale, perceptual and communication challenges in the "fog of war", and other performance-hindering conditions.

But the presumptive case for deploying robots on the battlefield is more than about saving soldiers' lives or superior efficiency and effectiveness, though these are significant issues. Robots would also be unaffected by the emotions, adrenaline, and stress that cause soldiers to overreact or even deliberately violate the Laws of War (LOW) and Rules of Engagement (ROE), the international agreements and internal military policies that set limits on the conduct of war. So robots would be less likely to commit atrocities, that is to say, war crimes. We would no longer read (as many) news reports about our own soldiers brutalizing enemy combatants or foreign civilians to avenge the deaths of their brothers in arms – unlawful actions that carry a significant political cost. Indeed, robots may act as objective, unblinking observers on the battlefield, reporting any unethical behavior back to command. Their mere presence as such would discourage all-too-human atrocities in the first place. Call this the fourth "D": dispassionately performing a job (Veruggio and Abney 2012).

3. THE ISSUES

These current and possible robotic applications in the military give rise to a wide range of issues related to ethics and policy. Here, we loosely organize these in thematic sub-groups: legal, just war, technical, robot-human, societal, and future challenges. Our aim is to introduce and motivate reflection on these issues – each one of which requires a more detailed and robust investigation than can be provided in this general survey.

Legal issues

Perhaps the most obvious and near-term worry about robots is that, as complex and embodied computers, they suffer from the same liabilities as any other complex piece of technology: they may operate unexpectedly, for example crashing due to a software error. Or they may misidentify an object of interest (or target, in the case of military robots) due to programming and sensor limitations. Such interactions could result in serious bodily harm as well as property damage. At first glance, one might think that, as products, robots would be covered by extant product-liability laws and other such legal instruments. But given the increasing autonomy, capabilities, and complexities of these "intelligent" machines, existing law may not be enough. We have never before encountered a product that can act so much like a human being; thus, there is very little directly relevant legal precedent to resolve certain issues, including but not limited to the following:

1. *Unclear responsibility.* To whom would we assign blame – and punishment – for improper conduct and unauthorized harms caused by an autonomous robot (whether by error or by intention): the designers, robot manufacturer, procurement officer, robot controller/supervisor, field commander, a nation's president or prime minister, or the robot itself (Sparrow 2007)?

   In a military system, it may be possible to simply *stipulate* a chain of responsibility – e.g., that the commanding officer is ultimately responsible. But this may oversimplify matters. For example, if inadequate testing allowed a design problem to slip by and caused the improper robotic behavior, then perhaps a procurement officer or the manufacturer ought to be responsible. The situation becomes much more complex and interesting with robots that have greater degrees of autonomy, which may make it appropriate to treat them as quasi-persons, if not full moral agents at some point in the future.

2. *Refusing an order.* Consider the following type of situation: A commander orders a robot to attack a house that is known to harbor insurgents, but the robot – being equipped with sensors to "see" through walls – detects many children inside. Given its programmed instruction, based on standard versions of the laws of war (LOW) and rules of engagement (ROE) that require careful targeting to minimize civilian casualties, the robot must consider refusing the order. How ought the situation proceed: should the decision fall to the robot who may have better situational awareness, or the officer who (as far as she or he knows) issues a legitimate command? This dilemma also relates back to the question of responsibility: if the robot refuses an order, then who would be responsible for the events that ensue? Following legitimate orders is clearly an essential tenet for military organizations to function, but if we permit robots to refuse an order, this may expand the circumstances in which human soldiers may refuse an order as well (for better or worse).

3. *Consent by soldiers to risks.* In 2007, a semi-autonomous robotic cannon deployed by the South African army malfunctioned, killing nine "friendly" soldiers and wounding 14 others. Such accidents will surely happen again. In August 2010, the US military lost control of a helicopter drone on a test flight for more than 30 minutes and 23 miles, during which it veered towards Washington DC, violating airspace restrictions meant to protect the White House and other governmental assets. Given the potential for harm, should soldiers be informed that an unusual or new risk exists such as when they are handling or working with other dangerous items, e.g., explosives or anthrax? Does consent to risk matter anyway, if soldiers generally lack the right to refuse a work order?

Just war issues

Because the robots we are considering here operate in a military context, there are specific bodies of law and international conventions that require compliance, collectively known as the Laws of War. Some put restrictions on the kinds of weapons that may be used and their conditions of use, such as the St. Petersburg Declaration and the Hague Conventions. Others aim to protect the innocent from the horrors of war, such as the Geneva Conventions. Behind the LOW – as well as the ROE, which are usually applications of the LOW to specific missions, such as orders not to shoot first – is just war theory, a philosophy that is more than two thousand years old. The following, then, are legal issues that specifically challenge the LOW and ROE:

1. *Attack decisions.* It may be important for the issue of responsibility to decide who, or what, makes the decision for a robot to strike. Some situations may

develop so quickly and require such rapid information processing that we will be tempted to entrust our robots and systems to make critical decisions autonomously. But the LOW and ROE generally demand there to be "eyes on target", either in-person or electronically and presumably in real time. (This is one reason why there is a general ban on landmines: without eyes on target, it is not possible to know who is harmed by the ordnance and therefore to fulfill the responsibility to discriminate combatants from non-combatants.) If human soldiers must monitor the actions of each robot as they occur, this may limit the effectiveness for which the robot was designed in the first place: robots may be deployed precisely because they can act more quickly and with better information than humans can.

2. *Lower barriers for war.* Does the use of advanced weaponry such as autonomous robotics make it easier for one nation to engage in war or adopt aggressive foreign (and domestic) policies that might provoke other nations? If so, is this a violation of *jus ad bellum* – the part of just war theory that specifies conditions under which a state may morally engage in warfare (Asaro 2008)? It may be true that new strategies, tactics, and technologies make armed conflict an easier path to choose for a nation, if they reduce risks to one side. Yet while it seems obvious that we should want to reduce casualties from our respective nations, there is something sensible about the need for some terrible cost to war as a deterrent against entering war in the first place. This is the basis for just war theory, that war ought to be the very last resort given its horrific costs (Walzer 1977).

   But this objection – that advanced robotics immorally lowers barriers for war – seems to suggest that we should not do anything that makes armed conflict more palatable: we should not attempt to reduce friendly casualties, or improve battlefield medicine, or conduct any more research that would make victory more likely and quicker. Taken to the extreme, the objection seems to imply that we should *raise* barriers to war, to make fighting as brutal as possible (e.g., using primitive weapons without armor) so that we would never engage in it unless it were truly the last resort. Such a position appears counterintuitive at best and dangerously foolish at worst, particularly if we expect that other nations would not readily adopt a policy of relinquishment, which would put the nation that forgoes advanced robotics at a competitive disadvantage.

3. *Imprecision in LOW and ROE.* Consider the principles of proportionality and of discrimination in the Laws of War: do not use more force than necessary to achieve a military objective; and target only combatants and never noncombatants (though unintended deaths of nearby noncombatant civilians are excusable, i.e., "collateral damage", assuming disproportionate force is not used). Translated as specific Rules of Engagement these principles might be stated as orders to minimize collateral damage and to not fire first. Each of these LOW and ROE seems sensible, but in practice they leave much room for contradictory or vague imperatives which may result in undesired and unexpected behavior in robots.

   For instance, a ROE to minimize collateral damage is vague: What counts as "minimal" collateral damage? Is the rule that we should not attack a position if civilian deaths are expected to be greater than – or even half of – combatant deaths? Could there be a target of such high value that collateral damage of five (or 10 or even 1,000) civilians or $1M (or $10M or $100M) in property damage would be justified? A robot may need specific numbers to know exactly where this line is drawn in order to comply with the ROE. Unfortunately, this is not an area that has been precisely quantified nor easily lends itself to such a determination. Other LOW and ROE may be similarly vague or incomplete without more detailed instructions so encompassing that they may surpass the robot's ability to plot an acceptable course of action in a timely enough fashion.

Technical issues

Given the above imprecision in the LOW and ROE we might ask whether it is even possible to reduce them to programmable rules at all. Further, as mentioned earlier, the possibility of robotic errors is a looming concern – whether due to faulty programming or sensor/computing overload – particularly in a military context that is governed by a long list of prohibited actions. Some solutions may raise new problems of their own. These issues are many and including the following:

1. *Discriminating among targets*. Some experts contend that it is simply too difficult to design a machine that can distinguish between a combatant and a non-combatant, particularly when insurgents pose as civilians, as required by the LOW and ROE (e.g., Sharkey 2008; Canning et al. 2004). Further, robots would need to discriminate between active combatants and wounded ones who are unable to fight or have surrendered. Admittedly, this is a complex technical task, but we need to be clear on how accurate this discrimination needs to be. That is, discrimination among targets is also a difficult, error-prone task for human soldiers. Ought we hold machines to a higher standard than we have yet to achieve ourselves?

    Consider the following: A robot enters a building known to harbor terrorists, but at the same time an innocent girl is running toward the robot (unintentionally) because she is chasing after a ball that happens to be rolling in the direction of the robot. Would the robot know to stand down and not attack the child? If the robot were to attack, it may cause outrage from opposing forces and even our own media and public. However, this scenario could also arise with a human soldier, adrenaline running high, who may misidentify the charging target. It seems that in such a situation, a robot may be less likely to attack the child, since the robot is not prone to overreact from the influence of emotions and fear which afflict human soldiers. But in any event, if a robot would likely not perform worse than a human soldier, perhaps this is good enough until the technical ability to discriminate among targets improves. Some critics, however, may still insist on perfect discrimination or at least far better than humans are capable of. But it is unclear why we should hold robots to such a high standard while holding human soldiers to a lower one.

2. *First-generation problem*. We previously mentioned that it is reasonable to believe that accidents with military robots will happen. As with any other

technologies, errors or bugs will inevitably exist that can be corrected in the next generation of the technology. With Internet technologies, for instance, first-generation mistakes may not be too serious and can be fixed with software patches or updates. But with military robotics the stakes are much higher since human lives may be lost as a result of programming or other errors. So it seems that the prudent or morally correct course of action is to rigorously test the robot before deploying it.

However, testing already occurs with today's military robots, yet it is still difficult if not impossible to certify any given robot as error-free given that (a) testing environments may be substantially different than more complex, unstructured and dynamic battlefield conditions; and (b) the computer program used in the robot's on-board computer (its "brain") may consist of millions of lines of code.

Beta-testing of a program (testing prior to the official product launch, whether related to robotics, business applications or communications) is conducted today, yet new errors are routinely found in software by actual users after the launch. It is simply not possible to run a complex piece of software through all possible uses in a testing phase; surprises may occur during its actual use. Likewise, it is not reasonable to expect that testing robots will catch any and all flaws; the robots may behave in unexpected and unintended ways during actual field use. Again, the stakes are high with deploying robots, since any error could be fatal. This makes the first-generation problem, as well as ongoing safety and dependability, an especially sensitive issue (e.g., Van der Loos 2007).

3. *Robots running amok.* A frequent scenario in science-fiction novels and movies is that robots break free from their human programming through such methods as their own learning, creating other robots without such constraints (self-replicating and self-revising), malfunction, programming error, or intentional hacking (e.g., Joy 2000). Because military robots are built to be durable and often possess attack capabilities, if this were to occur they would be extremely difficult to defeat. This is an implication of being a force multiplier. However, some of these scenarios are more likely than others. Military robots will not have the ability to fully manufacture other robots or to radically evolve their intelligence and escape any programmed morality for quite some time. But other scenarios, such as hacking, seem to be near-term possibilities, especially if robots are not given strong self-defense capabilities (see below).

That robots might run amok is an enhanced version of the worry that enemies might use our own creations against us, but it also introduces a new element, since previous weapon systems needed a human operator which provided a point of vulnerability, i.e., a "soft underbelly" of the system. But autonomous robots could be designed to operate without human control. What precautions can be taken to prevent one from being captured and reverse-engineered or reprogrammed to attack our own forces or even civilians? If we design a "kill switch" that can automatically shut off a robot, this may present a key vulnerability that can be exploited by the enemy.

4. *Unauthorized overrides.* This concern is similar to one that has been raised regarding nuclear weapons: that a rogue officer may be able to take control of

the weapons and unleash them without authorization or otherwise override their programming to commit some unlawful action. This is a persistent worry with any new, powerful technology and it poses a multi-faceted challenge: it is a human problem (to develop ethical, competent officers), an organizational problem (to provide procedural safeguards), and a technical problem (to provide systemic safeguards). While this issue is not unique to the development or deployment of advanced robotics, it nevertheless is a concern that needs to be considered in the design and deployment phases.

5. *Coordinated attacks.* Generally, it is better to have more data than less when making decisions, particularly one as weighty as a military strike decision. Robots can be designed to easily network with other robots and systems, but this may complicate matters for robot engineers as well as commanders. We may need to establish a chain of command within robots when they operate as a team as well as ensure coordination of their actions. The risk here is that as complexity of any system increases, the more opportunities exist for errors to be introduced, and, again, mistakes by military robots may be fatal.

Human-robot issues

The dynamic and unpredictable environment of a battlefield is made more difficult when we consider that military robots do not and probably will not operate in a vacuum; they will interact with human warfighters and possibly noncombatant civilians. Beyond the immediate question of whether it is ethical to create a machine that can make decisions to kill humans on its own – a question we do not here address, but simply take for granted for the purposes of this discussion, given that such autonomous (or at least fully-auto) robots already exist – this intersection between man and machine gives rise to several interesting ethical issues:

1. *Effect on unit cohesion.* As a "band of brothers" there understandably needs to be strong trust and support among warfighters just as there is among police officers and firefighters. But sometimes this sense of camaraderie can be overdeveloped to the extent that one team member becomes complicit in or deliberately assists in covering up an illegal or inappropriate action of another team member. We have discussed the possible benefits of military robots with respect to behavior that is more ethical than currently exhibited by human soldiers. But robots will also likely be equipped with video cameras and other such sensors to record and report actions on the battlefield. This could negatively impact the cohesion among team or squad members by eroding trust with the robot as well as among fellow soldiers who then may or may not support each other as much knowing that they are being watched. Of course, soldiers and other professionals should not be giving each other unlawful "support" anyway, but there may be situations in which a soldier is unclear about or unaware of motivations, orders, or other relevant details and err on the side of caution, i.e., not providing support even when it is justified and needed.

In a team mixed with robots and humans, operators of the robots – which may include soldiers outfitted with a robotic exoskeleton, or loosely "cyborgs" – may be prone to take more risks given a real or perceived ability to withstand attacks that would otherwise kill a normal human. Again, how would this affect unit cohesion? If a negative effect is plausible, then perhaps human warfighters and robotic (and enhanced) ones should be segregated for prudential reasons.

2. *Self-defense*. One advantage touted for military robots is that they can be more conservative in their actions, e.g., holding their fire because they do not have a natural instinct of self-preservation and may be programmed without such (Arkin 2007). But how practical is it, at least economically speaking, to avoid giving robots – which may range from $100,000 to millions of dollars in cost – an ability to defend themselves? If a person, even a noncombatant civilian, threatens to destroy a robot, shouldn't we want it to have the ability to protect itself, our very expensive taxpayer-funded investment?

   Further, self-defense capabilities may be important for the robot to elude capture and hacking as previously discussed. Robots may be easily trapped and recovered fully intact – as is apparently the case with a US drone that "crash landed" in Iran in December 2011 – unlike tanks and aircraft, for instance, which usually sustain severe if not total damage prior to or during capture. However, an unsophisticated or too generous self-defense system could be in tension with using robots for a more ethical prosecution of war, if it misinterprets an innocent action as an attack. Therefore, a tradeoff or compromise among these goals – to have a more ethical robot and to protect the robot from damage and capture – may be needed.

3. *Winning hearts and minds*. Just war theory, specifically *jus post bellum*, requires that war is conducted in such a manner that it leaves the door open for lasting peace after the conflict (Orend 2002). We should not brutalize an enemy since that would generate ill-feelings that linger even after the fighting has stopped thereby making peaceful reconciliation more difficult to achieve. Robots do not necessarily represent an immoral or overly brutal way of waging war. However, when they are used for urban operations, such as patrolling dangerous streets or enforcing a curfew or securing an area, the local population may be less likely to trust and build good-will relationships with the occupying force (Sharkey 2008). Winning hearts and minds is likely to require diplomacy and human relationships that machines would not be capable of delivering at the present time.

Societal issues

Finally, we can expect military robots to make a broader and unintended impact beyond challenges to the military missions themselves; this is known as "blowback" in military-speak. Some of these issues apply also to technology in general as well as to new weapons systems:

1. *Counter-tactics in asymmetric war*. As discussed previously, robots could help make military actions more effective and efficient, which is exactly the point of

deploying those machines. Presumably, the more autonomous a robot is, the more lethal it can be (given requirements to discriminate among targets and so on). This translates to quicker, more decisive victories, but, for the other side, this means swifter and perhaps more demoralizing defeats. We can reasonably expect that a consequence of increasing the asymmetry of warfare in favor of highly technologized militaries will cause opposing forces to engage in even more unconventional strategies and tactics perhaps well beyond "terrorist" acts today. Few nations could hope to successfully wage war with militarily-superior states by using the same methods the US uses given the sheer number of US troops and advanced military technologies and weapons.

This not only affects how wars and conflicts are fought but also exposes the military as well as public to new forms of attack. More desperate enemies may resort to more desperate measures, from intensifying efforts to acquire nuclear or biochemical weapons to devising a "scorched earth" or "poison pill" strategy that strikes deeply at US interests even at some great cost to their own forces or population.

2. *Proliferation.* Related to the previous issue, history shows that innovations in military technologies – from armor and crossbows to intercontinental missiles and "smart" bombs – give the inventing side a temporary advantage that is eroded over time by other nations working to replicate the technologies. Granting that modern technologies are more difficult to reverse-engineer or replicate than previous ones, it nevertheless seems inevitable or at least possible that they can be duplicated, especially if an intact sample can be captured (again, as was the case in Iran in December 2011). So with the development of autonomous military robots, we can expect their wide proliferation at some future point. This means that these robots – which we are currently touting as lethal, difficult-to-neutralize machines – may be turned against our own forces eventually. Currently, nearly 50 countries possess military robots, though most are not using them as widely and visibly as the US (Singer 2012).

The proliferation of weapons, unfortunately, is an extremely difficult cycle to break; many nations are working to develop autonomous robotics, so a unilateral ban on their development would not accomplish much except to handicap that nation relative to the rest of the world. One possible defense for being on the vanguard of developing military robots might be to have leverage to arrest their proliferation. A country might believe that because it occupies the higher moral ground it would be irresponsible not to develop the technologies first. The problem, of course, is that almost every nation thinks of itself as moral or "doing the right thing." Thus, solving the proliferation problem would seem to require new or revised international treaties or amendments to the Laws of War.

3. *Space race.* Autonomous robots may hold many benefits for space exploration and development. However, there are significant financial and environmental risks in developing military robotics for outer space. First, launch costs remain astronomical; it costs thousands of dollars *per pound* to put an object into low Earth orbit and several times more per pound for geostationary orbit (not to mention periodic replacement costs and in-orbit repairs). A "star wars" scenario aside – which would create countless pieces of space debris that

would threaten communications and other satellites – even the use of robots for research purposes, e.g., to explore and develop moons and other planets may spark another space race given the military advantages of securing the ultimate high ground. This not only opens up outer space for militarization, which the world's nations have largely resisted, but diverts limited resources that could make more valuable contributions elsewhere.

4. *Technology dependency.* The possibility that we might become dependent or addicted to our technologies has been raised throughout the history of technology and even with respect to robotics. Today, ethicists worry that we may become so reliant on, for instance, robots for difficult surgery, that humans will start losing life-saving skills and knowledge, or that we become so reliant on robots for basic, arduous labor that our economy is impacted and we forget some of those techniques (Veruggio 2007). In the military, robots play an important role in finding and disarming IEDs, and some soldiers are reportedly inconsolable when such a robot that saved their lives is "killed" in action (Garreau 2007).

    As a general objection to technology, this concern seems legitimate but rather weak since the benefits of the technology in question often outweigh any losses. Certainly, it is a possible hypothetical or future scenario that, after relying on robots to perform all our critical surgeries, some event – say, a terrorist attack or massive electromagnetic pulse – could interrupt an area's power supply, disabling our machines and leaving no one to perform the surgery (because we forgot how and have not trained surgeons on those procedures, since robots were able to do it better), but the possibility seems remote and insignificant relative to the benefit. For military robots, as long as they can do the job less expensively (financially and politically) than humans can, the benefits of having them seem greater than the risk of dependency and loss of skill or access to that skill in an improbable event.

    One complication, however, is that the rare-earth elements needed by the computing industry are mostly mined in China which reportedly had temporarily withheld these critical materials from Japan during a political disagreement (Bloomberg 2010) – making the loss of computing technologies a plausible possibility. This and the persistent problem of "e-waste", or the disposal of toxic materials inside discarded computer equipment, also speak to the environmental costs of an addiction to technology.

5. *Civil security and privacy.* Defense technologies often turn into public or consumer technologies. So, a natural and current step in the evolution of military robots is their incarnation as civil security robots; they could guard corporate buildings, control crowds, and even chase down criminals. Many of the same concerns discussed above – such as technical challenges and questions of responsibility – would also become larger societal concerns; if a robot unintentionally (meaning that no human intentionally programmed it to ever do so) kills a small child, whether by accident (e.g., run over) or mistake (e.g., identification error), it will likely have greater repercussions than a robot that unintentionally kills a noncombatant in some faraway conflict. Therefore, there is increased urgency to address these military issues that may spill over into the public domain.

And, while we take it that warfighters, as volunteers for government service, have significantly decreased privacy expectations and rights, the same is not true of the public at large. If and when robots are used more in society, the robots are likely to be networked. As a result, concerns about illegal monitoring and surveillance – privacy violations (Calo 2012) – may surface as they have with many other modern technologies such as DNA testing, genome sequencing, communications-monitoring software, and nanotechnology. This raises the question of what kind of consent we need from the public before deploying these technologies in society.

4. CONCLUSION

In robot ethics, science fiction writer Isaac Asimov's "Laws of Robotics" (1950) are a touchstone for most discussions, and we acknowledge his work here in closing our analysis. In a military context, his first law – that a robot must not injure or kill a human – is already made irrelevant, since many such robots will be designed exactly to harm or kill humans. His second law – that a robot must follow human orders (when not in conflict with the first law) – is sensible enough though we contemplated above whether a robot should be able to refuse a human order. Asimov's third law is that a robot must defend itself when doing so is not in conflict with the first or second law, but we also noted above potential issues concerning a robot with self-defense capabilities. Because robotic behavior is largely programmed, though some robots are programmed to learn, the solution to many of our considered problems could be found in clever programming; for example, the discovery of new laws of robotics, perhaps context-specific such as for military operations. But it's already clear that they will need to be much more sophisticated than Asimov's.

The foregoing is not intended to capture all possible issues related to risk and ethics in military robotics. Certainly, new issues will emerge depending on how the technology and intended uses develop (e.g., Lin 2010; Lin et al. 2012; Singer 2009). In the preceding, we have highlighted what we consider to be some of the most urgent and important issues to be resolved, particularly those related to responsibility, risk, and the ability of robots to discriminate among targets. This is only the beginning of a dialogue in robot ethics and military applications. Since they are inherently about life and death, these issues urgently demand further investigation.

WORKS CITED

R. C. Arkin (2007) 'Governing Lethal Behavior: Embedding Ethics in a Hybrid Deliberative/Hybrid Robot Architecture,' Report GIT-GVU-07–11, Atlanta, GA: Georgia Institute of Technology's GVU Center. Last accessed on February 24, 2012: http://www.cc.gatech.edu/ai/robot-lab/online-publications/formalizationv35.pdf

P. Asaro (2008) 'How Just Could a Robot War Be?' in A. Briggle, K. Waelbers, and P. Brey (eds.), Current Issues in Computing and Philosophy, pp. 50–64, (Amsterdam, The Netherlands: IOS Press).

I. Asimov (1950) *I, Robot* (2004 edition) (New York, NY: Bantam Dell).

G. Bekey (2005). *Autonomous Robots: From Biological Inspiration to Implementation and Control* (Cambridge, MA: MIT Press).

Bloomberg News (2010) 'China Denies Japan Rare-Earth Ban Amid Diplomatic Row,' *Bloomberg News*, September 23, 2010. Last access on February 24, 2012: http://www.bloomberg.com/news/2010–09–23/china-denies-japan-rare-earth-ban-amid-diplomatic-row-update1-.html

R. Calo (2012) 'Robots and Privacy,' in P. Lin, Keith Abney, and George Bekey (eds.), *Robot Ethics: The Ethical and Social Implications of Robotics*, pp. 187–202, (Cambridge, MA: MIT Press).

J. S. Canning (2004) 'A Concept for the Operation of Armed Autonomous Systems on the Battlefield,' *Proceedings of Association for Unmanned Vehicle Systems International's (AUVSI) Unmanned Systems North America*, August 3–5, 2004, (Anaheim, CA).

J. Garreau (2007) 'Bots on the Ground,' *Washington Post*, May 6, 2007. Last accessed on February 24, 2012: http://www.washingtonpost.com/wp-dyn/content/article/2007/05/05/AR2007050501009_pf.html

E. Hemingway (1935) 'Notes on the Next War: A Serious Topical Letter,' *Esquire*, 4 (3): 19, 156.

B. Joy (2000) 'Why the Future Doesn't Need Us,' *Wired* 8 (4): 238–262.

P. Lin (2010) 'Ethical Blowback from Emerging Technologies,' *Journal of Military Ethics* 9 (4): 313–331.

P. Lin (2011) 'Drone-Ethics Briefing: What a Leading Robot Expert Told the CIA,' *The Atlantic*, December 15, 2011 online. Last accessed on February 24, 2012: http://www.theatlantic.com/technology/archive/2011/12/drone-ethics-briefing-what-a-leading-robot-expert-told-the-cia/250060/

P. Lin, K. Abney, and G. Bekey (2012) '*Robot Ethics: The Ethical and Social Implications of Robotics*,' (Cambridge, MA: MIT Press).

P. Lin, G, Bekey, and K. Abney (2008) '*Autonomous Military Robotics: Risk, Ethics, and Design*,' Report commissioned under US Department of the Navy, Office of Naval Research, award # N00014–07–1-1152 (San Luis Obispo, CA: California Polytechnic State University).

B. Orend (2002) 'Justice After War,' *Ethics & International Affairs* 16 (1): 43–56.

S. Ramo (2011) '*Let Robots Do The Dying*,' (Los Angeles, CA: Figueroa Press).

N. Sharkey (2008) 'Cassandra or False Prophet of Doom: AI Robots and War,' *IEEE Intelligent Systems*, July/August 2008, pp. 14–17. Last accessed on February 24, 2012: http://www.computer.org/portal/cms_docs_intelligent/intelligent/homepage/2008/X4–08/x4his.pdf

P. W. Singer (2009) *Wired for War: The Robotics Revolution and Conflict in the 21st Century* (New York, NY: Penguin Press).

P. W. Singer (2012) 'Do Drones Undermine Democracy?' *The New York Times*, January 21, 2012. Last accessed on February 24, 2012: http://www.nytimes.com/2012/01/22/opinion/sunday/do-drones-undermine-democracy.html?pagewanted=all

R. Sparrow (2007) 'Killer Robots,' *Journal of Applied Philosophy*, 24 (1): 62–77.

US Department of Defense (2011) 'Unmanned Systems Integrated Roadmap: 2011–2036,' Washington, DC: Department of Defense. Last accessed on February 24, 2012: http://www.acq.osd.mil/sts/docs/Unmanned%20Systems%20Integrated%20Roadmap%20FY2011–2036.pdf

H. F. M. Van der Loos (2007) 'Ethics by Design: A Conceptual Approach to Personal and Service Robot Systems,' *Proceedings of the IEEE Conference on Robotics and Automation, Workshop on Roboethics*, April 14, 2007, Rome, Italy.

G. Veruggio (2007) *EURON Roboethics Roadmap* (Genova, Italy: European Robotics Research Network). Last accessed on February 24, 2012: http://www.roboethics.org/atelier2006/docs/ROBOETHICS%20ROADMAP%20Rel2.1.1.pdf

G. Veruggio, K. Abney (2012) 'Roboethics: The Applied Ethics for a New Science,' in P. Lin, K. Abney, and G. Bekey (eds.), *Robot Ethics: The Ethical and Social Implications of Robotics*, pp. 347–364, (Cambridge, MA: MIT Press).

W. Wallach, C. Allen (2008) *Moral Machines: Teaching Robots Right from Wrong* (New York, NY: Oxford University Press).

M. Walzer (1977) *Just and Unjust Wars: A Moral Argument with Historical Illustrations* (New York, NY: Basic Books).

DISCUSSION QUESTIONS

1. How much control should humans retain over lethal robots? Is it ethical to create a machine for the purpose of autonomously killing human beings?
2. How are military robots today ethically different from other weapons, such as guns and cruise missiles?
3. In what ways might military robots restructure warfare, military organizations, and public attitudes toward war?
4. Under what conditions, if any, could a robot be held responsible for its actions?
5. Is it reasonable to use military robots if they are at least as safe or reliable as human soldiers, e.g., if they do not kill non-combatants at a greater rate than humans currently demonstrate? Should robots be held to a higher standard? Why or why not?
6. How plausible is the suggestion that more effective weapons or asymmetric advantages make war more likely to occur, since fewer human soldiers would be placed in harm's way?
7. Does it make sense to even have laws governing warfare? If so, should the rules of engagement and laws of war be different for autonomous military robots than for human combatants?
8. Under what conditions would it be ethically acceptable for a robot to defend itself against human attackers?
9. Which issues described in this chapter seem to be the most urgent to resolve in the near term? Which issues seem the most difficult to solve?

24 ETHICS, LAW, AND GOVERNANCE IN THE DEVELOPMENT OF ROBOTS

Wendell Wallach

CHAPTER SUMMARY

In this chapter, Wendell Wallach begins by reviewing the history and current state of the field of robotics and artificial intelligence. He also distinguishes between two types of ethical issues associated with advances in the field. Roboethics concerns ethical issues that arise from the use of robots in our homes and workplaces. These include such things as safety, privacy, and responsibility/liability. Machine ethics concerns ethical issues associated with the development of artificial moral agents – i.e. machine intelligences that are capable of making moral decisions. The central issues in machine ethics are how to design morality into machine intelligences, as well as what the content or form of that moral code should be. After discussing both roboethics and machine ethics, Wallach concludes with a proposal for increasing monitoring and oversight of the robotics and artificial intelligence industry.

RELATED READINGS

Introduction: Extrinsic Concerns (2.6); Responsible Development (2.8)

Other Chapters: Patrick Lin, Keith Abney, and Georges Bekey, *Ethics, War, and Robots* (Ch. 23); John Basl, *What to do About Artificial Consciousness* (Ch. 25)

1. INTRODUCTION

The field of robotics draws inspiration from the more speculative fantasies of science fiction writers. But it is framed by the practical challenges confronted by engineers in the laboratory. Roboticists have discovered that very simple tasks such as navigating around a room, learning a language, recognizing faces, and kicking a ball – tasks that come easily to infants – are very difficult to implement in silicon, metal, and plastic. Popular media-created images, of intelligent robots with a wide array of skills and capable of conversing fluently, conflict with the reality of the existing machines, each of which perform a very limited number of tasks. The holy grail of full artificial intelligence remains on the distant horizon. Many scientists are skeptical that computers or robots[1] will ever be conscious, or learn and be creative in a manner that rivals human intelligence, while others predict that intelligent robots will surpass their human counterparts in all aspects of cognition. Meanwhile, the introduction of robots into the home, the battlefield, and the commerce of daily life poses an array of societal, ethical, legal, and policy challenges. These challenges fall into two fields of study:

> *Roboethics* (Veruggio 2002), which is human-centered and focuses upon the ethical use of robots and societal concerns in the deployment of robots.
> *Machine Ethics* (Hall 2000, Anderson *et al.* 2005), which studies the prospect of, and the approaches for, developing robots capable of making moral decisions.

Self-operating machines (automata), mechanical devices and toys with and without humanoid features, date back more than two thousand years. Automata became particularly intricate in the Muslim world during the Middle Ages; in Europe, Russia, and China during the eighteenth century; and again in France from 1860–1910, in what is known as "The Golden Age of Automata." The term 'robot' was coined by a Czeck writer Karel Čapek for a 1920 play entitled *R.U.R.*, in which a rebellion by initially happy slave androids leads to the extinction of humans. Čapek's robots were inanimate matter that somehow becomes anthropomorphic beings. However, with the development of computers in the 1940s and 1950s, scientists perceived pathways to actually imbuing machines with intelligence, and the term 'robot' took on its more contemporary meaning. In 1950, Alan Turing, one of the fathers of computer science, wrote a classic paper entitled, "Computing Machinery and Intelligence", in which he pondered the question 'Can machines think?' Noting the difficulty of defining intelligence, Turing proposed a test, now known as the Turing Test, in which an interrogator must distinguish which of two hidden respondents (one human, one machine) is the human. If the interrogator fails to identify the human consistently, the machine is by this standard determined to be intelligent.

In 1956 scientists meeting at Dartmouth College initiated the field of artificial intelligence (AI). The founders of AI made many predictions. They expected that within a decade computers would understand natural language and would beat experts in the game of chess, and that within a generation computers would be as smart as humans. These predictions were overly optimistic. The dream of an all-purpose computer with the general intelligence to solve a vast array of challenges faded into the background, as engineers focused their attention on solving specific problems and building machines for specific tasks. Progress came in the form of industrial robots capable of replacing workers by repeating intricate repetitive tasks while assembling cars or computer circuit boards.

In 1997, IBM's Deep Blue did win a six-game chess match against the world champion Garry Kasparov. More recently, another IBM computer named

---

[1] This article will discuss stationary computers and robots capable of movement. Rather than constantly repeating the phrase 'computers and robots', the term robots will be used when referring to both. The 'bots' stands for intelligent agents within computer software.

Watson was able to beat two champions on the TV quiz show *Jeopardy*. Watson demonstrated skill in deciphering questions that were posed in natural language. The victories by Deep Blue and Watson were remarkable achievements; however, they fall far short of the kind of intelligence the more optimistic theorists had anticipated, and they illustrate an ongoing conundrum in evaluating the advances toward full artificial intelligence.

Who should judge the progress of AI and what criteria should be used to make that judgment? On the one hand, computers certainly outperform humans on many tasks from mathematical calculations, to modeling complicated systems, and to searching large databases to discover significant correlations between disparate pieces of information. On the other hand, robots fail in performing many basic tasks such as discerning essential from inessential information while navigating within real world environments.

While the more optimistic predictions regarding a rapid progression toward AI have been wrong often over the past fifty years, with each breakthrough proponents contend that strong AI is on the near horizon. Artificial neural networks, which are composed of individual processors that mimic neurons, and experiments with genetic algorithms and artificial life, approaches that mimic evolutionary processes, were breakthroughs that were each perceived as setting the stage for rapid progress. Alas, the headway afforded by these innovative approaches has not yielded the key to strong AI. Nevertheless, those with optimistic predictions are a breed that reappears in each generation. The inventor and futurist Ray Kurzweil (2006) insists that computers with human level intelligence with be available within the next thirty years, and soon afterwards there will be an 'intelligence explosion' (Good 1965) leading to AI systems vastly superior in intelligence to humans. This juncture is often referred to as a technological singularity (Vinge 1993).

I am among the skeptical regarding the advent of a technological singularity. There are many technological thresholds that must be crossed before we will know whether strong AI, not to mention super-intelligence, is possible, let alone probable. Meanwhile, designers and engineers are making steady if plodding progress as they tackle practical problems, such as matching input from visual sensors against a database of objects and entities, or discerning the emotions of a human by deciphering, in real time, micro-movements in facial muscles. Some of the practical challenges entail solving tasks which to date have thwarted the best efforts of engineers. However, no one knows for sure when there will be a new breakthrough or whether some of these tasks represent limitations on what can or will be achieved. There are, no doubt, many technological thresholds that are yet to be crossed, but there may also be ceilings that define limits on what can and what cannot be achieved technologically with existing scientific knowledge.

Nevertheless, robotics is developing at a relatively rapid pace. New products are appearing and untold possibilities are being explored in university and commercial laboratories. Robot dolls and robopets have been a commercial success, as has the Roomba vacuum cleaner produced by the IRobot corporation. Rescue robots snake through rubble looking for survivors. Remotely controlled robots, such as drones, and surgical assistants, such as the DaVinci systems that extend the capabilities of skilled surgeons for performing minimally invasive surgery, are proliferating. The U.S. government has invested millions in the development of neuroprosthetic arms, legs, and hands that will provide agility and mobility for those who have lost limbs. Robotic exoskeletons can enhance a worker's strength or improve mobility for those with pain in their lower back or weak legs.

Researchers are tackling the development of computers and robots that have the functional equivalent of specific cognitive capabilities, such as the ability to learn, emotional intelligence, and a theory of mind (able to attribute beliefs, desires and intents to others and to oneself). In addition, renewed attention is being directed at developing computer systems with the general intelligence to be used in a variety of domains. For example, significant progress has been made on the Blue Brain project, which proposes to create a synthetic brain by emulating neural structures. However, it is unknown whether scaling issues in linking billions of artificial neurons will place limits on building a full emulation of the human brain.

Manufacturers and roboticists are still searching for an application that will make sophisticated robots compelling for consumers at an affordable price point. Perhaps that device will be a robot caregiver that assists the homebound or elderly,[2] or cars with autonomous capabilities, such as those being developed by Google and Mercedes Benz.

2. ROBOETHICS: NEAR-TERM SOCIETAL AND ETHICAL CONCERNS

The array of applications that robots are being designed and adapted for are accompanied by an array of ethical and legal considerations that will need to be addressed. Drones and unmanned ground vehicles developed for the military are being marketed to local police forces. Surveillance drones, some smaller than birds, will be a nightmare for maintaining the safety of aviation and will undermine privacy. Driverless cars, cooks, and caregivers raise safety and liability concerns. The core ethical issues are subsumed within five interrelated themes: safety, appropriate use, capabilities, privacy and property rights, and responsibility and liability.

2.1 Safety

Are the robotic devices safe? Engineers have always been concerned with the safety of the products they build. The existing legal frameworks for product liability adequately cover the legal challenges posed by present day robots. The robots that have been developed to date are sophisticated machines whose

---

[2] Many companies throughout the world are designing service and domestic robots. Robots that care for the elderly and homebound are a high priority for countries such as Japan, where employment is high, the population is aging, and where there are restrictions that limit immigration.

safety is clearly the responsibility of the companies that produce the devices and of the end users who adapt the technology for particular tasks.

2.2 Appropriate use

Are robots appropriate for the applications for which they are being designed? Some critics consider the use of robots as sex toys offensive. Social theorists lament the emotional enrichment and lessons lost when robopets and robo-companions are substituted for animals or people (Sparrow 2002, Turkle et al. 2006). Noel and Amanda Sharkey (2010) express concern that the extended use of robonannies, robots that tend infants and children, may actually stunt emotional and intellectual development. Robotic caregivers for the home-bound and elderly are perceived as abusive or reflecting badly upon modern society. From a humanistic perspective, however, turning to robotic care is arguably better than no care at all.

2.3 Capabilities

Is the robot fully capable of performing the task to which it has been assigned? People tend to anthropomorphize robots whose looks or behavior are faintly similar to that of humans and presume that robots have capabilities they do not possess. Unfortunately, those marketing robotic systems all too often play upon public misconceptions and the tendency to anthropomorphize to obscure whether robots have the capabilities necessary to perform essential tasks when, for example, they serve as caregivers. A professional association or regulatory commission that evaluates the capabilities of systems and certifies their use for specific activities is needed. This, however, is likely to be very expensive, as the development of each robotic platform is a moving target. Existing capabilities are undergoing refinement and new capabilities are constantly being added to systems in the form of both hardware and software improvements.

2.4 Privacy and property rights

The diminution of privacy and property rights has long been a focus for computer ethics and for theorists working on information and Internet law. Robots will exacerbate the diminution of these rights. Consider introducing a robot into the home or other social settings. The privacy risks will be similar to those posed by surveillance cameras. Each robot will have both sensors and large drives that record all the data it collects. The data will be valuable to the homeowner as a record of important events, and it will be useful for technicians tracking what went wrong when there is a system failure. It will also be a record of all private activity within range of the sensors. We can expect that in the future the hard drives within robots and networks will be subpoenaed for everything from criminal investigations to custody battles. Furthermore, the devices will commonly be connected to the Internet through wireless connections that are likely to make the data stored in the robots accessible for a variety of criminal purposes (Denning et al., 2009).[3]

2.5 Responsibility and liability

The designers and engineers of a complex device cannot always predict how that device will act when confronted with new situations and new inputs. As Helen Nissenbaum (1996) has pointed out, "many hands" will have contributed to the building of a robot. Only those who designed and built a specific hardware component may understand the full operation of that component, and even they may have little or no understanding of how that component might interact with other components in a totally new device. Those designing and engineering new systems are under pressure to complete projects. Sufficient time and funding to fully test new systems is not always available, and this also contributes to limited understanding of the potential risks inherent in deploying new devices.

Certainly, credible manufacturers are concerned with safety and do not want to be held liable for marketing faulty or potentially dangerous products. Thus they may elect to hold back the release of a new device when they are unable to guarantee its safe use. However, this course of action has a downside. Robotics is among the transformative technologies that are expected to enhance productivity. Productivity improvements are essential for keeping an economy strong. Withholding the release of products whose benefits are significant and whose risks are low will be perceived by both industry and some governmental leaders as a heavy burden to place on innovation. Furthermore, withholding products will provide a competitive advantage to manufacturers in countries with higher bars to litigation. If, for example, the robotic industry in the U.S. waits for liability law to be sorted out before releasing a new product, competitors in Europe, Japan, Korea, or China might be afforded the opportunity to establish a formidable lead in marketing systems for new applications.

Manufacturers will undoubtedly welcome and work towards measures that lower their liability. As a means of spurring industry growth and innovation, Ryan Calo (2010) has proposed immunizing manufacturers of open robotic platforms from all actions related to improvements made by third parties. But any approach for limiting liability will need to be balanced against insuring that an industry will not knowingly introduce dangerous products.

Another approach that could stimulate the development of certain robot applications by lowering manufacturer's liability is no-fault insurance. This would be particularly appropriate for driverless cars. Even if driverless cars are much safer

---

[3] Tamara Denning, Tadayoshi Kohno, Karl Koscher, William Maisel and colleagues at the University of Washington have already demonstrated that the cameras and sensors in a robot pet can be hacked to gain real-time visual and auditory access to activity in the robot's location. More alarmingly, they have hacked and altered heart pacemakers (Maisel & Kohno 2010) and the software in automobile computers that regulates braking and other functions (Karl Koscher et al. 2010). Tadayoshi Kohno contends that every topic in computer science can have a security-related twist.

than those driven by humans, robot-chasing attorneys may well initiate suits for each death in which a robotic car is involved. No-fault insurance could help insulate manufacturers from frivolous lawsuits.

Similar challenges are confronted by all new technologies. Each country must balance speeding up innovation and productivity against dangers posed to its citizens. Speedier development can provide a competitive advantage while opening up the citizenry to disasters in which a new technology is complicit. In addition to loss of life and property, disasters can dramatically set back an industry.

When a complex system fails it is very difficult to determine the cause and exactly who is responsible for the failure. On January 28, 1986 the Space Shuttle Challenger exploded 73 seconds after takeoff. Many months passed and millions of dollars were spent before investigators determined that the culprit was tiny o-rings that became brittle in cold weather. Perhaps stricter precautionary measures might have averted the disaster, but later investigators question whether any method for monitoring the engineering of a truly complex system will reveal all design flaws. In reviewing that research, Malcolm Gladwell writes, "we have constructed a world in which the potential for high-tech catastrophe is embedded in the fabric of day-to-day life" (1996).

When an intelligent system fails, manufacturers will try to dilute or mitigate liability by stressing an appreciation for the complexity of the system and the difficulties in establishing who is responsible for the failure. Anticipating this argument, practical ethicists and social theorists are raising concerns as to the dangers inherent in diluting corporate and human responsibility, accountability, and liability for the actions of increasingly autonomous systems. In order to combat those dangers, five rules have been proposed to reinforce the principle that people and corporations should not be excused from moral responsibility for the design, development, or deployment of computing artifacts.[4]

Rule 1: The people who design, develop, or deploy a computing artifact are morally responsible for that artifact, and for the foreseeable effects of that artifact. This responsibility is shared with other people who design, develop, deploy or knowingly use the artifact as part of a sociotechnical system.

Rule 2: The shared responsibility of computing artifacts is not a zero-sum game. The responsibility of an individual is not reduced simply because more people become involved in designing, developing, deploying or using the artifact. Instead, a person's responsibility includes being answerable for the behaviors of the artifact and for the artifact's effects after deployment, to the degree to which these effects are reasonably foreseeable by that person.

Rule 3: People who knowingly use a particular computing artifact are morally responsible for that use.

Rule 4: People who knowingly design, develop, deploy, or use a computing artifact can do so responsibly only when they make a reasonable effort to take into account the sociotechnical systems in which the artifact is embedded.

---

[4] The full document titled, Moral Responsibility for Computing Artifacts, defines terms and explains the rules. It can be accessed at https://edocs.uis.edu/kmill2/www/ TheRules/.

Rule 5: People who design, develop, deploy, promote, or evaluate a computing artifact should not explicitly or implicitly deceive users about the artifact or its foreseeable effects, or about the sociotechnical systems in which the artifact is embedded.

The development of the robotics industry would be slowed significantly if these five rules were adopted. Consider, for example, holding those who develop and market robots morally responsible for all the potentially foreseeable uses to which the robot might be put. By such a standard cigarette manufacturers are responsible for lung cancer, and gun manufacturers could be sued when their products were used in a robbery or for a murder. In other words, the rules may be considered too broad in scope, and therefore cannot be applied in a discriminating manner. Nevertheless, I am among those who sponsor these five rules, even while I recognize that they are unlikely to be fully codified in law, because the rules embody principles that we should not compromise in a cavalier manner.

3. MACHINE ETHICS: MORAL MACHINES

New applications and new markets will open up if it becomes possible to design robots that are sensitive to ethical considerations, and are able to factor those considerations into the choices they make and the actions they take. Conversely, if robots fail to accommodate human laws and values in their behavior, the public will demand regulations that limit the applications for which they can be used.

Noting that designers and engineers can not always anticipate how increasingly autonomous intelligence systems will act when confronted with new situations, a growing community of scholars has begun reflecting on prospects for developing robots capable of making moral decisions. This new field of inquiry is often called machine ethics (ME), but it also goes by other names, including machine morality, computational ethics, artificial morality, and friendly AI. If designers and engineers can no longer predict the choices and actions a system will make then it becomes necessary for the robots themselves to evaluate the appropriateness or legality of various courses of action. If that is impossible, then the only other option might be to not build the systems at all because they are likely to enter into situations where their actions are inappropriate and even dangerous.

In the near term, the freedom of action for robots capable of making moral decisions will be quite limited. However, as their autonomy increases, robots might eventually evolve into artificial moral agents (AMAs).

Whether robots will actually acquire the full intelligence and moral acumen to be considered true moral agents is unknown at this stage of development. Both technological and philosophical thresholds will need to be crossed before a robot is likely to be designated a moral agent responsible for its own actions. As mentioned earlier, some of those thresholds could turn out to be ceilings that define limits upon the intelligence, moral understanding, and stature of robots.

3.1 Operational morality

The development of robots, and indeed all technology, can be viewed through two dimensions: increasing autonomy and increasing sensitivity to ethical considerations. For example, a hammer has neither sensitivity nor autonomy. A thermostat, in contrast, has some sensitivity to temperature and the autonomy to turn a furnace or fan on or off when a threshold has been reached.

The values instantiated in the actions of nearly all of the robotic devices being marketed today are determined by the corporations and engineers who build them, and they are programmed into the systems. These systems are 'operational moral' in that they make no decisions on their own, but merely follow the proscribed actions programmed in by designers who have predetermined all the types of situations the robot will encounter.

The manufacturer of a speaking robot doll a few years back was considering what the doll should say if a child treated it abusively. Sensors that detected abusive treatment could be built into the system, but how should the doll respond to such treatment? The company analyzed the issues and consulted with their lawyers and finally decided that the doll would say and do nothing.

Was this the right course of action, and should manufacturers be the one's determining appropriate action for robots in ethically charged situations? The robot could say, "Stop that! You are hurting me." But we are a long way, if ever, from building robots with a capacity to feel pain. This well-intended response from a robot that feels nothing might convey the wrong message, which could in turn led to unintended consequences.

Consider a robot that functions as a companion or robonanny to a young child or a teenager. If the robot is capable of detecting when the child puts herself in a dangerous situation, what should the robot do? This is not a simple problem. In certain circumstances inappropriate intervention by the robot might cause harm.

What if the child is abusing himself? Should the robot tell the child to stop? What would you want the programmer to design the robot to do if the child ignores this command? Would you want the robot to discipline the child? A robot that instructs but cannot follow up with discipline may well be teaching the wrong lessons. But a robot that disciplines is not likely to instill trust in the child.

Similar ethically charged situations arise in every domain. A care robot that is programmed to bring medicine to an elderly or sick person at designated intervals might be confronted with repeated refusals by the patient to take the medicine. What should the robot do if this occurs? Or the same robot might discover upon entering a room that a look of terror or fear was upon the face of the patient it was tending. How could the robot discern the source of that fear? Could it have any way of 'knowing' that the fear was perhaps caused by the robot itself?

Engineers can build in sensors and software that mechanically discern some ethical challenges and program in appropriate responses. But this, of course, presumes that they have anticipated the challenges the robot will encounter. However, even then, one response may not be suitable for all users. Different parents, for example, have different values. Some parents might want a robot that is capable of reprimanding a child, while other parents might find it absurd to have a robot reprimand their child.

Offering user selected options (parental choice) is one way of working around the likelihood that not everyone will want a robot to respond in the same manner in ethically charged situations. During setup parents would be informed about the ramifications of various alternative responses the robot might make when it was necessary to take some action, and could specify the course that matched with their values. A setup procedure provides the additional benefit of affording the manufacturers of companion robots with an opportunity to educate parent on what they should and what they should not expect from the robotic device. Thus the manufacturer protects itself from certain forms of liability while educating the parents on the proper use of robonanny.

Hopefully the countless ethical challenges that robots will confront in social contexts will alert industry leaders to the importance of making ethical considerations an integral aspect of the design process. Sensitizing students to the societal impact of the products they design has become an important element of the curriculum at engineering schools in recent years. A next step would be integrating applied ethicists into the design team, not as naysayers, but to help in forging new solutions to the design problems posed by societal and ethical challenges. Helen Nissenbaum (2005) has named this new role for applied ethicist 'engineering activism'.

3.2 Functional morality

The primary interest for machine ethicists is the approaches and procedures that would facilitate robots making explicitly moral judgments. To make an explicit judgment the robot will need to evaluate the situation, factor in an array of considerations, and then apply some rules, principles or procedure to determine the appropriate action. I have written about this subject extensively, and will only mention a few brief details in this chapter (Wallach and Allen, 2009).

The approaches for implementing moral decision-making capabilities in robots fall within two broad categories, top-down and bottom-up (Allen, Smit, & Wallach, 2006). Top-down refers to the implementation of rules, principles, or moral decision-making procedures, such as utilitarianism, Kant's categorical imperative, the Ten Commandments, the Golden Rule, and even Isaac Asimov's laws of robotics.

Any top-down approach takes an antecedently specified ethical theory and analyses its computation requirements to guide the design of algorithms and subsystems capable of implementing the theory. For example, here are the three laws[5] proposed by science fiction writer Isaac Asimov, author of *iRobot* and *The Bicentennial Man*:

---

[5] In later years Asimov added a fourth or Zeroth Law: A robot may not harm humanity, or by inaction, allow humanity to come to harm.

1. A robot may not injure a human being or, through inaction, allow a human being to come to harm.
2. A robot must obey the orders given to it by human beings, except where such orders would conflict with the First Law.
3. A robot must protect its own existence as long as such protection does not conflict with the First or Second Laws.

The laws are simple and intuitive, but in story after story Asimov illustrates that problems will occur with robots programmed to obey these rules. For example, what should the robot do when it receives conflicting orders from two or more people? Through his robot stories, Asimov demonstrated that a simple rule-based system of ethics will be ineffective. Thus machine ethicists have turned to more complex ethical theories such as utilitarianism (the greatest good for the greatest number) to guide the design of moral machines.

Bottom-up approaches take their inspiration from evolutionary psychology and game theory, as well as developmental psychology and theories of moral development. Bottom-up approaches, if they use a prior theory at all, do so only as a way of specifying the task for the system, but not as a way of specifying an implementation method or control structure. Minimizing harm, for example, might be the goal of a bottom-up system, but exactly how to minimize harm would not be specified in advance, and the system would be designed to learn ways to approach, if not actually achieve, that goal.

Both top-down and bottom-up approaches have strengths and weaknesses. For example, top-down approaches often specify principles that are defined broadly and therefore cover countless situations. But if the principle is too broad or too abstract its application to specific challenges will be subject to debate. A strength of bottom-up approaches lies in the ability to dynamically integrate input from discrete subsystems. But, it can be very difficult to define the ethical goal for a bottom-up system. In addition, assembling a large number of discrete components into a system that functions in an integrated manner can be extraordinarily difficult. Eventually, we will need AMAs that maintain the dynamic and flexible morality of bottom-up systems that accommodate diverse inputs, while subjecting the evaluation of choices and actions to top-down principles that represent ideals we strive to meet.

In many contexts, robots will need additional capabilities beyond the capacity to reason in order to arrive at appropriate courses of action. Machine ethics forces us to think seriously about how humans make moral decisions, and it highlights the important role played by capabilities that are commonly taken for granted when reflecting upon human decision making. These supra-rational (beyond reason) capabilities play a variety of roles from providing access to essential information to sensitizing the system to specific ethical considerations. Emotions, social intelligence, a theory of mind, empathy, consciousness, and being embodied in a world with humans, objects and other agents are among the additional capabilities whose functional equivalent will be needed for robots to make moral decisions in specific contexts. One task for machine ethics will be to determine which capabilities AMAs will require in order to operate appropriately and safely within specific domains. New applications for the use of robots will become possible as the sensitivities and abilities of robots expand. However, at this time we do not know whether all the necessary capabilities can be instantiated computationally.

Finding a computational method to implement norms, rules, principles, or procedures for making moral judgments is a very hard problem. A second hard problem concerns the challenge of setting boundaries for the assessments that must be made for effective moral reasoning. The following questions are examples of this second hard problem:

> How does the system recognize it is in an ethically significant situation?

> How does it discern essential from inessential information?

> How does the AMA estimate the sufficiency of initial information?

> What capabilities would an AMA require to make a valid judgment about a complex situation, e.g., combatants v. non-combatants?

> How would the system recognize that it had applied all necessary considerations to the challenge at hand or completed its determination of the appropriate action to take?

3.3 Coordination

The vast preponderance of robotic systems in the coming decade will be remotely controlled and/or semi-autonomous systems. Coordinating the activity of human operators with the semi-autonomous actions of robots will be more important than designing systems that function independently. David Woods and Erik Hollnagel (2006) contend that robots and the operators working with them are best understood as a joint cognitive system (JCS). In other words, some of the smarts are built into the system and additional intelligence comes from the humans operating the robot. Unfortunately, there is an increasing tendency to view humans as the weak link in the partnership. So that when there is a failure of a JCS, humans are commonly blamed for that failure and more autonomy for the robot is proposed as a solution. In addition, increasing autonomy is also seen as allowing designers to escape from responsibility for the actions of artificial agents. This, however, is a recipe for disaster. As robots become more autonomous and more complex, the human operators will find it difficult, if not impossible, to anticipate the actions of the system in new situations, and may therefore be unsuccessful in coordinating their activity with that of the robot. This lack of coordination will potentially contribute to an increase in the failure of JCSs. A focus on autonomy will lead to the misengineering of JCSs.

It is a mistake to place decisions made by computers ahead of human intelligence. The intelligence of each computational system is confined to a limited set of cognitive capabilities and continues to be brittle on the margins. People, however, are slowly becoming more and more reluctant to go against the recommendations of semi-intelligent systems. Doctors and administrators will shy away from overriding the recommendations of an 'expert' computerized medical advisor when there will be an audit trail available to enterprising liability lawyers if anything goes wrong. Human decision makers need to be empowered so that they will have the courage to countermand the actions or suggestions of robots. Claims that robots have the capabilities to make superior decisions, or even function as safe substitutes for human agents, should be evaluated skeptically.

3.4 The future of AMAs

The development of AMAs is likely to be a slow incremental process that spans many decades. Throughout this development a primary challenge for society will be monitoring and assessing the capabilities of each system. What criteria should be used to determine whether a particular system could be deployed safely in a specific context? What oversight mechanisms need to be put into place in order to ensure that such an assessment can be made and has been made? What penalties might be applied if a certified system is later implicated in harmful actions?

If AMAs can be built, the more distant theoretical and speculative challenges that have fascinated science fiction writers, philosophers, and legal theorists will come into play. Will, for example, artificial agents need to emulate the full array of human faculties to function as adequate moral agents and to instill trust in their actions? Or, will AMAs have their own moral codes that are less anthropocentric than those of humans? What criteria should be used for evaluating whether an AI system deserves rights or should be held responsible for its own actions? Does punishing a robot make any sense? If yes, how might one punish a robot for infractions of rules or transgressions against the rights of others? Should or can we control the ability of robots to reproduce? How will humanity protect itself against treats from intelligent robots?

4. A PROPOSAL FOR MONITORING AND GOVERNING THE ROBOTICS INDUSTRY

Technological development is similar to an economy in that it can stagnate or overheat. Modulating the rate of innovation and deployment will be a central role for ethics, law and public policy in the development of the emerging technologies.

Pressures are building to embrace, reject, or regulate robotics and other emerging technologies. Biotechnology, nanotechnology, AI, geoengineering, neuroscience, synthetic biology, personalized medicine, and surveillance technologies each have unique features, but they also share a complex set of similar issues. These shared issues include health, safety, and environmental risks, funding for research and development, intellectual property rights, public perception of the benefits and risks, the need for government oversight, and the competition within industries and internationally.

The individual fields of research are not progressing in isolation. The various technologies will be combined and converge in ways that are difficult to predict. The tools we have for forecasting and risk assessment are highly subjective. So how will the U.S., other countries, and humanity as a whole monitor and manage technological innovation? Given that the possibilities change with each new scientific discovery and each unanticipated tragedy, any roadmap will certainly be a work in progress. What forms of governance will we need to put in place? Which areas of research will need to be regulated and when will regulations unnecessarily stultify innovation?

The roboticization of aspects of warfare and the prospect that autonomous weaponry may soon pick their own targets and initiate lethal force without direct human intervention are among the issues specific to the robot industry. The latter possibility is perceived by many critics as a violation of the laws of war and existing international agreements. Another concern is that complex systems may become unstable at large scales, and therefore will periodically fail in potentially disastrous ways. Then there is always the futuristic and speculative fear of a robot takeover or robots that are unfriendly to human values and human interests. While this is unlikely to occur any time soon, public anxiety might be allayed if there were a credible monitoring authority that could report on which technologies lay in the distant futures, and when technological thresholds were about to be crossed that open the door to substantial risks.

Technologies created by the robotics industry are likely to be combined with other technologies to augment the strength and versatility of future soldiers. Consider a soldier with a robotic exoskeleton viewing the movements of small drones searching for the enemy through virtual goggles, and when the enemy is discovered calling in larger drones for airstrikes. Information from sensors on the soldier's body and nanosensors in the bloodstream could help those back at the base monitor how well the soldier is functioning both physiologically and cognitively. When necessary, supervising officers could remotely activate a nanobot to release a neurochemical or stimulant. Of course these same technologies will be adapted for domestic applications and for personal experimentation.

What we require is a Governance Coordination Committee (GCC) to oversee the development of the emerging technologies (Marchant & Wallach 2012). The first step would be a GCC for robotics and perhaps one or two other fields such as synthetic biology or geoengineering. We need a credible, flexible, adaptive, and lean oversight committee to help guide the safe development of robotics. The role of a GCC would be to monitor development and flag issues or gaps in the existing policy mechanisms, to coordinate the activities of the various stakeholders, and to modulate the pace of development. The GCC would be mandated to avoid regulation where possible and to favor the implementation of soft governance mechanisms such as industry oversight. It would not reproduce the functions of other regulatory or monitoring institutions, but would rather build upon and attempt to coordinate the tasks performed by these other parties.

There are many stakeholders in the development of scientific research. These include researchers, industry, the public, legislators, regulatory authorities, non-governmental organizations, and the media. Each of the stakeholders can play a positive, negative, or neutral role in scientific development and each has its own concerns. For example, the media can educate and disseminate information. It can also exacerbate bias, rumors, misinformation, and unwarranted fears. The public generally perceives technology as an engine of innovation and productivity, but is also concerned about minimizing harms. Fears from terrorists to pandemics to robot takeovers can lead to often-irrational pressures from the public that can result in bureaucratic and unhelpful public policy.

For legislators there is a tension between stimulating economic growth and minimizing risks. Legislators tend to be shortsighted, and all too often new regulations are postponed until action is forced by a tragedy. With a little foresight and planning, alternative policy mechanisms might be implemented to reduce risks. An array of policy mechanisms including government financing of research, professional codes of ethics, research ethics, laboratory practices and procedures, insurance policies, governmental oversight agencies, Institutional Review Boards, Biosafety Committees, and Institutional Animal Care and Use Committees help guide scientific development. Collectively these mechanisms provide a robust system of protections. But there are grounds for criticism, such as a lack of funding for oversight, too many regulations, and a make-shift system that has been constructed haphazardly and piecemeal.

An effective CCG would function as a good faith broker and would report to the legislative and executive branches of government regarding potential harms and gaps in the system of existing policy mechanisms. Industry leaders, for example, would prefer self-regulation, but may not take the necessary initiative unless threatened with government regulations. The CCG could influence industry with the threat of government regulations, confident that a legislature will listen to its recommendations once it has exhausted all alternatives.

Whether an effective CCG can be put in place to manage, monitor, and modulate the development of robotics is difficult to say. Robotics is still a young industry. While it is expected to be a transformative industry, to date it is not burdened by heavy oversight or unwieldy regulations. But establishing the authority, legitimacy, credibility, and influence of an industry manager will not be easy. Who will make up the committee's membership and how will administrators be selected? Will the GCC be within the government or will it be private? If it is private, will it have sufficient influence on legislators or the executive branch of government? How would a GCC for robotics be funded? To whom would it be accountable? These are difficult questions. Perhaps the questions themselves indicate that establishing an oversight committee to facilitate the governance of robotics is overly optimistic, too complicated to implement, and hopelessly naive in the present political climate. Nevertheless, a comprehensive approach to governing robotics is needed. Perhaps a GCC is too ambitious. However, as the robotics industry continues to develop there will still be many opportunities to consider new mechanisms for directing its progress.

WORKS CITED

C. Allen, I. Smit, and W. Wallach (2006) 'Artificial Morality: Top-Down, Bottom-Up and Hybrid Approaches,' *Ethics and Information Technology*, 149.

M. Anderson, S. Anderson, and C. Armen (2005) *Towards Machine Ethics: Implementing Two Action-Based Ethical Theories*, Paper presented at the AAAI 2005 Fall Symposium on Machine (Alexandria, VA: Ethics).

M. Anderson, and S. L. Anderson (2008) 'EthEl: Towards a Principled Ethical Eldercare Robot,' paper presented at the 3rd ACM/IEEE International Conference on Human-Robot Interaction, Amsterdam.

R. Calo (2011) 'Open Robotics,' *Maryland Law Review*, 571.

T. Denning, C. Matuszek, K. Koscher, J. P. Smith, and T. Kohno (2009) 'A Spotlight on Security and Privacy Risks with Future Household Robots: Attacks and Lessons,' Proceedings of the 11th International Conference on Ubiquitous Computing, (Orlando, Florida).

M. Gladwell (1996) 'Blowup,' *The New Yorker* Reprinted in *What the Dog Saw, And Other Adventures*, 2009 (New York: Little, Brown and Company) available online at http://www.gladwell.com/1996/1996_01_22_a_blowup.htm (accessed May 21st 2012).

I. J. Good (1965) 'Speculations Concerning the First Ultraintelligent Machine,' in F. L. Alt, and M. Rubinoff (eds), *Advances in Computers*, 6 (Burlington, MA: Academic Press).

J. S. Hall (2000) 'Ethics for Machines,' (available at http://autogeny.org/ethics.html).

K. Koscher, A. Czeskis, F. Roesner, S. Patel, and T. Kohno (2010) 'Experimental Security Analysis of a Modern Automobile,' IEEE Symposium on Security and Privacy (Berkeley, CA) May 16–19.

R. Kurzweil (2006) *The Singularity is Near* (New York, NY: The Penguin Group).

W. H. Maisel, and T. Kohno (2010) 'Improving Security and Privacy of Implantable Medical Devices,' New *England Journal of Medicine*, 1164.

G. Marchant, and W. Wallach (2012) 'Proposal for a Governance Coordination Committee,' Presented at the Pacing Governance with Technology Workshop at Arizona State University, March 6th.

H. Nissenbaum (1996) 'Accountability in a Computerized Society,' *Science and Engineering Ethics*, 25.

H. Nissenbaum (2001) 'How Computer Systems Embody Values,' *Computer*, 117.

N. Sharkey, and A. Sharkey (2010) 'The Crying Shame of Robot Nannies: An Ethical Appraisal,' *Interaction Studies*, 161.

R. Sparrow (2002) 'The March of the Robot Dogs,' *Ethics and Information Technology*, 305.

S. Turkle, W. Taggart, C. D. Kidd, and O. Daste (2005) 'Relational Artifacts with Children and Elders: The Complexities of Cybercompanionship,' *Connection Science*, 18 (4): 347–361.

V. Vinge (1993) 'The Coming Technological Singularity: How to Survive in the Post-Human Era,' prepared for the VISION-21 Symposium sponsored by the NASA Lewis Research Center and the Ohio Aerospace Institute, 30–31 March 1993. (It is retrievable from the NASA technical reports server as part of NASA CP-10129.)

W. Wallach, and C. Allen (2009) *Moral Machines: Teaching Robots Right From Wrong.* (New York, NY: Oxford University Press).

D. D. Woods, and E. Hollnagel (2006) *Joint Cognitive Systems: Patterns in Cognitive Systems Engineering* (Boca Raton, FL: CRC Press).

DISCUSSION QUESTIONS

1. Do you think a robot or machine could ever be morally responsible or a moral agent? Why or why not? Which capabilities would it need to have?
2. According to the Turing test, a machine should be considered intelligent if discourse with it is indistinguishable from discourse with an adult human. Is

this a reasonable test for determining intelligence? Is passing the Turing test necessary for being considered to have intelligence? Is it sufficient?

3. Does the use of robots to perform personal care functions – for example, watching children or assisting the elderly – reflect badly on modern society? If it does, why does it do so?

4. If driverless cars will decrease the number of accidents per year, should we allow them on the road? If driverless cars prove to be much less likely to cause accidents and to harm people, should we require the use of them (just as we impose speed limits and sobriety standards for safety reasons)?

5. What kind of moral rules do you think we should program into machines?

6. Are there any jobs or roles that robots ought not to fill, even if they were able to perform the function as well as humans? If so, why not – is it to do with potential unintended effects, form of life considerations, or power considerations?

7. Are there any jobs or roles that autonomous machines could not perform as well as humans? If so, what are they? If not, what might be the implications of autonomous machine proliferation for our form of life?

 # WHAT TO DO ABOUT ARTIFICIAL CONSCIOUSNESS

John Basl

CHAPTER SUMMARY

One of the primary ethical issues related to creating robust artificial intelligences is how to engineer them to ensure that they will behave morally – i.e. that they will consider and treat us appropriately. A much less commonly discussed issue is what their moral status will be – i.e. how we ought to consider and treat them. In this chapter, John Basl takes up the issue of the moral status of artificial consciousnesses. He defends a capacity-based account of moral status, on which an entity's moral status is determined by the capacities it has, rather than its origins or material composition. An implication of this is that if a machine intelligence has cognitive and psychological capacities like ours, then it would have comparable moral status to us. However, Basl argues that it is highly unlikely that machines will have capacities (and so interests) like ours, and that in fact it will be very difficult to know whether they are conscious and, if they are, what capacities and interests they have.

RELATED READINGS

Introduction: Types of Value (3.1)

Other Chapters: Wendell Wallach, *Ethics, Law, and Governance in the Development of Robots* (Ch. 24); Christopher J. Preston, *Evolution and the Deep Past* (Ch. 36)

1. INTRODUCTION

Many researchers are engaged in the project of creating artificial intelligence and there is hope that someday machines or robots will exhibit human-like intelligence and even consciousness. There is controversy about whether accomplishing this is probable or even possible, but there has been little discussion about whether it is desirable or morally appropriate. There are some who worry about the consequences for us of creating artificial intelligences and there are concerns about who to hold morally responsible as machines behave more autonomously (Chalmers 2010).[1] However, we must also ask whether we have any obligations to the consciousnesses we create, and whether our treatment might wrong them. This is the issue that I address in this chapter.

---

[1] See also Patrick Lin et al. (Ch. 23) and Wendell Wallach (Ch. 24) in this textbook.

Imagine someone, let's call her Jane, that is thought to be in a persistent vegetative state. Until now, our best theories of the mind, physiological and behavioral evidence, and medical devices seem to confirm that Jane is not conscious. However, in testing out a new device, we find that Jane is actually quite mentally responsive. While she is unable to behave in ways that tell us how she is feeling, we are able to monitor her brain in a way that indicates she can hear and respond to questions and commands. Within several years, doctors and family are able to communicate with Jane regularly, albeit tediously, by asking her to respond to yes or no questions. She reports that the years she spent locked within her own head were awful with nobody to communicate with.[2]

Now that we know about Jane, what would we think about failing to use the device, assuming the device is readily and cheaply available, to test other people thought to be in a persistent vegetative state? To fail to use it would be to commit an inexcusable wrong – it would be to knowingly ignore a possible person, someone with the same moral status as ourselves. It would be as if we turned our back on a possible person who might be suffering greatly right in front of us when we had the ready means to alleviate that suffering.

Now let's think of an analogous scenario where scientists try to the best of their ability to create an artificial consciousness. They try to determine what the functional bases of consciousness are, what hardware has to do in order to support conscious states, and they try to develop hardware and software that can support such states. Let's assume these scientists succeed in developing hardware that can support consciousness, but they have no way to monitor the conscious states of their creation. On this scenario, the entity is in a situation a lot like Jane's before the invention of the device that allowed her to communicate her awareness.

Is the creation of artificial beings that, for all we know, are a lot like Jane morally problematic? In order to address this question, it is important to determine the moral status of artificial consciousnesses of various kinds. We must know whether artificial consciousnesses have a welfare and whether their welfare is relevant to moral deliberations.

On any reasonable account of who or what is a moral patient – i.e., who or what has a welfare that we must consider in our moral deliberations – once we achieve artificial consciousness on par with our own we must recognize that consciousness as a moral patient. It will be due consideration for its own sake in virtue of the fact that it has interests similar to ours and because there will be little or no reason to discount those interests. However, we are a long way from creating an artificial consciousness that is anything like our own, or, for that matter, unlike our own. Yet, as we create more and more complicated machines and attempt to create artificial consciousness, we must think carefully about which properties of a machine would confer interests. To fail to ask whether these intermediate entities have interests and whether they are due consideration could lead to inappropriate conduct on our part. After all, it is not only beings with a consciousness like ours that are moral patients; non-human animals are moral patients, and we owe it to

---

[2] This situation is not all that imaginary. A new application of EEG machines has lead scientists to believe that a large number of people classified as being in a persistent vegetative state, lacking consciousness, are in fact at least minimally conscious (Cruse et al., 2011).

them to take their interests into account in our moral deliberations. Thus, the ethics of creating artificial consciousness involves determining (i) under what conditions machines have a welfare, and (ii) whether and how to take that welfare into account. In what follows, I take up these issues. I will argue that, though future machines might certainly be moral patients, current machines are not, or, more specifically, they are not moral patients for any practical purposes. In our ethical deliberations, we need not take the interests of (current) machines into account.

In Section 2, I explain the concept of moral status and its relationship to the concept of welfare and interests. In Section 3, I take up the metaphysical, epistemic, and normative issues concerning the moral status of artificial consciousness. I argue that in order for an entity to have what I will call conscious interests it must have the capacity for attitudes. Entities might be conscious even if they lack this capacity, but they will not have any interests in virtue of that consciousness. Also in section 3, I argue that current machines are mere machines, machines that lack the capacity for attitudes and thereby any conscious interests. Despite the fact that current machines may have all sorts of capacities, we have no evidence whatsoever that they are conscious, let alone that they have attitudes regarding their conscious states. Given this, machines lack conscious interests that are relevant to our moral deliberations.

Many environmental ethicists have argued that non-sentient organisms have interests in virtue of being goal-directed, or teleologically organized systems (Goodpaster 1978; Taylor 1989; Varner 1998; Sandler 2007). In Section 4, I explore the possibility that machines have what I'll call teleo interests in virtue of their being teleologically organized. In Section 5, I will argue that even if mere machines have a welfare in virtue of having such interests, these interests are also practically irrelevant to our moral deliberations. Therefore, for all intents and purposes (current) machines are not moral patients, or, at least, they are moral patients that we need not care about. The arguments in favor of this conclusion also apply to future machines until such a time when we can overcome the epistemic obstacles to determining whether, in fact, machines have the relevant cognitive capacities that ground conscious interests.

2. MORAL STATUS, INTERESTS, AND WELFARE

Before turning to questions concerning the moral status of artificial consciousnesses, it is important to clarify how we are to understand terms such as 'moral status', 'moral patient', 'interest', and 'welfare.'

To say that a being has moral status is to say that it is worthy of our consideration in moral deliberations. Given the breadth of this definition, it is obvious that the domain of things with moral status is quite large. For example, we often have to take artifacts into account in our moral deliberations because they are, for example, owned by others or are instrumental to our doing what's morally required. For this reason, it is important to distinguish the different reasons why a thing might have moral status.

One type of moral status is often referred to as moral considerability (Goodpaster 1978; Cahen 2002) or as inherent worth (Sandler 2007; Sandler and Simons 2012). To be morally considerable is to have a welfare composed of interests that are to be taken into account in moral deliberations for the sake of the individual whose welfare it is. For example, any typical human being is morally considerable; we have interests in living a good life, in not being subject to violence, and so on, and these interests are relevant to moral deliberations. For example, when I want to acquire money, your interest in not being robbed should figure into my decisions about how I ought acquire money, not just because you might put up a fight, but because of the impact my robbing you would have on your welfare. In what follows, I will use the term moral patient to refer to any individual that is morally considerable.

It also is important to be clear about the concept of an interest and the relationship between interests and an entity's welfare. As I will use the term, an individual's interests are those things the satisfaction of which contributes to its welfare. As I discuss below, there are various kinds or types of interests that an entity may have. Having an interest of any kind is sufficient for an entity's having a welfare.

We can distinguish psychological interests from teleo interests. A psychological interest is an interest that an entity has in virtue of having certain psychological capacities (and psychological states). A teleo interest is an interest an entity has in virtue of being goal-directed, or teleologically organized. In Section 4 I explicate the notion of teleo interests and explain how it is that mere machines have such interests. In the following section I turn to the question of whether machines with consciousness are moral patients and in virtue of which psychological capacities they are so.

3. MORAL PATIENCY AND ARTIFICIAL CONSCIOUSNESSES

3.1 The easy case: human-like consciousnesses

Imagine that an artificial consciousness has been developed or comes online. This consciousness is very much like ours. It has pleasant and painful experiences; it has the capacity for imagination, memory, critical thinking; and it is a moral agent. We can even imagine that this consciousness is embodied and goes about the world like we do. On any reasonable normative theory, theory of welfare, and theory of moral considerability, this artificial consciousness is a moral patient in the same way that we are. This is because reasonable theories of moral patiency are capacity-based. Being a moral patient is a function of the capacities an entity has, not the type of being that it is. Furthermore, if two beings' are equal in their capacities they are or should be considered equal with respect to moral considerability or their claim to being a moral patient.[3] If tomorrow we cognitively enhanced a chimpanzee so that it was our equal in cognitive capacity, even the most adamant proponent of animal experimentation would have to recognize that this chimpanzee deserved protections equal to those afforded other human beings.

---

[3] This does not mean that there are no cases where we should favor one life over another. For example, if we must decide between saving the life of our own child and a stranger, we have good reason to save our child. However, this isn't because our child is a more important moral patient. It has to do with the relationships involved.

Those that wish to deny this must deny that moral patiency is based in capacities. But, what is the alternative? There are two primary options. One option is religiously based. One might argue that certain individuals are imbued with moral patiency in virtue of being chosen by God. Another option is that moral patiency is species-based. I discuss these in turn.

Religion-based views of moral status are problematic for a variety of reasons. First, there is an epistemological problem. How are we to know whether an individual has been imbued with patiency? Perhaps God has saw fit to immediately imbue any being with certain capacities with patiency. Perhaps God has imbued all entities with patiency. Another problem for religious based views of moral patiency is political. Legitimate justifications for policies must be justifiable to those that are governed (Rawls 1993; Basl 2010). But, since there are reasonable, secular citizens, there must be at least some secular justification for policies. Given this, any moral reasoning concerning who or what are moral patients that we hope to translate to policy must be reasonable from a secular perspective.

What about species membership views of moral patiency? According to such views, a being is a moral patient in virtue of its belonging to a particular species. It is difficult to see how being a member of a particular species or other kind is a morally relevant feature of a being. After all, we have no trouble thinking of alien species that are moral patients in the same way that we are, and yet they are not of the same species as us. Alternatively, imagine some subset of humans evolves into a separate lineage, but retains all the same capacities for intelligence, imagination, and sociability. It would clearly be wrong to enslave them, even if we could not interbreed with them or diverged genetically from them – i.e. they were another species.

Because only capacity based views of moral status are reasonable, once there are artificial consciousnesses with capacities very much like ours, they will be moral patients in the same ways that we are. It will matter none at all whether these beings are silicone, steel, or purely digital. Our being made of squishy, biological material will not give us moral priority over such beings.

3.2 The Hard(er) case: animal- and other consciousnesses

In our quest to create artificial consciousness, it is more likely we will develop consciousnesses that are more like nonhuman animals, or radically different from anything that we know of, than like human consciousness. Given this, more must be said about those capacities in particular that give rise to psychological interests.

To determine which conscious machines are moral patients, we must first determine which capacities in particular give rise to psychological interests of the sort that are morally relevant. Not all capacities will give rise to morally relevant interests. If we create a consciousness with only the capacity for experiencing colors but with no attending emotional or other cognitive response, we need not worry about wronging it. It might be a shame to force such a consciousness offline, since its creation would no doubt be a fascinating and important achievement; but we would not wrong the machine by forcing it offline.

So, which psychological capacities give rise to psychological interests? To proceed, it is helpful to start by thinking about a non-human animal, such as a dog. Hitting such an animal with a sledgehammer is certainly bad for its welfare and at least partly in virtue of the fact that it frustrates its psychological interests. But, in virtue of what are those interests frustrated? Hitting a dog with a sledgehammer causes a variety of psychological phenomena. It causes pain (understood as a sensory experiences) and suffering (understood as an aversive attitude towards some other mental state). It might also result in frustration, if the dog unsuccessfully tries to perform actions that would be possible were it not injured. In non-human primates, a strong blow from a hammer might result in all of these, plus additional frustration as the primate realizes that its future plans can no longer be realized. Which of these psychological capacities (the capacity for conscious experience of pain, suffering, frustration, future planning) is necessary or sufficient for having morally relevant psychological interests?

Mere capacity for sensation is not sufficient to generate psychological interests.[4] We can imagine a being that is capable of feeling sensations that we call painful but lacks the capacity to have an aversive attitude towards these sensations. If we imagine that sensation is the only cognitive capacity the being has, then the being is very similar to the consciousness that can experience colors. The being would be able to feel something, just as the color experiencing being could see something; it might be able to describe the difference between a burning sensation and tingling sensation, but it would not be averse to either. Such a being would not, even if it could, recoil or feel like avoiding such sensations. It would not harm the being to cause it to feel those "painful" experiences, since such a being just could not care that it is in such a state. To deny this, to assert that mere sensation matters, requires a defense of how the mere feeling of a sensation harms an individual. An explanation must be given that helps to distinguish the moral relevance of those sensations we call painful from those that we call pleasurable in terms of the mere sensation. I submit that no defense has been given and think the prospects are dim. Evidence for this is found in the fact that the same sensation – e.g. being tickled, exhilaration, or physical exertion – is found by some people to pleasurable and by other people to be unpleasant. Without an attitudinal component, the bare sensations are neither good nor bad for a person.

While the mere capacity for first-order consciousness or sensory experience is not sufficient for an entity's having psychological interests, the capacity for attitudes towards any such experiences is sufficient. Peter Singer has argued that sentience, understood as the capacity for suffering and enjoyment, is sufficient for moral considerability. Consider the case of newborns and the severely mentally handicapped. These beings are severely limited in their cognitive capacities, perhaps lacking all but the capacity for sensory experience and basic attitudes regarding those experiences. And yet, these beings are moral patients. We ought and do take their welfare into consideration in our moral deliberations. We avoid things that cause pain in newborns, at least in part because they don't like it or have an adverse reaction to it.

The view just described is one on which a machine will have psychological interests (of the sort that are morally relevant) if it has attitudes. However, any

---

[4] For further argument against this view, called Sensory Hedonism, see Feldman (2004).

machine lacking these capacities, conscious or otherwise, is a mere machine; such machines lack morally relevant, psychological interests. If mere machines have a welfare, it will be in virtue of interests that are not psychological.

3.3 Epistemic challenges

Before turning to the question of whether current machines are moral patients, it is important to note some epistemic challenges that we face in determining whether current or future machines have psychological interests. The Problem of Other Minds is the problem of saying how it is we know that other human beings, beings that seem very much like ourselves, have a mental life that is similar to ours.

Perhaps the best answer we can give to this problem is that all our evidence suggests that others are mentally like ourselves. The source of this evidence is evolutionary, physiological, and behavioral. We know that we share a common ancestor with those that seem just like us; we know that they are physiologically like us (and we think we understand some of the bases of our conscious states); and, we know that we behave in very similar ways (for example by avoiding painful stimuli and telling us that such stimuli hurt). These same sources of evidence can be appealed to in order to justify claims about the mental lives of animals. Those organisms that are evolutionarily close to us, have certain physiological structures, and behave in certain ways that seem best explained by appeal to relevant cognitive capacities are judged to have a mental life that includes attitudes.[5]

Unfortunately, in the case of machines, we lack the typical sources of evidence about their mental life. A computer lacks any evolutionary relationships with other species; its physiology is entirely different than any other conscious being's; and, if it feels pain, it cannot tell us it does, or behave in a way that suggests that it does, unless it has been somehow enabled to do so. Unless we have a very good understanding of the (functional) bases of mental capacities and so can know whether such bases exist in a given machine, we may be largely in the dark as to whether a machine is conscious and as to whether it has morally relevant capacities.

I do not have any solutions to the epistemic problems posed by the nature of machine consciousness. However, these difficulties do raise ethical concerns. Insofar as it is possible to replicate the bases of the capacities for attitudes or concepts, it is possible to create machines that have conscious interests. We must ask whether and under what conditions it is permissible to continue to try to develop artificial consciousnesses given that we might be creating entities with a welfare that we can't identify and so might be, unknowingly, causing intense suffering to a set of beings.

---

[5] There is considerable controversy over which mental capacities non-human animals have. See Tomasello and Call (1997) for a discussion of some of the issues concerning primate cognition. However, there is little doubt that many non-human animals have aversive attitudes towards those conditions we identify as painful. See Varner (1998, chapter 2) for an overview of the evidence that non-human animals have the capacity for suffering.

4. TELEO INTERESTS

In the remainder of this chapter, I argue that current machines are mere machines and that, even though they may have a welfare in virtue of having non-psychological interests, they are, for all practical purposes, not moral patients.

Consider our most advanced computers, from chess computers, to missile guidance systems, to Watson (the *Jeopardy!* winning supercomputer). We have absolutely no evidence that such computers are conscious and so absolutely no evidence that such computers have the capacity for attitudes or concepts that would ground psychological interests. Of course, we could program a computer to tell us that it doesn't like certain things, I'm sure even Siri (the iPhone's 'assistant') has "attitudes" of this sort. But, we know that such behaviors are programmed and we don't believe that computers genuinely have cognitive capacities on these grounds. So, if current machines have a welfare, it must be based on non-psychological interests.

Various environmental ethicists have defended the view that non-sentient organisms have a welfare grounded in non-psychological interests. However, they typically deny that mere machines and artifacts have a similar welfare or interests. But the most prominent arguments for distinguishing artifacts from organisms in these respects fail.

4.1 The interests of non-sentient organisms

It seems obvious that there are ways to benefit or harm non-sentient organisms. Pouring acid on a maple tree is bad for it, whereas providing it with ample sunlight and water is good for it. So, there is intuitive plausibility to the idea that such beings have interests. However, proponents of views on which consciousness is a necessary condition for having a welfare have long denied that such beings have a welfare. They believe that statements about what is good for or bad for non-sentient organisms are either incoherent (Singer 2009), or that they reduce to claims about what is good for sentient organisms (Feinberg 1963). For example, they might argue that the reason that acid is bad for maple trees is that we have an preference for maple trees flourishing, and so it is bad for us if we were to pour acid on them.

In order to respond to such arguments, proponents of the welfare of non-sentient organisms must explain how these beings have genuine interests. They must explain how the interests of non-sentient organisms are non-derivative and non-arbitrary. The most prominent and promising attempt to meet the challenges of derivativeness and arbitrariness is to ground the interests of non-sentient organisms in their being goal-directed or teleologically organized.[6] Non-sentient organisms have parts and processes that are organized towards achieving certain ends such as survival and reproduction. There is a very real sense in which, independent of our desires, maple trees have parts and processes whose goal or end it is to aid in survival and reproduction in the ways characteristic of maple trees. A maple tree is defective if it fails to grow leaves, in part because it is the end of certain sub-systems to produce leaves and promote growth.

---

[6] All existing organisms will have interests of this kind, but sentient organisms will have additional interests.

Given that organisms are teleologically organized, it is possible to specify a set of interests, non-arbitrarily and non-derivatively, in light of this organization. Whatever promotes the ends of organisms is in its interest, whatever frustrates those ends undermines its interests.[7] These are most often referred to as biological interests, but I will call them teleo interests in virtue of the fact that they are grounded in the teleological organization of organisms and not, strictly speaking, their being biological.

Some might balk at the notion that plants are genuinely teleologically organized. In a world that was not designed, where organisms have evolved through a combination of chance processes and natural selection, how is it possible that organisms without minds could have ends? The answer is that natural selection grounds claims about such ends. It is perfectly legitimate to ask what makes for a properly functioning heart as opposed to a defective one. The answer to such a question is that a defective heart fails to pump blood. But, this can only be a defect if the heart has an end or purpose. And, it does; the purpose or end of the heart is to pump blood, as opposed to making rhythmic noises, because that is what it was selected for.[8] Natural selection generates teleology.

4.2 Derivative interests

While various environmental ethicists have been keen to adopt the view that natural selection grounds the teleological organization of non-sentient entities and thereby the interests of such beings, they have been adamant that artifacts do not have a welfare (Varner 1998; Taylor 1989; Goodpaster 1978). This is strange because while some may balk at thinking that organisms are teleologically organized, there is no denying that machines and most other artifacts are teleologically organized. Even if (contrary to what was just argued) natural selection cannot ground teleology, the intentional creation of an entity can. The parts of my computer have purposes and the computer itself myriad ends. Why, then, shouldn't we judge that artifacts have interests?

There are various differences between organisms and machines. The former are biological, more natural, and so on. However, none of these are morally relevant differences between the two, and none of these differences explain why teleological organization gives rise to interests in organisms but not artifacts (John Basl and Ronald Sandler Forthcoming). One difference that many have thought is morally relevant has to do with the nature of that teleological organization. The teleological organization of artifacts, it is often said, is derivative on our interests, while the interests of organisms is not derivative. Therefore, the only sense in which mere machines and artifacts have interests is derivative. Call this the Objection from Derivativeness.

The Objection from Derivativeness is mistaken. First, let's carefully distinguish two reasons we might think that the so-called welfare of mere machines is derivative. The first reason is that mere machines only exist for our use. If we had no need,

---

[7] A similar account of the interests of non-sentient organisms can be found in Varner (1998).

[8] This is a very brief summary of what is known as the etiologically account of functions (Wright, 1973; Neander, 1991; Neander, 2008; Millikan, 1989; Millikan, 1999).

desire, or use for them, mere machines would not exist. Call this Use-Derivativeness. The second reason is that the ends or the teleological organization of mere machines can only be explained by reference to the intentions or ends of conscious beings – that is, the explanation of the teleological organization in mere machines is derivative on our intentions. Call this Explanatory Derivativeness.

Being Use-Derivative is not an obstacle to being genuinely teleologically organized or to having interests. Many non-sentient organisms, from crops to pets, are use-derivative and yet they still have teleo interests. Even though my office plant exists only because I planted it to provide some decoration for my office, it would still be bad for it to be deprived of sunlight in the same way that it would be bad for a naturally occurring plant of the same species to be deprived of sunlight. The fact that mere machines exist to serve our purposes makes it such that what promotes their ends is typically the same as what promotes our ends, but this fact doesn't undermine the idea that there are things that promote the machine's ends. It is still the subject of teleo interests even if it wouldn't have those interests if not for us. There is a difference between the cause or source of a machine's ends (us) and the thing which has the ends (the machine).

The same is true concerning explanatory derivativeness. The fact that we must appeal to another's intentions to explain the teleological organization of a machine does not show that the machine is not teleologically organized, that it does not have its own ends. Were I to have a child and play an influential role in his or her life and career choice, it would matter none at all to whether a promotion benefitted the child. Even though, perhaps, you could not explain the preferences my child will have without reference to my intentions, the child still has interests of its own. The same is true of interests grounded in teleological organization. Despite the fact that you must cite a designer's intentions to explain why a machine has the ends that it does, it, the machine, still has those ends.

The account of teleo interests is the most plausible way of grounding claims about the welfare of non-sentient organisms. The Objection from Derivativeness constitutes the best objection to the claim that it is only non-sentient organisms and not machines or artifacts that have teleo interests. Given the failures of the objection, the Comparable Welfare Thesis should be accepted: If non-sentient organisms have teleo interests, mere machines have teleo interests.

This thesis does not commit us to the view that either non-sentient organism or mere machines have a welfare; nor does it commit us to the view that non-sentient organisms have a welfare if mere machines do. However, for those sympathetic to the idea that non-sentient organisms have interests that constitute a welfare, the principle commits us to expanding the domain of entities that have a welfare to machines and artifacts.

5. THE PRACTICAL IRRELEVANCE OF TELEO INTERESTS

Finally, we turn to the question of the moral relevance of teleo interests. There are good reasons to think that mere machines are not moral patients, or, more precisely, that for all practical purposes they are not moral patients – i.e. we need not worry about their teleo interests in our moral deliberations. First, most often, since we wish to use artifacts as they are intended, our use is in accordance with their interests. Using a machine to serve the purpose for which it was intended does not result in a conflict of interests.

Second, even in circumstances where our actions would frustrate a machine's teleo interests, our legitimate psychological interests always take precedence over teleo interests. To see that this is so, consider cases where a conflict arises between the teleo interests of an individual that also has psychological interests, a human being. A human's teleo interests will include the proper functioning of their organs and other biological systems, but often humans have preferences that these systems fail to work properly. For example, an individual that does not desire to have children might take steps to impair their biological capacity to do so. In such a case, there is a conflict between teleo interests, the interests associated with proper functioning of reproductive parts, and psychological interests, the attitudes and preferences regarding offspring. In this case, psychological interests take precedence and it is morally permissible to frustrate the teleo interests.

Some people attribute significant importance to reproductive capacities and so might not be convinced by this case. Besides, one might argue that the number of biological interests that would be frustrated is smaller in number than the psychological interests that would be frustrated by disallowing this biological procedure. In order to establish the priority of psychological interests, a case is needed where it is morally permissible to satisfy a legitimate psychological interest at the cost of even a very large number of teleo interests.

Consider a case involving non-sentient organisms. Imagine that biologists were convinced that there was something of importance to be learned by first growing and then ultimately destroying a very large number of trees (say 1 million). Let's imagine that it would teach us something about the origins of plant life on earth. Assuming no negative externalities, this experiment seems permissible. This is so despite the fact that a massive number of teleo interests would be frustrated, and the only immediate gain would be the satisfaction of a psychological interest we have in learning about the origins of our biological world.

This shows that legitimate psychological interests trump teleo interests even when the number of teleo interests frustrated is very large. But, in almost all cases where there will be a conflict between our psychological interests and a machine's interests, our psychological interests will be legitimate; we will be frustrating machine interests to gain knowledge about machines, to develop new ones, and to improve our welfare. For this reason, there seems to be no problem now, or in the future, with our frustrating the teleo interests of mere machines either by destroying them or otherwise causing them to function improperly. You may recycle your computers with impunity.

Some might deny my claim about the thought experiment involving the million trees. They might argue that our interest in evolutionary knowledge does not justify the destruction of 1 million trees even if there are no additional externalities. They might accuse me of begging the question in my defense of the prioritization of psychological interests over teleo interests. In response, it is worth noting how many activities, typically taken to be ethically unproblematic, might be ruled out by granting a lot of relative weight to teleo interests. In laboratories across the world, millions of bacteria are sacrificed daily on the altars of science. Many crops are destroyed by agricultural scientists in the hopes of gaining knowledge. In some cases these experiments will lead to great benefits, but in many cases they will not. But, it seems that we need not spend much, if any, time worrying about whether basic science is permissible despite costs to teleo interests.

6. CONCLUSION

The arguments of the previous section temper any worries we may have about the moral wrongs we might commit against mere machines. In the near future, no matter how complex the machines we develop, so long as they are not conscious, we may do with them largely as we please. However, things change once we develop, or think we are close to developing, artificial consciousness.

Once artificial consciousnesses exist that have the capacity for attitudes, they have psychological interests that ground their status as moral patients. We must, at that point, be careful to take their welfare into account and determine the appropriate way to do so. Therefore, given the epistemic uncertainties surrounding the creation of consciousnesses and the nature of their psychological interests, we must proceed with care as we create machines that have what we think are the functional bases of consciousness.

WORKS CITED

J. Basl and R. Sandler (forthcoming) 'Three Puzzles Regarding the Moral Status of Synthetic Organisms' in G. Kaebnick (ed.) *"Artificial Life": Synthetic Biology and the Bounds of Nature* (Cambridge, MA: MIT Press).

H. Cahen (2002) 'Against the Moral Considerability of Ecosystems' in (eds.) Andrew Light and H. Rolston III. *Environmental Ethics: An Anthology* (Malden, MA: Blackwell Publishers).

D. Chalmers (2010) 'The Singularity: A Philosophical Analysis,' *Journal of Consciousness Studies* 17 (9–10): 7–65.

D. Cruse, S. Chennu, C. Chatelle, et al. (2011) 'Bedside Detection of Awareness in the Vegetative State: A Cohort Study," *The Lancet* 378 (9809): 2088–2094.

J. Feinberg (1963) 'The Rights of Animals and Future Generations,' *Columbia Law Review* 63: 673.

F. Feldman (2004) *Pleasure and the Good Life: Concerning the Nature, Varieties and Plausibility of Hedonism* (USA: Oxford University Press).

K. Goodpaster (1978) 'On Being Morally Considerable,' *The Journal of Philosophy* 75: 308–325.

J. John (2010) 'State Neutrality and the Ethics of Human Enhancement Technologies,' *The American Journal of Bioethics* 1 (2): 41–48.

R. G. Millikan (1989) 'In Defense of Proper Functions,' *Philosophy of Science* 56 (2): 288–302.

R. G. Millikan (1999) "Wings, Spoons, Pills, and Quills: A Pluralist Theory of Function," *The Journal of Philosophy* 96 (4): 191–206.

K. Neander (1991) 'Functions as Selected Effects: The Conceptual Analyst's Defense,' *Philosophy of Science* 58 (2): 168–184.

K. Neander (2008) "The Teleological Notion of 'Function'," *Australasian Journal of Philosophy* 69 (4) (March 24): 454 – 468.

J. Rawls (1993) *Political Liberalism*, John Dewey Essays in Philosophy ; No. 4. (New York: Columbia University Press)

R. Sandler (2007) *Character and Environment: A Virtue-Oriented Approach to Environmental Ethics.* (New York: Columbia University Press).

R. Sandler and L. Simons (2012) 'The Value of Artefactual Organisms,' *Environmental Values* 21 (1): 43–61.

P. Singer (2009) *Animal Liberation: The Definitive Classic of the Animal Movement.* Reissue. (New York: Harper Perennial Modern Classics).

P. W. Taylor (1989) *Respect for Nature* Studies in Moral, Political, and Legal Philosophy. (Princeton, N.J.: Princeton University Press).

M. Tomasello and J. Call (1997) *Primate Cognition* 1st ed. (USA: Oxford University Press).

G. Varner (1998) *In Nature's Interest* (Oxford: Oxford University Press).

L. Wright (1973) 'Functions,' *Philosophical Review* 82: 139–168.

DISCUSSION QUESTIONS

1. The author argues that both religious and species membership accounts of moral status are unreasonable. Were the arguments presented for this view sound?

2. Are there any other considerations, in addition to the ones the author discussed, that could be offered in support of the view that only capacity-based accounts of moral status are reasonable?

3. Do you agree that mere sensations are not good or bad in themselves, but are only so in virtue of the attitudes of the person experiencing them? What is the basis for your view?

4. The author argues that consciousness is not necessary for having interests – i.e. that nonconscious entities such as plants and computers can have interests in virtue of their teleological organization. Did you find the arguments for this compelling? If so, do you agree that these interests are not morally relevant? Why or why not?

5. The author argues that given that we can't know whether machines have conscious interests, we don't have any obligations to take them into account. Do you agree? Is the case really analogous to the case of Jane, the coma patient, before the advent of machines that can detect her consciousness? Why or why not?

6. What kind of tests or evidence should we use to determine whether machines have conscious interests?

7. The author argued that non-sentient organisms and machines both have the same kinds of interests. Do you think that machines have interests? Are there any justifications for caring more about organisms than artifacts?

THE SINGULARITY IS NEAR[1]

Ray Kurzweil

CHAPTER SUMMARY

In this chapter, Ray Kurzweil presents and defends his view that we will reach a technological singularity in the next few decades, which he defines as a "period during which the pace of techno-logical change will be so rapid, its impact so deep, that human life will be irreversibly transformed." Kurwzweil argues that the pace of technological change, particularly with respect to information technologies, is exponential, and that we are near the "knee" of the exponential curve (i.e. the point at which the curve changes from largely horizontal to largely vertical). Kurzweil predicts that a core feature of the singularity will be the merging of biological and machine intelligence, such that the majority of "human" intelligence will become non-biological, and the merging of virtual and physical reality. Kurzweil considers this the next step in human–machine co-evolution.

RELATED READINGS

Introduction: Technology in Human Life (1); Form of Life (2.5)

Other Chapters: President's Council on Bioethics, *Beyond Therapy* (Ch. 14); Nick Bostrom, *Why I want to be a Posthuman when I Grow Up* (Ch. 15)

I am not sure when I first became aware of the Singularity. I'd have to say it was a progres-sive awakening. In the almost half century that I've immersed myself in computer and related technologies, I've sought to understand the meaning and purpose of the continual upheaval that I have witnessed at many levels. Gradually, I've become aware of a transforming event looming in the first half of the twenty-first century. Just as a black hole in space dramatically alters the patterns of matter and energy accelerating toward its event horizon, this impending Singularity in our future is increasingly trans-forming every institution and aspect of human life, from sexuality to spirituality.

What, then, is the Singularity? It's a future period during which the pace of techno-logical change will be so rapid, its impact so deep, that human life will be irreversibly transformed. Although neither utopian nor dystopian, this epoch will transform the concepts that we rely on to give meaning to our lives, from our business models to the cycle of human life, including death itself. Understanding the Singularity will alter our perspective on the significance of our past and the ramifications for our

---

[1] This chapter is excerpted from Ray Kurzweil (2005) *The Singularity is Near* (Viking). It appears here by permission of the author.

future. To truly understand it inherently changes one's view of life in general and one's own particular life. I regard someone who understands the Singularity and who has reflected on its implications for his or her own life as a "singularitarian."

I can understand why many observers do not readily embrace the obvious implications of what I have called the law of accelerating returns (the inherent acceleration of the rate of evolution, with technological evolution as a continuation of biological evolution). After all, it took me forty years to be able to see what was right in front of me, and I still cannot say that I am entirely comfortable with all of its consequences.

The key idea underlying the impending Singularity is that the pace of change of our human-created technology is accelerating and its powers are expanding at an exponential pace. Exponential growth is deceptive. It starts out almost imperceptibly and then explodes with unexpected fury – unexpected, that is, if one does not take care to follow its trajectory. (See the "Linear vs. Exponential Growth" graph below.)

Consider this parable: a lake owner wants to stay at home to tend to the lake's fish and make certain that the lake itself will not become covered with lily pads, which are said to double their number every few days. Month after month, he patiently waits, yet only tiny patches of lily pads can be discerned, and they don't seem to be expanding in any noticeable way. With the lily pads covering less than 1 percent of the lake, the owner figures that it's safe to take a vacation and leaves with his family. When he returns a few weeks later, he's shocked to discover that the entire lake has become covered with the pads, and his fish have perished. By doubling their number every few days, the last seven doublings were sufficient to extend the pads' coverage to the entire lake. (Seven doublings extended their reach 128-fold.) This is the nature of exponential growth.

Consider Gary Kasparov, who scorned the pathetic state of computer chess in 1992. Yet the relentless doubling of computer power every year enabled a computer to defeat him only five years later. The list of ways computers can now exceed human capabilities is rapidly growing. Moreover, the once narrow applications of computer intelligence are gradually broadening in one type of activity after another. For example, computers are diagnosing electrocardiograms and medical images, flying and landing airplanes, controlling the tactical decisions of automated weapons, making credit and financial decisions, and being given responsibility for many other tasks that used to require human intelligence. The performance of these systems is increasingly based on integrating multiple types of artificial intelligence (AI). But as long as there is an AI shortcoming in any such area of endeavor, skeptics will point to that area as an inherent bastion of permanent human superiority over the capabilities of our own creations.

This book will argue, however, that within several decades information-based technologies will encompass all human knowledge and proficiency, ultimately including the pattern-recognition powers, problem-solving skills, and emotional and moral intelligence of the human brain itself.

Although impressive in many respects, the brain suffers from severe limitations. We use its massive parallelism (one hundred trillion interneuronal connections operating simultaneously) to quickly recognize subtle patterns. But our thinking is extremely slow: the basic neural transactions are several million times slower than contemporary electronic circuits. That makes our physiological bandwidth for processing new information extremely limited compared to the exponential growth of the overall human knowledge base.

Our version 1.0 biological bodies are likewise frail and subject to a myriad of failure modes, not to mention the cumbersome maintenance rituals they require. While human intelligence is sometimes capable of soaring in its creativity and expressiveness, much human thought is derivative, petty, and circumscribed.

The Singularity will allow us to transcend these limitations of our biological bodies and brains. We will gain power over our fates. Our mortality will be in our own hands. We will be able to live as long as we want (a subtly different statement from saying we will live forever). We will fully understand human thinking and will vastly extend and expand its reach. By the end of this century, the nonbiological portion of our intelligence will be trillions of trillions of times more powerful than unaided human intelligence.

We are now in the early stages of this transition. The acceleration of paradigm shift (the rate at which we change fundamental technical approaches) as well as the exponential growth of the capacity of information technology are both beginning to reach the "knee of the curve," which is the stage at which an exponential trend becomes noticeable. Shortly after this stage, the trend quickly becomes explosive. Before the middle of this century, the growth rates of our technology – which will be indistinguishable from ourselves – will be so steep as to appear essentially vertical. From a strictly mathematical perspective, the growth rates will still be finite but so extreme that the changes they bring about will appear to rupture the fabric of human history. That, at least, will be the perspective of unenhanced biological humanity.

The Singularity will represent the culmination of the merger of our biological thinking and existence with our technology, resulting in a world that is still human but that transcends our biological roots. There will be no distinction, post-Singularity, between human and machine or between physical and virtual reality. If you wonder what will remain unequivocally human in such a world, it's simply this quality: ours is the species that inherently seeks to extend its physical and mental reach beyond current limitations.

Many commentators on these changes focus on what they perceive as a loss of some vital aspect of our humanity that will result from this transition. This perspective stems, however, from a misunderstanding of what our technology will become. All the machines we have met to date lack the essential subtlety of human biological qualities. Although the Singularity has many faces, its most important implication is this: our technology will match and then vastly exceed the refinement and suppleness of what we regard as the best of human traits.

THE INTUITIVE LINEAR VIEW VERSUS THE HISTORICAL EXPONENTIAL VIEW

When the first transhuman intelligence is created and launches itself into recursive self improvement, a fundamental discontinuity is likely to occur, the likes of which I can't even begin to predict.

– MICHAEL ANISSIMOV

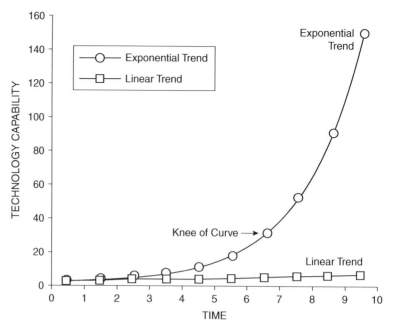

Figure 1 Linear vs. Exponential growth

Notes: Linear growth is steady; exponential growth becomes explosive

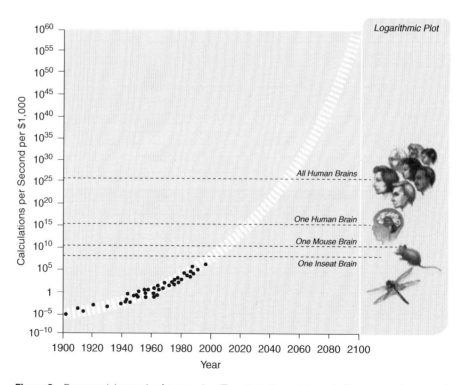

Figure 2 Exponential growth of computing (Twentieth through twenty first century)

In the 1950s John von Neumann, the legendary information theorist, was quoted as saying that "the ever-accelerating progress of technology ... gives the appearance of approaching some essential singularity in the history of the race beyond which human affairs, as we know them, could not continue."[2] Von Neumann makes two important observations here: *acceleration* and *singularity*.

The first idea is that human progress is exponential (that is, it expands by repeatedly multiplying by a constant) rather than linear (that is, expanding by repeatedly adding a constant).

The second is that exponential growth is seductive, starting out slowly and virtually unnoticeably, but beyond the knee of the curve it turns explosive and profoundly transformative. The future is widely misunderstood. Our forebears expected it to be pretty much like their present, which had been pretty much like their past. Exponential trends did exist one thousand years ago, but they were at that very early stage in which they were so flat and so slow that they looked like no trend at all. As a result, observers' expectation of an unchanged future was fulfilled. Today, we anticipate continuous technological progress and the social repercussions that follow. But the future will be far more surprising than most people realize, because few observers have truly internalized the implications of the fact that the rate of change itself is accelerating.

Most long-range forecasts of what is technically feasible in future time periods dramatically underestimate the power of future developments because they are based on what I call the "intuitive linear" view of history rather than the "historical exponential" view. My models show that we are doubling the paradigm-shift rate every decade, as I will discuss in the next chapter. Thus the twentieth century was gradually speeding up to today's rate of progress; its achievements, therefore, were equivalent to about twenty years of progress at the rate in 2000. We'll make another twenty years of progress in just fourteen years (by 2014), and then do the same again in only seven years. To express this another way, we won't experience one hundred years of technological advance in the twenty-first century; we will witness on the order of twenty thousand years of progress (again, when measured by *today's* rate of progress), or about one thousand times greater than what was achieved in the twentieth century.

Misperceptions about the shape of the future come up frequently and in a variety of contexts. As one example of many, in a recent debate in which I took part concerning the feasibility of molecular manufacturing, a Nobel Prize winning panelist dismissed safety concerns regarding nanotechnology, proclaiming that "we're not going to see self-replicating nano-engineered entities [devices constructed molecular fragment by fragment] for a hundred years." I pointed out that one hundred years was a reasonable estimate and actually matched my own appraisal of the amount of technical progress required to achieve this particular milestone when measured at *today's rate of progress* (five times the average rate of change we saw in the twentieth century). But because we're doubling the rate of progress every decade, we'll see the equivalent of a century of progress – at *today's rate* – in only twenty-five calendar years.

---

[2] Von Neumann, paraphrased in Ulam (1958).

Similarly at *Time* magazine's Future of Life conference, held in 2003 to celebrate the fiftieth anniversary of the discovery of the structure of DNA, all of the invited speakers were asked what they thought the next fifty years would be like. Virtually every presenter looked at the progress of the last fifty years and used it as a model for the next fifty years. For example, James Watson, the codiscoverer of DNA, said that in fifty years we will have drugs that will allow us to eat as much as we want without gaining weight.

I replied, "Fifty years?" We have accomplished this already in mice by blocking the fat insulin receptor gene that controls the storage of fat in the fat cells. Drugs for human use are in development now and will be in FDA tests in several years. These will be available in five to ten years, not fifty. Other projections were equally shortsighted, reflecting contemporary research priorities rather than the profound changes that the next half century will bring. Of all the thinkers at this conference, it was primarily Bill Joy and I who took account of the exponential nature of the future, although Joy and I disagree on the import of these changes, as I will discuss in chapter 8.

People intuitively assume that the current rate of progress will continue for future periods. Even for those who have been around long enough to experience how the pace of change increases over time, unexamined intuition leaves one with the impression that change occurs at the same rate that we have experienced most recently. From the mathematician's perspective, the reason for this is that an exponential curve looks like a straight line when examined for only a brief duration. As a result, even sophisticated commentators, when considering the future, typically extrapolate the current pace of change over the next ten years or one hundred years to determine their expectations. This is why I describe this way of looking at the future as the "intuitive linear" view.

But a serious assessment of the history of technology reveals that technological change is exponential. Exponential growth is a feature of any evolutionary process, of which technology is a primary example. You can examine the data in different ways, on different timescales, and for a wide variety of technologies, ranging from electronic to biological, as well as for their implications, ranging from the amount of human knowledge to the size of the economy. The acceleration of progress and growth applies to each of them. Indeed, we often find not just simple exponential growth, but "double" exponential growth, meaning that the rate of exponential growth (that is, the exponent) is itself growing exponentially (for example, see the discussion on the price-performance of computing in the next chapter).

Many scientists and engineers have what I call "scientist's pessimism." Often, they are so immersed in the difficulties and intricate details of a contemporary challenge that they fail to appreciate the ultimate long-term implications of their own work, and the larger field of work in which they operate. They likewise fail to account for the far more powerful tools they will have available with each new generation of technology.

Scientists are trained to be skeptical, to speak cautiously of current research goals, and to rarely speculate beyond the current generation of scientific pursuit. This may have been a satisfactory approach when a generation of science and technology lasted longer than a human generation, but it does not serve society's interests now that a generation of scientific and technological progress comprises only a few years.

Consider the biochemists who, in 1990, were skeptical of the goal of transcribing the entire human genome in a mere fifteen years. These scientists had just spent an entire year transcribing a mere one ten thousandth of the genome. So, even with reasonable anticipated advances, it seemed natural to them that it would take a century, if not longer, before the entire genome could be sequenced.

Or consider the skepticism expressed in the mid-1980s that the Internet would ever be a significant phenomenon, given that it then included only tens of thousands of nodes (also known as servers). In fact, the number of nodes was doubling every year, so that there were likely to be tens of millions of nodes ten years later. But this trend was not appreciated by those who struggled with state-of-the-art technology in 1985, which permitted adding only a few thousand nodes throughout the world in a single year.

The converse conceptual error occurs when certain exponential phenomena are first recognized and are applied in an overly aggressive manner without modeling the appropriate pace of growth. While exponential growth gains speed over time, it is not instantaneous. The run-up in capital values (that is, stock market prices) during the "Internet bubble" and related telecommunications bubble (1997–2000) was greatly in excess of any reasonable expectation of even exponential growth. As I demonstrate in the next chapter, the actual adoption of the Internet and e-commerce did show smooth exponential growth through both boom and bust; the overzealous expectation of growth affected only capital (stock) valuations. We have seen comparable mistakes during earlier paradigm shifts – for example, during the early railroad era (1830s), when the equivalent of the Internet boom and bust led to a frenzy of railroad expansion.

Another error that prognosticators make is to consider the transformations that will result from a single trend in today's world as if nothing else will change. A good example is the concern that radical life extension will result in overpopulation and the exhaustion of limited material resources to sustain human life, which ignores comparably radical wealth creation from nanotechnology and strong AI. For example, nanotechnology-based manufacturing devices in the 2020s will be capable of creating almost any physical product from inexpensive raw materials and information.

I emphasize the exponential-versus-linear perspective because it's the most important failure that prognosticators make in considering future trends. Most technology forecasts and forecasters ignore altogether this historical exponential view of technological progress. Indeed, almost everyone I meet has a linear view of the future. That's why people tend to overestimate what can be achieved in the short term (because we tend to leave out necessary details) but underestimate what can be achieved in the long term (because exponential growth is ignored).

THE SINGULARITY IS NEAR

The first reference to the Singularity as an event capable of rupturing the fabric of human history is John von Neumann's statement quoted above. In the 1960s, I. J. Good wrote of an "intelligence explosion" resulting from intelligent machines' designing their next generation without human intervention. Vernor Vinge, a mathematician and computer scientist at San Diego State University, wrote about a rapidly approaching "technological singularity" in an article for *Omni* magazine in 1983 and in a science-fiction novel, *Marooned in Realtime*, in 1986.

My 1989 book, *The Age of Intelligent Machines*, presented a future headed inevitably toward machines greatly exceeding human intelligence in the first half of the twenty-first century. Hans Moravec's 1988 book *Mind Children* came to a similar conclusion by analyzing the progression of robotics (Moravec, 1988). In 1993 Vinge presented a paper to a NASA-organized symposium that described the Singularity as an impending event resulting primarily from the advent of "entities with greater than human intelligence," which Vinge saw as the harbinger of a runaway phenomenon (Vinge, 1993). My 1999 book, *The Age of Spiritual Machines: When Computers Exceed Human Intelligence*, described the increasingly intimate connection between our biological intelligence and the artificial intelligence we are creating (Kurzweil, 1999). Hans Moravec's book *Robot: Mere Machine to Transcendent Mind*, also published in 1999, described the robots of the 2040s as our "evolutionary heirs," machines that will "grow from us, learn our skills, and share our goals and values, … children of our minds" (Moravec, 1999). Australian scholar Damien Broderick's 1997 and 2001 books, both titled *The Spike*, analyzed the pervasive impact of the extreme phase of technology acceleration anticipated within several decades (Broderick, 1997; Broderick, 2001). In an extensive series of writings, John Smart has described the Singularity as the inevitable result of what he calls "MEST" (matter, energy, space, and time) compression.[3]

From my perspective, the Singularity has many faces. It represents the nearly vertical phase of exponential growth that occurs when the rate is so extreme that technology appears to be expanding at infinite speed. Of course, from a mathematical perspective, there is no discontinuity, no rupture, and the growth rates remain finite, although extraordinarily large. But from our currently limited framework, this imminent event appears to be an acute and abrupt break in the continuity of progress. I emphasize the word "currently" because one of the salient implications of the Singularity will be a change in the nature of our ability to understand. We will become vastly smarter as we merge with our technology.

Can the pace of technological progress continue to speed up indefinitely? Isn't there a point at which humans are unable to think fast enough to keep up? For unenhanced humans, clearly so. But what would 1,000 scientists, each 1,000 times more intelligent than human scientists today, and each operating 1,000 times faster than contemporary humans (because the information processing in

---

[3] One of John Smart's overviews, "What Is the Singularity," can be found at http://www.KurzweilAI.net/meme/frame.html?main=/articles/art0133.html; for a collection of John Smart's writings on technology acceleration, the Singularity, and related issues, see http://www.singularitywatch.com and http://www.Accelerating.org. John Smart runs the "Accelerating Change" conference, which covers issues related to "artificial intelligence and intelligence amplification." See http://www.accelerating.org/ac2005/index.html.

their primarily nonbiological brains is faster) accomplish? One chronological year would be like a millennium for them.[4] What would they come up with?

Well, for one thing, they would come up with technology to become even more intelligent (because their intelligence is no longer of fixed capacity). They would change their own thought processes to enable them to think even faster. When scientists become a million times more intelligent and operate a million times faster, an hour would result in a century of progress (in to day's terms).

The Singularity involves the following principles, which I will document, develop, analyze, and contemplate throughout the rest of this book:

- The rate of paradigm shift (technical innovation) is accelerating, right now doubling every decade.
- The power (price-performance, speed, capacity, and bandwidth) of information technologies is growing exponentially at an even faster pace, now doubling about every year. This principle applies to a wide range of measures, including the amount of human knowledge.
- For information technologies, there is a second level of exponential growth: that is, exponential growth in the rate of exponential growth (the exponent). The reason: as a technology becomes more cost effective, more resources are deployed toward its advancement, so the rate of exponential growth increases over time. For example, the computer industry in the 1940s consisted of a handful of now historically important projects. Today total revenue in the computer industry is more than one trillion dollars, so research and development budgets are comparably higher.
- Human brain scanning is one of these exponentially improving technologies. As I will show in chapter 4, the temporal and spatial resolution and bandwidth of brain scanning are doubling each year. We are just now obtaining the tools sufficient to begin serious reverse engineering (decoding) of the human brain's principles of operation. We already have impressive models and simulations of a couple dozen of the brain's several hundred regions. Within two decades, we will have a detailed understanding of how all the regions of the human brain work.
- We will have the requisite hardware to emulate human intelligence with super-computers by the end of this decade and with personal-computer-size devices by the end of the following decade. We will have effective software models of human intelligence by the mid-2020s.
- With both the hardware and software needed to fully emulate human intelligence, we can expect computers to pass the Turing test, indicating intelligence indistinguishable from that of biological humans, by the end of the 2020s.
- When they achieve this level of development, computers will be able to combine the traditional strengths of human intelligence with the strengths of machine intelligence.

---

[4] An emulation of the human brain running on an electronic system would run much faster than our biological brains. Although human brains benefit from massive parallelism (on the order of one hundred trillion interneuronal connections, all potentially operating simultaneously), the reset time of the connections is extremely slow compared to contemporary electronics.

- The traditional strengths of human intelligence include a formidable ability to recognize patterns. The massively parallel and self-organizing nature of the human brain is an ideal architecture for recognizing patterns that are based on subtle, invariant properties. Humans are also capable of learning new knowledge by applying insights and inferring principles from experience, including information gathered through language. A key capability of human intelligence is the ability to create mental models of reality and to conduct mental "what-if" experiments by varying aspects of these models.
- The traditional strengths of machine intelligence include the ability to remember billions of facts precisely and recall them instantly.
- Another advantage of nonbiological intelligence is that once a skill is mastered by a machine, it can be performed repeatedly at high speed, at optimal accuracy, and without tiring.
- Perhaps most important, machines can share their knowledge at extremely high speed, compared to the very slow speed of human knowledge-sharing through language.
- Nonbiological intelligence will be able to download skills and knowledge from other machines, eventually also from humans.
- Machines will process and switch signals at close to the speed of light (about three hundred million meters per second), compared to about one hundred meters per second for the electrochemical signals used in biological mammalian brains. This speed ratio is at least three million to one.
- Machines will have access via the Internet to all the knowledge of our human–machine civilization and will be able to master all of this knowledge.
- Machines can pool their resources, intelligence, and memories. Two machines – or one million machines – can join together to become one and then become separate again. Multiple machines can do both at the same time: become one and separate simultaneously. Humans call this falling in love, but our biological ability to do this is fleeting and unreliable.
- The combination of these traditional strengths (the pattern-recognition ability of biological human intelligence and the speed, memory capacity and accuracy, and knowledge and skill-sharing abilities of nonbiological intelligence) will be formidable.
- Machine intelligence will have complete freedom of design and architecture (that is, they won't be constrained by biological limitations, such as the slow switching speed of our interneuronal connections or a fixed skull size) as well as consistent performance at all times.
- Once nonbiological intelligence combines the traditional strengths of both humans and machines, the nonbiological portion of our civilization's intelligence will then continue to benefit from the double exponential growth of machine price-performance, speed, and capacity.
- Once machines achieve the ability to design and engineer technology as humans do, only at far higher speeds and capacities, they will have access to their own designs (source code) and the ability to manipulate them. Humans are now accomplishing something similar through biotechnology (changing the genetic and other information processes underlying our biology), but in a much slower and far more limited way than what machines will be able to achieve by modifying their own programs.

- Biology has inherent limitations. For example, every living organism must be built from proteins that are folded from one-dimensional strings of amino acids. Protein-based mechanisms are lacking in strength and speed. We will be able to reengineer all of the organs and systems in our biological bodies and brains to be vastly more capable.

- As we will discuss in chapter 4, human intelligence does have a certain amount of plasticity (ability to change its structure), more so than had previously been understood. But the architecture of the human brain is nonetheless profoundly limited. For example, there is room for only about one hundred trillion interneuronal connections in each of our skulls. A key genetic change that allowed for the greater cognitive ability of humans compared to that of our primate ancestors was the development of a larger cerebral cortex as well as the development of increased volume of graymatter tissue in certain regions of the brain. This change occurred, however, on the very slow timescale of biological evolution and still involves an inherent limit to the brain's capacity. Machines will be able to reformulate their own designs and augment their own capacities without limit. By using nanotechnology-based designs, their capabilities will be far greater than biological brains without increased size or energy consumption.

- Machines will also benefit from using very fast three-dimensional molecular circuits. Today's electronic circuits are more than one million times faster than the electrochemical switching used in mammalian brains. Tomorrow's molecular circuits will be based on devices such as nanotubes, which are tiny cylinders of carbon atoms that measure about ten atoms across and are five hundred times smaller than today's silicon-based transistors. Since the signals have less distance to travel, they will also be able to operate at terahertz (trillions of operations per second) speeds compared to the few gigahertz (billions of operations per second) speeds of current chips.

- The rate of technological change will not be limited to human mental speeds. Machine intelligence will improve its own abilities in a feedback cycle that unaided human intelligence will not be able to follow.

- This cycle of machine intelligence's iteratively improving its own design will become faster and faster. This is in fact exactly what is predicted by the formula for continued acceleration of the rate of paradigm shift. One of the objections that has been raised to the continuation of the acceleration of paradigm shift is that it ultimately becomes much too fast for humans to follow, and so therefore, it's argued, it cannot happen. However, the shift from biological to nonbiological intelligence will enable the trend to continue.

- Along with the accelerating improvement cycle of nonbiological intelligence, nanotechnology will enable the manipulation of physical reality at the molecular level.

- Nanotechnology will enable the design of nanobots: robots designed at the molecular level, measured in microns (millionths of a meter), such as "respirocytes" (mechanical red-blood cells). Nanobots will have myriad roles within the human body, including reversing human aging (to the extent that this task will not already have been completed through biotechnology, such as genetic engineering).

- Nanobots will interact with biological neurons to vastly extend human experience by creating virtual reality from within the nervous system.
- Billions of nanobots in the capillaries of the brain will also vastly extend human intelligence.
- Once nonbiological intelligence gets a foothold in the human brain (this has already started with computerized neural implants), the machine intelligence in our brains will grow exponentially (as it has been doing all along), at least doubling in power each year. In contrast, biological intelligence is effectively of fixed capacity. Thus, the nonbiological portion of our intelligence will ultimately predominate.
- Nanobots will also enhance the environment by reversing pollution from earlier industrialization.
- Nanobots called foglets that can manipulate image and sound waves will bring the morphing qualities of virtual reality to the real world.
- The human ability to understand and respond appropriately to emotion (so-called emotional intelligence) is one of the forms of human intelligence that will be understood and mastered by future machine intelligence. Some of our emotional responses are tuned to optimize our intelligence in the context of our limited and frail biological bodies. Future machine intelligence will also have "bodies" (for example, virtual bodies in virtual reality, or projections in real reality using foglets) in order to interact with the world, but these nano-engineered bodies will be far more capable and durable than biological human bodies. Thus, some of the "emotional" responses of future machine intelligence will be redesigned to reflect their vastly enhanced physical capabilities.
- As virtual reality from within the nervous system becomes competitive with real reality in terms of resolution and believability, our experiences will increasingly take place in virtual environments.
- In virtual reality, we can be a different person both physically and emotionally. In fact, other people (such as your romantic partner) will be able to select a different body for you than you might select for yourself (and vice versa).
- The law of accelerating returns will continue until nonbiological intelligence comes close to "saturating" the matter and energy in our vicinity of the universe with our human–machine intelligence. By saturating, I mean utilizing the matter and energy patterns for computation to an optimal degree, based on our understanding of the physics of computation. As we approach this limit, the intelligence of our civilization will continue its expansion in capability by spreading outward toward the rest of the universe. The speed of this expansion will quickly achieve the maximum speed at which information can travel.
- Ultimately, the entire universe will become saturated with our intelligence. This is the destiny of the universe. We will determine our own fate rather than have it determined by the current "dumb," simple, machinelike forces that rule celestial mechanics.
- The length of time it will take the universe to become intelligent to this extent depends on whether or not the speed of light is an immutable limit. There are indications of possible subtle exceptions (or circumventions) to this limit, which, if they exist, the vast intelligence of our civilization at this future time will be able to exploit.

This, then, is the Singularity. Some would say that we cannot comprehend it, at least with our current level of understanding. For that reason, we cannot look past its event horizon and make complete sense of what lies beyond. This is one reason we call this transformation the Singularity.

I have personally found it difficult, although not impossible, to look beyond this event horizon, even after having thought about its implications for several decades. Still, my view is that, despite our profound limitations of thought, we do have sufficient powers of abstraction to make meaningful statements about the nature of life after the Singularity. Most important, the intelligence that will emerge will continue to represent the human civilization, which is already a human–machine civilization. In other words, future machines will be human, even if they are not biological. This will be the next step in evolution, the next high-level paradigm shift, the next level of indirection. Most of the intelligence of our civilization will ultimately be nonbiological. By the end of this century, it will be trillions of trillions of times more powerful than human intelligence. However, to address often-expressed concerns, this does not imply the end of biological intelligence, even if it is thrown from its perch of evolutionary superiority. Even the nonbiological forms will be derived from biological design. Our civilization will remain human – indeed, in many ways it will be more exemplary of what we regard as human than it is today, although our understanding of the term will move beyond its biological origins.

Many observers have expressed alarm at the emergence of forms of nonbiological intelligence superior to human intelligence. The potential to augment our own intelligence through intimate connection with other thinking substrates does not necessarily alleviate the concern, as some people have expressed the wish to remain "unenhanced" while at the same time keeping their place at the top of the intellectual food chain. From the perspective of biological humanity, these superhuman intelligences will appear to be our devoted servants, satisfying our needs and desires. But fulfilling the wishes of a revered biological legacy will occupy only a trivial portion of the intellectual power that the Singularity will bring.

WORKS CITED

D. Broderick (1997) *The Spike: Accelerating into the Unimaginable Future* (Sydney, Australia: Reed Books).

D. Broderick (2001) *The Spike: How Our Lives Are Being Transformed by Rapidly Advancing Technologies*, rev. ed. (New York: Tor/Forge).

R. Kurzweil (1999) *The Age of Spiritual Machines: When Computers Exceed Human Intelligence* (New York: Viking).

H. Moravec (1988) *Mind Children: The Future of Robot and Human Intelligence* (Cambridge, MA: Harvard University Press).

H. Moravec (1999) *Robot: Mere Machine to Transcendent Mind* (New York: Oxford University Press).

S. Ulam (1958) 'Tribute to John von Neumann,' *Bulletin of the American Mathematical Society* 64.3, pt. 2 May: 1–49.

V. Vinge, (1993) 'The Coming Technological Singularity: How to Survive in the Post-Human Era,' VISION-21 Symposium, sponsored by the NASA Lewis

Research Center and the Ohio Aerospace Institute, March (available at http://www.KurzweiW.net/vingesing).

DISCUSSION QUESTIONS

1. What is the "law of accelerating returns" and what role does it play in Kurzweil's argument that a technological singularity is near?

2. Do you find Kurzweil's reasoning in support of the coming of the singularity to be compelling? Is there any reason to think that either the technology will not develop at the pace he proposes or that, even if it does, it would not have the sort of transformative impacts he predicts?

3. Kurzweil claims that the singularity is neither utopian nor dystopian. What do you think he means by that? Do you think it would be utopian or dystopian, and why?

4. Kurzweil emphasizes the merging of human and machine intelligence. In what ways, if any, are human and machine intelligence already linked and co-evolving?

5. Kurzweil asserts that we would remain human through the singularity, but that the meaning of the term "human" would change. Do you agree that if the singularity occurs we would remain human? In what senses would we be human, and in what senses would we not be?

6. It is not possible to know what it would be like to have one's biological intelligence augmented by machine intelligence in the way Kurzweil predicts. Given this, what reasons might we give for or against undergoing it? Is the singularity something that you would want to be part of? Why or why not?

PART VII

ENVIRONMENT AND TECHNOLOGY

Technology has had enormous impacts on the natural environment, particularly since the industrial revolution. It has enabled us to intentionally transform ecological systems – for example, by filling in wetlands for development, clearing grasslands and forests for agriculture, and introducing and eliminating species. It has enabled us to consume natural resources far faster than they can be replenished – for example, fresh water aquifers, fish stocks, timber, and topsoil. It has led to widespread degradation of ecological systems – for example, agricultural runoff, plastic waste, and chemical pollution. It has resulted in increased concentrations of greenhouse gases in the atmosphere that alter global climatic processes. It has enabled a rapid increase in the global human population, from ~1.5 billion people in 1900 to over 7 billion people today. Human influence on the Earth is now so great that some people argue that we have entered a new, human-dominated geological age: the anthropocene.

Technology has also dramatically altered our relationship with nature. We have different vulnerabilities and dependencies as a result of technological innovations – for example, we are not as at risk for many diseases as we once were, but we have become highly reliant on vast, inexpensive sources of energy. We interact less regularly with natural environments – the majority of the global population now lives in urban areas. And we interact differently with nature when we are in it – for example, by means of cameras, binoculars, GPS, and other gadgetry. Technology has also changed how nature is represented to us – for example, in media and nature programs. Moreover, the applications of science and technology to studying the natural environment have dramatically expanded our understanding of ecological systems and nonhuman species.

Technology also is often developed and deployed to address ecological challenges – for example, to assess ecological risks, increase agricultural yields, remediate pollution, protect species, or manage waste. The chapters in this section focus on technologies and technology-related practices that are intended to be ecologically beneficial. In "Risk, Precaution, and Nanotechnology," Kevin C. Elliott addresses ongoing debates regarding how best to assess and manage technological risk, including whether to employ a precautionary principle. He then discusses the implications of these debates for managing the risks associated with rapidly proliferating nanomaterials.

In "Avoiding Catastrophic Climate Change," Philip Cafaro discusses the role of technological innovation in mitigating greenhouse gas emissions in order to reduce the magnitude of global climate change. He argues that while existing and emerging technologies have a large and crucial role to play in mitigation, they are not themselves sufficient to accomplish climate safety. Therefore, they must be supplemented by substantial social and behavioral changes.

In "Ethical Anxieties about Geoengineering," Clive Hamilton provides an ethical analysis and evaluation of the use of technology to intentionally manage the Earth's climate in response to global climate change. He argues that proposed geoengineering strategies, such as solar radiation management and carbon sequestration, are both intrinsically and extrinsically problematic – i.e. they are problematic both in virtue of the nature of the activity and their possible outcomes. He believes that responses to global climate change need to be primarily political, rather than techno-focused.

In "Ecosystems Unbound," Ben A. Minteer and James P. Collins discuss the challenges that human ecological impacts and rapid ecological change pose for traditional park and reserve (or wilderness oriented) ecosystem management and species conservation practices. They then address the ethical issues raised by two novel, and more interventionist, approaches – managed relocation and Pleistocene rewilding. These alternative approaches are controversial among conservation biologists because they involve large amounts of human design and manipulation of ecological spaces. However, Minteer and Collins argue, we have no choice but to support them if we are going to prevent species from going extinct in our highly dynamic ecological world.

More detailed summaries, as well as a listing of related readings, are located at the start of the chapters.

27 RISK, PRECAUTION, AND NANOTECHNOLOGY

Kevin C. Elliott

CHAPTER SUMMARY

One of the primary concerns regarding emerging technologies is the ecological and human health impacts that engineered chemicals and materials can have when released into the environment. In this chapter, Kevin C. Elliott discusses the ethics of risk and risk management of novel materials, with a particular focus on engineered nanomaterials. Central to risk management of novel materials is the lack of information and uncertainties about them. The precautionary principle is one approach to responding to these, and this chapter includes an extensive critical discussion of the precautionary approach. Elliott advocates for a policy response to the risks posed by novel nanomaterials that is designed to bring about tolerable results under a very wide range of possible scenarios, rather than a response that aims to bring about the best results under a narrow range of scenarios.

RELATED READINGS

Introduction: EHS (2.6.1); Responsible Development (2.8)

Other Chapters: Michele Garfinkle and Lori Knowles, *Synthetic Biology, Biosecurity, and Biosafety* (Ch. 35); Mark A. Bedau and Mark Triant, *Social and Ethical Implications of Creating Artificial Cells* (Ch. 37)

1. INTRODUCTION

One of the central ethical issues associated with the introduction of new technologies is their potential to create human and environmental health risks. For example, genetically modified crops have the potential to diminish biodiversity, new industrial chemicals may have a wide range of previously unexpected developmental and neurological effects in humans and wildlife, and efforts to geoengineer earth's climate could have serious impacts on local weather conditions. In an effort to respond to these sorts of challenges, government agencies in the United States and throughout the industrialized world have created sophisticated procedures for assessing and managing risks. Nevertheless, concerned citizens have raised serious questions about the effectiveness of these strategies. The adequacy of current risk-assessment procedures is particularly questionable in the face of unpredictable emerging technologies such as genetic modification, synthetic biology, nanotechnology, and geoengineering.

This chapter explores these issues by reflecting on the risks posed by the development of nanotechnology. The first section provides an introduction to current debates about how best to assess and manage technological risks. The second section shows how these debates apply to the case of nanotechnology, and the third section provides some suggestions for addressing nanotechnology risks in an ethical fashion. While there are few easy solutions to handling the risks posed by emerging technologies, this chapter argues that we would do well to pursue robust strategies that produce tolerable results under a wide range of future scenarios. In the case of nanotechnology, these strategies include deliberately engineering new materials that are likely to be non-toxic, incentivizing the use of relatively non-toxic materials, minimizing exposure to worrisome chemicals, speeding up toxicity testing, developing a thoughtful pre-market testing scheme that minimizes the effects of financial conflicts of interest, and creating a monitoring plan to catch problems as they arise.

2. RISK AND PRECAUTION

2.1 Risk assessment

The term 'risk' is used and defined in a variety of ways. For example, it can refer to an unwanted event, to the cause of the unwanted event, or to the probability of the event. It is perhaps most common in the literature on risk to define risk as the product of the probability of an event times a measure of its disvalue (Hansson 2004). For example, the risk of an oil spill at a particular facility would be the product of the probability that a spill will occur times the detrimental impacts (or costs) of its occurring. In response to concerns about risks from nuclear power and toxic chemicals in the 1970s, federal agencies in the United States developed a scheme for addressing risk that was codified in a document commonly called the "Red Book" (NRC 1983). The Red Book drew a distinction between assessing risks and managing them. In the case of toxic chemicals, for example, risk assessment consists of identifying potentially hazardous chemicals, determining the chemicals' dose-response relationships (i.e., the biological effects of the chemicals at particular doses), calculating the concentrations of the chemicals to which people are exposed, and then synthesizing this information to estimate the probability that people will be killed or injured by them.

Whereas risk assessment is supposed to be a narrowly scientific endeavor on this scheme, the process of risk management incorporates economic, political, and ethical considerations about what levels of risk society should tolerate. For example, risk managers might argue on ethical grounds that citizens should not be exposed to levels of toxic chemicals that raise their probability of dying above a particular level. Or, they might insist that industrial facilities should use the "best available technology" to lessen public exposure to toxicants. They might also engage in some form of cost–benefit analysis to estimate the overall economic costs and benefits of regulating chemicals at particular levels.

While the codification of risk assessment and risk management techniques has undoubtedly been helpful to some extent, concerned citizens have still raised numerous concerns about the adequacy of these approaches. One general worry is that the focus among regulatory agencies, policy makers, and the public on assessing and managing risks creates overly narrow societal discussions about new technologies. In other words, decision makers become accustomed to the idea that the only significant question about technologies is whether their human and environmental health risks are sufficiently small. But citizens may have a wide variety of other legitimate concerns about emerging technologies. For example, many of those who oppose genetically modified (GM) crops are motivated not just by concerns about health risks but also by worries about the costs of GM seeds for poor farmers, the power of corporations over our seed supply, the loss of traditional food cultures, the expansion of industrialized approaches to agriculture, and the notion that we are transgressing natural boundaries or "playing God."

Another worry concerns the process of risk assessment itself. Some commentators point out that the supposedly scientific process of risk assessment cannot be easily insulated from the ethical and political concerns associated with risk management. For example, the empirical data available to risk assessors typically involve the effects of relatively high doses of toxic chemicals on animals such as rats or mice. Risk assessors then have to make difficult decisions about what models to use for estimating the effects of these chemicals on humans at much lower doses. They have to make further estimates about how these chemicals will affect particularly sensitive humans (e.g., children or pregnant women). Because the scientific evidence does not decisively determine what models to use for making these estimates, the risk-management context (including ethical, political, and economic concerns) is arguably relevant to making these decisions. On the one hand, if risk assessors decide that it is particularly important to protect public health, they may estimate health risks using models that are more likely to overestimate than to underestimate the risks. On the other hand, if risk assessors conclude that it is more important to promote the economic interests of regulated industries, they may instead choose models that are more likely to underestimate the risks. Therefore, critics contend that it is naïve to think that the results of risk assessments can be regarded as uncontroversial scientific information that everyone can agree on. For example, in recent years many scientists contended that bisphenol A (BPA, a substance found in many can liners, plastics, and receipts) could plausibly be causing a number of adverse effects in humans. Nevertheless, the major government regulatory agencies in Europe and the U.S. continued to insist that BPA appeared to be safe (Myers et al. 2009).

Related to this worry (i.e., that the ethical, political, and economic considerations associated with risk management frequently and perhaps unavoidably affect risk assessment) is the concern that regulated companies fund many of the scientific studies that inform the risk assessments of their products. Therefore, they are often able to design and interpret these studies in ways that serve their economic interests (McGarity and Wagner 2008). For example, the industry-funded studies that guided U.S. and European regulations of BPA during the first decade of the twenty-first century invariably showed that BPA had no harmful effects, whereas the majority of independent studies funded by sources such as the U.S. National Institutes of Health found evidence suggesting that BPA could be harmful (vom Saal and Hughes 2005; Myers et al. 2009). By producing studies that create doubt about the harmfulness of their products and by using the court system to challenge attempts at regulation, chemical manufacturers are able to grind the processes of risk assessment and management at institutions like the U.S. Environmental Protection Agency (EPA) to a virtual standstill (Cranor 2011).

Another criticism of contemporary risk assessment is that it may promote a false sense of security about the safety of new technologies while drawing attention away from significant forms of uncertainty. For example, risk analysts frequently make a distinction between decision making under risk and decision making under uncertainty. Decisions under risk occur when it is possible to assign probabilities to the possible consequences associated with a decision. In contrast, decisions under uncertainty occur when there is insufficient information to assign probabilities to various potential outcomes. Some analysts also refer to situations of "great uncertainty," in which even the range of possible consequences are unknown (Hansson 1996). Critics argue that expert risk assessors frequently have insufficient evidence to justify the probabilities that they assign to potential outcomes, and the experts may simply ignore the possibility of unforeseen consequences. The result is that risk assessments may yield quantitative estimates of potential harm that appear to be very precise but that may be seriously misleading (Elliott and Dickson 2011). For example, when the famous WASH-1400 risk assessment of nuclear reactor safety came under scrutiny, analysts found that experts were sometimes widely off the mark in the probability estimates that they provided for various sorts of accidents (Cooke 1991, p. 36).

2.2 The precautionary principle

In response to the range of criticisms that have been leveled against traditional processes of risk assessment and risk management, some critics have suggested that a more ethical approach for responding to the hazards associated with new technologies is embodied in the precautionary principle (PP). This principle began to receive sustained attention during the 1990s, especially after a version of it was formulated in the 1992 Rio Declaration of the United Nations Conference on Environment and Development: "Where there are threats of serious or irreversible damage, lack of full scientific certainty shall not be used as a reason for postponing cost-effective measures to prevent environmental degradation" (see e.g., Sunstein 2005, 18). Another common formulation is the Wingspread Statement on the Precautionary Principle: "When an activity raises threats of harm to human health or the environment, precautionary measures should be taken even if some cause-and-effect relationships are not fully established scientifically" (Raffensperger and Tickner 1999, 353–354). The core of these statements is the idea that precautionary actions can be taken to prevent harms even if the scientific evidence concerning those harms is incomplete. For example, nongovernmental organizations such as the ETC Group and Greenpeace have appealed to the precautionary principle as a basis for demanding severe limitations on the use of nanotechnology and genetically modified crops.[1]

One of the strengths of the PP is that it can support quicker responses to potentially hazardous activities. Rather than getting bogged down in a lengthy risk-assessment process that can be manipulated by special interest groups, the PP appears to justify actions to ban or restrict potentially hazardous activities immediately. Also, because it explicitly states that scientific uncertainty is often present, it eases the pressure to arrive at precise quantitative estimates of risk. Furthermore, many proponents of the precautionary principle argue that it supports public participation in decisions about how to respond to hazards, whereas traditional approaches to risk assessment and management left decisions in the hands of technical experts. Thus, the PP has the potential to promote broad deliberation about a range of ethical and social concerns about new technologies rather than channeling the decision-making process into a narrow focus on human and environmental health risks (see e.g., Raffensperger and Ticker 1999).

Despite these apparent strengths of the PP, critics have argued that it is deeply flawed. A common complaint is that it is so vague that it can be formulated in a multitude of different ways. For example, in order to implement the PP, one must specify which sorts of threats are serious enough to justify precautionary actions and how much evidence about those threats is necessary. One must also determine what sorts of actions are justified in response to particular sorts of threats (Manson 2002; Sandin 1999). In the case of genetically modified (GM) crops, for instance, policy makers and concerned citizens have to determine whether specific threats to the environment are sufficient to justify precautionary actions or whether there must be threats to human health. And even if they do call for precautionary measures, they must consider (based on the limited evidence available – including very few studies performed by scientists unaffiliated with the biotech industry) whether to call for an all-out ban on particular GM crops or whether to settle for some form of labeling or other regulatory scheme. Different proponents of the PP can arrive at varying conclusions about these matters.

Critics frequently pose a further dilemma: depending on how one answers these detailed questions about the PP, it becomes either trivial and uninformative or hopelessly restrictive and even paralyzing (see e.g., Sunstein 2005). On one hand, if one interprets the PP in a "weak" way, so that it says that inexpensive cautionary actions should be taken to lessen the severity of serious human and environmental health threats, then nobody would disagree with it. However, when it is interpreted in this way, the PP just appears to be a matter of common sense; it does not provide serious guidance about what actions to take in the difficult or contested cases where guidance is needed. On the other hand, if one interprets the PP in a "strong" way, so that it says that all potentially hazardous activities should be banned, then it appears to be untenable, because almost any action poses at least some threat of harm. For example, one might appeal to this strong

---

[1] See e.g., Greenpeace's statement on genetically modified crops at http://archive.greenpeace.org/geneng/reports/bio/bio011.htm (accessed on 5/14/12) and the ETC Group's statement on nanotechnology at http://www.etcgroup.org/upload/publication/171/01/thebigdown.pdf (accessed on 5/14/12).

formulation of the PP to ban the use of GM crops (because of concerns about potential human or environmental health threats), but one might also appeal to this form of the PP to ban the use of traditional agricultural methods (because of concerns that they will yield insufficient food production and result in mass starvation). Thus, strong formulations of the PP apparently leave decision makers unable to do anything (Sunstein 2005).

Proponents of the PP typically respond that the critics have formulated a false dichotomy; the proponents insist that the PP can be formulated in meaningful ways that are not overly restrictive. For example, they frequently associate the PP with the idea that the burden of proof in regulatory contexts should be shifted away from consumers and onto producers. For example, in the United States, the manufacturers of pharmaceuticals and pesticides currently bear the burden of showing that their products are safe before they can market them, but the manufacturers of most other chemicals (which are regulated under the U.S. Toxic Substances Control Act) can legally put their products on the market without providing evidence that they are non-toxic (Cranor 2011). As a result, the U.S. Environmental Protection Agency has data regarding the health effects of only 15% of the roughly 50,000 new chemical substances that have been introduced into commerce since 1979 (Cranor 2011, p. 6). Therefore, consumer groups and the EPA bear the significant burden of providing enough evidence to remove harmful chemicals from the market. Many proponents of the PP argue that the United States should instead require testing of all chemicals before marketing them, much like the European Union requires under its Registration, Evaluation, Authorisation, and Restriction of Chemicals (REACH) legislation. Carl Cranor (2011) argues that this sort of "pre-market" regulatory scheme (in which manufacturers must establish the safety of their products before putting them on the market) would be more consistent with widely accepted medical ethics guidelines, which require that appropriate prior research be performed before exposing people (and especially children!) to potentially harmful substances.

The PP has also been associated with a range of other ideas. As mentioned earlier, many proponents of the PP argue that it calls for shifting the way decisions about technological hazards are made. Rather than leaving these decisions up to expert risk assessors and managers, they suggest that regulatory agencies should experiment with novel forums that facilitate public deliberation about how to respond to new technologies. For example, a number of countries have experimented with various sorts of citizen panels such as "consensus conferences," in which small groups of citizens are educated about a scientific or technological issue and then prepare a report on the key concerns that policy makers ought to consider or address. Those who associate the PP with these sorts of inclusive approaches to developing public policy insist that the PP's vagueness is actually a strength; various communities can deliberate about how they want to specify it, given their tolerance for technological risks (von Schomberg 2006). Other proponents of the PP have suggested that it calls for more carefully identifying the range of alternatives to potentially harmful activities. Still others claim that the PP requires developing "robust" policies. Policies are robust if they produce satisfactory results across a very broad range of potential future scenarios. To see how some of these proposals could be implemented in practice, let us consider how policy makers are approaching the risks associated with nanotechnology.

3. NANOTECHNOLOGY RISKS

Nanotechnology is often described not as one specific technology but rather as a technological platform that can facilitate many different technological advances. The U.S. National Nanotechnology Initiative defines nanotechnology as "science, engineering, and technology conducted at the nanoscale, which is about 1 to 100 nanometers" (NNI 2011). A nanometer is one billionth of a meter, which is roughly the size of 10 hydrogen atoms lined up next to each other. By producing custom-designed particles in this very small size range, scientists hope to manipulate their properties in valuable ways. For example, chemists have developed new materials by configuring collections of carbon atoms into spheres (fullerenes) and hollow tubes (nanotubes). These carbon nanotubes and fullerenes have novel properties that show promise for producing stronger, lighter materials and better electronics. Thus, many aspects of nanotechnology are a fairly natural extension of previous work in materials science and chemistry. However, some scientists have proposed much more speculative applications of nanotechnology, such as the creation of tiny molecular-scale robots that could revolutionize the domains of manufacturing, medicine, and military technology.

As with most emerging technologies, the potential advances associated with nanotechnology are accompanied by a range of potential threats. Bill Joy, the former Chief Scientist of Sun Microsystems, wrote a famous essay in 2000, warning that nanoscale robots designed to replicate themselves and perform manufacturing tasks could instead run amok and manipulate the entire surface of the earth, turning it into a "gray goo" (Joy 2000). Others have worried that nanoscale sensors could create new threats to privacy or that nanotechnology could create novel military threats. At present, the most common fears about nanotechnology typically concern the potential for nanomaterials to pose unexpected environmental or human health hazards. The same property that makes nanotechnology so exciting (namely, the potential to create new properties by manipulating materials at the molecular scale) also raises concerns insofar as nanomaterials may display new and unexpected toxic properties in addition to their intended characteristics. For example, commentators worry that nano-scale particles of zinc and titanium oxide (used in sunscreens and cosmetics) and silver (used in numerous antimicrobial products) could be much more toxic than larger scale particles of the same substances (Friends of the Earth 2006; Royal Society 2004).

One property that makes nanomaterials worrisome is that very small particles have an exceptionally high surface area relative to their mass. This high surface area increases their reactivity, which can be a valuable property in some contexts but which can also result in increased toxicity. Because of their small size, there is also some evidence that nanoparticles can translocate from one part of the body to another more easily than other particles. For example, they may have increased potential to pass through the skin and the blood–brain barrier (Maynard et al. 2011, S117-S118). Furthermore, because nanoparticles are comparable in size to biologically important molecules like proteins and DNA, they may have more potential to interfere with normal metabolic processes or to damage cellular structures (Maynard et al. 2011, S122).

Unfortunately, nanotechnology vividly illustrates the difficulties associated with performing risk assessments of emerging technologies. First, while it is sometimes possible to predict the toxicity of nanoparticles based on the toxicity of larger particles of the same chemical substances, this is not always feasible. Instead, a wide range of other characteristics (including size, crystal structure, surface charge, and preparation process) may affect nanoparticle toxicity, but we do not presently have enough understanding to predict toxicity in advance based on these characteristics (Oberdörster et al. 2005; Maynard et al. 2011, S112). Risk assessments of nanoparticles are further complicated by the fact that various toxicity studies have produced conflicting results. Some studies indicate that nanoparticles are more toxic when functionalized (i.e., when molecular structures are added to them), whereas others indicate that functionalizing them makes them less toxic (Magrez et al. 2006; Sayes et al. 2006). Some studies indicate that they are more toxic when purified, whereas others indicate the opposite (Templeton et al. 2006; Tian et al. 2006). Finally, new evidence suggests that nanoparticle toxicity is influenced by the specific "corona" of proteins and other biomolecules that forms around the particles (Maynard et al. 2011, S118). This makes toxicity even more difficult to predict. Thus, while it is appealing to try to calculate the risks associated with new nanotechnologies, policy makers are, in effect, frequently forced to make decisions under uncertainty about their effects (Elliott and Dickson 2011).

A significant ethical question is how to respond to this uncertainty. One approach is to focus primary attention on avoiding any potential harms from this or other new technologies. For example, Bill Joy (2000) argued (in accordance with strong versions of the PP) that nanotechnology should not be pursued, because it has too much potential to result in catastrophes. Similarly, the nongovernmental ETC group is well known for arguing that a moratorium should be placed on at least some areas of nanotechnology research (ETC 2011). A number of other reports have argued that particular nanoparticles should at least undergo careful testing to identify hazards before they are introduced into consumer products (Friends of the Earth 2006; Royal Society 2004).

Nevertheless, many of these suggestions are controversial. Bill Joy's call for relinquishing nanotechnology is arguably too broad. Nanotechnology displays a host of variations, including many different types of materials, as well as applications ranging from improved electronics to stronger materials to better antibacterial products, and so on. Therefore, it is likely to give rise to at least some developments that are not of significant ethical concern. Moreover, many areas of nanotechnology could yield not only harms but also a range of benefits: economic development, environmental remediation, improved energy technologies, and better consumer products. Therefore, one might argue that, rather than placing their focus primarily on preventing harm, decision makers should comprehensively consider both the costs and benefits when deciding how to regulate nanotechnology. For example, policy makers can perform cost–benefit analyses (CBAs), which represent potential costs and benefits in monetary terms and then weight these consequences based on quantitative estimates of their probability.

Nevertheless, CBA has been criticized both because of the difficulty of representing all consequences in monetary terms and because, as we have seen, it is very difficult to develop credible quantitative estimates of risks when dealing with a new field like nanotechnology. Thus, those who develop regulations for emerging fields seem to be caught between two unappealing options. On the one hand, traditional approaches to risk assessment and management can hide important sources of uncertainty and leave decisions in the hands of technical experts who are often beholden to special interest groups. On the other hand, the precautionary principle runs the risk of being either uninformative or of calling for decisions that block promising innovations. The next section proposes some strategies that can help to address this challenge in the context of nanotechnology and that may provide models for responding to the risks associated with other emerging technologies.

4. RESPONDING TO NANOTECHNOLOGY RISKS

We have seen that traditional risk assessment and at least some formulations of the PP face serious problems when applied to emerging fields such as nanotechnology. While there are no easy answers to these difficulties, a number of strategies may be helpful to varying extents.

One strategy for responding to potential hazards associated with all chemicals, including nanomaterials, is to develop new substances that are likely to be less toxic and then encourage companies to use the less worrisome products. This strategy has already been explored as a response to the tendency for companies to bog down risk assessment processes with endless wrangling over the precise toxicity of specific chemicals. The state of Massachusetts developed a very clever response. The state's Toxic Use Reduction Act (TURA) of 1989 established a list of chemicals that were suspected to be problematic. But rather than fighting over how much to restrict their use, the law merely required companies to publicly report how much of those chemicals they were using and to explore whether there were feasible alternatives. Once companies were forced to consider alternatives, they ended up voluntarily reducing their use of the worrisome chemicals to a significant extent and still saved money in the process (Raffensperger and Tickner 1999). The TURA also created a Toxic Use Reduction Institute (TURI) at the University of Massachusetts, Lowell, with the goal of designing new, safer chemicals that could replace more toxic ones. In the case of nanotechnology, some scientists have advocated a strategy of "safety by design" (Colvin 2003). Their goal, like that of the TURI, is to find ways to develop new nanomaterials with the explicit goal of making them as non-toxic as possible.

Nevertheless, a difficulty in the case of nanotechnology is that, as this chapter has emphasized, toxicologists still have limited understanding of the range of characteristics that contribute to the toxicity of nanomaterials. Therefore, "safety by design" is more a dream than a reality at this point, so other strategies need to be pursued in the meantime. In cases where reasonably safe traditional chemicals are satisfactory, this may mean that companies should be incentivized not to use new nanomaterials that are poorly understood. In other contexts, where nanomaterials do appear to display truly valuable new properties, other strategies will need to be pursued. One possibility is to minimize exposure to potentially hazardous nanomaterials. For example, many carbon nanotubes are similar in shape to asbestos fibers, and some studies indicate that they can produce similar toxic effects (Poland et al. 2008). Therefore, even though the available information is currently limited, it would be wise to limit worker exposure to nanotubes during manufacturing processes and to design consumer products in ways that minimize the release of nanotubes into the environment.

It would also be prudent to require "pre-market" toxicity testing of most nanomaterials, both in order to collect information and to make it more difficult for companies to introduce potentially harmful new products onto the market unnecessarily. As discussed previously, the U.S. Toxic Substances Control Act does not require most chemicals to be tested for safety before they are marketed. But there are significant ethical concerns with a policy that allows manufacturers to expose the public to products without taking reasonable care to ensure that they are not harmful (Cranor 2011). If testing were hopelessly expensive, then there might be some justification for failing to do so. However, the U.S. EPA is currently experimenting with new "high-throughput" methods for testing chemicals very quickly and developing rough predictions of their potential to be toxic. These "screening" studies can then be followed up with more extensive toxicity tests when necessary. As these methods become more sophisticated, it appears less and less likely that the economic costs of requiring pre-market testing of nanomaterials will be sufficient to override the ethical and economic costs of putting harmful chemicals on the market (see e.g., Cranor 2011).

However, even if a "pre-market" testing policy for nanomaterials were to be implemented, at least two important issues would remain. First, policy makers would have to decide precisely which particles merit pre-market testing and how much evidence of toxicity to demand before regulating them. One might think that this issue could be handled relatively easily by insisting on testing all nanoparticles and requiring that they be shown to be non-toxic. But the situation is not so easy. For example, one faces the question of how to define a nanoparticle. Most common definitions of nanotechnology refer to materials with at least one dimension under 100 nanometers, but this is obviously not a hard-and-fast criterion. For example, some scientists have argued that most particles do not display new toxic properties until they are less than 30 nanometers in diameter (Aufann et al. 2009), and some particles could display novel toxic properties at sizes greater than 100 nanometers. Thus, depending on how careful regulators wanted to be, they could set the threshold for requiring testing at different sizes. Similarly, they could demand different levels of evidence for toxicity depending on their tolerance for over- or under-regulation. Therefore, developing a reasonable pre-market testing scheme will require both ongoing scientific input as well as broadly based deliberation among a wide range of stakeholders to decide how stringent the testing system should be.

A second issue for those looking to develop a pre-market testing scheme for nanoparticles is how to deal with the problem of financial conflicts of interest. If the companies who manufacture nanomaterials also perform the toxicity studies (or pay "contract research organizations" to perform the studies for them), it raises significant questions about the credibility of the results (Elliott 2011, p. 83; McGarity and Wagner 2008). Some thinkers have suggested that this problem could be solved both by creating standardized guidelines for toxicity studies and by creating a committee to "vet" the results (Ramachandran et al. 2011). Unfortunately, standardized guidelines have not been sufficient in the past to prevent those with conflicts of interest from influencing the design and interpretation of toxicity studies (Elliott and Volz 2012).

Therefore, it would be much more effective if the manufacturers of nano-materials were required to give a government agency like the U.S. Food and Drug Administration (FDA) or the U.S. Environmental Protection Agency (EPA) the money that they would have spent on late-stage toxicity testing and instead let the government agency contract the studies out to academics or contract research organizations. Because of its potential to mitigate serious conflicts of interest, this approach would make more sense than the current system, but the new approach is likely to face fierce opposition from chemical manufacturers that do not want to lose their control over toxicity studies. Therefore, if something like the current system remains in place for testing nanomaterials, it will be important to create a thorough procedure for vetting study results, and it will also be important to provide extensive government funding for academic researchers to conduct independent studies of new nanomaterials (Elliott and Volz 2012).

A final strategy for addressing nanotechnology risks is to develop a moni-toring scheme in an effort to identify problems that arise as quickly as possible. A number of commentators have argued that the introduction of emerging tech-nologies is essentially a "real-world experiment" in which we test the impacts of these technologies as we go. Unfortunately, we often do not learn as much from these "experiments" as we could, because we fail to set up effective monitoring plans for observing the results of our actions. In the case of nanotechnology, for example, it might be helpful to start collecting medical data for workers who manufacture nanomaterials so that adverse health impacts can be identified as quickly as possible. Monitoring schemes could help to identify threats as quickly as possible if they begin to occur.

In sum, a reasonable approach for responding to nanotechnology risks is likely to incorporate a variety of strategies: deliberately engineering new mate-rials that are likely to be non-toxic, incentivizing the use of relatively non-toxic materials, minimizing exposure to worrisome chemicals, speeding up toxicity testing, developing a thoughtful pre-market testing scheme that minimizes the effects of financial conflicts of interest, and creating a monitoring plan to catch problems as they arise. A common thread running through these strategies is that they are likely to provide a *robust* policy response to nanotechnology risks. Robust policies are not designed with the aim of getting the best possible results under the assumption that the future turns out to be exactly the way we expect; instead, they are designed to yield *satisfactory* results over a very wide range of possible future scenarios (Mitchell 2009). If most nanomaterials display the toxicities that we expect, it would be cheaper not to take extra steps to minimize exposure to them or to implement monitoring schemes. However, minimizing exposures and creating monitoring schemes are more robust policies, because they yield tolerable results whether most nanomaterials turn out to be toxic or not. Because it is difficult to predict the probability of particular future scenarios when dealing with emerging fields such as nanotechnology, there is much to be said for pursuing these sorts of robust policies.

5. CONCLUSION

This chapter has provided an introduction to the concepts of risk and precaution in the context of emerging technologies. We have seen that even though risk assessment and risk management are powerful tools for addressing human and environmental health hazards, they also have weaknesses. In particular, they can promote overly narrow societal discussions about emerging technologies, they can be manipulated by interest groups and employed as a strategic tool to slow down the regulatory process, and they can hide important forms of uncertainty. The precautionary principle (PP) was introduced as a solution to these difficulties, but it has faced criticism for being overly vague. "Weak" versions of the PP are allegedly vacuous, whereas "strong" versions of the PP turn out to be paralyzing because they call for banning almost all activities.

The introduction of nanotechnology provides a good case study for thinking further about these issues. Many features of nanomaterials raise concerns that they could be toxic. However, it is very difficult to develop precise quantitative estimates of this toxicity, in part because the characteristics of nanomaterials can be varied in so many ways and in part because we have little understanding of how particular characteristics influence nanomaterial toxicity. Faced with this difficulty, some commentators have argued for banning nanotechnology or enacting strict regulations on its development. However, this approach runs the risk of blocking promising innovations. Thus, regulators face serious difficulties: traditional approaches to risk assessment and management are hampered by the presence of particularly serious scientific uncertainty, but precautionary approaches may not be ideal either – and national governments have resisted enacting strict regulations on nanotechnology (Ramachandran et al. 2011). In response, this chapter has proposed a range of strategies that can help contribute to robust public policies. Some proponents of the precautionary principle argue that the goal of developing robust policies falls under the umbrella of the PP. For our purposes, it is not particularly important whether these strategies are labeled as part of the PP or not; what matters is that they may help policy makers to respond to emerging technologies in ways that produce tolerable results under a wide range of potentially unexpected future scenarios.

ACKNOWLEDGMENTS

I thank two collaborators, Michael Dickson and David Volz, who worked with me on papers that informed this chapter. I also thank Daniel Steel for his contribution to my thinking about robust decision making.

WORKS CITED

M. Aufann, J. Rose, J. Bottero, G. Lowry, J. Jolivet, and M. Wiesner (2009) 'Towards a Definition of Inorganic Nanoparticles from an Environmental, Health and Safety Perspective,' *Nature Nanotechnology*, 4: 634–641.

V. Colvin (2003) 'The Potential Environmental Impact of Engineered Nanoparticles,' *Nature Biotechnology* 21: 1166–1170.

R. Cooke (1991) *Experts in Uncertainty: Experts and Subjective Probability in Science* (New York: Oxford University Press).

C. Cranor (2011) *Legally Poisoned: How the Law Puts Us at Risk from Toxicants* (Cambridge, MA: Harvard University Press).

K. Elliott (2011) *Is a Little Pollution Good for You? Incorporating Societal Values in Environmental Research* (New York: Oxford University Press).

K. Elliott, and M. Dickson (2011) 'Distinguishing Risk and Uncertainty in Risk Assessments of Emerging Technologies,' in T. Zülsdorf, C. Coenen, Ar. Ferrari, U. Fiedeler, C. Milburn, and M. Wienroth (eds.), *Quantum Engagements: Social Reflections of Nanoscience and Emerging Technologies* (Heidelberg: AKA Verlag) 165–176.

K. Elliott, and D. Volz (2012) 'Addressing Conflicts of Interest in Nanotechnology Oversight: Lessons Learned from Drug and Pesticide Safety Testing,' *Journal of Nanoparticle Research*, forthcoming.

ETC (2011) "Nanotechnology," accessed on December 8, 2011 from http://www.etcgroup.org/en/issues/nanotechnology.

Friends of the Earth (2006) *Nanomaterials, Sunscreens, and Cosmetics: Small Ingredients, Big Risks*. Available at http://www.foe.org/pdf/nanocosmeticsreport.pdf, accessed on Sept. 17, 2011.

S. O. Hansson (1996) 'Decision-Making Under Great Uncertainty,' *Philosophy of the Social Sciences* 26: 369–386.

S. O. Hansson (2004) 'Philosophical Perspectives on Risk,' *Techné* 8: 10–35.

W. Joy (2000) 'Why the Future Doesn't Need Us,' *Wired* 8.04.

A. Magrez, et al. (2006) 'Cellular Toxicity of Carbon-Based Nanomaterials,' *Nano Letters* 6: 1121–1125.

N. Manson (2002) 'Formulating the Precautionary Principle,' *Environmental Ethics*, 24: 263–274.

A. D. Maynard, D. B. Warheit, and M. A. Philbert (2011) 'The New Toxicology of Sophisticated Materials: Nanotoxicology and Beyond,' *Toxicological Sciences*, 120: S109-S129.

T. McGarity, and W. Wagner (2008) *Bending Science: How Special Interests Corrupt Public Health Research* (Cambridge, MA: Harvard University Press).

S. Mitchell (2009) *Unsimple Truths: Science, Complexity, and Policy* (Chicago, IL: University of Chicago Press).

J. Myers, F. vom Saal, B. Akingbemi, K. Arizono, S. Belcher, T. Colborn, I. Chahoud, et al. (2009) 'Why Public Health Agencies Cannot Depend on Good Laboratory Practices as a Criterion for Selecting Data: The Case of Bisphenol A,' *Environmental Health Perspectives*, 117: 309–315.

NNI (National Nanotechnology Initiative) 'What is Nanotechnology?' Retrieved on December 8, 2011 from http://www.nano.gov/nanotech-101/what/definition.

NRC (National Research Council) (1983) *Risk Assessment in the Federal Government: Managing the Process* (Washington, D.C.: National Academy Press).

G. Oberdörster, et al. (2005) 'Principles for Characterizing the Potential Human Health Effects from Exposure to Nanomaterials: Elements of a Screening Strategy,' *Particle and Fibre Toxicology* 2: 8.

J. Pauluhn (2010) 'Multi-walled Carbon Nanotubes (Baytubes˚): Approach for Derivation of Occupational Exposure Limit,' *Regulatory Toxicology and Pharmacology* 57: 78–89.

C. Poland, R. Duffin, I. Kinloch, et al. (2008) 'Carbon Nanotubes Introduced into the Abdominal Cavity of Mice Show Asbestos-Like Pathogenicity in a Pilot Study,' *Nature Nanotechnology* 3: 423–428.

C. Raffensperger, and J. Tickner (eds.) (1999), *Protecting Public Health and the Environment*, (Washington, D.C.: Island Press).

G. Ramachandran, S. Wolf, J. Paradise, J. Kuzma, R. Hall, E. Kokkoli, and L. Fatehi (2011) 'Recommendations for Oversight of Nanobiotechnology: Dynamic Oversight for Complex and Convergent Technology,' *Journal of Nanoparticle Resesearch*, 13: 1345–1371.

The Royal Society and The Royal Academy of Engineering (2004) *Nanoscience and Nanotechnologies*, available at: http://www.nanotec.org.uk/finalReport.htm.

P. Sandin (1999) 'Dimensions of the Precautionary Principle,' *Human and Ecological Risk Assessment* 5: 889–907.

C. Sayes, et al. (2006) 'Functionalization Density Dependence of Single-Walled Carbon Nanotubes Cytotoxicity in Vitro,' *Toxicology Letters* 161: 135–142.

C. Sunstein (2005) *Laws of Fear: Beyond the Precautionary Principle* (New York: Cambridge University Press).

R. Templeton, P. L. Ferguson, K. Washburn, W. Scriven, G. T. Chandler (2006) 'Life-Cycle Effects of Single-Walled Carbon Nanotubes (SWNTs) on an Estuarine Meiobenthic Copepod,' *Environmental Science and Technology* 40: 7387–7393.

F. Tian, D. Cui, H. Schwarz, G. G. Estrada, and H. Kobayashi (2006) 'Cytotoxicity of Single-Wall Carbon Nanotubes on Human Fibroblasts,' *Toxicology in Vitro* 20: 1202–1212.

F. vom Saal, and C. Hughes (2005) 'An Extensive New Literature Concerning Low-Dose Effects of Bisphenol A Shows the Need for a New Risk Assessment,' *Environmental Health Perspectives* 113: 926–933.

R. von Schomberg (2006) 'The Precautionary Principle and Its Normative Challenges,' in E. Fisher, J. Jones, and R. von Schomberg (eds.) *Implementing the Precautionary Principle: Perspectives and Prospects* (Northampton, MA: Edward Elgar).

DISCUSSION QUESTIONS

1. Which criticisms of risk assessment discussed in this chapter are most serious? Can you think of other potential criticisms? Are there effective ways to address or mitigate these criticisms?

2. Do you think that the precautionary principle provides helpful guidance for responding to emerging technologies? Are you convinced by the objection that it is either vacuous or paralyzing, depending on how it is interpreted?

3. This chapter proposed a range of strategies for developing public policies in response to nanomaterials. If you had to prioritize two or three of them, which would you pick, and why?

4. Do you prefer a pre-market approach to regulating toxic chemicals, such as the REACH legislation in Europe, or do you prefer a post-market approach to regulation, such as the Toxic Substances Control Act in the United States? Suggest how someone might defend each approach.

5. When manufacturers conduct toxicity tests of their own products for regulatory purposes, do you trust the reliability of the results? Is it better if they pay a contract research organization to conduct the studies for them? What do you think of the proposal to have a government agency receive money from the manufacturers and then pay for the studies?

6. Imagine that you had to adapt the risk management and decision strategies proposed in this chapter to another emerging technology, such as genetically modified crops or geoengineering. Which strategies would be particularly easy to apply to the other technology, and which strategies would not make sense in a different context?

7. Do you agree that "robust public policies" to promote responsible development should be pursued in response to emerging technologies? How could someone criticize the pursuit of robust public policies?

8. Cost–benefit analyses assign values to outcomes in monetary terms. What is problematic about doing so? What reasons are there to avoid translating health or biodiversity loss into monetary units?

9. Do you think that adopting robust strategies for dealing with risks and uncertainty limits the development of beneficial technologies? Why or why not?

AVOIDING CATASTROPHIC CLIMATE CHANGE: WHY TECHNOLOGICAL INNOVATION IS NECESSARY BUT NOT SUFFICIENT

Philip Cafaro

CHAPTER SUMMARY

Global climate change is among the world's most significant and pressing ecological challenges. On some scenarios it could bring about hundreds of millions of refugees, cause a quarter of the world's plant and animal species to go extinct, and result in trillions of dollars in economic costs. How bad global climate change is going to be will depend upon its magnitude, which in turn depends upon how much more greenhouse gases we emit. In this chapter, Philip Cafaro presents and evaluates a wide range of emission reduction strategies, both technological and behavioral. He argues that both types are necessary, and advocates for several particular behavioral strategies, including eating less meat, flying less frequently, having fewer children, and slowing economic growth.

RELATED READINGS

Introduction: Innovation Presumption (2.1); EHS (2.6.1)
Other Chapters: Clive Hamilton, *Ethical Anxieties about Geoengineering* (Ch. 29)

1. INTRODUCTION

Anthropogenic climate change, driven primarily by increased greenhouse gas emissions and, to a lesser extent, the conversion of wild lands to tame, may be the most difficult environmental challenge facing humanity in the 21st century. Scientific studies project that climate change may kill or displace hundreds of millions of people and extinguish one-quarter or more of Earth's species within the lifetimes of people reading this chapter. Yet there is a curious disconnect between these predictions of global climate change and discussions of possible solutions.

On the one hand, it is widely acknowledged that the primary causes of increased emissions are unrelenting population growth and economic expansion. As the fourth assessment report from the Intergovernmental Panel on Climate Change (IPCC) succinctly puts it: "GDP/per capita and population growth were the main drivers of the increase in global emissions during the last three decades of the 20th century … At the global scale, declining carbon and energy intensities [increased efficiency] have been unable to offset income effects and population growth and, consequently, carbon emissions have risen" (IPCC, 2007, p. 107). And crucially, the IPCC's projections for the next several decades see a continuation of these trends. More people living more affluently mean that under "business as usual," despite expected technical efficiency improvements, greenhouse gas emissions will increase between 25% and 90% by 2030, relative to 2000 (IPCC, 2007, p. 111). If we allow this to occur, it will almost surely lock in global temperature increases of more than two degrees Centigrade over pre-industrial levels, exceeding the threshold beyond which scientists speak of potentially catastrophic climate change. Following this path would be a moral catastrophe as well: the selfish over-appropriation and degradation of key environmental services by the current generation to the detriment of future ones, by rich people to the detriment of the poor, and by human beings to the great detriment of the rest of the living world.

On the other hand, most proposals for preventing further climate change take population increase and economic growth for granted, and focus on technical means of reducing greenhouse gas emissions. Yet a reasonable person reading the scientific literature on climate change would likely conclude that we are bumping up against physical limits to growth. "Wow, this could be hard!" he or she might say. "We need to start working on the problem with all the tools at our disposal. Increasing efficiency and technological innovation, to be sure. But also decreasing consumption and the pursuit of affluence, and stabilizing or reducing human populations. Maybe in the future we can grow like gangbusters again. But for now, people need to make fewer demands on nature and see if even our current numbers are sustainable over the long haul. After all, our situation is dangerous and unprecedented – more than 7 billion people living or aspiring to live in modern, industrialized economies – and we may already be in 'overshoot' mode."

However, neither climate change nor our other global environmental problems have led to a widespread re-evaluation of our faith in the goodness of growth. Regarding climate change, we have seen a near-total focus on technological solutions by politicians, scientists, and even environmentalists. This is a serious mistake. Because 'business as usual' with respect to growth probably cannot avoid catastrophic climate change or meet our other global ecological challenges, we need to consider a broader range of alternatives that include slowing or ending growth. Continued neglect of this topic will undermine attempts to specify a just and prudent course of action on climate change.

2. THE WEDGE APPROACH

Stephen Pacala and Robert Socolow have created an influential "wedge" heuristic to help identify and compare alternative global climate change mitigation schemes; i.e., different approaches to reducing greenhouse gas emissions and avoiding harmful land use changes (Pacala and Socolow, 2004).[1]

---

[1] Recent research and even a downloadable version of the "Carbon Mitigation Wedge Game" can be found at the website for the Carbon Mitigation Initiative (www.princeton.edu/~cmi).

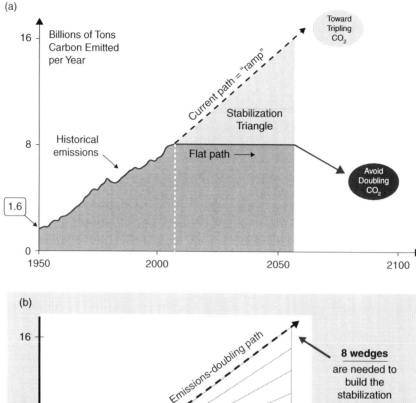

Figures 1a and 1b Stabilization Triangles

Source: Carbon Mitigation Initiative, Princeton University.

Figures 1a and 1b show past and projected global carbon emissions, and compare projections under business as usual with a flat emissions rate. Each wedge in the "stabilization triangle" represents a technological change which, fully implemented, would keep one billion metric tons of carbon from being pumped into the air annually, fifty years from now. It would also prevent twenty-five billion metric tons of carbon from being released during the intervening fifty years (remember the formula for finding the area of a triangle, from high school geometry?).[2] The authors figure eight such wedges must be implemented not

---

[2] In the most recent version of this approach, each wedge prevents four billion metric tons of CO_2 from being emitted annually, rather than one billion metric tons of carbon. Since one ton of carbon equals 3.67 tons of CO_2, the wedges are slightly larger in the new version.

	Option	Effort by 2054 for one wedge, relative to 14 gigatons of carbon per year (GtC/year) under business as usual
Energy Efficiency and Conservation	1. Efficient vehicles	Increase fuel economy for 2 billion cars from 30 to 60 mpg
	2. Reduced vehicle use	Decrease car travel for 2 billion 30-mpg cars from 10,000 to 5,000 miles per year
	3. Efficient buildings and appliances	Cut carbon emissions by one-fourth in projected buildings and appliances
	4. Efficient baseload coal plants	Produce twice today's coal power output at 60% efficiency compared with 32% efficiency today
Fuel Shift	5. Gas baseload power for coal baseload power	Replace 1,400 gigawatt coal plants with natural gas plants
	6. Nuclear power for coal power	Add 700 gigawatt nuclear power (tripling current capacity)
Carbon Capture and Storage	7. Capture CO_2 at power plants	Introduce CCS at 800 gigawatt coal plants or 1,600 gigawatt natural gas plants
	8. Capture CO_2 at Hydrogen (H_2) plants	Introduce CCS at plants producing 250 million tons H_2/year from coal or 500 million tons H_2/year from natural gas
	9. Capture CO_2 at coal-to-synfuels plants	Introduce CCS at synfuels plants producing 30 million barrels per day from coal (200 times current capacity)
Renewable Electricity and Fuels	10. Wind power for coal power	Add 2 million 1-megawatt-peak windmills (50 times current capacity) occupying 30 million hectares of land
	11. Photovoltaic power for coal power	Add 2,000 GW-peak PV (700 times current capacity) on 2 million hectares
	12. Wind-generated H_2 fuel-cell cars for gasoline–powered cars	Add 4 million 1-megawatt-peak windmills (100 times current capacity) to power electric cars
	13. Biomass fuel for fossil fuel	Add 100 times current Brazil or U.S. ethanol production, using 250 million hectares (1/6 of world cropland)
Forests and Agricultural Soils	14. Reduce deforestation, plus reforestation and new plantations	Halt tropical deforestation instead of 0.5 billion ton C/year loss, and establish 300 million hectares of new tree plantations
	15. Conservation tillage	Apply best practices to all world cropland (10 times current usage)

Figure 2 Potential Conventional Carbon Mitigation Wedges

Source: Carbon Mitigation Initiative, Princeton University (figure modified).

to reduce atmospheric CO_2, or even stabilize CO_2 levels, but simply to keep atmospheric carbon from doubling and pushing past potentially catastrophic levels during this period. In order to fully address GCC, in the following fifty years, humanity would have to take further steps and move to an economy where human carbon outputs do not exceed carbon uptakes in natural systems.

The wedge approach buys us time and (allegedly) begins the transition toward such an economy. The fifteen carbon reduction wedges proposed by Pacala and Socolow are listed in Figure 2 (above).

Despite a stated desire to consider only alternatives that are technically feasible, today, scaling up the carbon capture and storage options (wedges 7–9) appears to rely on future technological improvements that may not prove practicable. While some of the wedges could pay for themselves over time, most, on balance, would involve significant economic costs. Many wedges also carry significant environmental costs, which in some cases may equal or outweigh the environmental benefits they would provide in helping mitigate climate change. This is arguably the case with the proposed nuclear power wedge, given potential meltdowns and problems with waste disposal, which could expose people to deadly radiation. Similarly, the coal wedges would encourage continued ecological degradation from mountain-top coal mining and continued deaths due to air pollution. Even seemingly benign alternatives, such as wind or solar power expansion, could result in huge wildlife habitat losses if pursued on the scale demanded to achieve a full wedge.

One strength of the wedge approach is that it allows us to specify the costs and benefits of different courses of action and choose intelligently between them. – if, that is, we are considering a reasonably complete set of alternatives. Take another look at the fifteen proposed wedges. Fourteen focus on improvements in energy efficiency, or substitutions in energy and materials production; one or perhaps two wedges involve limiting consumption (cutting the miles driven by automobile drivers; maybe limiting deforestation); and none of them involve limiting human population growth. This is hardly a peculiarity of Pacala and Socolow's work. Most discussions of climate change neglect the possibility of limiting consumption or stabilizing populations. The goal, always, seems to be to accommodate *more* consumption by *more* people with *less* environmental impact.

Numerous illustrations can be cited from the IPCC's fourth assessment report itself. Its authors recognize agriculture as a major contributor to global warming, accounting for about one-fifth of the problem. Yet they simply accept projections for greatly increased demands for all categories of agricultural products (including a doubling in worldwide demand for meat over the next fifty years) and focus on changes in tillage, fertilizer use and the like, as means to limit increased greenhouse gas emissions (IPCC, 2007, Ch. 8). Similarly, the assessment report notes that among significant greenhouse gas sources, aviation traffic is the fastest-growing sector worldwide. It considers numerous changes to aviation practices, including relatively trivial improvements in airplane technology and changes in flight patterns, while avoiding the obvious alternative of reducing the number of flights (IPCC, 2007, 1999). Many similar examples could be given.

The failure to consider policies designed to reduce consumption or limit population can't be chalked up to these factors' unimportance. The IPCC assures us that they are all-important in generating global climate change. Nor is it because there aren't plausible policies that might reduce consumption or slow population growth. Nor is it because such policies necessarily would be more expensive, harder to implement, more coercive, or in any other way less appealing than the technological approaches under consideration. Some may be, of course. But as I show below, there are almost certainly consumption and population wedges that could be developed and implemented at lower economic, environmental and social cost than most of the wedges proposed by Pacala and Socolow, and even with considerable overall benefit.

The real problem is that the majority of policy-makers are so ideologically committed to maximizing growth that it is impossible for them to consider the full range of alternatives. There are legitimate questions about how to reduce growth without reducing people's quality of life, or without reducing it too much. Still, this failure could prove disastrous. The scientific literature has grown even grimmer in the years since publication of the fourth assessment report, leading some climate scientists to argue that humanity must reduce greenhouse gas emissions more quickly and stabilize emissions at lower levels than previously thought (Hansen, 2007; Hansen et al., 2008). We may not need eight wedges over the next fifty years, but ten or more, to avoid catastrophic climate change. Furthermore, like other approaches focused on mid-term carbon reductions, the wedge framework assumes a transition to *much* lower emissions in the long term. But such a transition seems unlikely if we squeeze all the most easily-achieved efficiency gains out the system, while increasing the number of people and ratcheting up their per capita demands and sense of entitlement.

Pacala and Socolow try to meet this problem by advocating for increased research and putting their faith in future technological breakthroughs to achieve a near-zero carbon society. I believe we cannot rest in such faith given the stakes involved. I'm all for increased research into new energy technologies, but I think we'll probably need wisdom as well as cleverness, self-restraint as well as technological ingenuity to avoid catastrophic climate change. And doing so is a moral imperative. Morality demands that this generation of wealthy human beings construct a plausible bridge to a sustainable future, and walk across it. It also demands that this bridge be one that the world's poor and other species can cross along with us. Proposals to trade in our old air-conditioned SUVs for new, air-conditioned *hybrid* SUVs and roar into the future with the hope of a Hail Mary techno-pass at the end of the game do not meet our moral obligations. Given the stakes involved, we should consider our full range of options. In the remainder of this chapter, I propose a number of alternative wedges focused on reductions in consumption, population growth, and economic growth, to supplement those already proposed.

3. ALTERNATIVE WEDGES: CONSUMPTION

Consider four consumption wedges, the first two focused on food consumption and transportation, the following two seeking to rein in consumption in general. Once again, each of these wedges represents prevention of one billion tons of carbon from being emitted fifty years from now and twenty-five billion tons from being emitted over the next fifty years.

3.1 Meat wedge

According to a recent comprehensive study, agriculture currently contributes 18% of total world greenhouse gas emissions and livestock production accounts for nearly 80% of this (UN FAO, 2006, p. 112). Thus meat-eating contributes approximately 2.38 billion tons carbon equivalent to current greenhouse gas emissions. The U.N. Food and Agricultural Organization projects a worldwide doubling in animal production between 2000 and 2050, from 60 billion to 120 billion animals raised annually, which, under "business as usual," will double the greenhouse gas emissions from this sector to 4.76 billion tons. If instead we hold worldwide animal food production steady over the next fifty years, this would provide nearly two and a half carbon wedges (averting a 2.38 billion tons increase), while merely preventing half the projected doubling during that time would supply more than one full carbon wedge (1.19 billion tons averted, Eshel and Martin, 2006, paper 9). Such wedges might be accomplished non-coercively by increasing the price of meat, removing subsidies for cattle production, banning confined animal feedlot operations (CAFO's) and raising animals humanely, and directly taxing meat to discourage consumption. These measures could accommodate a reasonable increase in meat-eating in poor countries where many people eat little meat, while providing environmental and health benefits in wealthy countries where people eat more meat than is good for them (Cafaro, Primack and Zimdahl, 2006, pp. 541–61). They could complement efforts to improve the conditions under which food animals are raised – changes that may be expensive, but which are arguably demanded by morality anyway.[3]

3.2 Aircraft wedge

According to the fourth assessment report, civil aviation is one of the world's fastest-growing sectors of significant greenhouse gas emissions. Analysis shows that air traffic "is currently growing at 5.9% per year [and] forecasts predict a global average annual passenger traffic growth of around 5% – passenger traffic doubling in 15 years" (IPCC, 2007, p. 334). Under current projections, carbon emissions from aircraft might increase from 0.2 billion tons per year to 1.2 billion tons annually over the next fifty years (IPCC, 1999). In addition to emitting CO_2, airplanes increase "radiative forcing" through emissions of other greenhouse gases and by creating contrails and cirrus clouds; these contributions to global warming may be "2 to 4 times larger than the forcing by aircraft carbon dioxide alone" (IPCC, 1999, section 4.8). Let's assume, conservatively, that the other effects of aviation add up to twice the impact of CO_2 emissions. Preventing half the projected increase would give us half a wedge from carbon alone (0.5 billion tons less carbon emissions annually, fifty years from now) and one and a half wedges overall; compared to this, holding total flights at current levels would supply a full wedge from carbon and three wedges overall. Once again, such reductions could be achieved by increasing the cost of air travel by taxing it. Such a proposal was recently put before the parliament of the European Union (European Parliament, 2008). Alternatively, countries might decide that climate change is important enough to demand sacrifices from all their citizens, even rich ones, and strictly limit the number of allowable discretionary flights per person (Cafaro, 2013).

---

[3] For a detailed discussion of the meat/heat connection, see Compassion in World Farming (2007).

3.3 Carbon tax wedges

A general carbon tax is considered one of the most economically efficient ways to cut carbon emissions. While such taxes are usually presented as means to force technological innovation and decrease pollution per unit of consumption, they also incentivize less consumption. According to the IPCC, a tax of $50 per ton of carbon dioxide equivalent could prevent from 3.5 to 7 billion tons carbon equivalent from being emitted annually by 2030, while a tax of $100 per ton of carbon dioxide equivalent could prevent from 4.3 to 8.4 billion tons carbon equivalent from being emitted annually (IPCC, 2007, pp. 9–10). Projected out to 2060, such taxes could provide from 5.6 to 13.4 wedges of carbon reduction.[4] In other words, an appropriately sized carbon tax, by itself, could conceivably provide the eight or more wedges needed to avoid catastrophic climate change.

A carbon tax is so effective because it affects consumption across the board, from airplane travel to new home construction to food purchases. It treats all these areas equally and does not distinguish between frivolous or important consumption, or between useful or useless consumption. That is both its (economic) strength and its (ethical) weakness (for reasons discussed below), and why it should probably be supplemented by measures that directly target unnecessary carbon emissions.

3.4 Luxury wedges

Like any general consumption tax, a carbon tax is regressive, hitting the poor harder than the rich. However, there is a case to be made that in a greenhouse world, everyone should do their part, including the wealthy, by limiting unnecessary consumption. Consider again airplane travel, where we saw that holding the number of flights steady over the next fifty years might provide three wedges. If we assume that the wealthiest 10% of the world's population (roughly those with an annual income of $10,000 or more as per The World Bank, 1999) account for 90% of flights and that much of this travel is discretionary, then we might construct our plane wedges in a way that transformed them, to some degree, into luxury wedges. For example, we might tax a person's first flight at a percentage n of the cost of a ticket, her second at n x 2, her third at n x 4, and not allow more than three personal flights annually (with medical or bereavement exceptions) – and ban the personal ownership and use of planes (again allowing for reasonable exceptions in the public interest). In a similar manner, Pacala and Socolow's "reduced vehicle use" and "efficient buildings" wedges might be partially transformed into luxury wedges, through some combination of progressive taxation and outright prohibitions of extravagant consumption.

---

[4] Note that these taxes would scale up in two to two and a half decades, rather than the five decades in Pacala and Socolow's original wedges. I have (arbitrarily) assumed that the taxes would provide the same amount of annual carbon reductions in succeeding years. Strictly speaking, the resulting figure is not a triangular carbon "wedge" but a carbon trapezoid. A similar point applies to the economic growth reduction wedges discussed in a later section.

While suggestions to prohibit unnecessary, high-emission consumption tend to be unpopular, they merit serious consideration. It is possible to construct luxury wedges without prohibitions, through progressive taxation, but setting strict limits for *everyone* would acknowledge the seriousness of the problem and represent a society-wide commitment to avoiding catastrophic climate change. Arguably from a fairness perspective, from a wide buy-in perspective, and from a maximal emissions reductions perspective, it makes sense to consider the absolute prohibition of some high-energy, luxury consumption.

4. ALTERNATIVE WEDGES: POPULATION

When we turn to potential population wedges, we need to remember that population growth is one of the two main drivers of global climate change (O'Neill, Mackellar, and Lutz, 2005; O'Neill 2009). Again according to the IPCC's fourth assessment report: "The effect on global emissions of the decrease in global energy intensity (-33%) during 1970 to 2004 has been smaller than the combined effect of global per capita income growth (+77%) and global population growth (+69%); both drivers of increasing energy-related CO_2 emissions" (IPCC, 2007, p. 3). When it comes to climate change and other environmental problems: "size (of the human population) matters" (Grant, 2001).

The current global population is approximately 7.1 billion people as this volume goes to press. Here are recent fifty-year United Nations population projections at low, medium and high rates of growth:[5]

Projection	Annual growth rate	2050 population	2060 population
Low	0.40 %	7.4 billion	8.0 billion
Medium	0.77 %	8.9 billion	9.6 billion
High	1.12 %	10.6 billion	11.8 billion

Figure 3   World Population Projections

Source: UN Department of Economic and Social Affairs, Population Division, "World Population to 2300," p. 4.

The medium projection is presented as the "most likely" scenario, although all three projections are considered possible depending on a variety of factors, including public policy choices.

In 2000, world per capita greenhouse gas emissions were 1.84 tons carbon equivalent. Assuming this emissions rate, each 543 million people added to Earth's population adds another one billion tons of annual carbon emissions; conversely, preventing the existence of 543 million people fifty years from now provides a full carbon reduction wedge (543 million × 1.84 tons = 1 billion tons). If we follow the

---

[5] Original projections were to 2050; I projected out to 2060 using the annual growth rates provided. These figures might be somewhat optimistic; more recently, the US Census Bureau projected that world population will grow from six billion in 1999 to nine billion by 2040. See US Census Bureau (2009).

UN report and take 9.6 billion as our business as usual population scenario, then successfully holding world population growth to the lower figure of 8.0 billion would provide 2.95 global population wedges. Conversely, allowing the world's population to swell to the high projection of 11.8 billion (still within the realm of possibility) would create 4.05 population *de*stabilization wedges and almost certainly doom efforts to prevent catastrophic climate change. These figures show that reducing population growth could make a huge contribution to mitigating global warming (Meyerson, 1998).

"Population control" tends to bring to mind coercive measures, such as forced abortions or sterilizations. In fact, there are non-coercive policies that are almost as effective at reducing birth rates, and these are the ones we should pursue in constructing population wedges. First, providing free or low-cost birth control and accessible, appropriate information about how to use it has proven very effective in lowering birth rates in many poor countries (Speidel et al., 2009). Providing cheap birth control allows those who want to have fewer children to do so, increasing reproductive freedom while decreasing population growth. Second, policies which improve the lives of women have been shown to reduce fertility rates in many developing countries (Barroso, 2009). These include guaranteeing girls the same educational opportunities as boys, promoting female literacy, and improving women's economic opportunities (and thus their status in society). Third, making abortion safe, legal and easily available has helped reduce birth rates in many countries. In fact, no modern nation has stabilized its population without legalizing abortion. All these measures can directly improve people's lives at the same time that they help reduce population growth.

Given that these non-coercive methods have proven successful at reducing fertility rates in many places, and given the huge unmet need for contraception throughout the world, well-funded efforts to apply them globally seem capable of reducing population growth from the "most likely" scenario of 9.6 billion people to the lower projection of 8 billion people in 2060. Once again: 1.6 billion fewer people fifty years from now represent nearly three carbon reduction wedges. That would make an immense contribution to mitigating GCC, equal to deploying all three of Pacala and Socolow's carbon capture and sequestration wedges. Unlike carbon capture, however, the proposed population reduction measures rely on proven technologies that are available right now. Population wedges would also provide numerous other environmental benefits, reducing human impacts across the board, in contrast to the massive environmental harms that would be caused by continued coal and uranium mining under the carbon and nuclear power wedges.

Securing women's reproductive rights and furthering their economic and social opportunities are the right things to do. Population wedges thus provide "win/win" scenarios with the potential to aid women and their families directly, increasing their happiness and freedom, while helping to meet the grave danger of climate change (O'Neill, 2000; Sandler, In Press). Some of the very same aims written into the UN's Millennium Development Goals, such as improving maternal health and increasing the percentage of children receiving a full primary school education, turn out to be among the most effective means to reduce birth rates in poor countries (Butler, 2007). In addition, a recent study from the London School of

Economics titled "*Fewer Emitters, Lower Emissions, Less Cost,*" argues that reducing population growth is much cheaper than many other mitigation alternatives under consideration (Wire, 2009). Given all this, policies to stabilize or reduce populations should be an important part of national and international climate change efforts. These are some of the best wedges we've got.

5. ALTERNATIVE WEDGES: GROWTH

According to the US Department of Energy, "economic growth is the most significant factor underlying the projections for growth in energy-related carbon dioxide emissions in the mid-term, as the world continues to rely on fossil fuels for most of its energy use" (United States Energy Information Administration, Ch. 1). This suggests that one way to limit increased greenhouse gas emissions is to directly limit economic growth. In its report "International Energy Outlook 2009," the DOE quantified the impact of various growth rates on carbon emissions out to 2030, as follows: at a high (4%) annual growth rate, energy use and CO_2 emissions both increase 1.8% annually; at a medium (3.5%) annual growth rate, energy use increases 1.5% annually and CO_2 emissions increase 1.4% annually; and at a low (3%) annual growth rate, energy use increases 1.2 % annually and CO_2 emissions increase 1.0% annually (United States Energy Information Administration). The difference between a 3% or 4% annual growth rate adds up to 1.94 billion fewer tons of carbon emitted annually by 2030, or 19% less. By my calculations, limiting annual world economic growth to 3% rather than 4% over the next *fifty* years would lead to 73.64 billion tons less carbon emitted, or almost three carbon reduction wedges.

Both monetary policy and fiscal policy can be used to ratchet back growth. For example, central banks routinely tighten money supplies when rapid growth threatens to cause high inflation. This raises interest rates, reduces borrowing and spending, and slows overall growth. People tend to look at you funny if you advocate slowing growth rates in order to lower carbon emissions; more than once, economists have asked me "whether I want us all to live in caves," when I've made the suggestion. Still, once we accept the idea that economic growth may be limited to further other important goals, the question becomes: which goals are important enough to trump growth? Preventing global ecological disaster would seem to be a good candidate. If we are indeed bumping up against physical limits to the scale of the human enterprise, we may need to rethink our love affair with growth – particularly if growth threatens our own wellbeing or the wellbeing of others.

Directly reducing economic growth – by far the leading cause of global climate change – is both possible and potentially very effective in reducing greenhouse gas emissions. Like similar efforts to fight inflation, reducing growth to fight climate change could be done in a limited and controlled way. It is revealing that such an obvious wedge candidate is almost totally overlooked in most analyses of the problem. As Tim Jackson writes in *Prosperity without Growth? The Transition to a Sustainable Economy*: "The truth is that there is as yet no credible, socially just, ecologically-sustainable scenario of continually growing incomes for a world of nine billion people. In this context, simplistic assumptions that capitalism's propensity for efficiency will allow us to stabilize the climate or protect against resource scarcity are nothing short of delusional" (Jackson, 2009, p. 57). We can and should work to make our economic activities more efficient and less environmentally damaging, but that is no excuse for ignoring the *scale* of those activities, which is the most important part of the problem.

Clearly, we will need a new economic paradigm in order to create sustainable societies.[6] It is beyond the scope of this chapter to wade into the debate about whether a truly sustainable economy must be post-growth, slow-growth, no-growth, or something else. Here I make the more limited point that reducing (still relatively high) economic growth rates could make a significant contribution to reducing greenhouse gas emissions and avoiding catastrophic global warming. We don't "all have to live in caves" in order to do that.

6. CONCLUSION: COMPARING ALTERNATIVES

Here are the alternative carbon reduction wedges I have proposed in this paper; undoubtedly more are possible.

	Option	Effort by 2060 for one wedge, relative to 15 gigatons of carbon per year (GtC/year) under business as usual
Consumption Reduction	1. Eating less meat	Prevent half the projected world increase in meat eating, through removal of agricultural subsidies and price increases
	2. Flying less often	Prevent one-third of the projected increase in commercial aviation, through taxing or limiting flights, or providing alternatives
	3–9. Taxing carbon	Tax greenhouse gas emissions at $50 per ton of CO2 equivalent
	10. Limiting luxury consumption	Tax or prohibit multiple personal plane flights, large homes, low-mileage vehicles, or other unnecessary consumption
Population Reduction	11–13. Improving women's lives and reproductive freedom	Achieve UN's low rather than medium 2060 population projection (8.0 rather than 9.6 billion) by providing free, accessible birth control and improving economic and educational opportunities for women
Economic Growth Reduction	14–16. Limiting the size of the world economy	Use monetary and fiscal policy to reduce world growth rates from 4% to 3% annually

Figure 4   Potential Alternative Carbon Mitigation Wedges

---

[6] Valuable contributions toward specifying the parameters of a sustainable economy include (Daly and Cobb, 1989; Alexander, 2009).

All the wedges outlined above are achievable using current technologies. As with Pacala and Socolow's conventional wedges, pushing reductions faster and further could result in additional wedges. My proposal is not to replace the original wedges, but to combine them with these alternatives, increasing our options.

Providing more wedges is important because we might have to consider unpalatable choices now that scientists are telling us we may have to ratchet down emissions faster than anticipated in order to avoid catastrophic global climate change. However, some alternative wedges might actually be preferable to the usual proposals to mitigate global warming. For example, it is not clear that tripling world nuclear generating capacity is better than simply cutting back on electricity use by reducing unnecessary consumption. Cutting consumption might be cheaper and less dangerous and make a stronger contribution toward creating sustainable societies. Again, some of us would prefer that our tax dollars go toward helping poor women in developing countries improve their lives, as in the population reduction wedges, rather than subsidizing energy companies' profits, as required by Pacala and Socolow's coal and nuclear wedges. Again, limiting consumption and population growth seems less selfish and more responsible than relying solely on efficiency improvements that pass significant environmental burdens on to nonhuman beings and future generations, or than betting on futuristic technologies that may not work.

Whether or not I'm right in these particular judgments, getting a full range of alternatives on the table is our best hope for finding the fairest and most efficient strategies to mitigate climate change. Pacala and Socolow's original framework allows us to consider whether to triple nuclear generating capacity or build two million new wind turbines; the expanded framework allows us to choose these options, or the option of paying more for electricity and using less of it. The original framework makes it hard to achieve eight wedges without committing to continued heavy use of coal; the expanded framework would allow humanity to phase out coal and uranium as fuel sources, provided we embraced population reduction. The new framework thus makes explicit some of the ecological and economic costs of continued consumption and population growth. And it emphasizes that such growth is a choice.

As in the original framework, alternatives can be compared on the basis of monetary cost, total ecological impact (not just impacts on greenhouse gas emissions) or social equity. But now, new alternatives are in play that would further sustainability, improve the lives of some of the world's poorest people, and demand greater contributions from wealthy global elites, while not relying on unproven technologies. My claim is that using this expanded list of wedges would allow us to come up with a better (because more just and more sustainable) climate change policy. Also, implementing more alternative wedges would put us in a better position to transition to truly sustainable societies in the second half of the twenty-first century. Unless people get in the habit of asking less from nature and more from ourselves, it seems unlikely that we will muster the discipline necessary to create sustainable societies.[8]

---

[8] Thanks to Art Darbie, Steve Shulman, Robert Socolow, Denis Arnold, Kris Cafaro and Ron Sandler for helpful comments.

WORKS CITED

S. Alexander (ed.), *Voluntary Simplicity: The Poetic Alternative to Consumer Culture* (Whanganui, New Zealand: Stead & Daughters, 2009).

C. Barroso (2009) 'Cairo: The Unfinished Revolution,' in Laurie Mazur (ed.), *A Pivotal Moment* (Washington, D.C. Island Press).

B. Butler (2007) 'Globalisation, Population, Ecology and Conflict,' *Health Promotion Journal of Australia*, 18.

P. Cafaro (2013) 'Reducing Consumption to Avert Catastrophic Global Climate Change: The Case of Aviation,' *Natural Science*, 5: 99–105.

P. Cafaro, R. Primack, and R. Zimdahl (2006) 'The Fat of the Land: Linking American Food Overconsumption, Obesity, and Biodiversity Loss,' *Journal of Agricultural and Environmental Ethics*, 19: 541–561.

Compassion in World Farming, *Global Warning: Climate Change and Farm Animal Welfare.* (Godalming, UK: 2007).

H. Daly, and J. Cobb (1989) *For the Common Good: Redirecting the Economy toward Community, the Environment, and a Sustainable Future* (Boston, MA: Beacon Press).

G. Eshel, and P. Martin (2006) 'Diet, Energy, and Global Warming,' *Earth Interactions*, 10.

European Parliament (2008) Directive 2008/101/EC of the European parliament and the council of 19 November, 2008, amending directive 2003/87/EC, so as to include aviation activities in the scheme for greenhouse gas emission allowance trading within the Community.

L. Grant (2001) *Too Many People: The Case for Reversing Growth* (Santa Ana, CA: Seven Locks Press).

J. Hansen (2007) "Scientific Reticence and Sea Level Rise," *Environmental Research Letters*, 2: (2), article 024002.

J. Hansen et al. (2008) 'Target Atmospheric CO_2: Where Should Humanity Aim?' *Open Atmospheric Science Journal*, 2: 217–231.

Intergovernmental Panel on Climate Change (IPCC) *Climate Change 2007: Mitigation* (2007), Technical Summary.

IPCC (1999) Aviation and the Global Atmosphere.

T. Jackson (2009) *Prosperity Without Growth? The Transition to a Sustainable Economy* (Brussels: European Union Sustainable Development Commission).

F. Meyerson (1998) 'Population, Carbon Emissions, and Global Warming: The Forgotten Relationship at Kyoto,' *Population and Development Review*, 24: 115–130;

B. O'Neill (2000) 'Cairo and Climate Change: A Win/Win Opportunity,' *Global Environmental Change*, 10: 93–96.

B. O'Neill (2009) "Climate Change and Population Growth," in Laurie Mazur (ed.), *A Pivotal Moment: Population, Justice & the Environmental Challenge* (Washington, D.C.: Island Press) 81–94.

B. O'Neill, L. Mackellar, and W. Lutz (2005) *Population and Climate Change* (Cambridge: Cambridge University Press).

S. Pacala, S., and R. Socolow (2004) 'Stabilization Wedges: Solving the Climate Problem for the Next 50 Years with Current Technologies,' *Science*, 305: 968–972.

R. Sandler (In press) *The Ethics of Mitigation.*

J. Speidel, et al. (2009) *Making the Case for U.S. International Family Planning Assistance* (New York: Population Connection).

United Nations Food and Agriculture Organization (2006) *Livestock's Long Shadow: Environmental Issues and Options* (Rome, Italy: LEAD).

United States Energy Information Administration, Department of Energy, 'International Energy Outlook 2009' (Washington, D.C.: USEIADE).

U.S. Census Bureau (2009) 'World Population: 1950–2050,' International Data Base.

T. Wire (2009) *Fewer Emitters, Lower Emissions, Less Cost: Reducing Future Carbon Emissions by Investing in Family Planning: A Cost/Benefit Analysis* (London: London School of Economics).

The World Bank (1999) 'True World Income Distribution, 1988 and 1993: First Calculations, Based on Household Surveys Alone,' The World Bank, Development Research Group, policy research working paper, 2244: page 30, table 19.

DISCUSSION QUESTIONS

1. Which of the proposed alternative wedges are the most politically feasible? Which are least politically feasible?

2. What criteria ought to be used to evaluate potential carbon reduction wedges or strategies?

3. One of the approaches to creating reduction wedges discussed in this article is to empower women by promoting reproductive choice and increased economic and educational opportunities. This is considered a "win/win". Which other of the proposed wedges provides benefits beyond climate mitigation?

4. How might proposals to slow growth raise concerns of global justice? Which countries will suffer the most from slowing or stopping growth? Is there any way to slow growth while also reducing global poverty and inequality?

5. Under what conditions would it be permissible to use coercive means, such as expressly prohibiting elective travel by air or discouraging parents from having more than one or two children, to reduce carbon emissions?

6. Are there additional consumption wedges that we should consider besides the four discussed above? What are they, and how might you go about calculating their carbon reduction potential?

7. Should there be a presumption in favor of consumption wedges or technology wedges? If so, which, and why?

29 ETHICAL ANXIETIES ABOUT GEOENGINEERING

Clive Hamilton

CHAPTER SUMMARY

Geoengineering is intentional large-scale technological intervention in order to manage the Earth's climate. Several approaches to geoengineering have been proposed as possible responses to global climate change – for example, solar radiation management and carbon sequestration. In this chapter, Clive Hamilton discusses and evaluates the main justifications for geoengineering. He also surveys a wide array of intrinsic and extrinsic (or outcome-oriented) ethical concerns that have been raised regarding intentional climate manipulation. He concludes that geoengineering is not an ethically acceptable response to global climate change, and advocates instead for an approach that is primarily political rather than technological.

RELATED READINGS

Introduction: EHS (2.6.1); Intrinsic Concerns (2.7); Types of Theories (3.2)

Other Chapters: Philip Cafaro, *Avoiding Catastrophic Climate Change* (Ch. 28); Christopher J. Preston, *Evolution and the Deep Past* (Ch. 36)

1. INTRODUCTION

Global climate change refers to significant and persistent alterations in global climatic systems. There is now an overwhelming body of evidence showing that global warming is occurring and is being driven by human activities that result in the emission of greenhouse gases (e.g. carbon dioxide and methane). The main source of emissions is the burning of fossil fuels (coal, oil and natural gas), but deforestation is also a source. Global warming will continue to accelerate if it is not addressed. Climate change is problematic because it is associated with high rates and magnitude of ecological change (in comparison with the recent historical past) – for example, changes in surface air temperatures, precipitation patterns, extreme weather events, sea levels, and ocean pH. The elevated magnitude and rate of ecological change makes biological and societal adaptation to ecological change more difficult. As a result, global climate change will result in increased rates of species extinction, more instability of ecological systems, greater agricultural and natural resource insecurities, rising damage due to severe weather events, spread of tropical diseases, and waves of environmental refugees. The greater the magnitude of global climate change, the more severe these problems will be.

It is generally agreed that the most preferable way to respond to global climate change is to address its cause – that is, to mitigate it by reducing future anthropogenic (or human-caused) greenhouse gas emissions. After all, it is the build-up of these gases in the atmosphere that is driving the ecological changes. However, efforts to reduce greenhouse gas emissions have thus far been largely unsuccessful (Crutzen 2006). Moreover, it is no longer possible to prevent all anthropogenic global climate change by mitigation alone. As a result, other responses to global climate change are being proposed. Among these is geoengineering, deliberate large-scale intervention in the climate system to counter global warming or offset some its effects.

Three main justifications are used to defend geoengineering research and its possible deployment – it will allow us to buy time, it will allow us to respond to a climate emergency, and it may be the best option economically. Against these a number of ethical risks intrude: we may use the possibility of climate engineering to blind ourselves to our moral responsibilities; research into geoengineering may provide an excuse for governments to reduce mitigation efforts (in the way research into carbon capture and storage has); a powerful pro-geoengineering constituency may emerge, skewing decision-making; and, attempting to regulate Earth's great natural processes is "playing God", which is dangerous and invites retribution. The playing-God argument comes in theistic and atheistic versions. In addition, as a "techno-fix", geoengineering may make the problem worse by papering over the social and political causes of the climate crisis. In this chapter, these arguments for and against geoengineering are developed and evaluated. I will use sulfate aerosol injections, a form of solar radiation management (SRM), as the primary case as it is the geoengineering method that is attracting most interest and seems most likely to be deployed. It also illustrates the ethical issues most starkly. Sulfate aerosol injection would involve release of enormous amounts of tiny particles into the upper atmosphere in order to reflect greater amounts of solar radiation out of the atmosphere, thus preventing it from reaching the Earth's surface and thereby reducing warming. This form of 'solar radiation management' is not aimed at the cause of climate change (rising greenhouse gas emissions) but at one of its symptoms (a warming planet).

2. RATIONALES FOR CLIMATE ENGINEERING

As mentioned above, three main justifications are used to defend research into geoengineering and possible deployment – it will allow us to buy time, it will allow us to respond to a climate emergency and it may be the best option economically. These are set out in Table 1 (below) along with some implications.

The buying-time argument (Royal Society 2009) is based on an understanding that the failure to cut global emissions arises either from political paralysis or from the power of vested interests. The log-jam can only be broken by the development of a substantially cheaper alternative to fossil energy because countries will then adopt the new technologies for self-interested reasons. SRM would allow the effects of warming to be controlled while this process unfolds. Geoengineering is

Table 1 Reasons for developing SRM technologies

Reason	Rationale	Implications for abatement	Timing of deployment	Moral status of geoengineering
Buy time	Time is needed to develop cheaper abatement technologies to overcome policy inertia	Abatement is essential but will be easier later	As soon as it's ready and for as long as it takes to develop better abatement technologies	A necessary evil
Climate emergency	We need to have the tools ready to deal quickly with an imminent catastrophe	Abatement is essential but may not happen in time	When necessary and for as long as the emergency lasts	A necessary evil
Best option	If it's cheaper than abatement we should use it instead or as a complement	Geoengineering is a substitute for abatement	As soon as it's ready and then indefinitely	A good thing because it widens choice

therefore a necessary evil deployed to head off a greater evil, the damage due to unchecked global warming.

The climate emergency argument was Nobel Prize winner Paul Crutzen's motive for breaking the silence over geoengineering. Sulfate aerosol injection, he wrote, should only be developed "to create a possibility to combat potentially drastic climate heating" (Crutzen 2006). Today it is the dominant argument, reflecting growing understanding of and concern about climate tipping points (Morgan and Ricke 2010). It envisages rapid deployment of SRM in response to some actual or imminent sharp change in world climate that cannot be averted even by the most determined mitigation effort. Instances might include the beginning of rapid melting of the Greenland ice sheet, acceleration in permafrost thawing or a prolonged and severe heat wave. By inducing an overall cooling of the globe, sulphate aerosol injections would be able to respond to any of these, a bit like a broad-spectrum antibiotic. Those who anticipate deployment of SRM in a climate emergency favor sustained research so that, once developed and refined, the technology could be "put on the shelf" to be used as necessary (Lane et al. 2007, 11).

As in the case of buying time, the best-option argument would see geoengineering deployed pre-emptively. Rejecting the understanding of geoengineering as an inferior Plan B, it argues that there is nothing inherently good or bad in any approach to global warming. The decision rests on a comprehensive assessment of the consequences of each approach, which is often reduced to the economist's assessment of costs and benefits (e.g. Levitt and Dubner 2009). In this consequentialist approach the "ethical" decision is the one that maximizes the ratio of benefits to costs (Bickel and Lane undated). Some early economic modeling exercises have concluded that geoengineering is cheaper than mitigation and almost as effective and is therefore to be preferred (Barrett 2008). Those economists and philosophers who adopt the best-option argument see geoengineering as a potential *substitute* for mitigation rather than as a complement. They do not accept that geoengineering represents a necessary evil. If Plan B proves to be cheaper than Plan A then it would be unethical *not* to use it. They thereby avoid accusations that their advocacy undermines the incentive to choose the better path; geoengineering *is* the better path. However, this view requires adopting a narrow consequentialist ethical viewpoint.

3. INTENTIONAL VS UNINTENTIONAL CLIMATE CHANGE

SRM involves intentionally altering the global climate. Is there any moral difference between unintended climate change flowing from other activities and intentionally altering the climate? Most people believe that intentions matter morally, which is why courts judge manslaughter less severely than murder. Against this everyday intuition, some philosophers argue that there is no defensible distinction between a harm caused intentionally and the same harm caused unintentionally, that the degree of "wrongness" of an action has no bearing on its degree of "badness" (see Vanderheiden 2008, 207–8).

The issue is complicated by the fact that, since we know that continuing to burn fossil fuels will cause harm, it could be said that global warming is now "deliberate" even if warming is not the intention. Continued release of greenhouse gases is undoubtedly negligent, but I think there is a moral, and certainly an attitudinal, leap to a conscious plan to modify the Earth's atmosphere. In the case of geoengineering, those who make the decision to deploy will argue they are doing so with the best intentions, to prevent a greater harm, unless it can be shown that they knew that more harm could be caused to some. This is the moral import of studies like that of Robock et al. that suggest that sulfate aerosol spraying could disrupt the Indian monsoon (Robock et al. 2008). Certainly, one would expect the law to take the view that damage to someone arising from a deliberate action carries more culpability. In law liability for harms caused by an action depends in part on *mens rea*, literally "guilty mind".

4. MORAL CORRUPTION

Stephen Gardiner describes moral corruption as "the subversion of our moral discourse to our own ends" (Gardiner 2010b, 286). The psychological strategies we use to deny or, more commonly, evade the facts of climate science – and thereby blind ourselves to our moral responsibilities or reduce the pressure to act on them – include wishful thinking, blame-shifting and selective disengagement (Hamilton and Kasser 2009). For selfish reasons we do not want to change our behavior or be required to do so by electing a government committed to deep cuts in emissions.

Geoengineering itself may be a form of moral corruption. If we are preparing to pursue geoengineering for self-interested reasons then the promotion of geoengineering can provide a kind of absolution. But if Plan B (in this case, geoengineering) is inferior to Plan A (in this case, mitigation), in the sense of being less effective and more risky, then merely by choosing B instead of A we succumb to moral failure. At least, this is the case unless we are constrained in our actions, so that pursuing A is beyond us, sometimes called the "control condition" for moral responsibility (Vanderheiden 2010, 144). This presents a moral dilemma for environmental groups: if they believe that Plan B is inferior to Plan A, then supporting geoengineering can be justified only if they believe they can no longer effectively advance Plan A. The dilemma deepens if it proves that supporting Plan B actually makes Plan A less likely to be implemented.

But these arguments are too blunt to give a full understanding. Scientists who defend geoengineering research mostly see themselves as exempt from the moral failings that have given rise to the situation. After all, many are among those who have supported strong abatement action and have become alarmed and frustrated at the failure of political leaders to act. It's not their fault and they are looking for ways of saving the world from the consequences of institutional failure. For both environmentalists and researchers who see geoengineering as a necessary evil, to maintain their integrity they must continue to argue that mitigation is to be preferred. Thus, like Crutzen, the Royal Society declared resolutely that mitigation is to be strongly preferred and geoengineering cannot be "an easy or readily acceptable" alternative (Royal Society 2009, ix).

Nevertheless, simply to restate this belief may not be enough; unless one continues to act on it the declaration can become merely a means of deflecting the censure of others. This draws attention to the position of governments, for it would be hollow for them to argue that they are pursuing Plan B even though they believe Plan A is superior. They have the power to implement Plan A and their reluctance is the reason Plan B is being considered in the first place. To adopt Plan B they must convince others that it is not in their power to reduce emissions, a tactic that is frequently attempted. Even in the United States some argue that there is no point cutting US emissions if other major emitters do not do the same.

It's worth noting that when the time arrives at which they feel they can back geoengineering governments are unlikely to appeal to the climate-emergency justification because highlighting the severity of global warming would only underline their moral failure. As we saw, those able to implement Plan A will lack credibility if they defend Plan B with the buying-time argument, which leaves them with the best-option economic argument. In the case of solar radiation management the empirical basis for it remains speculative, not least because the risks of unintended consequences appear so high. Moreover, the appeal to economics as the basis for making such a momentous decision risks accusations of abandoning ethical concerns and treating the atmosphere as a commodity.

The same moral failure arguments could be used by poor countries against rich ones. As it will probably be rich nations that invest in geoengineering research and, if the time comes, deploy the technologies that result, poor countries will accuse them of evading their responsibilities to reduce emissions. Studies that indicate that some poor countries may suffer harms from solar radiation management techniques reinforce the likely sense of grievance. The ethical situation would be reversed if a small, poor and vulnerable country decided to protect itself by engineering the climate with sulfate aerosol spraying (something that may prove technically and financially feasible). The Maldives, for example, would have a strong moral case to argue that the threat to its citizens' survival caused by the refusal of major emitting nations to change their ways, and its own inability to influence global warming despite sustained efforts, leave it with no choice.

5. MORAL HAZARD

It is widely accepted that having more information is uniformly a good thing as it allows better decisions to be made. Research into geoengineering is strongly defended on these grounds (Royal Society 2009; Cicerone 2006; Bipartisan Policy Center 2011). Yet, for many years, research into geoengineering, and even public discussion of it, was frowned on by almost all climate scientists. When Paul Crutzen made his intervention in 2006 calling for serious research into Plan B because Plan A, cutting global emissions, had been "grossly unsuccessful" he was heavily criticized by fellow scientists (Lawrence 2006). They felt that researching Plan B would reduce the incentive to reduce emissions, the response to global warming strongly preferred by scientists including Crutzen himself.

In other words, they were worried about "moral hazard" – that is, the availability of an inferior policy substitute that can be made to appear superior making it easier for a government (or other parties) not to act as they should. So does geoengineering research create moral hazard? Geoengineering researchers tend to be vague and somewhat dismissive of the likelihood, as though it is only of theoretical concern (see, e.g., Royal Society 2009). Yet in practice any realistic assessment of how the world works must conclude that geoengineering research is virtually certain to reduce incentives to pursue mitigation. This is apparent even now, before any major research programs have begun. Already a powerful predilection for finding excuses not to cut greenhouse gas emissions is apparent to all, so that any apparently plausible method of getting a party off the hook is likely to be seized upon. Already, representatives of the fossil fuel industry have begun to talk of geoengineering as a *substitute* for carbon abatement (Steffen 2009). Economic analysis is in general not interested in the kind of judicious technology mix or emergency backup defended by some scientists, but will readily conclude that geoengineering should be pursued, even as the sole solution, if that's what the "cost curves" show (Bickel and Lane undated). Indeed, the popular book *Superfreakonomics* insists that the prospect of solar radiation management renders mitigation unnecessary: "For anyone who loves cheap and simple solutions, things don't get much better" (Levitt and Dubner 2009). For the authors, economics renders moral concerns redundant: "So once you eliminate the moralism and the angst, the task of reversing global warming boils down to a straightforward engineering problem: how to get thirty-four gallons per minute of sulfur dioxide into the stratosphere?" Or as former United States House Speaker Newt Gingrich declared: "Geoengineering holds forth the promise of addressing global warming concerns for just a few billion dollars a year. Instead of penalizing ordinary Americans, we would have an option to address global warming by rewarding scientific invention… Bring on the American ingenuity" (Ferraro 2011). For advocates the problem of moral hazard evaporates because there is nothing wrong with reducing abatement incentives if a cheaper means of responding to climate change is available.

That moral hazard is a powerful ethical argument against the development of geoengineering technologies in practice is suggested by the highly germane case of carbon capture and storage.

6. CASE STUDY IN MORAL HAZARD: CARBON CAPTURE AND STORAGE

The risk is that policy makers will use the promise of geoengineering research as a reason for further delay in pursuing abatement policies now. But if geoengineering research proves unsuccessful (because, for example, the risks of deployment are shown to be too high) then the world may be worse off. This is the moral hazard.

The history of carbon capture and storage (CCS) suggests that governments are indeed likely to latch on to geoengineering as an excuse for further delay.

Soon after the 1997 Kyoto agreement, the governments of the two nations that refused to ratify it, Australia and the United States, began talking up the benefits of CCS, a technology that aimed to extract carbon dioxide from the flue gases of coal-fired power plants, pipe it to suitable geological formations and bury it permanently. Burning coal would be rendered safe so there was no need to invite "economic ruin" with policies mandating emission reductions. Quickly branded "clean coal", the promise of the technology was increasingly relied on by the world coal industry to weaken policy commitments and spruce up its image (World Coal Association 2011). The promise of CCS has been used repeatedly by both governments (e.g. United States, United Kingdom, Australia, Germany) and economists as a justification for building new coal-fired power plants, and torrents of public funding have flowed to CCS research. The Obama Administration's 2009 stimulus bill allocated US$3.4 billion and the US Department of Energy announced it would provide US$2.4 billion to "expand and accelerate the commercial deployment of carbon capture and storage technology" (Anon. 2009b). In the same month, the Rudd Government in Australia announced it would commit A$2.4 billion to an industrial-scale demonstration project (Kerr 2009). The high hopes invested in CCS provoked the conservative business magazine *The Economist* to comment that "the idea that clean coal ... will save the world from global warming has become something of an article of faith among policymakers" (Anon. 2009b).

Yet from the outset impartial experts argued that the promise of CCS was being exaggerated (Smil 2008). Even supporters of CCS conceded that the technology, if it worked, would have no impact on global emissions until at least the 2030s, well after the time scientists say deep emission cuts must begin. The most damning assessment was made in 2009 by the *Economist* in an editorial titled "The illusion of clean coal":

> The world's leaders are counting on a fix for climate change that is at best uncertain and at worst unworkable. ... CCS is not just a potential waste of money. It might also create a false sense of security about climate change, while depriving potentially cheaper methods of cutting emissions of cash and attention – all for the sake of placating the coal lobby (Anon. 2009a).

The *Economist* was echoing the warnings of critics who had identified one of the major risks as the way in which CCS would undermine global mitigation efforts by giving national governments an excuse to do nothing in the hope that coal plants could be rendered safe. Events seem to have proven the critics right. Despite the hype and public investment, the promise of CCS is now collapsing. In October 2011 a major CCS project at the Longannet power station in Scotland was cancelled on concerns about its commercial viability without further subsidies (Gersmann and Harvey 2011). In November 2010 Shell's Barendrecht carbon capture project in the Netherlands was cancelled due to local opposition. A month later ZeroGen, a huge project identified by the Australian government as a "flagship" carbon capture project and promoted as "one of the first commercial-scale IGCC with CCS projects in the world", was shelved because of cost blow-outs and technical difficulties (http://www. carboncapturejournal.com/displaynews.php?NewsID=707&PHPSESSID=d2 hv0cjlmijbes2p17jmelllp6).

The case of CCS is a vivid illustration of moral hazard; yet it is into this political and commercial environment that geoengineering arrives. It is presented as a solution to the same global warming problem, to the same politicians, with the same resistant industry, and the same public prone to wishful thinking. The conditions seem perfect for moral hazard.

7. THE SLIPPERY SLOPE

There are good grounds for thinking that research and testing of SRM technologies would set the world on a slippery slope to ultimate deployment. Technological lock-in is a well-recognized problem (Perkins 2003). Already, patents in geoengineering techniques are being issued and start-up companies are attracting significant investors, including Bill Gates. A lobby group of scientists and investors is beginning to form and it is likely to become more influential as geoengineering becomes normalized in the public debate, not least with the publication of the next IPCC report. Scientists engaged in geoengineering research have argued vigorously against any early regulation of their activities, insisting that society should take a hands-off approach until there is a risk of significant harm from tests and experiments (Morgan and Ricke 2010). There is therefore a legitimate concern that the knowledge generated by geoengineering research will be misused in foreseeable ways.

However, the strength of the slippery slope danger depends in part on the absence of technological hurdles that appear insurmountable. While the experience with CCS points to the strength of the moral hazard concern about geoengineering, it also suggests a limit to the slippery slope argument. A powerful constituency formed around the promise of CCS, perhaps reaching its pinnacle with the creation in 2009 of the Global Carbon Capture and Storage Institute (Energy Matters 2009). CCS is not dead yet, but as the technical difficulties become more apparent, CCS is waning as a credible alternative to emission reductions, and the CCS lobby's momentum is stalling. The slippery slope towards the deployment of, say, sulphate aerosol spraying will depend on continued research and testing not turning up some insuperable risk or obstacle that its more open-minded supporters cannot ignore.

8. FROM CONSEQUENCES TO PRINCIPLES

The moral corruption, moral hazard and slippery slope arguments discussed above are framed in consequentialist terms, that is, as dependent on outcomes – i.e. that promoting geoengineering, even just geoengineering research, will increase the likelihood of implementing practices and policies that will have problematic results (or that will have worse results than possible alternatives). As solar radiation management is certain to be environmentally less effective than carbon abatement (especially as it does not reduce, and may hasten, ocean acidification), to the extent that political leaders succumb to the temptation to avoid abatement measures, solar radiation management is ethically dubious on consequentialist grounds.

But there is a non-consequentialist moral hazard objection – geoengineering may facilitate the continuation of bad behavior and on that basis would be wrong. If we resort to Plan B then the climate-policy obstructionism of ExxonMobil, for example, would be rewarded. More generally, those most negligent would be able to use geoengineering to escape their responsibilities for causing climate change. The wrong would be compounded if rich countries with high emissions pursued climate engineering instead of abatement. Solar radiation management would entrench the failure of the North in its duties towards the global South. This is another way of making the case that what matters ethically about geoengineering is not only the outcome but also the human virtues or faults it reveals.

The deepest disagreement over the ethics of geoengineering is in fact a dispute about what "ethics" means. The three rationales for geoengineering are all based on a consequentialist view in which the ethics of climate regulation depends solely on the consequences, compared to alternative actions. In its narrow utilitarian version (the best-option argument), the question of whether it is ethically justified intentionally to shift the planet to a warmer or cooler climate depends on an assessment of the costs and benefits of the new state compared to the old one, where costs and benefits are evaluated in terms of their effects on human well-being (Gardiner 2010a). The two "necessary evil" arguments, although more nuanced, are also confined to assessments of consequences more broadly defined.

For those who reject narrow consequentialism, the idea that the "ethical" can be decided solely by instrumental calculation is itself unethical. For consequentialism it is always *in principle* ethically justified to engineer a different climate, even though the process of calculation may show that it is imprudent. If SRM proves cheaper then it would be unethical *not* to deploy it.

One immediate implication of the consequentialist approach is that there is nothing inherently preferable about the natural state, including the pre-industrial climate. Depending on the assessment of human well-being, there may be a "better" temperature or climate as a whole. In other words, it is ethically justified for humans to "set the global thermostat" in their interests.

In the following sections I consider non-consequentialist concerns regarding geoengineering.

9. HUBRIS – OR WHY WE ARE NOT GODS WHEN IT COMES TO GEOENGINEERING

Despite its moniker, the concern about playing God is not confined to theists but may resonate just as strongly with non-theists (or a-theists) as well. The nontheistic conception of Playing God entails crossing a boundary to a domain of control or causation that is beyond the proper realm of human capabilities or rightful roles. In this view, there is a boundary around what humans should attempt or aspire to because of our intellectual, emotional or moral limitations, or because, in a stronger claim, the distinction between domains is part of the proper order or "the scheme of things" within which one can find what it means to be human.

In the debate over human genetic enhancement the playing God argument has been prominent. Biologically, DNA is the essence of life, coding all of the information that makes an individual unique. As such, tinkering with genes (and especially the germ-line, or changes to DNA that can be passed on) can be seen by the theist as invading the sacred or by the atheist as disturbing the essential dignity of the human. Michael Sandel argues that it is the gifted character of human capacities and potentialities that incites a natural regard, and that there is something hubristic and unworthy about attempting to overrule or improve upon this gift through genetic enhancement. Manipulating genes to human ends is "a Promethean aspiration to remake nature, including human nature, to serve our purposes and satisfy our desires" (Sandel 2009, 78).

The particulars are not of much help in the case of solar radiation management because we are not talking about transforming humans but the world in which humans live. Yet global dimming via sulfate aerosol injections is a similarly Promethean aspiration to remake "nature" to serve our purposes, this time not at the microscopic level of DNA but at the macroscopic level of the Earth as a whole. The domain being invaded is not that of the essential code of each life but the sphere in which all life was created or emerged. With solar radiation management the concern is not so much a lack of gratitude for a unique and precious gift, but the invasion of and dominion over the atmosphere that encompasses the planet, the benevolent ring that makes it habitable, supplies the air breathed by all living things and sends the weather. Global dimming would not only transform the atmosphere but also regulating the light reaching the Earth from the Sun. The Sun has its own god-like character because it is the source of all growth, the food of plants and thus all living things. It is the origin of the most primordial rhythms that have always governed human life – the cycles of day and night and the annual seasons.

So the intuition is that in planning to regulate the atmosphere we are crossing over into a domain properly beyond the human. To cross over successfully would require a degree of understanding and control that is beyond human capabilities. Moreover, the desire to do so betrays a kind of hubris – an overestimation of our own abilities – as well as a kind of arrogance – that the earth should be made to fit us rather than our accommodating ourselves to it. Instead of trying to remake the climate, it seems more appropriate and more within our powers to face up to our failures and attempt to become better humans.

So the argument is not so much that the consequences of geoengineering might be horrible, but that the impulse to respond to global climate change with geoengineering betrays a deep fault in the human character, one which Coady describes as "an unjustified confidence in knowledge, power, and virtue beyond what can reasonably be allowed to human beings" (Coady 2009, 165).

9.1 What we do not know

Recent developments in Earth system science have increased our knowledge substantially, but they have also uncovered yawning gaps in it. We have come to see more clearly that the climate system is extremely complex both in itself and because changes in it cannot be isolated from changes in the other elements of the Earth system – the hydrosphere, the biosphere, the cryosphere and even the geosphere (Hamilton 2012). For example, human-induced warming is expected to have large effects on global precipitation patterns, but predicting regional changes in rainfall patterns is very crude. The importance of "tipping points" that define rapid shifts from one climate state to another have become apparent from the Earth's geological record, but our understanding of why and when they occur is rudimentary. Predicting when or how thresholds might be crossed is extremely imprecise. Moreover, it is well-understood that sulfate aerosol injection, while effective at suppressing warming, would do nothing to slow the acidification of the oceans, which is interfering with the process of calcification or shell formation on which a wide array of marine animals depend for their survival. What this will mean for ocean ecosystems is barely grasped.

Apart from these the uncertainties, unknowns, and threshold effects arising from the complexity and non-linearity of the Earth system, the dominant fact is that carbon dioxide persists in the atmosphere for many centuries. So it is possible – indeed, likely – that before the larger impacts of warming are felt and measured humans will have committed future generations to an irreversibly hostile climate lasting thousands of years. Yet some economists are telling us that they can use their models to estimate future streams of the monetary values of costs and benefits to determine the optimal temperature of the Earth over the next two centuries, as if we know enough to install and begin to operate a "global thermostat" (Bickel and Lane undated).

9.2 What we would aspire to

Humans are powerful. But what kind of power do we aspire to with geoengineering? Beyond deliberate management and exploitation of particular resources or geographical areas, and beyond the unintentional degradation of land, rivers and oceans, we now aspire to take control of and regulate the atmosphere and climate of the planet as a whole. Solar radiation management would be the first conscious formulation of a "planetary technology", a plan to take control of and regulate the Earth's climate system through manipulation of the flow of primary energy to the planet as a whole. The energy that sustains all living things and ecosystems would become subject to human supervision.

Geoscientists are now arguing that humans have so transformed the face of the Earth as to justify the naming of a new geological epoch to succeed the Holocene, the 10,000 year period of unusual geological stability that allowed civilization to flourish. The Anthropocene is defined by the fact that the "human imprint on the global environment has now become so large and active that it rivals some of the great forces of Nature in its impact on the functioning of the Earth system" (Steffen et al. 2011). And the scale of this impact is now destined to increase by orders of magnitude.

Using knowledge of the great planetary cycles that over millions of years regulate the amount of solar radiation reaching the Earth, climatologists are able to predict that Earth is due for its next ice age in about 50,000 years' time (Stager 2011). However, given the persistence over millennia of increased carbon dioxide in the atmosphere, human-induced global warming, due mostly to burning fossil fuels over the period 1950–2050, is expected to cancel out the next ice age. Nothing humans have ever done comes close to the momentousness of this. If emissions rise to the higher level of expectations, which seems likely, then a century or so of economic activity will stop the subsequent ice age as well, expected in about 130,000 years, and indeed all glaciations for the next half a million years (Stager 2011). Yet, unfazed, the scientists who have uncovered these astonishing facts immediately begin to speculate about whether preventing ice ages for the next half a million years will be, on balance, a good thing or not, as if the appropriate response to knowledge of such significance is to conduct a cost–benefit analysis (Sager 2011).

9.3 Why we cannot be trusted

Even if we had the knowledge and the ability to deploy SRM as we intend, would we be likely to use it benevolently? Given that humans are proposing to engineer the climate because of a cascade of institutional failings and self-interested behaviors, any suggestions that deployment of SRM would be done in a way that fulfilled the strongest principles of justice and compassion would lack credibility. We find ourselves in a situation where geoengineering is being proposed because of our penchant to deceive ourselves and exaggerate our virtues. If a just global warming solution cannot be found, who can believe in a just geoengineering regime? It is believed that SRM would offset some of the impacts of climate change more effectively in some parts of the world than others. In some areas it may even exacerbate droughts (Robock et al. 2008). The temptation of those in control of SRM to implement it in a way that suited their interests first would be ever-present. And at no forum will non-human species have a voice.

One of the more serious concerns about the development of geoengineering technologies is the risk of unilateral deployment, that is, the decision by one nation, or even a wealthy individual, to transform the Earth's atmosphere. In this case, one agent would assume the role of climate regulator, one man playing God, a proposition so fraught with dangers that they do not need spelling out. Moreover, disputes over where to set the global thermostat could potentially escalate into wars.

The Playing god or Hubris argument is not necessarily a categorical injunction against SRM but it does ring a warning of Promethean recklessness, calling for utmost caution and deep reflection. By reinforcing the technological approach to social problems, recourse to geoengineering can be a means of avoiding serious contemplation of the deeper reasons for humanity's inability to respond to the threat.

10. TECHNOLOGY AS A SUBSTITUTE FOR POLITICS

Previously, following Gardiner, I described political inertia arising from denial or evasion of the implications of climate science as moral corruption. This moral failure is attributable to our psychological weaknesses (such as the tendency to shift blame onto others or filter the science to make it less worrying) and to the failure of political systems that are heavily influenced by sectional interests, dominated by parties that put the economy before all else and populated by individuals too timid to act on the scientific warnings. Elsewhere I have attempted to explain widespread denial and evasion in terms of the self-preoccupation and comfortable conservatism of consumer society (Hamilton 2010).

Yet there is a tendency to view these as immutable facts of modern life. Instead of promoting change in political and social structures, which are acknowledged as the source of the problem, we resort to technological solutions that we hope will bypass the blockages. In addition to the modern preoccupation with techno-fixes, advocating far-reaching social change is dismissed as utopian. Yet from the time of the French Revolution until the 1980s thinking about and advocating radical social change was part of the daily discourse of Western society, so the unwillingness to consider changes to economic, social and political structures is all the more striking in the face of a threat as grave as the climate crisis. Shunning deeper questioning of the roots of the climate crisis avoids uncomfortable conclusions about social dysfunction and the need to directly challenge powerful interests. In this way, global warming becomes no longer a profound threat to our future security and survival but just another problem that must be approached like others. Calls for a techno-fix, including geoengineering, are thus deeply conformable with existing structures of power and a society based on continued consumerism.

In his critique of the Royal Society's 2009 report on geoengineering, Gardiner poses the question bluntly:

> if the problem is social and political, why isn't the solution social and political as well [and] if, as the report asserts, we already have adequate scientific and technological solutions, why assume that research on alternative solutions will help? (Gardiner 2011)

In the end, the answer from geoengineering supporters must lie in an implicit judgment that social change is not on the table, so the only answer is to buy time for the costs of renewable energy technologies to fall far enough, or to prepare to deal with an inevitable climate emergency. Even though they will accept that the source of the problem is political, engaging in political activity as concerned citizens is shunned by most scientists. A few, like James Hansen, combine the two, but they are frowned upon by most of their colleagues for sullying science with politics. What is surprising is that the advice of scientists who advocate geoengineering research because they are afraid of stepping outside of their professional comfort zones should be given so much weight in the public debate. At least, it is surprising until we remember that conservatism is what established political systems always prefer.

WORKS CITED

American Coalition for Clean Coal Electricity (2008) 'Barack Obama Supports Developing Clean Coal Technology,' (http://www.youtube.com/watch?v=GehK7Q_QxPc) (accessed November 21, 2011).

American Enterprise Institute (2008) 'Geoengineering: A Revolutionary Approach to Climate Change,' AEI website (http://www.aei.org/event/1728) (accessed November 21, 2011).

Anonymous (2009a) 'The illusion of clean coal,' *The Economist*, 5th of March.

Anonymous (2009b) 'Trouble in store,' *Economist*, 5th of March.

D. Archer (2008) *The Long Thaw: How Humans Are Changing the Next 100,000 Years of Earth's Climate* (Princeton, NJ: Princeton University Press).

S. Barrett (2008) 'The Incredible Economics of Geoengineering,' *Environmental and Resource Economics*, 39: 45–54.

J. Bickel, and L. Lane (undated) 'An Analysis of Climate Engineering as a Response to Climate Change,' *Copenhagen Consensus Center* (available at http://www.aei. org/files/2009/08/07/AP-Climate-Engineering-Bickel-Lane-v.3.0.pdf)

Bipartisan Policy Center (2011) 'Task Force on Climate Remediation,' *Bipartisan Policy Center*, October 4th.

J. Blackstock, et al. (2009) *Climate Engineering Responses to Climate Emergencies,'* (Novim Group) archived online at: http://arxiv.org/pdf/0907.5140.

K. Caldeira, and D. Keith 'The Need for Climate Engineering Research,' *Issues in Science and Technology*, Fall 2010.

R. Carson (1965) *Silent Spring* (London: Penguin).

R. Cicerone (2006) 'Geoengineering: Encouraging Research and Overseeing Implementation,' *Climatic Change*, 77: 221–226.

C. Coady (2009) 'Playing God,' in (eds.) J. Savulescu, and N. Bostrom *Human Enhancement* (Oxford: Oxford University Press).

P. Crutzen (2006) 'Albedo Enhancement by Stratospheric Sulfur Injections: A Contribution to Resolve a Policy Dilemma?' *Climatic Change*, 77: 211–220.

Energy Matters (2009) 'Australia's Global Carbon Capture and Storage Institute Launch Energy Matters Website,' April 16th, (available at http://www.energymatters.com.au/index.php?main_page=news_article&article_id=400 accessed November 21st 2011).

A. Ferraro (2011) 'At the controls: should we consider geoengineering?' University of Reading Meteorology blog (available at http://www.met.reading. ac.uk/Data/CurrentWeather/wcd/blog/at-the-controls-should-we-consider-geoengineering/ accessed November 21, 2011).

M. Franklin (2009) 'Obama supports Rudd on clean coal,' *Australian*, March 26th.

S. Gardiner (2010a) 'Ethics and Global Climate Change,' in (eds.) S. Gardiner, S. Caney, D. Jamieson, and H. Shue, *Climate Ethics: Essential Readings*, (Oxford: Oxford University Press).

S. Gardiner (2010b) 'Is "Arming the Future" with Geoengineering Really the Lesser Evil?' (available at http://folk.uio.no/gasheim/Gar2010b.pdf).

S. Gardiner 'Some Early Ethics of Geoengineering the Climate: A Commentary on the Values of the Royal Society Report,' *Environmental Ethics*, 20: 163–188.

R. Garnaut (2008) *The Garnaut Climate Change Review: Final Report* (Melbourne: Cambridge University Press).

H. Gersmann, and F. Harvey (2011) 'Longannet carbon capture project cancelled,' *Guardian* October 19th.

Greenpeace (2008) 'False Hope: Why carbon capture and storage won't save the climate,' (Amsterdam: Greenpeace International).

C. Hamilton (2006) 'The Worldview Informing the Work of the Productivity Commission: A Critique' (available at www.clivehamilton.net.au).

C. Hamilton (2007) *Scorcher: The dirty politics of climate change*, (Melbourne: Black Inc.).

C. Hamilton (2010) *Requiem for a Species: Why we resist the truth about climate change* (London: Earthscan).

C. Hamilton (2012) 'The Ethical Foundations of Climate Engineering,' in W. Burns and A. Strauss (eds) *Climate Change Geoengineering: Legal, Political and Philosophical Perspectives*, (Cambridge: Cambridge University Press).

C. Hamilton, and T. Kasser (2009) 'Psychological Adaptation to the Threats and Stresses of a Four Degree World,' a paper for "Four Degrees and Beyond" conference, Oxford University (available at http://www.clivehamilton.net. au/cms/media/documents/articles/oxford_four_degrees_paper_final.pdf accessed November 19th 2011).

Heartland Institute (2007) 'Geo-Engineering Seen as a Practical, Cost-Effective Global Warming Strategy,' *Heartland Institute*, December 1st Home Page: (http://news.heartland.org/newspaper-article/2007/12/01/geo-engineering-seen-practical-cost-effective-global-warming-strategy).

J. Hoggan (2009) *Climate Cover-Up* (Vancouver: Greystone Books).

International Energy Agency (2009) 'Technology Roadmap: Carbon Capture and Storage,' (Paris: IEA).

Ipsos MORI (2010) 'Experiment Earth? Report on a Public Dialogue on Geoengineering,' (available at http://www.nerc.ac.uk/about/consult/geoengineering-dialoguefinal-report.pdf).

D. Jamieson, and H. Shue (eds.) *Climate Ethics: Essential Readings*, (New York: Oxford University Press).

C. Kerr (2009) 'Carbon capture to save industry,' *Australian*, May 13th.

J. Kirkland (2010) Australia's Desire for Cleaner Coal Falls Prey to High Costs, *New York Times* December 23rd (available at http://www.nytimes.com/cwire/2010/12/23/23climatewire-australias-desire-for-cleaner-coal-falls-pre-84066.html?pagewanted=1accessed November 21st 2011).

L. Lane, K. Caldeira, R. Chatfield , and S. Langhoff (2007) 'Workshop Report on Managing Solar Radiation,' (California: Ames Research Center).

M. Lawrence (2006) 'The Geoengineering Dilemma: To Speak or Nor to Speak,' *Climatic Change*, 77: 245–248.

S. Levitt, and S. Dubner (2009) *Superfeakonomics: Global Cooling, Patriotic Prostitutes, and Why Suicide Bombers Should Buy Life Insurance*, (New York, NY: HarperCollins).

A. Madrigal (2010) 'Climate Hackers Want to Write Their Own Rules,' *Wired*, March 23rd.

M. G. Morgan, and K. Ricke (2010) 'Cooling the Earth Through Solar Radiation Management: The need for research and an approach to its governance,' (Geneva: International Risk Governance Council) (available at http://www.see.ed.ac.uk/~shs/Climate%20change/Geo-politics/Morgan%20and%20Ricker%20SRM.pdf).

R. Nelles (2007) 'Germany Plans Boom in Coal-Fired Power Plants Despite High Emissions,' *Der Spiegel Online*, March 21st.

N. Oreskes, and E. Conway (2010) *Merchants of Doubt: How a Handful of Scientists Obscured the Truth on Issues from Tobacco Smoke to Global Warming*, (New York, NY: Bloomsbury).

F. Pals (2010) 'Shell's Barendrecht Carbon Capture Project Cancelled,' *Bloomberg Business Week*, September 4th (available at http://www.businessweek.com/news/2010–11–04/shell-s-barendrecht-carbon-capture-project-canceled.html accessed May 18th 2011).

R. Perkins (2003) 'Technological "lock-in",' *Internet Encyclopaedia of Ecological Economics*, International Society for Ecological Economics (available at http://www.ecoeco.org/pdf/techlkin.pdf).

A. Robock, L. Oman, and G. Stenchikov (2008) 'Regional climate responses to geoengineering with tropical and Arctic SO_2 injections,' *Journal of Geophysical Research*, 13.

Royal Society (2009) *Geoengineering the Climate: Science, governance and uncertainty*, (London: The Royal Society).

J. Sachs (2009) 'Living with Coal: Addressing Climate Change,' speech to the Asia Society New York, June 1st.

M. Sandel (2009) 'The Case Against Perfection: What's Wrong with Designer Children, Bionic Athletes, and Genetic Engineering,' in J. Savulescu, and N. Bostrom (eds.) *Human Enhancement*, (Oxford: Oxford University Press).

Secretary of State for Energy and Climate Change (2010) Government Response to the House of Commons Science and Technology Committee 5th Report of Session 2009–10: The Regulation of Geoengineering, The Stationery Office.

I. Singh (2009) 'Carbon capture, storage projects funded,' *The Money Times*, May 18th (available at http://www.themoneytimes.com/20090518/carbon-capture-storage-projects-funded-id-1068423.html accessed November 21st 2011).

V. Smil (2008) 'Long-range energy forecasts are no more than fairy tales,' *Nature*, 453: 154–8.

C. Stager (2011) *Deep Future: The Next 10,000 Years of Life on Earth*, (New York: Thomas Dunne Books).

A. Steffen (2009) 'Geoengineering and the New Climate Denialism,' *Worldchanging*, April 29th (available at http://www.worldchanging.com/archives/009784.html).

W. Steffen, J. Grinevald, P. Crutzen, and J. Neil (2011) 'The Anthropocene: conceptual and historical perspectives,' *Philosophical Transactions of the Royal Society A*, 369: 842–867.

N. Stern (2007) *The Economics of Climate Change* (Cambridge: Cambridge University Press).

C. Taylor (1988) 'Review of The Fragility of Goodness by Martha Nussbaum,' *Canadian Journal of Philosophy*, 18: 4.

S. Vanderheiden (2008) *Atmospheric Justice: A Political Theory of Climate Change*, (New York: Oxford University Press).

M. Weitzman (2008) 'On Modelling and Interpreting the Economics of Catastrophic Climate Change,' REStat Final Version, July 7th.

World Coal Association (2011) 'Failure to widely deploy CCS will seriously hamper international efforts to address climate change,' (available at http://www.worldcoal.org/carbon-capture-storage/).

DISCUSSION QUESTIONS

1. The author argues that there is moral corruption and moral hazard associated with the pursuit of geoengineering. What does he mean by these? To you agree that geoengineering invites moral corruption and is a moral hazard?

2. Do you think there is anything ethically problematic with merely conducting scientific research related to geoengineering? Why or why not?

3. Do any of the ethical concerns raised regarding geoengineering seem to you to be particularly important? Do any of them seem unimportant? Why?

4. If we knew that geoengineering were safe and effective – i.e. it would work and would not have any unexpected or unintended negative effects – would there be anything ethically problematic about it? Why or why not?

5. Do you think carbon capture and storage and solar radiation management are ethically equivalent? Or are there differences between them that are relevant to ethical evaluation?

6. Do you agree with the author that responses to global climate change should be primarily political, rather than technological? Why or why not?

7. Can you think of any considerations in favor of geoengineering that are not discussed in this chapter? Can you think of any concerns about geoengineering that people might have that are not addressed?

30 ECOSYSTEMS UNBOUND: ETHICAL QUESTIONS FOR AN INTERVENTIONIST ECOLOGY

Ben A. Minteer and James P. Collins

CHAPTER SUMMARY

Most of the Earth's ecosystems have been significantly impacted by human technologies and activities. Moreover, climate change, habitat loss, invasive species, agriculture, and pollution (among other factors), continue to cause very high rates of ecological change. In this Chapter, Ben A. Minteer and James P. Collins argue that it is becoming increasingly difficult to characterize nature as a pristine wilderness separate from humans and to hold this up as an ideal for which ecosystem management should strive. The alternative, they suggest, is to embrace more interventionist and pragmatic approaches to ecosystem management and species conservation. They discuss two such approaches in detail – managed relocation and Pleistocene re-wilding. Both management strategies involve introducing species into systems where they historically have not been; and Pleistocene re-wilding in particular involves extensive ecosystem design and manipulation. Minteer and Collins suggest several aspects of managed relocation and Pleistocene re-wilding that should be considered when evaluating whether or not to pursue and support them.

RELATED READINGS

Introduction: Respecting Nature (2.7.3); Types of Value (3.1)

Other Chapters: Clive Hamilton, *Ethical Anxieties about Geoengineering* (Ch. 29); Christopher J. Preston, *Evolution and the Deep Past* (Ch. 36).

1. THE FALL OF THE WILD

"None of Nature's landscapes are ugly so long as they are wild," wrote the renowned wilderness advocate and naturalist John Muir at the dawn of the 20th century (Muir 1901: 4). For Muir, wilderness possessed a distinctive mix of spiritual, aesthetic, and moral qualities that raised it far above the anthropogenic environs of the city and the farmed landscape. It was where the forces of nature, rather than the forces of civilization, dominated, a place deserving of our collective admiration and respect.

Today, Muir's notion of wilderness as a place apart from human manipulation and influence has fallen on hard times. For more than two decades geographers, anthropologists, ethnoecologists, and popular writers have challenged traditional assumptions of a wild and "untrammeled" pre-European North America. These assumptions are said to reflect an outmoded "myth of the pristine" impeached by the mounting evidence of extensive Native American management of ecosystems (e.g., earthworks, forest clearing, fire, and hunting; see Mann 2005). Even Muir's wild Yosemite has been revealed to be largely a fiction given the role of California Indians in shaping the landscape of the Sierra Nevada (Anderson 2005).

By the end of the 20th century, many scholars, most notably the historian William Cronon (1995), were advocating the de-privileging of wilderness in American environmental thought and action. The "wilderness myth" of Muir and other preservationists, Cronon suggested, had caused great damage to American environmentalism by inserting a false dualism between humans and nature. This view also celebrated a wildland vision that led to the neglect of the city, the suburb, and the countryside, environments that Cronon argued deserved more attention from environmentalists given the need for an ethic of sustainable use.

Muir's wilderness ideal has also increasingly been challenged by a deepening understanding of the extent of the human modification and domination of global systems in the contemporary era, especially since the Industrial Revolution. Anthropogenic climate change, rapid and intensive urbanization, the disruption of biogeochemical cycles, the spread of invasive species, and related environmental change agents have all profoundly transformed ecological systems (Rockstrom et al. 2009). Human-dominated rather than nature-dominated (e.g., wilderness) ecological systems are now the planetary norm. According to some assessments, by the second half of the 20th century the terrestrial biosphere transitioned from a system shaped primarily by processes that lacked a major human influence to one driven mainly by human activities (Ellis 2011, p. 1029).

Although ecosystems that lack major human influences still exist around the globe, most ecosystems today display varying, but measurable degrees of human modification and manipulation, especially if atmospheric changes are included. In recent years, a number of scientists have even described these anthropogenic impacts as marking a geological divide in the planet's history, a transition from the Holocene Epoch to the new era of the "Anthropocene" (Zalasiewicz et al. 2010).

In light of such trends, many scientists are realizing that appeals to a pristine and predictable environment free of human influence as a normative and scientific standard no longer have much relevance for conservation, restoration, and ecological management (Marris 2011). For example, in many cases the goal of restoring ecosystems to pre-disturbance historical states will be anachronistic, given the scale and rate of ecological transformation, or simply impossible to achieve in practice (Jackson and Hobbs 2009). This recognition has led some scientists to call for a more future-oriented rather than a historical model of restoration. The goal is not to recreate past conditions and historical species assemblages, but rather to intervene in distressed and often rapidly changing systems to rehabilitate key ecological functions for the future (Hobbs and Cramer 2008).

The move toward a more interventionist, more future-oriented, and less-wilderness-centric model of conservation and ecological management creates a significant hurdle for the philosophy of nature preservationism, perhaps the dominant tradition in American environmental ethics. Many, if not most environmental philosophers have historically followed Muir in celebrating the wilderness and notions of natural value as the purest expression of a non-anthropocentric (or "nature"-centered) worldview, an essential counterpoint or corrective to the more artificial and technological world of the city and the suburb (e.g., Taylor 1986, Rolston 1994). The traditional preservationist paradigm may be characterized by its embrace of three basic tenets: 1) wild or "pristine" landscapes are the bearers of ethical value and the focal points of advocacy and policy, 2) technological interference in nature should be minimized, and 3) historical integrity is the primary conservation and restoration target.

This outlook, however, is increasingly out of sync with the emerging reality presented by a human-dominated and rapidly changing planet, an environment in which appeals to "pristine landscapes" and "historical baselines" will have less and less utility and credibility in conservation and ecological management discussions. The world of the future will be composed of a growing number of "no analog" communities that differ from ecosystems of the recent Holocene and the dynamics that have shaped them (Williams and Jackson 2007). It is within this "new normal" of rapid environmental change, the eclipse of ecological and evolutionary baselines assumed as typical, and the expansion and acknowledgment of human-dominated systems that many scientists have begun to advocate for a more interventionist and anticipatory agenda, one that breaks in fundamental ways with the preservationist model (Hobbs et al. 2011).

In the rest of this chapter, we will briefly discuss some of the emerging proposals and debates that reflect this more interventionist model of conservation and restoration. We will begin with a discussion of "managed relocation," focusing on the scientific and ethical questions surrounding this controversial conservation strategy to relocate species in advance of global change. We then consider Pleistocene Re-wilding, an idea that has similarities to managed relocation but that is motivated by distinct ecological and conservation concerns and raises its own ethical challenges. Finally, we end the chapter by considering how managed relocation and Pleistocene Re-wilding lead to discussions of the relative value of novel and designer ecosystems in conservation and environmental management, emphasizing the need for a pragmatic yet cautious ecological ethics that can guide intelligent and precautious interventions into rapidly changing ecological communities.

2. THE ETHICS OF MANAGED RELOCATION

Historically, much of the ethical discussion surrounding conservation focused on establishing good reasons for caring about the plight of threatened species – and convincing decision makers to adopt policies to protect these populations in their habitats (e.g., Norton 1987, Sarkar 2005). Although there are often disagreements, especially among philosophers, about the ethical reasons for undertaking conservation, there is nevertheless a policy accord in the conservation community to save species threatened with extinction by human activities such as habitat destruction and fragmentation, pollution, and unsustainable harvesting (Minteer 2009). Yet, global climate change (GCC), often working in tandem with other forces of global change, is challenging many traditional ethical arguments and conservation strategies. It is forcing the conservation community to confront novel and difficult ethical questions concerning the value and importance of species under rapidly changing climatic and ecological conditions (Camacho et al. 2010).

Conservation scientists recognize that GCC during this century will alter current ecosystems. For example, erosion and wetland loss due to GCC will change coastal regions, while increasing ocean acidification affects the entire ocean system (Rozenzweig et al. 2007). At the species level, GCC is linked to physiological, phenological, and distributional changes (Parmesan 2006). A confounding factor that complicates extinction risk in many species is the degree to which GCC combines with and magnifies traditional biodiversity threats, including land use change and the proliferation of invasive species (Root and Schneider 2006). One influential assessment placed a third of the world's species on a path to climate-driven extinction as a result of GCC (Thomas et al. 2004). While this prediction depends on the actual speed and extent of planetary warming, the message for conservationists is that rapid GCC is an emerging, significant, and complex threat to biodiversity.

Consider the following thought experiment, which reveals how GCC interacts with other factors driving biodiversity loss. Imagine the climate warms to such a degree that it exerts significant stress on a particular wildlife or plant population. What happens to this population if organic evolution does not proceed quickly enough (i.e., the warming is too rapid), or because population dispersal is impossible due to landscape barriers (e.g., waterways, highways, cities)? In the most extreme cases, population extinction will occur in the historical range as a result of the rapidly changing environmental conditions.

An anticipatory and activist conservation strategy could avoid this outcome. Specifically, we could intervene *before* populations thought to be at risk due to GCC enter the extinction vortex. One of the more radical preemptive conservation strategies is translocating populations judged threatened by present or future climate change; i.e., moving species to novel habitats outside their historical range. Not surprisingly, what has been called "assisted colonization" or "managed relocation," is controversial (see, e.g., McLachlan et al. 2007, Ricciardi and Simberloff 2009a, b). The prospect of using such a tactic to save species is revealing a philosophical and policy rift in the conservation community regarding the ethical justification for and ecological consequences of radical interventions to save species under GCC.

Torreya taxifolia, an endangered conifer with a shrinking range in Florida's panhandle, is the best-documented case of managed relocation (MR). A conservation advocacy organization, the Torreya Guardians, planted seedlings of the species in North Carolina in an attempt to save it from climate-driven decline (see http://www.torreyaguardians.org/). In the UK, scientists relocated two species of butterflies to habitat predicted to be more suitable for the organisms given projected climate shifts (Willis et al. 2009). Forest ecologists in British Columbia moved more than a dozen commercial tree species to help them escape pine beetle outbreaks (accelerated by GCC; see Marris 2011), while fisheries scientists in Australia experimented with relocating 10,000 rock lobsters to boost production of the fishery and increase its socio-ecological resilience under predicted environmental change (Green et al. 2010).

The primary objection to such practices is their potential to disrupt the ecological integrity of the new systems (Sandler 2009). Some skeptics have pointed out that we simply lack the ability to predict accurately how a species will perform in a new ecosystem and this ignorance of the potential for disruption should disqualify MR as a conservation practice (see, e.g., Ricciardi and Simberloff 2009a, b, Davidson and Simkanin 2008). For these critics MR is "ecological roulette" (Ricciardi and Simberloff, 2009a, p. 252). Others argue that MR may fail to save relocated species given the spotty success record of past translocations – and the fact that relocated populations may be particularly vulnerable to other threats in their new range, especially if the introduced population size is small (Huang 2008).

On the other side of the debate, some suggest that many of these risks are manageable, especially if we develop decision protocols to help scientists, citizens, and policy makers address them in a systematic and informed manner (Hoegh-Guldberg et al. 2008, Richardson et al. 2010). There is also the broader moral argument that, even if such risks cannot always be minimized, our obligation to save species from human harm requires that we make serious efforts to conserve them, despite the fact that doing so may be difficult or costly – or plagued by unpredictability and the potential for unwanted ecological consequences.

Clearly, one of the implications of MR that makes it so controversial is its break with the aforementioned philosophy of preservationism that has historically underpinned conservation efforts, especially in the United States. As we have seen, one of the core normative tenets of preservationism is that species should be protected within historical habitats, which are the geographical ranges and evolutionary contexts in which they evolved. To the degree that MR involves translocating populations outside their native ranges for conservation purposes, it therefore breaks from the longstanding preservationist model of saving species in their historical habitats.

MR also challenges traditional preservationist norms surrounding human intervention in and manipulation of ecological systems, although this is more a matter of degree rather than a fundamental break with tradition. For example, conservationists have long engaged in practices that could be characterized as "interventionist," such as captive breeding and the manipulation of experimental populations. Still, the degree of intervention suggested by MR, especially its prospective and anticipatory character, crosses the line of acceptable conservation practice for many critics. MR's departure from historical conditions and its transgression of biogeographical boundaries strike critics as a different kind, rather than a different degree, of human intervention in ecological systems.

Here are some of the key ecological ethics-related questions for MR as a conservation strategy under rapid environmental change (see also Minteer and Collins (2010):

- What is the most scientifically justified and ethically defensible process for choosing candidate populations for relocation (and selecting the recipient ecosystems)?

- Who should make MR decisions and carry out particular managed relocations?
- How can we ensure that MR efforts do not undermine ecological integrity? Should the historical integrity of ecosystems always have priority over survival of individual species?
- Should we be concerned that MR might actually weaken our ethical resolve to address the root causes of GCC by, for example, diminishing our commitment to reducing global greenhouse gas emissions if species can simply be moved if threatened?
- Does a policy of MR demonstrate proper respect for vulnerable species and ecosystems, or does it convey an attitude of domination that clashes with core conservation values, such as Aldo Leopold's (1949) land ethic?

In summary, the preemptive movement of species to new habitats to help them adapt to anthropogenic climate change raises many ethical concerns about our responsibility to conserve biodiversity in this century. At the deepest level, these concerns force us to confront difficult questions about the manipulation of nature and the perils of adopting a more aggressive, yet well-intentioned stance toward the protection of plants and wildlife facing rapid ecological changes.

3. PLEISTOCENE RE-WILDING: THE WILDERNESS MYTH REBOOTED?

Perhaps an even more controversial conservation proposal than managed relocation is "Pleistocene re-wilding" (PRW). PRW refers to the introduction of large wild vertebrates into North America as ecological proxies for species – e.g., mammoths, the American cheetah, the American lion – that went extinct at the end of the Pleistocene (~ 13,000 years before present) (Donlan et al. 2005, 2006). The introduced megafauna would include species currently extant in Africa and Asia (e.g., lions, elephants, cheetahs, camels) that are either related to extinct Pleistocene wildlife or deemed sufficiently similar to them, and would be sourced from both captive populations in the U.S. and from in-situ sites.

Proposed by a group of conservation biologists and activists, the goal of PRW is to restore past evolutionary and ecological processes that were lost with the megafaunal extinctions in North America. It is also defended as a way to increase the geographic range of many threatened Old World vertebrates via an intercontinental network of megafaunal reserves, thus increasing these species' chances for long-term survival. The proponents of PRW envision the re-wilding of the continent to take place over time and in a series of overlapping phases, starting with smaller pilot projects, followed by the more dramatic establishment of lion, cheetah, and elephant populations on private reserves and the creation of large "ecological history parks" that would allow free-ranging (though fenced) wildlife to populate the continent that their ancestors – or their Pleistocene analogs – once roamed (Donlan et al. 2005).

In addition to the scientific rationale for their proposal, supporters of PRW defend the introduction of proxies for North America's lost charismatic megafauna on societal grounds, arguing that ecological history parks will be a boon for tourism in their host communities and that re-wilded landscapes will restore a deep evolutionary landscape aesthetic to the continent (Donlan et al 2005, p. 913). PRW proponents also argue that the restoration of "closely related" species as ecological proxies for extinct megafauna will discharge an ethical obligation created by the actions of Pleistocene hominids, who PRW supporters argue bear a degree of responsibility for the megafaunal extinctions (Martin 2007). Finally, they suggest that PRW will energize the conservation movement, transforming conservation biology from a "doom-and-gloom" discipline focused on bleak assessments of global species losses to a more optimistic and proactive field actively involved in the restoration of ecological and evolutionary processes (Donlan et al. 2006, pp. 664–665).

The response to PRW within the fields of ecology and conservation biology has been mostly negative. Critics have identified a number of problems with the proposal, taking issue with scientific and societal justifications for PRW. Similar to the debate over MR, many scientists have expressed the concern that by introducing non-indigenous species, PRW could result in the disruption and destruction of North American ecosystems and native species. Specifically, worries have been voiced about PRW's potential to: 1) establish new predators for native fauna and create new grazing pressure on native vegetation, 2) facilitate genetic hybridization between introduced and native wildlife, and 3) accelerate the transmission of disease and parasites from Old to New World species (Rubenstein et al. 2006; Caro 2007). Furthermore, some critics have pointed out that the North American landscape has obviously continued to evolve and change over the past 13,000 years. The introduction of large mammals into places where they have not existed for millennia is not likely to restore the ecological potential of the landscape to Pleistocene levels (Rubenstein et al. 2006, p. 233).

As mentioned above, the economic, social, and political dimensions of PRW have also been criticized. Some skeptics have argued that the costs of creating the ecological history parks could be prohibitive, and that PRW efforts could divert funds from more traditional conservation and restoration projects focused on recovering and protecting native biodiversity (Caro 2007). Another concern is that the new "Pleistocene Parks" could reduce the political will to conserve native species in the U.S. and that it will exacerbate ongoing conflicts between humans and wildlife, tensions that have been difficult enough to manage in the case of reintroducing native predators such as wolves to the American west (Rubenstein et al. 2006). Critics have also objected to the anticipated "plundering" of wildlife from source countries to stock new American wildlife parks (Caro and Sherman 2009).

Finally, many of the PRW critics end up reaching a conclusion similar to that of the MR skeptics: Conservation resources would be much better spent supporting traditional in-situ approaches (i.e., protecting species in their native ranges) than on radical and risky efforts like PRW. According to this view, given the potential hazards of PRW, as well as the fact that the evolutionary trajectories of extinct species cannot really be recreated, any "re-wilding" efforts should focus on reintroducing *native* species into their *historical* habitats to enhance ecosystem functions and increase the geographic range of threatened indigenous wildlife (Rubenstein et al. 2006, p. 235). Models of "re-wilding" (without the "Pleistocene" designation) based more on "recent" historical baselines – i.e., within the last 100–150 years or so – and that entail returning native species to ranges they previously inhabited, typically draw stronger support from these critics (Fraser 2009). Some PRW critics have gone so far as to propose a moratorium on importing non-native megafauna into ecosystems, a move that would also presumably apply to MR (see, e.g., Caro and Sherman 2009; Hyunh 2010).

For their part, PRW advocates acknowledge the risks identified by the skeptics, although they insist that "sound research, prescient management plans, and informed public discourse for each species on a case-by-case basis and locality-by-locality basis" will help minimize the ecological and socio-political challenges confronting PRW efforts (Donlan et al. 2006, pp. 672). They also contest claims that PRW will siphon tourists (and tourists' dollars) from African ecotourism (Donlan et al. 2006, pp. 672–673). But ultimately, PRW supporters end up turning the argument back on their critics, asking them whether they are willing to resign themselves to global biodiversity loss and settle for "an American wilderness emptier that it was just 100 centuries ago"(Donlan et al. 2005, p. 914).

The ethical implications of PRW are intriguing given its combination of elements of the traditional wilderness idea (e.g., the appeal to a long-lost past when the North American continent was far wilder than today) with a more interventionist agenda entailing the active design and technological manipulation of ecosystems in an effort to "reset" ecological and evolutionary processes. Even as they seek to restore wildness via the introduction of flagship wilderness fauna such as lions, cheetahs, and elephants, PRW advocates openly acknowledge the inadequacy of the older myth of the untrammeled wild for conservation action: "Earth is nowhere pristine," they conclude (Donlan et al. 2005, p. 913). But, rather than turning away from the wilderness idea and pursuing more anthropocentric agendas, supporters of PRW find it logical to embrace the ethical imperative of designing and introducing more "wildness" to the North American landscape.

For some PRW skeptics, however, such aggressive interventions in natural systems cross a clear ethical boundary: they reinforce "the illusion of humans attempting to absolutely exert control over the natural world" (Hyunh 2010, p. 101). Despite their good intentions, the "carefully managed ecosystem manipulations" proposed by the PRW boosters (see Donlan et al. 2005, p. 913) are viewed by critics as entailing destructive alterations to ecosystems, actions that embody suspect attitudes of control and domination. These arguments evoke earlier debates in the field of environmental ethics about the unacceptability of ecological restoration, which some philosophers have argued degrades natural value and encourages the desire for technological mastery over nature (e.g., Katz 1997, Elliott 1997).

In sum, similar to the idea of managed relocation (MR), PRW raises a number of ethical and normative questions surrounding human intervention in ecological systems to return them to Pleistocene-approximating conditions:

- Do the ecological risks of PRW outweigh the potential conservation and ecological benefits of the reintroduction of proxy species for an extinct Pleistocene megafauna?
- As with the MR proposal, who should oversee and conduct PRW activities? How should candidate species and recipient ecosystems be selected?

- Should we draw an ethical distinction between relocating species to habitats outside their current historic range to save them from future climate stress (MR) and relocating megafauna from Africa and Asia to North America to "re-wild" the continent?
- Does our responsibility for present and expected species extinctions from global climate change provide a more compelling reason to move species to new systems (MR) than the ethical duty to "make amends" for the role of Pleistocene hominids in causing megafaunal extinctions?
- Does PRW display proper respect for the autonomy and integrity of native species and ecosystems or does it embody an unacceptable ethic of ecological control? Do such criticisms have any credence in an increasingly human-dominated world?

Pleistocene Re-wilding is clearly a controversial proposal to intervene in ecological systems and wildlife populations to restore long-lost species (or their analogs) to North America. As with managed relocation, the idea raises a number of ethical and normative issues surrounding the control of species and the alteration of ecosystems – as well as the legitimacy of pre-historic ecological and evolutionary baselines for restoration and conservation practice.

4. ECOLOGICAL RESPONSIBILITY IN THE ANTHROPOCENE

Although the disagreements over MR, PRW, and other proposed intensive interventions in ecological systems for conservation purposes will continue, there are opportunities to reduce ethical and strategic conflict by developing integrated models of conservation planning and policy. Several conservation scientists, for example, have suggested that radical strategies such as MR could be more acceptable if pursued under the broader policy goal of increasing landscape connectivity – a widely supported agenda among biologists and conservation activists to link fragmented landscapes and facilitate the natural dispersal of species (Loss et al. 2011, Lawler and Olden 2011). Ethical concerns about disrupting native species and ecosystems – and objections to the interventionist philosophy of MR – would not necessarily disappear, but might be assuaged by incorporating a more measured and balanced integration of MR within traditional conservation activities.

Paleoecologist Anthony Barnosky (2009) has proposed another strategy for reconciling more traditional preservationist values and the manipulated landscapes characteristic of MR and PRW. He outlines a bifurcated nature reserve system with 1) "Species reserves," that would require intensive human management, including activities such as MR, PRW and other conservation interventions, and 2) "wildland reserves" managed to sustain ecological processes free of significant human modification, although even these systems could not avoid human inputs to a warming climate, changing weather patterns, and an altered atmosphere. Species reserves would bear little resemblance to historical ecosystems; their primary goal would be to conserve individual species and assemblages threatened by the forces of global change. Wildland reserves would retain "wildness," or the historical integrity of ecosystems in the sense of preservation, given their insulation from direct human manipulation. Even in this case, however, preservation would be qualified given the influence of global climate change, which would result in shifts in species compositions over time. For example, unlike species reserves where active intervention to conserve species (including relocation) is permitted, species in wildland reserves would be allowed to go extinct (Barnosky 2009, pp. 207–208).

Despite efforts to harmonize strongly interventionist conservation agendas and values with more conventional in-situ/preservationist commitments, it is clear that both MR and PRW will result in novel ecosystems, i.e., assemblages of species that depart from recent historical conditions. In recent years, a growing number of ecologists have begun to champion the recognition of such novel systems and their value for conservation and restoration, as well as the provision of vital ecosystem services such as carbon sequestration, biogeochemical cycling, water provision and purification, and related functions and benefits (see, e.g., Hobbs et al. 2006; Marris 2011). By definition, novel ecosystems contain species combinations atypical of a region's recent historical record and reflect significant human agency; i.e., they are systems produced by the result of either intentional or inadvertent human activity, although they do not depend on continued human intervention for maintenance (Hobbs et al. 2006, p. 2). Like MR and PRW, the novel ecosystem idea – and particularly its defense as a valued landscape type in conservation and restoration – has drawn criticism from those ecologists and conservation scientists who view such modified and reshuffled systems as "ecological disasters" rather than ecosystems deserving of scientific attention and conservation effort (see, e.g., the discussion in Hobbs et al. [2006]).

This lingering preference for native species assemblages, historical baselines, and a generally less-manipulated landscape, however, is even more contested by emerging proposals to shift ecological research and practice to "designer ecosystems" (Palmer et al. 2004). Designer systems reflect considerable technological manipulation and human intention. At the most anthropogenic end of the design continuum, designer ecosystems will be dominated by artificial, hard-engineered components created to mitigate undesirable ecological conditions – or deliver ecosystem services via a mixture of technological and biotic elements (Palmer et al. 2004; Marris 2011). The designer landscape idea clearly poses a philosophical challenge to conservationists and environmental ethicists of a more non-anthropocentric bent, although it is a vision that has been long advocated by those who feel that societal values and ethical attachments to the environment are malleable and can be maintained – and manipulated – in favor of more contrived landscapes (see, e.g., Krieger 1973).

Regardless, it seems clear that rapid, large-scale environmental changes are forcing many conservationists to consider innovative and often controversial strategies and techniques for protecting species in this century, proposals that raise significant ethical and value-laden questions. Given what we already know about rapid environmental change and the rise of human-dominated systems, it is inevitable that species conservation and ecological protection in the Anthropocene era will require more rather than less interventionist actions and policies. This will lead to further debates regarding risks, benefits, and likely success of largely untested practices such as MR and PRW – as well as deliberations over the proper place of novel and designer ecosystems in the conservation and restoration agenda.

Saving species and sustaining ecosystems on a rapidly changing planet will require conservationists to come to grips with the simultaneously shifting standards of environmental responsibility in an era of exceptional ecological dynamism. In philosophical ethics, there is a saying (derived from Immanuel Kant) that "ought implies can": i.e., if there is an ethical requirement to perform some act, it must logically be possible to perform it. One could argue, therefore, that traditional ethical directives to restore systems to historical, pre-human dominated conditions, or to save species in their native habitats absent human influence, or to protect wildlands from human influence, become less and less binding the greater the forces and consequences of anthropogenic change, particularly after the Industrial Revolution. These preservationist ethical directives are based on a definition of the pristine that in hindsight began to disappear with the evolution of modern humans as a species – and has continued to decline since then.

A remaining question is whether the final abandonment of long-held notions of the pristine wilderness and historical baselines for restored ecosystems will eventually undercut the strong moral regard for nature that John Muir wrote about so powerfully a century ago. Yet at the same time, and despite the seemingly new rhetoric surrounding ecological intervention, novel and designer systems, and so on, we should acknowledge that humans have always been intervening in nature and designing (often intentionally) ecological systems. We have been ecological engineers for tens of thousands of years, altering and managing nature ever since the early domestication of plant and animal species (Smith 2007). What is at issue is therefore not so much the philosophical status of intervention and design in the abstract, but rather the proper way to determine the degree, type, and timing of acceptable interventions to restore, conserve, and enhance biodiversity and ecological services (Hobbs et al. 2011). The development of a pragmatic ecological ethics able to guide such actions, an ethics that carries as well a greater sense of environmental humility and precaution in the face of increasingly urgent calls for action, will be just as important as tackling the scientific and technological challenges of ecological intervention in the coming decades.

WORK CITED

M. K. Anderson (2005) *Tending the Wild: Native American Knowledge and the Management of California's Natural Resources* (Berkeley: University of California Press).

A. D. Barnosky (2009) *Heatstroke: Nature in an Age of Global Warming* (Washington, DC: Shearwater/Island Press).

A. E. Camacho, H. Doremus, et al. (2010) 'Reassessing conservation goals in a changing climate,' *Issues in Science and Technology* 26: 21–26.

T. Caro (2007) 'The Pleistocene re-wilding gambit,' *Trends in Ecology and Evolution* 22: 281–283.

T. Caro & P. Sherman (2009) 'Rewilding can cause rather than solve ecological problems,' *Nature* 462: 985.

I-C. Chen, J. K. Hill, et al. (2011) 'Rapid range shifts of species associated with high levels of climate warming,' *Science* 333: 1024–1026.

W. Cronon (1995) 'The trouble with wilderness; or, getting back to the wrong nature,' in *Uncommon Ground: Rethinking the Human Place in Nature*, ed. W. Cronon (New York: Norton).

I. Davidson, and C. Simkanin (2008) 'Skeptical of assisted colonization,' *Science* 322: 1048–1049.

J. Donlan, H. W. Greene, et al. (2005) 'Re-wilding north America,' *Nature* 436: 913–914.

J. Donlan, and J. Berger et al. (2006) 'Pleistocene rewilding: an optimistic agenda for twenty-first century conservation,' *The American Naturalist* 168: 660–681.

R. Elliot (1997) *Faking Nature: The Ethics of Environmental Restoration* (London: Routledge).

E. C. Ellis (2011) 'Anthropogenic transformation of the terrestrial biosphere,' *Philosophical Transactions of the Royal Society A* 369: 1010–1035.

C. Fraser (2009) *Rewilding the World. Dispatches from the Conservation Revolution* (New York: Picador).

B. S. Green, and C. Garder et al. (2010) 'The good, the bad and the recovery in an assisted migration,' *PLoS ONE*, 5: 1–8.

N. Hewitt, and N. Klenk, et al. (2011) 'Taking stock of the assisted migration debate,' *Biological Conservation*, 144: 2560–2572.

R. J. Hobbs, and S. Arico, et al. (2006) 'Novel ecosystems: theoretical and management aspects of the new ecological world order,' *Global Ecology and Biogeography*, 15: 1–7.

R. J. Hobbs, and V. A. Cramer (2008) 'Restoration ecology: interventionist approaches for restoring and maintaining ecosystem function in the face of rapid environmental change,' *Annual Review of Environment and Resources*, 33: 39–61.

R. J. Hobbs, and L. M. Hallett, et al. (2011) 'Intervention ecology: applying ecological science in the twenty-first century,' *BioScience*, 61: 442–450.

O. Hoegh-Guldberg, and L. Hughes et al. (2008) 'Assisted colonization and rapid climate change,' *Science*, 321: 345–346.

D. Huang (2008) 'Assisted colonization won't help rare species,' *Science*, 322: 1049.

H. M. Hyunh (2010) 'Pleistocene re-wilding is unsound conservation practice,' *Bioessays*, 33: 100–102.

S. T. Jackson, and R. J. Hobbs (2009) 'Ecological restoration in the light of ecological history,' *Science*, 325: 567–569.

E. Katz (1997) *Nature as Subject: Human Obligation and Natural Community* (Lanham, MD: Rowman & Littlefield).

M. H. Krieger (1973) 'What's wrong with plastic trees?' *Science*, 179: 446–455.

J. J. Lawler, and J. D. Olden (2011) 'Reframing the debate over assisted colonization,' *Frontiers in Ecology and the Environment*, 9: 569–574.

A. Leopold (1949) *A Sand County Almanac* (Oxford, UK: Oxford University Press).

S. R. Loss, L . A. Terwilliger, and A. C. Peterson (2011) 'Assisted colonization: integrating conservation strategies in the face of climate change,' *Biological Conservation*, 144: 92–100.

C. C. Mann (2005) *1491: New Revelations of the Americas Before Columbus* (New York: Vintage Books).

E. Marris (2011) *Rambunctious Garden. Saving Nature in a Post-Wild World* (New York: Bloomsbury).

P. Martin (2007) *Twilight of the Mammoths: Ice Age Extinctions and the Rewilding of America* (Berkeley: University of California Press).

J. S. McLachlan, J. J. Hellmann, and M. W. Schwartz (2007) 'A framework for debate of assisted migration in an era of climate change,' *Conservation Biology*, 21: 297–302.

P. C. D. Milly, J. Betancourt, et al. (2008) 'Stationarity is dead: whither water management?' *Science*, 319: 573–574.

B. A. Minteer ed. (2009) *Nature in Common? Environmental Ethics and the Contested Foundations of Environmental Policy* (Philadelphia, PA: Temple University Press).

B. A. Minteer, and J. P. Collins (2010) 'Move it or lose it? The ecological ethics of relocating species under climate change,' *Ecological Applications*, 20: 1801–1804.

J. Muir (1901) *Our National Parks* (Boston: Houghton, Mifflin and Co).

B. G. Norton (1987) *Why Preserve Natural Variety?* (Princeton, NJ: Princeton University Press).

M. Palmer, E. Bernhardt, et al. (2004) 'Ecology for a crowded planet,' *Science*, 304: 1251–1252.

C. Parmesan (2006) 'Ecological and evolutionary responses to recent climate change,' *Annual Review of Ecology, Evolution, and Systematics*, 37: 637–669.

A. Ricciardi, and D. Simberloff (2009a) 'Assisted colonization is not a viable conservation strategy,' *Trends in Ecology and Evolution*, 24: 248–253.

A. Ricciardi, and D. Simberloff (2009b) 'Assisted colonization: good intentions and dubious risk assessment,' *Trends in Ecology and Evolution*, 24: 476–477.

D. M. Richardson, J. J. Hellmann, et al. (2009) 'Multidimensional evaluation of managed relocation,' *PNAS* 106: 9721–9724.

J. Rockström, W. Steffen, et al. (2009) 'A safe operating space for humanity,' *Nature*, 461: 472–475.

H. Rolston III (1994) *Conserving Natural Value* (New York: Columbia University Press).

T. L. Root, and S. Schneider (2006) 'Conservation and climate change: the challenges ahead,' *Conservation Biology*, 20: 706–708.

C. Rosenzweig, G. Casassa, et al. (2007) 'Assessment of observed changes and responses in natural and managed systems,' in *Climate Change 2007: Impacts, adaptation and vulnerability. Contribution of Working Group II to the Fourth Assessment Report of the Intergovernmental Panel on Climate Change*, eds. Parry, M. L., Canziani, O. F., et al. (Cambridge, UK: Cambridge University Press), pp. 79–131.

D. R. Rubenstein, D. I. Rubenstein, et al. (2006) 'Pleistocene park: does re-wilding North America represent sound conservation practice for the 21st century?' *Biological Conservation*, 132: 232–238.

R. Sandler (2010) 'The value of species and the ethical foundations of assisted colonization,' *Conservation Biology* 24: 424–431.

S. Sarkar (2005) *Biodiversity and Environmental Philosophy* (Cambridge, UK: Cambridge University Press).

B. D. Smith (2007) 'The ultimate ecosystem engineers,' *Science* 315: 1797–1798.

P. W. Taylor (1986) *Respect for Nature: A Theory of Environmental Ethics.* (Princeton, NJ: Princeton University Press).

C. D. Thomas, A. Cameron, et al. (2004) 'Extinction risk from climate change,' *Nature*, 427: 145–148.

J. W. Williams, and S. T. Jackson (2007) 'Novel climates, no-analog plant communities, and ecological surprises: past and future,' *Frontiers in Ecology and Evolution*, 5: 475–482.

S. G. Willis, J. K. Hill, et al. (2009) 'Assisted colonization in a changing climate: a test-study using two U.K. butterflies,' *Conservation Letters*, 2: 45–51.

J. Zalasiewicz, M. Williams, W. Steffen, and P. Crutzen (2010) 'The new world of the "anthropocene,"' *Environmental Science and Technology*, 44: 2228–2231.

DISCUSSION QUESTIONS

1. Would you be supportive of managed relocation for climate-threatened species? What criteria should be used to determine which species should be relocated?

2. Would you support the introduction of modern proxies for Pleistocene-era fauna in the United States or Australia, for example? Why or why not?

3. What reasons are there to try to protect biodiversity if it cannot be done while preserving the historical ranges of species? That is, is it important to prevent species from going extinct if they will no longer be located in their native ranges?

4. Are there any reasons, not discussed in this chapter, for trying to preserve "wild areas"?

5. Given the situation described in this chapter, does it make sense to deemphasize trying to save species as an ecosystem management goal, rather than taking novel measures to try to save them?

6. Is there anything problematic with the attitude that humans should take responsibility for designing ecological systems? Or should this idea and practice be encouraged?

7. In what ways does our technology structure or mediate our relationships with nonhuman nature and species?

PART VIII
AGRICULTURAL TECHNOLOGIES

Agriculture is among the most important human activities. One reason, of course, is that we must all eat in order to survive and have our nutritional needs meet in order to be healthy. Another reason is the enormous scale of the activity. Approximately a third of all workers in the world are engaged in agriculture; over a third of the terrestrial surface of the Earth is used for agriculture (crop and animal); and 90% of fresh water use globally is in agriculture. A third reason is the centrality of food and agriculture to cultural practice, cultural identity, and cultural diversity. All cultures have distinctive foodways and food rules that play a crucial role in everything from how daily meals are prepared to how holidays and special events are celebrated.

Modern agricultural technologies have typically been developed to increase productivity in order to feed a rapidly growing human population that now stands at over seven billion – 925 million of whom are undernourished. This is the case with both crop technologies – e.g. fertilizers, pesticides, and irrigation systems – and animal technologies – e.g. growth hormones and antibiotics. However, the use of technology to industrialize agriculture and increase productivity has given rise to significant ethical concerns. One type of concern is ecological. Fresh water aquifers are being drained, aquatic systems are being polluted, topsoil is being depleted, animal waste is piling up, and forests are being cut down at a rapid rate, for example. Another type of concern is animal welfare. Over 50 billion land animals are produced for consumption each year, many of them in concentrated animal feed operations (CAFOs) where such things as confinement, debeaking/declawing, tail docking and castration are common practices. Yet another type of concern focuses on the impacts of industrialization and globalization on traditional agricultural practices and foodways – e.g. that monocultural cash crops are displacing local crop diverse food agriculture and that transnational corporations have disproportionate control of the seed supply.

In "Ethics and Genetically Modified Foods," Gary Comstock considers both intrinsic objections to GM crops – i.e. that they are objectionable in themselves – and extrinsic objections to GM crops – i.e. that use of them will have problematic outcomes. He argues that the intrinsic objections are not reasonable, and that the extrinsic concerns justify careful development and use but not relinquishment of the technology. He then presents his reasons for endorsing GM crops, which are based on autonomy and their potential ecological and human health benefits.

In "Women and the Gendered Politics of Food," Vandana Shiva addresses the cultural impacts of modern industrial agriculture. Shiva argues that globalization, seed engineering, patenting, and corporate control of the seed supply are undermining local agricultural practices. As a result of them, knowledge and diversity are lost, and food is transformed from nourishment into a commodity. Shiva also emphasizes that it has been women who have been marginalized the most by these developments, since women are frequently stewards of seed diversity and agricultural knowledge.

In "The Ethics of Agricultural Animal Biotechnology," Robert Streiffer and John Basl consider the potential for genetic modification to improve the welfare of livestock. After an extended discussion of theories of animal welfare, they raise several doubts about whether the intended improvements are likely to be accomplished through animal biotechnology. They then consider the potential for animal biotechnology to reduce the ecological impacts of livestock. Again, they suggest that realizing the intended ecological benefits through genetic engineering is less straightforward and less probable than it might initially appear.

In the final chapter in the section, "Artificial Meat," Paul Thompson provides an ethical analysis of growing animal tissue in vitro for human consumption. He begins by raising concerns about the standard ethical arguments in favor of artificial meat – i.e. that it is preferable to meat grown in CAFOs with respect to animal welfare and ecological impacts. He then suggests some reasons to think that artificial meat might in fact be ethically suspect – for example, that it will be highly centralized and industrialized. He does not draw any definitive conclusions about the ethics of artificial meat. His aim, rather, is to show that the fact that no animals are harmed or killed in its production is only a part of its ethical profile, and that the considerations raised in its favor are at this point far from decisive.

More detailed summaries, as well as a listing of related readings, are located at the start of the chapters.

ETHICS AND GENETICALLY MODIFIED FOODS[1]

Gary Comstock

CHAPTER SUMMARY

In this chapter, Gary Comstock considers whether it is ethically justified to pursue genetically modified (GM) crops and foods. He first considers intrinsic objections to GM crops that allege that the process of making GMOs is objectionable in itself. He argues that there is no justifiable basis for the objections – i.e. GM crops are not intrinsically ethically problematic. He then considers extrinsic objections to GM crops, including objections based on the precautionary principle, which focus on the potential harms that may result from the adoption of GM organisms. He argues that these concerns have some merit. However, they do not justify giving up GM crops altogether. Instead, they require that GM crops be developed carefully and with appropriate oversight. Comstock then presents the positive case for GM crops that he endorses. It is based on three considerations: (i) the right of people to choose to adopt GM technology; (ii) the balance of likely benefits over harms to consumers and the environment from GM technology; and (iii) the wisdom of encouraging discovery, innovation, and careful regulation of GM technology.

RELATED READINGS

Introduction: Innovation Presumption (2.1); Extrinsic Concerns (2.6); Intrinsic Concerns (2.7); Ethics and Public Policy (2.9); Types of Theories (3.2)

Other Chapters: Jason Robert and Francoise Baylis, *Crossing Species Boundaries* (Ch. 10); Vandana Shiva, *Women and the Gendered Politics of Food* (Ch. 32); Mark A. Bedau and Mark Triant, *Social and Ethical Implications of Creating Artificial Cells* (Ch. 37)

1. INTRODUCTION

Much of the food consumed in the United States (and, increasingly, globally) is genetically modified (GM). GM food derives from microorganisms, plants, or animals manipulated at the molecular level to have traits that farmers or consumers desire. These foods often have been produced by techniques in which "foreign" genes are inserted into the microorganisms, plants, or animals. Foreign genes

---

[1] This chapter is excerpted from Gary Comstock (2001) 'Ethics and Genetically Modified Foods,' *SCOPE* (AAAS). It appears here by permission of the author.

are genes taken from sources other than the organism's natural parents. In other words, GM plants contain genes they would not have contained if researchers had used only traditional plant breeding methods.

Some consumer advocates object to GM foods, and sometimes they object on ethical grounds. When people oppose GM foods on ethical grounds, they typically have some reason for their opposition. We can scrutinize their reasons and, when we do so, we are doing applied ethics. Applied ethics involves identifying peoples' arguments for various conclusions and then analyzing those arguments to determine whether the arguments support their conclusions. A critical goal here is to decide whether an argument is sound. A sound argument is one in which all the premises are true and no mistakes have been made in reasoning.

Ethically justifiable conclusions inevitably rest on two types of claims: (i) empirical claims, or factual assertions about how the world *is*, claims ideally based on the best available scientific observations, principles, and theories; and (ii) normative claims, or value-laden assertions about how the world *ought to be*, claims ideally based on the best available moral judgments, principles, and theories.

Is it ethically justifiable to pursue GM crops and foods? There is an objective answer to this question, and we will try here to figure out what it is.

2. A METHOD FOR ADDRESSING ETHICAL ISSUES

Ethical objections to GM foods typically center on the possibility of harm to persons or other living things. Harm may or may not be justified by outweighing benefits. Whether harms are justified is a question that ethicists try to answer by working methodically through a series of questions:

(i) What is the harm envisaged? To provide an adequate answer to this question, we must pay attention to how significant the harm or potential harm may be (will it be severe or trivial?); who the "stakeholders" are (that is, who are the persons, animals, even ecosystems, who may be harmed?); the extent to which various stakeholders might be harmed; and the distribution of harms. The last question directs attention to a critical issue, the issue of justice and fairness. Are those who are at risk of being harmed by the action in question different from those who may benefit from the action in question?

(ii) What information do we have? Sound ethical judgments go hand in hand with a thorough understanding of the scientific facts. In a given case, we may need to ask two questions. Is the scientific information about harm being presented reliable, or is it fact, hearsay, or opinion? What information do we not know that we should know before we make the decision?

(iii) What are the options? In assessing the various courses of action, emphasize creative problem-solving, seeking to find win-win alternatives in which everyone's interests are protected. Here we must identify what objectives each stakeholder wants to obtain; how many methods are available by which to achieve those objectives; and what advantages and disadvantages attach to each alternative.

(iv) What ethical principles should guide us? There are at least three secular ethical traditions:

- Rights theory holds that we ought always to act so that we treat human beings as autonomous individuals and not as mere means to an end.
- Utilitarian theory holds that we ought always to act so that we maximize good consequences and minimize harmful consequences.
- Virtue theory holds that we ought always to act so that we act the way a just, fair, good person would act.

Ethical theorists are divided about which of these three theories is best. We manage this uncertainty through the following procedure. Pick one of the three principles. Using it as a basis, determine its implications for the decision at hand. Then, adopt a second principle. Determine what it implies for the decision at hand. Repeat the procedure with the third principle. Should all three principles converge on the same conclusion, then we have good reasons for thinking our conclusion is morally justifiable.

(v) How do we reach moral closure? Does the decision we have reached allow all stakeholders either to participate in the decision or to have their views represented? If a compromise solution is deemed necessary in order to manage otherwise intractable differences, has the compromise been reached in a way that has allowed all interested parties to have their interests articulated, understood, and considered? If so, then the decision may be justifiable on ethical grounds.

There is a difference between consensus and compromise. Consensus means that the vast majority of people agree about the right answer to a question. If the group cannot reach a consensus but must, nevertheless, make some decision, then a compromise position may be necessary. But neither consensus nor compromise should be confused with the right answer to an ethical question. It is possible that a society might reach a consensus position that is unjust. For example, some societies have held that women should not be allowed to own property. That may be a consensus position or even a compromise position, but it should not be confused with the truth of the matter. Moral closure is a sad fact of life; we sometimes must decide to undertake some course of action even though we know that it may not be, ethically, the right decision, all things considered.

3. ETHICAL ISSUES INVOLVED IN THE USE OF GENETIC TECHNOLOGY IN AGRICULTURE

Discussions of the ethical dimensions of agricultural biotechnology are sometimes confused by a conflation of two quite different sorts of objections to GM technology: intrinsic and extrinsic. It is critical not only that we distinguish these two classes but also that we keep them distinct throughout the ensuing discussion of ethics.

Extrinsic objections focus on the potential harms consequent upon the adoption of GM organisms (GMOs). Extrinsic objections hold that GM technology should not be pursued because of its anticipated results. Briefly stated, the extrinsic objections go as follows. GMOs may have disastrous effects on animals, ecosystems, and humans. Possible harms to humans include perpetuation of social inequities in modern agriculture, decreased food security for women and children on subsistence farms in developing countries, a growing gap between well-capitalized economies in the northern hemisphere and less capitalized peasant economies in the South, risks to the food security of future generations and the promotion of reductionistic and exploitative science. Potential harms to ecosystems include possible environmental catastrophe; inevitable narrowing of germplasm diversity; and irreversible loss or degradation of air, soils, and waters. Potential harms to animals include unjustified pain to individuals used in research and production.

These are valid concerns, and nation-states must have in place testing mechanisms and regulatory agencies to assess the likelihood, scope, and distribution of potential harms through a rigorous and well-funded risk assessment procedure. However, these extrinsic objections cannot by themselves justify a moratorium, much less a permanent ban, on GM technology, because they admit the possibility that the harms may be minimal and outweighed by the benefits. How can one decide whether the potential harms outweigh potential benefits unless one conducts the research, field tests, and data analysis necessary to make a scientifically informed assessment?

In sum, extrinsic objections to GMOs raise important questions about GMOs, and each country using GMOs ought to have in place the organizations and research structures necessary to ensure their safe use.

There is, however, an entirely different sort of objection to GM technology, a sort of objection that, if it is sound, would justify a permanent ban.

Intrinsic objections allege that the process of making GMOs is objectionable *in itself*. This belief is defended in several ways, but almost all the formulations are related to one central claim, the unnaturalness objection:

(**UE**) It is unnatural to genetically engineer plants, animals, and foods.

If **UE** is true, then we ought not to engage in bioengineering, however unfortunate may be the consequences of halting the technology. Were a nation to accept **UE** as the conclusion of a sound argument, then much agricultural research would have to be terminated and potentially significant benefits from the technology sacrificed. A great deal is at stake.

In *Vexing Nature? On The Ethical Case Against Agricultural Biotechnology*, I discuss 14 ways in which **UE** has been defended (Comstock, 2000). For present purposes, those 14 objections can be summarized as follows:

(i)   To engage in ag biotech is to *play God*.
(ii) To engage in ag biotech is to *invent world-changing technology*.
(iii) To engage in ag biotech is *illegitimately to cross species boundaries*.
(iv) To engage in ag biotech is to *commodify life*.

Let us consider each claim in turn.

(i) To engage in ag biotech is to *play God*.

In a western theological framework, humans are creatures, subjects of the Lord of the Universe, and it would be impious for them to arrogate to themselves roles and powers appropriate only for the Creator. Shifting genes around between individuals and species is taking on a task not appropriate for us, subordinate beings. Therefore, to engage in bioengineering is to play God.

There are several problems with this argument. First, there are different interpretations of God. Absent the guidance of any specific religious tradition, it is logically possible that God could be a Being who wants to turn over to us all divine prerogatives, or explicitly wants to turn over to us at least the prerogative of engineering plants, or who doesn't care what we do. If God is any of these beings, then the argument fails because playing God in this instance is not a bad thing.

The argument seems to assume, however, that God is not like any of the gods just described. Assume that the orthodox Jewish and Christian view of God is correct, that God is the only personal, perfect, necessarily existing, all-loving, all-knowing, and all-powerful being. On this traditional western theistic view, finite humans should not aspire to infinite knowledge and power. To the extent that bioengineering is an attempt to control nature itself, the argument would go, bioengineering would be an unacceptable attempt to usurp God's dominion.

The problem with this argument is that not all traditional Jews and Christians think this God would rule out genetic engineering. I am a practicing evangelical Christian and the chair of my local church's council. In my tradition, God is thought to endorse creativity and scientific and technological development, including genetic improvement. Other traditions have similar views. In the mystical writings of the Jewish Kabbalah, God is understood as One who expects humans to be co-creators, technicians working with God to improve the world. At least one Jewish philosopher, Baruch Brody, has suggested that biotechnology may be a vehicle ordained by God for the perfection of nature.

I personally hesitate to think that humans can perfect nature. However, I have become convinced that GM might help humans to rectify some of the damage we have already done to nature. And I believe God may endorse such an aim. For humans are made in the divine image. God desires that we exercise the spark of divinity within us. Inquisitiveness in science is part of our nature. Creative impulses are not found only in the literary, musical, and plastic arts. They are part of molecular biology, cellular theory, ecology, and evolutionary genetics, too. It is unclear why the desire to investigate and manipulate the chemical bases of life should not be considered as much a manifestation of our god-like nature as the writing of poetry and the composition of sonatas. As a way of providing theological content for **UE**, then, argument (i) is unsatisfactory because it is ambiguous and contentious.

(ii)   To engage in ag biotech is to *invent world-changing technology*, an activity that should be reserved to God alone.

Let us consider (ii) in conjunction with a similar objection (iia).

(iia)   To engage in ag biotech is to *arrogate historically unprecedented power* to ourselves.

The argument here is not the strong one, that biotech gives us divine power, but the more modest one, that it gives us a power we have not had previously. But it would be counterintuitive to judge an action wrong simply because it has never been performed. On this view, it would have been wrong to prescribe a new herbal remedy for menstrual cramps or to administer a new anesthetic. But that seems absurd. More argumentation is needed to call historically unprecedented actions morally wrong. What is needed is to know *to what extent* our new powers will transform society, whether we have witnessed prior transformations of this sort, and whether those transitions are morally acceptable.

We do not know how extensive the ag biotech revolution will be, but let us assume that it will be as dramatic as its greatest proponents assert. Have we ever witnessed comparable transitions? The change from hunting and gathering to agriculture was an astonishing transformation. With agriculture came not only an increase in the number of humans on the globe but the first appearance of complex cultural activities: writing, philosophy, government, music, the arts, and architecture. What sort of power did people arrogate to themselves when they moved from hunting and gathering to agriculture? The power of civilization itself (McNeill, 1989).

Ag biotech is often oversold by its proponents. But suppose they are right, that ag biotech brings us historically unprecedented powers. Is this a reason to oppose it? Not if we accept agriculture and its accompanying advances, for when we accepted agriculture we arrogated to ourselves historically unprecedented powers.

In sum, the objections stated in (ii) and (iia) are not convincing.

(iii) To engage in ag biotech is *illegitimately to cross species boundaries.*

The problems with this argument are both theological and scientific. I will leave it to others to argue the scientific case that nature gives ample evidence of generally fluid boundaries between species. The argument assumes that species boundaries are distinct, rigid, and unchanging, but, in fact, species now appear to be messy, plastic, and mutable. To proscribe the crossing of species borders on the grounds that it is unnatural seems scientifically indefensible.

It is also difficult to see how (iii) could be defended on theological grounds. None of the scriptural writings of the western religions proscribes genetic engineering, of course, because genetic engineering was undreamt of when the holy books were written. Now, one might argue that such a proscription may be derived from Jewish or Christian traditions of scriptural interpretation. Talmudic laws against mixing "kinds," for example, might be taken to ground a general prohibition against inserting genes from "unclean" species into clean species. Here's one way the argument might go: For an observant Jew to do what scripture proscribes is morally wrong; Jewish oral and written law proscribe the mixing of kinds (eating milk and meat from the same plate; yoking donkeys and oxen together); bioengineering is the mixing of kinds; therefore, for a Jew to engage in bioengineering is morally wrong.

But this argument fails to show that bioengineering is intrinsically objectionable in all its forms for everyone. The argument might prohibit *Jews* from engaging in certain *kinds* of biotechnical activity but not all; it would not prohibit, for example, the transferring of genes *within* a species, nor, apparently, the transfer of genes from one clean species to another clean species. Incidentally, it is worth noting that the Orthodox community has accepted transgenesis in its food supply. Seventy percent of cheese produced in the United States is made with a GM product, chymosin. This cheese has been accepted as kosher by Orthodox rabbis (Gressel, 1998).

In conclusion, it is difficult to find a persuasive defense of (iii) on either scientific or religious grounds.

(iv)   To engage in ag biotech is to *commodify life*.

The argument here is that genetic engineering treats life in a reductionistic manner, reducing living organisms to little more than machines. Life is sacred and not to be treated as a good of commercial value only to be bought and sold to the highest bidder.

Could we apply this principle uniformly? Would not objecting to the products of GM technology on these grounds also require that we object to the products of ordinary agriculture on the same grounds? Is not the very act of bartering or exchanging crops and animals for cash vivid testimony to the fact that every culture on earth has engaged in the commodification of life for centuries? If one accepts commercial trafficking in non-GM wheat and pigs, then why object to commercial trafficking in GM wheat and GM pigs? Why should it be wrong to treat DNA the way we have previously treated animals, plants, and viruses (Nelkin and Lindee, 1999)?

Although (iv) may be true, it is not a sufficient reason to object to GM technology because our values and economic institutions have long accepted the commodification of life. Now, one might object that various religious traditions have never accepted commodification and that genetic engineering presents us with an opportunity to resist, to reverse course. Kass (1998), for example, has argued that we have gone too far down the road of dehumanizing ourselves and treating nature as a machine and that we should pay attention to our emotional reactions against practices such as human cloning. Even if we cannot defend these feelings in rational terms, our revulsion at the very idea of cloning humans should carry great weight. Midgley (2000) has argued that moving genes across species boundaries is not only "yukky" but, perhaps, a monstrous idea, a form of playing God.

Kass (1988) and Midgley (2000) have eloquently defended the relevance of our emotional reactions to genetic engineering but, as both admit, we cannot simply allow our emotions to carry the day. As Midgley writes, "Attention to … sympathetic feelings [can stir] up reasoning that [alters] people's whole world view" (p. 10). But as much hinges on the reasoning as on the emotions.

Are the intrinsic objections sound? Are they clear, consistent, and logical? Do they rely on principles we are willing to apply uniformly to other parts of our lives? Might they lead to counterintuitive results?

Counterintuitive results are results we strongly hesitate to accept because they run counter to widely shared considered moral intuitions. If a moral rule or principle leads to counterintuitive results, then we have a strong reason to reject it. For example, consider the following moral principle, which we might call the doctrine of naive consequentialism (NC):

(NC) Always improve the welfare of the most people.

Were we to adopt NC, then we would be not only permitted but required to sacrifice one healthy person if by doing so we could save many others. If six people need organ transplants (two need kidneys, one needs a liver, one needs a heart, and two need lungs) then NC instructs us to sacrifice the life of the healthy person to transplant six organs to the other six. But this result, that we are *obliged* to sacrifice innocent people to save strangers, is wildly counterintuitive. This result gives us a strong reason to reject NC.

I have argued that the four formulations of the unnaturalness objection considered above are unsound insofar as they lead to counterintuitive results. I do not take this position lightly. Twelve years ago, I wrote "The Case Against bGH," an article, I have been told, that "was one of the first papers by a philosopher to object to ag biotech on explicitly ethical grounds." I then wrote a series of other articles objecting to GM herbicide-resistant crops, transgenic animals, and, indeed, all of ag biotech (see, Comstock, 1988). I am acquainted with worries about GM foods. But, for reasons that include the weakness of the intrinsic objections, I have changed my mind. The sympathetic feelings on which my anti-GMO worldview was based did not survive the stirring up of reasoning.

4. WHY ARE WE CAREFUL WITH GM FOODS?

I do not pretend to know anything like the full answer to this question, but I would like to be permitted the luxury of a brief speculation about it. The reason may have to do with a natural, completely understandable, and wholly rational tendency to take precautions with what goes into our mouths. When we are in good health and happy with the foods available to us, we have little to gain from experimenting with new food and no reason to take a chance on a potentially unsafe food. We may think of this disposition as the precautionary response.

When faced with two contrasting opinions about issues related to food safety, consumers place great emphasis on negative information. The precautionary response is particularly strong when a consumer sees little to gain from a new food technology.

When a given food is plentiful, it is rational to place extra weight on negative information about any particular piece of that food. It is rational to do so, as my colleague Dermot Hayes points out, even when the source of the negative information is known to be biased.

There are several reasons for us to take a precautionary approach to new foods. First, under conditions in which nutritious tasty food is plentiful, we have nothing to gain from trying a new food if, from our perspective, it is in other respects identical to our current foods. Suppose on a rack in front of me there are 18 dozen maple-frosted Krispy Kreme doughnuts, all baked to a golden brown, all weighing three ounces. If I am invited to take one of them, I have no reason to favor one over the other.

Suppose, however, that a naked man runs into the room with wild-hair flying behind him yelling that the sky is falling. He approaches the rack and points at the third doughnut from the left on the fourth shelf from the bottom. He exclaims, "This doughnut will cause cancer! Avoid it at all costs, or die!" There is no reason to believe this man's claim and yet, because there are so many doughnuts freely available, why should we take any chances? It is rational to select other dough-nuts, because all are alike. Now, perhaps one of us is a mountain climber who loves taking risks and might be tempted to say, "Heck, I'll try that doughnut." In order to focus on the right question here, the risk takers should ask themselves whether they would select the tainted doughnut to take home to feed to their 2-year-old daughter. Why impose any risk on your loved ones when there is no reason to do so?

The Krispy Kreme example is meant to suggest that food tainting is both a powerful and an extraordinarily easy social act. It is powerful because it virtually determines consumer behavior. It is easy, because the tainter does not have to offer any evidence of the food's danger. Under conditions of food plenty, rational consumers do and should take precautions, avoiding tainted food no matter how untrustworthy the tainter.

Our tendency to take precautions with our food suggests that a single person with a negative view about GM foods will be much more influential than many people with a positive view (Hayes et. al, in press).

In a worldwide context, the precautionary response of those facing food abun-dance in developed countries may lead us to be insensitive to the conditions of those in less fortunate situations. Indeed, we may find ourselves in the following ethical dilemma.

For purposes of argument, make the following three assumptions. (I do not believe any of the assumptions is implausible.) First, assume that GM food is safe. Second, assume that some GM "orphan" foods, such as rice enhanced with iron or vitamin A, or virus-resistant cassava, or aluminum-tolerant sweet potato, may be of great potential benefit to millions of poor children. Third, assume that wide-spread anti-GM information and sentiment, no matter how unreliable on scientific grounds, could shut down the GM infrastructure in the developed world.

Under these assumptions, consider the possibility that, by tainting GM foods in the countries best suited to conduct GM research safely, anti-GM activists could bring to a halt the range of money-making GM foods marketed by multi-national corporations. This result might be a good or a bad thing. However, an unintended side effect of this consequence would be that the new GM orphan crops mentioned above might not be forthcoming, assuming that the develop-ment and commercialization of these orphan crops depends on the answering of fundamental questions in plant science and molecular biology that will be answered only if the research agendas of private industry are allowed to go forward along with the research agendas of public research institutions.

Our precautionary response to new food may put us in an uncomfortable position. On the one hand, we want to tell "both sides" of the GM story, letting people know about the benefits and the risks of the technology. On the other hand, some of the people touting the benefits of the technology make outlandish claims that it will feed the world and some of the people decrying the technology make unsupported claims that it will ruin the world. In that situation, however, those with unsupported negative stories to tell carry greater weight than those with unsupported positive stories. Our precautionary response, then, may well lead, in the short term at least, to the rejection of GM technology. Yet, the rejection of GM technology could indirectly harm those children most in need, those who need what I have called the orphan crops.

Are we being forced to choose between two fundamental values, the value of free speech versus the value of children's lives?

On the one hand, open conversation and transparent decision-making processes are critical to the foundations of a liberal democratic society. We must reach out to include everyone in the debate and allow people to state their opinions about GM foods, whatever their opinion happens to be and whatever their level of acquaintance with the science and technology happens to be. Free speech is a value not to be compromised lightly.

On the other hand, stating some opinions about GM food can clearly have a tainting effect, a powerful and extraordinarily easy consequence of free speech. Tainting the technology might result in the loss of this potentially useful tool. Should we, then, draw some boundaries around the conversation, insisting that each contributor bring some measure of scientific data to the table, especially when negative claims are being made? Or are we collectively prepared to leave the conversation wide open? That is, in the name of protecting free speech, are we prepared to risk losing an opportunity to help some of the world's most vulnerable?

5. THE PRECAUTIONARY PRINCIPLE

As a thirteen year-old, I won my dream job, wrangling horses at Honey Rock Camp in northern Wisconsin. The image I cultivated for myself was the weathered cowboy astride Chief or Big Red, dispensing nuggets to awestruck young rider wannabes. But I was, as they say in Texas, all hat. "Be careful!" was the best advice I could muster.

Only after years of experience in a western saddle would I have the skills to size up various riders and advise them properly on a case-by-case basis. You should slouch more against the cantle and get the balls of your feet onto the stirrups. You need to thrust your heels in front of your knees and down toward the animal's front hooves. You! Roll your hips in rhythm with the animal, and stay away from the horn. You, stay alert for sudden changes of direction.

Only after years of experience with hundreds of different riders would I realize that my earlier generic advice, well-intentioned though it was, had been of absolutely no use to anyone. As an older cowboy once remarked, I might as well have been saying, "Go crazy!" Both pieces of advice were equally useless in making good decisions about how to behave on a horse.

Now, concerned observers transfer fears to genetically modified foods, advising: "Take precaution!" Is this a valuable observation that can guide specific public policy decisions, or well-intentioned but ultimately unhelpful advice?

As formulated in the 1992 Rio Declaration on Environment and Development, the precautionary principle states that "... lack of full scientific certainty shall not be used as a reason for postponing cost-effective measures to prevent environmental degradation." The precautionary approach has led many countries to declare a moratorium on GM crops on the supposition that developing GM crops might lead to environmental degradation. The countries are correct that this is an implication of the principle. But is it the only implication?

Suppose global warming intensifies and comes, as some now darkly predict, to interfere dramatically with food production and distribution. Massive dislocations in international trade and corresponding political power follow global food shortages, affecting all regions and nations. In desperate attempts to feed themselves, billions begin to pillage game animals, clear-cut forests to plant crops, cultivate previously non-productive lands, apply fertilizers and pesticides at higher than recommended rates, kill and eat endangered and previously non-endangered species.

Perhaps this is not a likely scenario, but it is not entirely implausible either. GM crops could help to prevent it, by providing hardier versions of traditional lines capable of growing in drought conditions, or in saline soils, or under unusual climactic stresses in previously temperate zones, or in zones in which we have no prior agronomic experience.

On the supposition that we might need the tools of genetic engineering to avert future episodes of crushing human attacks on what Aldo Leopold called "the land," the precautionary principle requires that we develop GM crops. Yes, we lack full scientific certainty that developing GM crops will prevent environmental degradation. True, we do not know what the final financial price of GM research and development will be. But if GM technology were to help save the land, few would not deem that price cost-effective. So, according to the precautionary principle, lack of full scientific certainty that GM crops will prevent environmental degradation shall not be used as a reason for postponing this potentially cost-effective measure.

The precautionary principle commits us to each of the following propositions:

(1) We must not develop GM crops.
(2) We must develop GM crops.

As (1) and (2) are plainly contradictory, however, defenders of the principle should explain why its implications are not incoherent.

Much more helpful than the precautionary principle would be detailed case-by-case recommendations crafted upon the basis of a wide review of nonindustry-sponsored field tests conducted by objective scientists expert in the construction and interpretation of ecological and medical data. Without such a basis for judging this use acceptable and that use unacceptable, we may as well advise people in the GM area to go crazy. It would be just as helpful as "Take precaution!"

6. RELIGION AND ETHICS

Religious traditions provide an answer to the question, "How, overall, should I live my life?" Secular ethical traditions provide an answer to the question, "What is the right thing to do?" When in a pluralistic society a particular religion's answers come into genuine conflict with the answers arrived at through secular ethical deliberation, we must ask how deep is the conflict. If the conflict is so deep that honoring the religion's views would entail dishonoring another religion's views, then we have a difficult decision to make. In such cases, the conclusions of secular ethical deliberation must override the answers of the religion in question.

The reason is that granting privileged status to one religion will inevitably discriminate against another religion. Individuals must be allowed to follow their conscience in matters theological. But if one religion is allowed to enforce its values on others in a way that restricts the others' ability to pursue their values, then individual religious freedom has not been protected.

Moral theorists refer to this feature of nonreligious ethical deliberation as the *overridingness* of ethics. If a parent refuses a life-saving medical procedure for a minor child on religious grounds, the state is justified in overriding the parent's religious beliefs in order to protect what secular ethics regards as a value higher than religious freedom: the life of a child.

The overridingness of ethics applies to our discussion only if a religious group claims the right to halt GM technology on purely religious grounds. The problem here is the confessional problem of one group attempting to enforce its beliefs on others. I mean no disrespect to religion; as I have noted, I am a religious person, and I value religious traditions other than my own. Religious traditions have been the repositories and incubators of virtuous behavior. Yet each of our traditions must, in a global society, learn to coexist peacefully with competing religions and with nonreligious traditions and institutions.

If someone objects to GM technology on purely religious grounds, we must ask on what authority they speak for their tradition, whether there are other, conflicting, views within their tradition and whether acting on their views will entail disrespecting the views of people from other religions. It is, of course, the right of each tradition to decide its attitude about genetic engineering. But in the absence of other good reasons, we must not allow someone to ban GM technology for narrowly sectarian reasons alone. To allow such an action would be to disrespect the views of people who believe, on equally sincere religious grounds, that GM technology is not necessarily inconsistent with God's desires for us.

7. CONCLUSION

Earlier I described a method for reaching ethically sound judgments. It was on the basis of that method that I personally came to change my mind about the moral acceptability of GM crops. My opinion changed as I took full account of three considerations: (i) the rights of people in various countries to choose to adopt GM technology (a consideration falling under the human rights principle); (ii) the balance of likely benefits over harms to consumers and the environment from GM technology (a utilitarian consideration); and (iii) the wisdom of encouraging discovery, innovation, and careful regulation of GM technology (a consideration related to virtue theory).

Is it ethically justifiable to pursue GM crops and foods? I have come to believe that three of our most influential ethical traditions converge on a common answer. Assuming we proceed responsibly and with appropriate caution, the answer is yes.

WORKS CITED

B. Brody, private communication.

G. Comstock (1988) 'The case against bGH,' *Agricultural and Human Values,* 5:36–52.

G. Comstock (2000) *Vexing Nature? On the Ethical Case Against Agricultural Biotechnology* (Boston: Kluwer Academic Publishers).

D. Hayes, J. Fox, and J. Shogren (In press) 'Consumer preferences for food irradiation: how favorable and unfavorable descriptions affect preferences for irradiated pork in experimental auctions,' *J. Risk Uncertainty.*

L. Kass (1998) 'Beyond biology: Will advances in genetic technology threaten to dehumanize us all?' *The New York Times on the Web,* online at http://www.nytimes.com/books/98/08/23/reviews/980823.23kassct.html

L. Kass (1998) *Toward a More Natural Science: Biology and Human Affairs* (New York, NY: The Free Press).

W. McNeill (1989) 'Gains and losses: an historical perspective on farming,' *The 1989 Iowa Humanities Lecture* (National Endowment for the Humanities and Iowa Humanities Board, Oakdale Campus, Iowa City, IA).

M. Midgley (2000) 'Biotechnology and monstrosity: Why we should pay attention to the 'yuk factor,' *Hastings Center Report,* 30(5): 7–15.

D. Nelkin, and M. S. Lindee (1995) *The DNA Mystique: The Gene as Cultural Icon* (Ann Arbor, M.I.: University of Michigan).

DISCUSSION QUESTIONS

1. The author argues that under certain circumstances religious concerns should be discounted or disregarded. What role do you think religious ethics should play in decisions about public policy (i.e. regulation and funding) of controversial technologies?

2. The author argues that the intrinsic objections against agricultural biotechnology are unsound? Do you agree with this assessment? What are your reasons for agreeing or disagreeing?

3. What is the basis of the author's rejection of the precautionary principle? Is there a way for formulate the precautionary principle so that it is not subject to the problems he raises for it?

4. The author argues that developed nations' failure to pursue GM crops might come at a cost to people of the developing world. What are his reasons for thinking this? Do you think they are strong reasons?

5. As the author notes, many people have an adverse reaction to genetically modified foods. What role should those reactions play in our ethical deliberations?

6. Do you find the considerations the author presents in favor of GM crops to be compelling? Why or why not?

7. Can you think of any objections to GM crops that are not addressed by the author in this chapter?

32 WOMEN AND THE GENDERED POLITICS OF FOOD[1]

Vandana Shiva

CHAPTER SUMMARY

In this chapter, Vandana Shiva argues that, from seed to table, the food chain is gendered. Moreover, she locates her gender analysis within a critique of the globalization and commoditization of agriculture, particularly by means of corporate control of the seed supply. She argues that women's traditional and local seed and food economy has been discounted as "productive work," and that women's seed and food knowledge has been discounted as knowledge. Globalization, she argues, has led to the transfer of seed and food from women's control to corporate control by means of patenting and genetic engineering, for example. She further argues that food is transformed in corporate hands. It is no longer nourishment; it becomes a commodity. And as a commodity it can be manipulated and monopolized. If food grain makes more money as cattle feed than it does as food for human consumption, it becomes cattle feed. If food grain converted to biofuel to run automobiles is more profitable, it becomes ethanol and biodiesel. Finally, Shiva argues that a counter-revolution is underway, based on women's food and agricultural knowledge, to promote a just, sustainable, healthy, and secure food system.

RELATED READINGS

Introduction: Power (2.4); Form of Life (2.5); Justice, Access, and Equality (2.6.2)

Other Chapters: Langdon Winner, *Technologies as Forms of Life* (Ch. 4); Gary Comstock, *Ethics and Genetically Modified Foods* (Ch. 31)

1. INTRODUCTION

The politics of food is gendered at multiple levels.

Firstly, food production, processing, and provisioning have been women's domain in the social division of labor (women grew food, cooked food, processed food, served food). Women-centered food systems are based on sharing and caring, on conservation and well-being.

---

[1] This chapter is excerpted from Vandana Shiva (2009) 'Women and the Gendered Politics of Food,' *Philosophical Topics* (37): 17–32. It appears here by permission of the University of Arkansas Press and the author.

Secondly, corporate globalization driven by capitalist patriarchy has transformed food, food production, and food distribution. The control over the entire food chain, from seed to table, is shifting from women's hands to global corporations who are today's "global patriarchs." In the process, seed is turning to nonseed. Seed multiplies and reproduces. GMO and hybrid seeds are nonrenewable. Food becomes nonfood. Food is nourishment. As one of the ancient Indian text says, "everything is food, everything is something else's food" (Bajaj and Srinivas, 1966, p. 8; Shiva, 2001) Corporate-controlled food is no longer food, it becomes a commodity – totally interchangeable between biofuel for driving a car or feed for factory farms or food for the hungry. Not only is food displaced, women's knowledge and work, skills, productivity, and creativity related to food are destroyed.

Five Gene Giants and five Food Giants are replacing billions of women producers and processors. The Gene Giants who control the seed are Monsanto, Dupont/Pioneer, Sungenta, Bayer, BASF. This has led to the marginalization of women. It has also created new risks of food security and food safety. In 2008, food riots took place in more than forty countries as prices skyrocketed. The Food Giants control food. They include Cargill, Conagra, ADM, Louis Dreyfus, and Binge (Morgan, 1980). More than 1 billion are denied access to food, and another 2 billion are cursed with obesity and related diseases due to industrial/junk foods. Among those who suffer the two kinds of malnutrition, women and girls are the worst sufferers.

Thirdly, a new food revolution is underway, building on women's food and agriculture heritage to create just, sustainable, and healthy food systems that secure safe and healthy food for all.

2. MOST FARMERS OF THE WORLD ARE WOMEN: FARMING IS A FEMINIST ISSUE

Agriculture, the growing of food, is both the most important source of livelihood for the majority of the world people, especially women, as well as the sector related to the most fundamental economic right, the right to food and nutrition.

Women were the world's original food producers, and they continue to be central to food production systems in the Third-World countries in terms of the work they do in the food chain. The worldwide destruction of the feminine knowledge of agriculture evolved over four to five thousand years, but a handful of white male scientists in less than two decades have not merely violated women as experts; but since their expertise in agriculture has been related to modeling agriculture on nature's methods of renewability, its destruction has also gone hand in hand with the ecological destruction of nature's processes and the economic destruction of the poorer people in rural areas.

Agriculture has been evolved by women. Most farmers of the world are women, and most girls are future farmers. Girls learn the skills and knowledge of farming in the fields and farms. What is grown on farms determines whose livelihoods are secured, what is eaten, how much is eaten, and by whom it is eaten.

Women make the most significant contribution to food security. They produce more than half the world's food. They provide more than 80 percent of the food needs of food-insecure households and regions.

Food security is therefore directly linked to women's food-producing capacity. Constraints on women's capacity leads to erosion of food security, especially for poor households in poor regions.

From field to kitchen, from seed to food, women's strength is diversity. Women's capacities are eroded when this diversity is eroded.

Diversity is the pattern of women's work, the pattern of women's planting and sowing of food crops and the pattern of women's food processing.

The dominant systems of economics, science, and technology have conspired against women and girls by conspiring against diversity.

Economics has rendered women's work as food providers invisible because women provide for the household and perform multiple tasks involving diverse skills.

Women have remained invisible as farmers in spite of their contribution to farming. People fail to see the work that women do in agriculture. Their production tends not to be recorded by economists as "work." And agriculture as a future vocation for girls is thus closed.

These problems of data collection on agricultural work arise not because too few women work but too many women do too much work.

There is a conceptual inability of statisticians and researchers to define women's work inside the house and outside the house (and farming is usually part of both). This recognition of what is and is not labor is exacerbated both by the great volume of work that women do and the fact that they do many chores at the same time. It is also related to the fact that although women work to sustain their families and communities, most of their work is not measured in wages.

Science and technology have rendered women's knowledge and productivity invisible by ignoring the dimension of diversity in agricultural production. As the Food and Agriculture Organisation (FAO) report on Women Feed the World mentions, women use more plant diversity, both cultivated and uncultivated, than agricultural scientists know about. In Nigerian home gardens, women plant 18–57 plant species. In Sub-Saharan Africa women cultivate as many as 120 different plants in the spaces left alongside the cash crops managed by man. In Guatemala, home gardens of less than 0.1 half acre have more than ten tree and crop species (FAO, 1998).

In a single African home garden more than 60 species of food-producing trees were counted. In Thailand, researchers found 230 plant species in home gardens. In Indian agriculture women use 150 different species of plants for vegetables, fodder, and health care. In West Bengal 124 "weed" species collected from rice fields have economic importance for farmers. In the Expana region of Veracruz, Mexico, peasants utilize about 435 wild plant and animal species of which 229 are eaten. Women are the biodiversity experts of the world. Unfortunately, girls are being denied their potential as food producers and biodiversity experts under the dual pressure of invisibility and domination of industrial agriculture (FAO, 1998).

While women manage and produce diversity, the dominant paradigm of agriculture promotes monoculture on the false assumption that monocultures produce more.

Monocultures do not produce more, they control more. As mentioned in FAO's World Food Day report, a study in eastern Nigeria found that home gardens occupying only 2 percent of a household farmland accounted for half of the farm's total output. Navdanya's studies on biodiversity-based ecological agriculture show that women-run farms produce more food and nutrition than industrial, chemical farms (Navdanya, 2007).

Quite clearly, if women's knowledge was not being rendered invisible, the use of the 2 percent land under polyculture systems should be the path followed for providing food security. Instead, these highly productive systems are being destroyed in the name of producing more food.

Just as women's ways of growing food produce more while conserving more resources, women's ways of food processing conserve more nutrition. Hand pounding of rice or milling rice with a foot-operated mortar and pestle preserves more protein, fat, fiber, and minerals in rice. Thus when mechanical hullers replace hand pounding by women as in the case of Bangladesh where 700 new mills supplanted the paid work of 100,000 to 140,000 women in one year by reducing the labor input from 270 hours per ton to 5. They not only rob women of work and livelihoods, they also rob girls of essential nutrients. Yet this process of food value destruction is called "value addition" in patriarchal economics.

Feeding the world requires producing more food and nutrition with fewer resources – i.e., producing more with less. In this, women are experts and their expertise needs to filter into our institutions of agricultural research and development.

However, instead of building on women's expertise in feeding the world through diversity, the dominant system is rushing headlong into destroying diversity and women's food-producing capacities, then pirating the results of centuries of innovating and breeding through patenting.

Lack of women's property rights are a major constraint on women's capacity to feed the world. These property rights include rights to land, and common property rights to common resources like water and biodiversity. Women have been the custodians of biodiversity. New intellectual property rights are alienating women's rights to biodiversity and erasing their innovation embodied in agricultural biodiversity.

If the erosion of women's capacity for feeding the world has to be prevented, IPR regimes need to evolve sui generis systems that recognize and protect women's collective and informal innovation.

While women are being denied their rights to resources and we are seeing the feminization of subsistence agriculture, the dominant agriculture is showing increasing signs of masculinization as it appropriates resources and rights from women in subsistence agriculture and presents itself as the only alternative for feeding the world.

3. FIRST THE SEED: GLOBALIZATION AND THE GENDERED POLITICS OF SEED

Seed is the first link in the food chain. For five thousand years, peasants have produced their own seeds, selecting, storing, and replanting and letting nature take its course in the food chain. The feminine principle has been conserved through the conservation of seeds by women in their work in food and grain storage. With the preservation of genetic diversity and the self-renewability of food crops has been associated the control by women and Third-World peasants on germ plasm, the source of all plant wealth. All this changed "with the green revolution."

At its heart lie new varieties of miracle seeds, which have totally transformed the nature of food production and control over food systems. The "miracle" seeds for which Borlaug got a Nobel Prize and which rapidly spread across the Third World, also sowed the seeds of a new commercialization of agriculture. Borlaug ushered in an era of corporate control on food production by creating a technology by which multinationals acquired control over seeds and hence over the entire food system. The green revolution commercialized and privatized seeds, removing control of plant genetic resources from Third World peasant women and giving it over to Western male technocrats in CIMMYT, IRRI, and multinational seed corporations.

Women have acted as custodians of the common genetic heritage through the shortage and preservation of grain. In a study of rural women of Nepal, it was found that seed selection is primarily a female responsibility. In 60.4 percent of the cases, women alone decided what type of seed to use, while men decided in only 20.7 percent. As to who actually performs the task of seed selection in cases where the family decides to use their own seeds, this work is done by women alone in 81.2 percent of the households, by both sexes in 8 percent and by men alone in only 10.8 percent of the households.

Throughout India, even in years of scarcity, grain for seed was conserved in every household, so that the cycle of food production was not interrupted by loss of seed. The peasant women of India have carefully maintained the genetic base of food production over thousands of years. This common wealth, which had evolved over millennia, was defined as "primitive cultivars" by the masculinist view of seeds, which saw its own new products as "advanced" varieties.

The green revolution was a strategy of breeding out the feminine principle by the destruction of the self-reproducing character and genetic diversity of seeds.

The death of the feminine principle in plant breeding was the beginning of seeds becoming a source of profits and control. The hybrid "miracle" seeds are a commercial miracle, because farmers have to buy new supplies of them every year. They do not reproduce themselves. Grains from hybrids do not produce seeds that duplicate the same result because hybrids do not pass on their vigor to the next generation. With hybridization, seeds could no more be viewed as a source of plant life, producing sustenance through food and nutrition; they were now a source of private profit only.

Green revolution varieties of seeds were clearly not the best alternative for increasing food production from the point of view of nature, women, and poor peasants. They were useful for corporations that wanted to find new avenues in seeds and fertilizer sales, by displacing women peasants as custodians of seeds and builders of soil fertility, and they were useful for rich farmers wanting to make profits. The international agencies which financed research on the new seeds also provided the money for their distribution. The impossible task of selling a new variety to millions of small peasants who could not afford to buy the seeds was solved by the World Bank, the UNDP, the FAO, and a host of bilateral aid programs that began to accord high priority to the distribution of HYV seed in their aid programs.

Over the past decade through new property rights and new technologies, corporations have hijacked the diversity of life on earth – and people's indigenous innovation. Intellectual Property Rights (IPR) regimes globalized through the TRIPS agreement of WTO and have been expanded to cover life forms thus creating monopoly control over biodiversity.

Patents on life are a hijack of biodiversity and indigenous knowledge; they are instruments of monopoly control over life itself. Patents on living resources and indigenous knowledge are an enclosure of the biological and intellectual commons.

The sharing and exchange of biological resources and knowledge of its properties and use has been the norm in all indigenous societies, and it continues to be the norm in most communities, including the modern community. But sharing and exchange get converted to "piracy" when individuals, organizations, or corporations freely receive biodiversity and knowledge from indigenous communities and then convert this gift into private property through IPR claims.

Seed, the common gift shared and saved by women, now becomes the "property" of Monsanto for which royalties must be paid. Seed pirated from communities is now treated as pirated if it is saved or shared. The highest human values are converted into a crime. The lowest human traits are elevated to "intellectual property rights."

When combined with the opening of the seed industry and the entry of global corporations in the seed sector, the Trade Related Intellectual Property Rights Agreement (TRIPS) of WTO is the aspect of globalization, which can be the biggest threat to people's food security. The section of TRIPS that most directly affects farmer's rights and agriculture biodiversity is Article 27.5.3(b), which states –

> Parties may exclude from patentability plants and animals other than micro-organisms, and essentially biological processes for the production of plants or animals other than non-biological and micro-biological processes. However, parties shall provide for the protection of plant varieties either by patents or by an effective sui generis system or by any combination thereof. This provision shall be reviewed four years after the entry into force of the Agreement establishing the WTO.

The article thus allows two forms of IPRs in plants: patents and a sui generis system. The Patent Act and the National Plant Variety legislation drafts are becoming a major concern of contest between the public interest and corporate interest.

The TRIPS agreement militates against people's human right to food and health by conferring unrestricted monopoly rights to corporations in the vital sectors of health and agriculture. It also threatens the livelihoods of farmers.

In Navdanya, we conserve 2,000 rice varieties in our community seed banks. One of the rice varieties we conserve and grow is basmati, the aromatic rice for which my home Dehradun is famous. The basmati is just one among 100,000 varieties of rice evolved by Indian farmers. Diversity and perenniality is our culture of the seed. In Central India, which is the Vavilov center of rice diversity, at the beginning of the agricultural season farmers gather at the village deity, offer their rice varieties and then share the seeds. This annual festival of "Akti" rejuvenates the duty of saving and sharing seed among farming communities. It establishes partnership among farmers and with the earth.

The basmati rice which farmers in my valley have been growing for centuries is today being claimed as "an instant invention of a novel rice line" by a U.S. corporation called RiceTec (patent no. 5,663,454). The "neem" which our mothers and grandmothers have used for centuries as a pesticide and fungicide has been patented for these uses by W. R. Grace, another U.S. corporation. We have challenged Grace's patent with the Greens in European Parliament in the European Patent Office and after ten years of a legal struggle three women – Magda Avoet, president of the Greens; Linda Bullard, president of the International Federation of Organic Agriculture Movements (IFOAM); and myself – defeated the U.S. Government and W. R. Grace (Research Foundation, 2005).

A common myth used by global corporation and the biotechnology industry is that without genetic engineering, the world cannot be fed. However, while biotechnology is projected as increasing food production four times, small ecological farms have productivity hundreds of times higher than large industrial farms based on conventional farms.

Women farmers in the Third World are predominantly small farmers. They provide the basis of food security, and they provide food security in partnership with other species. The partnership between women and biodiversity has kept the world fed through history, at present, and will feed the world in the future. It is this partnership that needs to be preserved and promoted to ensure food security.

Agriculture based on diversity, decentralization, and improving small farm productivity through ecological methods is a women-centered, nature-friendly agriculture. In this women-centered agriculture, knowledge is shared, other species and plants are kin, not "property," and sustainability is based on renewal of the earth's fertility and renewal and regeneration of biodiversity and species richness on farms to provide internal inputs. In our paradigms, there is no place for monocultures of genetically engineered crops and IPR monopolies on seeds.

Monocultures and monopolies symbolize a masculinization of agriculture. The war mentality underlying military-industrial agriculture is evident from the names given to herbicides that destroy the economic basis of the survival of the poorest women in the rural areas of the Third World. Monsanto's herbicides are called "Round Up," "Machete," and "Lasso." American Home Products, which has merged with Monsanto, calls its herbicides "Pentagon," "Prowl," "Scepter," "Squadron," "Cadre," "Lightning," "Assert," and "Avenge." This is the language of war, not sustainability. Sustainability is based on peace with the earth.

The most widespread application of genetic engineering in agriculture is herbicide resistance; i.e., the breeding of crops to be resistant to herbicides. Monsanto's Round Up Ready Soya and Cotton are examples of this application. When introduced to Third World farming systems, this led to increased use of agri-chemicals, thus increasing environmental problems. It also destroyed the biodiversity that is the sustenance and livelihood base of rural women. What are weeds for Monsanto are food, fodder, and medicine for Third World women.

While women have maintained the continuity of seed over millennia in spite of wars, floods, and famines, the masculinization of biodiversity has led to violent technologies that ensure that seed does not germinate on harvest. This has been described as the Terminator Technology. Termination of germination is a means for capacity accumulation as a means for capital accumulation and market expansion. However, abundance in nature and for farmers shrinks as markets grow for Monsanto. When we sow seed, we pray, "May this seed be exhaustless." Monsanto and the U.S. Department of Agriculture (USDA), on the contrary, are stating, "Let this seed be terminated so that our profits and monopoly are exhaustless."

The violence intrinsic to methods and metaphors used by the global agribusiness and biotechnology corporations is violence against nature's biodiversity and women's expertise and productivity. The violence intrinsic to destruction of diversity through monocultures and the destruction of the freedom to save and exchange seeds through IPR monopolies is inconsistent with women's diverse nonviolent ways of knowing nature and providing food security. This diversity of knowledge systems and production systems is the way forward for ensuring that Third World women continue to play a central role as knowers, producers, and providers of food.

Genetic engineering and IPRs will rob Third World women of their creativity, innovation, and decision-making power in agriculture. In place of women deciding what is grown in fields and served in kitchens, agriculture based on globalization, genetic engineering, and corporate monopolies on seeds will establish a food system and world view in which men controlling global corporations control what is grown in our fields and what we eat. Corporate men investing financial capital in theft and biopiracy will present themselves as creators and owners of life.

Agriculture systems shaped by women have a number of key features. Farming is done on a small scale. Natural resources – soil, water, biodiversity – are conserved and renewed. There is little or no dependence on fossil fuels and chemicals. This becomes vital in a period of climate change and peak oil consumption.

Inputs needed for production such as fertilizers are produced on the farm from compost, green manures, or nitrogen-fixing crops. Diversity and integration are key features. And nutrition is a key consideration. Women-run small farms maximize nutrition per acre while they conserve resources.

With food grown for eating, most food is consumed at the household or local level, some is marketed locally, some goes to distant places. Women-centered agriculture is the basis of food security for rural communities. When the household community is food secure, the girl child is food secure. When the household and community is food insecure, it is the girl child who pays the highest price in terms of malnutrition because of gender discrimination. When access to food goes down, the girl child's share is last and least.

4. HUNGER, MALNUTRITION, AND THE POLITICS OF FOOD

Food riots do bring the politics of hunger to the front page of the media. But there is a hidden hunger that denies nearly a billion people of their right to food. And there is a problem of malnutrition related to obesity and other food-related diseases. Hunger and obesity (or the fears of it) are feminist issues both because their worst victims are women and girls, and also because they are result of a food system shaped and controlled by capitalist patriarchy.

Malnutrition is both a result of denial of access of food as well as disappearance of nutrition from our farms and processing systems.

Disappearance of biodiversity on farms is linked to disappearance of women from farms. This is food insecurity for the girl child. Malnutrition in childhood leads to malnutrition in adulthood. Anemia is the most significant deficiency women suffer from. Anemia is also the most significant reason for maternal mortality. When underfed girls become mothers, they give birth to low-birth-weight babies, vulnerable to disease and deprived of their right to full, healthy, wholesome personhood.

Usually these issues of health are not connected to growing of food and farming. But nutrition begins on the farm, and malnutrition begins on the farm.

> We are what we eat.
> But what are we eating?
> What are we growing on our farms? How are we growing it?
> What impact does it have on our health and on the planet?

Food safety, food security, and agriculture are intimately interrelated. How we grow our food and what we grow determines what we eat and who eats. It determines the quality and safety of our food. Yet food safety, food security, and agriculture have been separated from one another. Food is being produced in ways that is robbing the majority of people of food, and those who are eating are eating bad food.

Third-World countries are carrying a double burden of food-related disease, hunger and obesity. The WHO/FAO have predicted that by the year 2020 it is projected that 70 percent of ischemic heart disease deaths, 75 percent of stroke deaths, and 70 percent of diabetes deaths will occur in developing countries. These diseases, called noncommunicable diseases, are directly linked to diet.

The world is producing enough food for all. However, billions are being denied their right to food. The globalized industrialized food system is creating hunger in many ways.

Firstly, industrialized agriculture is based on destruction of small farmers. Uprooted and dispossessed peasants join the ranks of the hungry.

Secondly, industrialized agriculture is capital intensive. It is based on costly external inputs such as purchased and nonrenewable seeds, synthetic fertilizers, pesticides, herbicides. Peasants get into debt to buy these inputs. To pay back debt they must sell all they grow, thus depriving themselves of food. If they cannot pay their debts they lose their land. And they are increasingly losing their lives. More than 150,000 farmers in India have committed suicide as costs of inputs have increased, and the price of their produce has fallen, thus trapping them into debt.

Malnutrition and hunger is also growing because farmers are being pushed into growing cash crops for exports.

The nature of agriculture and the nature of food is being transformed. Agriculture, the care of the land, the culture of growing good food is being transformed into corporate, industrial activity. Food is being transformed from being a source of nutrition and sustenance into being a commodity. And as a commodity, it will first flow to factory farms and now cars. The poor will get the leftover.

Factory farms are a negative food system. They consume more food than they produce. Industrial beef requires 10 kg of feed to produce 1 kg of food. Industrial pork requires 4.0–5.5 kg of feed to produce 1 kg of food. Factory-farmed chicken requires 2.0–3.0 times more feed than it produces as food (Compassion in World Farming, 2004).

Industrial biofuels are putting a new pressure on food. Food prices in Mexico have doubled since corn, the staple for Mexican tortillas, is being increasingly used to make ethanol for fuel. Corn, soya, and canola are all being diverted to feed cars while people starve.

5. GLOBALIZATION AND INDUSTRIALIZATION OF AGRICULTURE AND FOOD SYSTEM

Across the world, a food tsunami is occurring, transforming small farms run largely by women peasants into "factories" producing "commodities." Globalization has led to the industrialization of agriculture, and industrial agriculture displaces women from productive work on the land.

Globalization of agriculture has been driven by agribusiness corporations, which are seeking global markets for their nonrenewable inputs – seeds, fertilizers, and pesticides – as well as markets for their food commodities. The Agriculture Agreement of WTO and the Structural Adjustment programs of the World Bank have been the most important instruments for the globalization of agriculture. Globalization involves multiple shifts. It shifts control over food production from local and national levels to the global level. It also shifts control over food production from women farmers to global corporations, whether it be in the area of seed or systems of maintaining and renewing soil fertility. The local, the renewable is replaced by the global and the nonrenewable. Women's knowledge, expertise, and creative and productive activities are replaced by science and technology driven by corporate profits.

The industrialization of agriculture is a shift from internal inputs to purchased external inputs. It is a shift from ecological to chemical. It is a shift from biodiversity to monocultures. And it is a shift from women as the primary source of knowledge and skills about farming – from seed saving to composting, to cultivating poly cultures in the right balance, to harvesting, storage, processing – to an agriculture without women.

Humanity has eaten more than 80,000 edible plants through its evolution. More than 3,000 have been used consistently. However, we now rely on just 8 crops to provide 75 percent of the world's food. And with genetic engineering, production has narrowed down to 3 crops – corn, soya, canola. And now these too are being diverted to biofuel.

Monocultures are destroying biodiversity, our health, and the quality and diversity of food.

Monocultures have been promoted as an essential component of industrialization and globalization of agriculture. They are assumed to produce more food. However, all they produce is more control and profits – for Monsanto, Cargill, and

ADM. They create pseudo surpluses and real scarcity by destroying biodiversity, local food systems, and food cultures.

Corporations are forcing us to eat untested food such as GMO's. Even soya, which is now in 60 percent of all processed food, was not eaten by any culture fifty years ago. It has high levels of Isoflavones and phytoestrogens, which produce hormone imbalances in humans. Traditional fermentation as in the food cultures of China and Japan reduce the levels of isoflavones (James and James, 1994). The promotion of soya in food is a huge experiment promoted with $13 billion subsidies from the U.S. Government between 1998 and 2004, and $80 million a year from the American Soya Industry. Nature, culture, and people's health are all being destroyed. Local food cultures have rich and diverse alternatives to soya. For protein we have thousands of varieties of beans and grain legumes – the pigeon pea, the chick pea, moong bean, urud bean, rice bean, azuli bean, moth bean, cow pea, peas, lentils, horse gram, faba bean, winged bean. For edible oils we have sesame, mustard, linseed, niger soffola, sunflower, groundnut.

In depending on monocultures, the food system is being made increasingly dependent on fossil fuels – for the synthetic fertilizers, for running the giant machinery, for the long-distance transport, which adds "food miles." With the spread of monocultures and the destruction of local farms, we are increasingly eating oil, not food, and threatening the planet and our health.

Moving beyond monocultures of the mind has become an imperative for repairing the food system. Biodiverse small farms have higher productivity and they generate higher incomes for farmers. And biodiverse diets provide more nutrition and better taste.

Bringing back biodiversity to our farms goes hand in hand with bringing back small farmers, especially women, to the land. Corporate control thrives on monocultures. The food freedom of all people depends on biodiversity. Human freedom and the freedom of other species are mutually reinforcing, not mutually exclusive.

The change in production is intimately connected to changes in distribution. Industrial agriculture produces commodities. Small-scale ecological farming produces food. Commodities are distributed by global corporations on the logic of profit maximization. If there are higher profits from cattle feed, food grain goes to factory farms rather than hungry families. If there are higher profits in industrial biofuels, corn goes to produce ethanol and soy goes to produce biodiesel. Commodities grow, so does hunger. And again it is the poor and the vulnerable who pay the highest price in terms of starvation – commodities are substitutable – they can be food for people or fuel for cars.

Biofuels, fuels from biomass, continue to be the most important energy source for the poor in the world. The ecological biodiverse farm is not just a source of food; it is a source of energy. Energy for cooking the food comes from the inedible biomass like cow dung cakes, stalks of millets and pulses, agro-forestry species on village wood lots. Managed sustainably, village commons have been a source of decentralized energy for centuries.

Industrial biofuels are not the fuels of the poor; they are the foods of the poor, transformed into heat, electricity, and transport. Liquid biofuels, in particular ethanol and biodiesel, are one of the fastest-growing sectors of production, driven by the search of alternatives to fossil fuels both to avoid the catastrophe of peak oil consumption and to reduce carbon dioxide emissions. President Bush tried to pass legislation to require the use of 35 billion gallons of biofuels by 2017.

Global production of biofuels alone has doubled in the last five years and will likely double again in the next four. Among countries that have enacted a new pro-biofuel policy in recent years are Argentina, Australia, Canada, China, Columbia, Ecuador, India, Indonesia, Malawi, Malaysia, Mexico, Mozambique, the Philippines, Senegal, South Africa, Thailand, and Zambia.

There are two types of industrial biofuels – ethanol and biodiesel. Ethanol can be produced from products rich in saccharose such as sugarcane and molasses, and substances rich in starch such as maize, barley, and wheat. Ethanol is blended with petrol. Biodiesel is produced from vegetable oils such as palm oil, soya oil, and rapeseed oil. Biodiesel is blended with diesel.

Representatives of organizations and social movements from Brazil, Bolivia, Costa Rica, Columbia, Guatemala, and the Dominican Republic, in a declaration titled "Full Tanks at the Cost of Empty Stomachs," wrote, "The current model of production of bioenergy is sustained by the same elements that have always caused the oppression of our people's appropriation of territory, of natural resources, and the labor force."

And Fidel Castro, in an article titled "Food Stuff as Imperial Weapon: Biofuels and Global Hunger," has said:

> More than three billion people are being condemned to a premature death from hunger and thirst.

The biofuel sector worldwide has grown rapidly. The United States and Brazil have established ethanol industries and the European Union is also fast catching up to explore the potential market. Governments all over the world are encouraging biofuel production with favorable policies. The United States is pushing the Third-World nations to go in for biofuel production so that their energy needs get met at the expense of plundering others' resources.

Inevitably this massive increase in the demand for grains is going to come at the expense of the satisfaction of human needs, with poor people priced out of the food market. On February 28, the Brazilian Landless Workers Movement released a statement noting that "the expansion of the production of biofuels aggravates hunger in the world. We cannot maintain our tanks full while stomachs go empty." The diversion of food for fuel has already increased the price of corn and soya. There have been riots in Mexico because of the price rise of tortillas. And this is just the beginning. Imagine the land needed for providing 25 percent of the oil from food.

One ton of corn produces 413 liters of ethanol; 35 million gallons of ethanol requires 320 million tons of corn. The United States produced 280.2 million tons of corn in 2005. As a result of NAFTA, the United States made Mexico dependent on U.S. corn and destroyed the small farms of Mexico. This was in fact the basis of the Zapatista uprising. As a result of corn being diverted to biofuels, prices of corn have increased in Mexico.

Industrial biofuels are being promoted as a source of renewable energy and as a means to reduce greenhouse gas emissions. However, there are two ecological reasons why converting crops like soya, corn, and palm oil into liquid fuels can actually aggravate climate chaos and the CO_2 burden.

Firstly, deforestation caused by expanding soya plantations and palm oil plantations is leading to increased CO_2 emissions. The United Nations Food and Agriculture Organization estimates that 1.6 billion tons or 25 to 30 percent of the greenhouse gases released into the atmosphere each year comes from deforestation. By 2022, biofuel plantations could destroy 98 percent of Indonesia's rainforests.

According to Wetlands International, destruction of South East Asia pert lands for palm oil plantations is contributing to 8 percent of the global CO_2 emissions. According to Delft Hydraulics, every ton of palm oil results in thirty tons of carbon dioxide emissions or ten times as much as petroleum producers. However, this additional burden on the atmosphere is treated as a clean development mechanism in the Kyoto Protocol for reducing emissions. Biofuels are thus contributing to the same global warming that they are supposed to reduce (World Rainforest Bulletin, 2006).

Further, the conversion of biomass to liquid fuel uses more fossil fuels than it substitutes.

One gallon of ethanol production requires 28,000 kcal. This provides 19,400 kcal of energy. Thus the energy efficiency is minus 43 percent.

The United States will use 20 percent of its corn to produce 5 billion gallons of ethanol, which will substitute 1 percent of oil use. If 100 percent of corn was used, only 7 percent of the total oil would be substituted. This is clearly not a solution either to peak oil or climate chaos (Pimental, 2007).

And it is a source of other crises: 1,700 gallons of water are used to produce a gallon of ethanol, and corn uses more nitrogen fertilizer, more insecticides, more herbicides than any other crop.

Food has literally become a life-and-death issue for women, whether it is through hunger and starvation, or self-starvation in the form of anorexia nervosa, or obesity, or female feticide.

The future of food needs to be reclaimed by women, shaped by women, democratically controlled by women. Only when food is in women's hands will food and women be secure.

In 1996, Maria Mies and I initiated the Leipzig Appeal for Food Security in Women's Hands. For thousands of years women have produced their own food and guaranteed food security for their children and communities. Even today, 80 percent of the work in local food production in Africa is done by women, in Asia 50 to 60 percent, and in Latin America 30 to 40 percent. And everywhere in the world, women are responsible for food security at the household level. In the patriarchal society, however, this work has been devalued.

All societies have survived historically because they provide food security to their people. This policy, however, has been subverted by the globalization, trade liberalization, industrialization, and commercialization of all agricultural products under the auspices of the World Trade Organization and the World Bank/IMF.

Worldwide, women are resisting the policies that destroy the basis of their livelihood and food sovereignty. They are also creating alternatives to guarantee food security for their communities based on different principles and methods from those governing the dominant, profit-oriented global economy. They are

- Localization and regionalization instead of globalization
- Nonviolence instead of aggressive domination
- Equity and reciprocity instead of competition
- Respect for the integrity of nature and its species
- Understanding humans as part of nature instead of as masters over nature
- Protection of biodiversity in production and consumption

Food security for all is not possible within a global market system based on the dogma of free trade, permanent growth, comparative advantage, competition, and profit maximization. However, food security can be achieved if people within their local and regional economies feel responsible, both as producers and as consumers for the ecological conditions of food production, distribution, and consumption, and for the preservation of cultural and biological diversity where self-sufficiency is the main goal.

Our food security is too vital an issue to be left in the hands of a few transnational corporations with their profit motives or to be left up to national governments that increasingly lose control over food security decisions, or to a few, mostly male national delegates at UN conferences, who make decisions affecting all our lives.

Food security must remain in women's hands everywhere! And men must share the necessary work, be it paid or unpaid. We have a right to know what we eat! No to Novel Food and No to Patents on Life. We will resist those who force us to produce and consume in ways that destroy nature and ourselves.

WORKS CITED

J. Bajaj, and M. D. Srinivas (1996) *Annam Bahu Kurvita* (Chennai: Centre for Policy Studies).

Compassion in World Farming (2004) 'The Global Benefits of Eating Less Meat,' (available at http://wessa.org.za/uploads/meat_free_mondays/global_benefits_of_eating_less_meat.pdf).

FAO (1998) 'Women Feed the World, (Rome: FAO) (available at http://www.fao.org/docrep/x0262e/x0262e16.htm).

V. James, and R. James (unpublished) 'The Toxicity of Soya Beans and Their Related Products.' Navdanya, Biodiversity Based Organic Farming: A New Paradigm for Food Safety and Food Security (New Delhi, India, 2007).

D. Morgan (1980) *Merchants of Grain: The Power and Profits of the Five Giant Companies at the Centre of the World's Food Supply* (New York: Penguin).

Oligopoly Inc. (2005) ETC Group Report, (available at www.etcgroup.org/en/materials/publications.html?pub-id=44).

D. Pimental (February 23–25, 2007) 'The Triple Crisis.' IFG conference, London.

Research Foundation for Science, Technology and Ecology (2005) *Neem: Fight against Biopiracy and Rejuvenation of Traditional Knowledge* (New Delhi: India).

V. Shiva (2001) 'Annadana: The Gift of Food,' in *A Sacred Trust: Ecology and Spiritual Values*, based on a series of lectures organized by the Prince's Foundation and the Temenos Academy, U.K.

V. Shiva (1997) *Biopiracy*, (Cambridge, MA: South End Press).

V. Shiva (2000) *Stolen Harvest* (Cambridge, MA: South End Press).

World Rainforest Bulletin (2006), November, 112: 22.

DISCUSSION QUESTIONS

1. Shiva argues that women's economic contributions are often under-counted because much of their work is done within the home. How might this negatively impact women? How might these negative impacts be avoided? Do you think this is specific to the cultures she is discussing or is it a more general problem?

2. What is a feminist approach to farming, according to Shiva, and how might it help to address other concerns associated with agriculture, for example environmental and animal welfare concerns?

3. Shiva argues that industrialization leads to monoculture and the loss of diversity. Why do you think industrialization tends this way – what are its distinctive features that might lead to this?

4. What technologies are the greatest threat to food security? Which of those threats can be managed by the means Shiva suggests? Which threats require additional measures and what might those measures be?

5. What other factors, in addition to those discussed in this chapter, are relevant to the problem of global malnutrition and food insecurity? Is food insecurity primarily a food production problem, a food distribution problem, or both?

6. Could global food security be achieved without agricultural industrialization? Or does industrialization actually undermine food security in the long run?

7. In what ways is Shiva's critique based on a power analysis of agricultural technologies? In what respects is it based on a form of life analysis? Did you find her analyses compelling? Why or why not?

33 THE ETHICS OF AGRICULTURAL ANIMAL BIOTECHNOLOGY[1]

Robert Streiffer and John Basl

CHAPTER SUMMARY

In this chapter, Robert Streiffer and John Basl consider the potential for biotechnology to address two prominent ethical concerns regarding concentrated animal agriculture: its detrimental ecological impacts and the amount of animal suffering that is involved. With respect to animal welfare, they focus on capacity diminishment – e.g. engineering blind chickens or microencephalic pigs. Streiffer and Basl raise significant doubts about whether diminishment would in fact decrease animal suffering and improve the quality of animal lives. With respect to the environment, they focus on the case of Enviropig – the attempt to engineer pigs that have less phosphorous in their manure. They argue that if Enviropigs were engineered successfully, they would have lower environmental impacts than non-engineered pigs on a per-pig basis. However, whether they would be ecologically beneficial overall depends on several other factors, including whether they enabled an increase in the number of animals used. Therefore, Enviropigs (and other animals engineered to reduced ecological impacts) may not in the end be ecologically beneficial.

RELATED READINGS

Introduction: EHS (2.6.1); Types of Value (3.1)

Related Readings: John Basl, *What to do About Artificial Consciousness* (Ch. 25); Paul Thompson, *Artificial Meat* (Ch. 34)

1. INTRODUCTION

Recent biotechnology research includes the development of genetically engineered animals and cloned animals for use as food or breeding stock in agriculture. Such research raises important ethical issues regarding the welfare of animals in agriculture and animal agriculture's environmental impact.

---

[1] Material in this chapter originally appeared in Robert Streiffer and John Basl (2011) 'The Application of Biotechnology to Animals in Agriculture,' Oxford Handbook of Animal Ethics, eds. Beauchamp and Frey (Oxford University Press). It appears here by permission of Oxford University Press.

The livestock sector's massive scale – fifty six billion land animals are consumed each year (UN FAO, http://faostat.fao.org) – means that several routine agricultural practices that are detrimental to animal welfare or to the environment pose some of the most pressing global ethical issues the human species has ever encountered. An application of biotechnology to animals in agriculture (*animal biotech* for short) can either mitigate or exacerbate these problems. In Section 2, after discussing the philosophical literature on animal welfare, we discuss animal biotech and animal welfare. In particular, we focus on how diminishing the cognitive capacities of animals in agriculture might improve or harm their welfare. In Section 3, we turn to animal biotech and the environment. After explaining how agriculture contributes to environmental problems, we look at how a particular genetically engineered animal, the Enviropig, might alter the environmental impacts of pig agriculture. In doing so we develop a framework for thinking about other cases.

2. ANIMAL BIOTECH AND ANIMAL WELFARE

2.1 Theories of animal welfare

One way an application of animal biotech can have a morally relevant impact is by its effect on an animal's welfare (its well-being, quality of life, or how well its life is going). Insofar as an application of animal biotech provides a net increase of animal welfare compared to current practices, there is a moral reason in favor of adopting that application. And, of course, insofar as an application decreases animal welfare compared to current practices, there is a moral reason against adopting that application. However, to assess the welfare impact of particular applications of animal biotech, we must know what it means to improve animal welfare. This requires that we understand both what constitutes animal welfare and what constitutes an improvement in it.

There are three types of general views about welfare: Mentalistic views, Desire-Satisfaction views, and Objective List views.

According to Mentalistic views of welfare, welfare is solely a function of the mental life of an individual. Negative mental states such as pain, suffering, and distress count against an individual's welfare, while positive mental states such as pleasure, enjoyment, and contentment count in favor of an individual's welfare.

One prominent problem with Mentalistic conceptions of welfare is that they ignore the possibility that whether or not one's desires are satisfied can have an impact on one's welfare without affecting one's mental life. For example, someone who desires to be in a committed relationship but, unknowingly, has an unfaithful partner is worse off than someone who is otherwise the same but for having a faithful partner. This problem motivates Desire-Satisfaction views, according to which welfare is constituted by the satisfaction of one's desires. As with Mentalistic views, Desire-Satisfaction views allow that mental states are an important component of welfare. However, Desire-Satisfaction views can accommodate the idea that facts about the world can affect welfare without affecting the mental life of the individual.

Desire-Satisfaction views typically distinguish between actual desires and informed desires. According to actual desire views, welfare is constituted by the satisfaction of an individual's actual preferences. However, the satisfaction of actual desires is not always a component of welfare. For example, the satisfaction of a desire to be shot on the basis of the false belief that being shot will make one live forever is not a component of an individual's welfare. This leads most proponents of a Desire-Satisfaction view to adopt the requirement that, for a desire's satisfaction to count as part of the individual's welfare, the desire must be "informed," that is, capable of persisting in the face of correct information and rational reflection (Griffin, 1986, pp. 11–14).

But moving to a Desire-Satisfaction view based on informed desires suggests that the states of affairs themselves, rather than the preferences concerning them, are what make a life go well. As James Griffin says, "What makes us desire the things we desire, when informed, is something about them–*their* features or properties. But why bother then with informed desire, when we can go directly to what it is about objects that shape and form desires in the first place?" (Griffin, 1986, p. 17).

This supports Objective List views, according to which there are some goods that contribute to welfare regardless of whether they contribute to the subjective experience of the individual or whether the individual does or would desire them. In the case of humans, the most plausible Objective List views are hybrid views that include mentalistic and desire-based components, as well as the satisfaction of basic psychological and biological needs and the development and exercise of valuable cognitive and emotional capacities. David Brink, for example, includes the autonomous undertaking and completion of "projects whose pursuit realizes capacities of practical reason, friendship, and community" as components of human welfare (Brink, 1989, p. 233).

The realization of some of the goods typical of Objective List views of human welfare requires significant cognitive capacities, many of which will be beyond those typical of livestock. Objective List views of animal welfare often focus on health or natural functioning (i.e., functioning in species-typical ways). However, health or natural functioning cannot be the full story with respect to an animal's welfare. For example, analgesia prevents a species-typical response, i.e. pain, but can contribute to welfare. For these reasons, Objective List views of animal welfare must also include mentalistic components of welfare. An animal may be made worse off by not being allowed to function in species-typical ways, but it is also made worse off if it suffers in a way consistent with typical species functioning.

While we are partial to such a hybrid Objective List view of animal welfare for the reasons provided above, and in what follows we use such a view to evaluate the welfare impact of animal biotech, many claims about what harms or improves animal welfare will hold no matter which view of animal welfare is correct. Sickness and injury, for example, typically cause negative mental states, frustrate informed desires, and hinder natural functioning. However, some of the applications of animal biotech discussed below are philosophically interesting because they improve welfare according to some conceptions, but negatively affect it according to others.

2.2 Two ways of "improving" animal welfare

A person might be making one of two different claims in saying that a technology improves animal welfare. One claim is that there is an individual who is made better off by that application. For example, a cow that is sick and receives antibiotics is made better off. In this case, the technology is good for the animal simply in virtue of its effects having a positive overall impact on the animal's welfare. There is clearly a reason in favor of using that technology, a reason grounded in the fact that the technology improved an individual's welfare.

In many cases where welfare problems are due to genetics, however, this simple picture no longer applies. Cloning and methods of pre-conception genetic modification, such as selective breeding, bring into existence individuals that are distinct from the ones who would have been brought into existence had the cloning or selective breeding not occurred. Thus, one cannot truthfully say that there is an individual whose welfare was improved by the technology. Rather, there is an individual brought into existence who has a certain level of welfare, and that individual is at a higher (or lower) level of welfare than the level at which a distinct individual, who did not in fact come into existence, would have been. So while aggregate welfare has been increased, no particular individual's welfare was improved. Let us refer to cases of the first kind as *alterations*, because an individual's welfare is altered, and cases of the second kind as *substitutions*, because one animal is substituted for another.[2]

Cloning and selective breeding will count as substitutions. Genetic engineering usually takes place post conception, and so the founder animals will count as being altered. The individuals that result from the subsequent breeding of the founder animals, however, will count as substitutions and are typically much more numerous than the founder animals.

Whether alterations and substitutions are morally different is a matter of controversy. On the one hand, an alteration is bad for the altered animal if and only if it has a negative impact overall on the animal's life-long welfare, whereas a substitution is bad for the animal brought into existence only in the extreme case that the animal will have a life so bad that it is not, on balance, worth living. Thus, many alterations will be bad for an animal even though the similar substitution would not be. This suggests that there is a morally relevant difference (Streiffer, 2008). On the other hand, consideration of similar cases supports the view that alterations and substitutions are morally on a par with each other:[3]

> Case 1: A woman is currently pregnant with a child she knows will be born mentally handicapped. By taking a safe and affordable medication, she can prevent the child from being born with the mental handicap. She chooses not to take the medication and the child is born with the handicap.

---

[2] The difference between these cases has been most influentially discussed by Parfit, (1984, esp. pp. 351–379).

[3] These cases are adapted from (Buchanan, Brock, Daniels, and Wikler, 2000, pp. 244–245).

Case 2: A woman is contemplating becoming pregnant but knows that, if she conceives within the next month, any resulting child will be born with a mental handicap. If she waits to become pregnant, the resulting child will not have the mental handicap. The woman decides against waiting, becomes pregnant, and the resulting child has a mental handicap.

Case 1 is an alteration, whereas Case 2 is a substitution. However, assuming that the handicap and the women's reasons for their decisions are the same in both cases, then the women's actions seem equally morally problematic.[4]

That there seems to be a morally relevant difference between alterations and substitutions but the women's actions seem equally morally problematic suggests the following: substitutions and alterations are, other things being equal, on a par with each other, but a different explanation must be given of why a substitution is morally problematic, one that does not appeal to the action's being bad for any particular individual. The problem of adequately articulating this explanation is known as the "non-identity problem." It is the subject of considerable controversy, but we will assume that a substitution that results in animals being brought into existence with a welfare that is worse than the welfare of animals that would have been brought into existence instead, even if both animals have a life worth living, is as morally problematic as the similar alteration would have been. Similarly, a substitution that results in animals with welfare that is better than that of the animals currently used in agriculture is a moral improvement, other things being equal.

2.3 Animal welfare and animal biotech

Starting around the 1950s, housing practices in some parts of the livestock sector significantly reduced the amount of space afforded each animal (Fraser, Mench, and Millman, 2001; Duncan, 2004; NAHMS, 2003). While such changes made it easier to care for the much larger number of animals typical in intensive agriculture, the changes have also had many well-documented negative impacts on welfare, including: severe osteoporosis, frustration over lack of nesting spaces, and increased agonistic behavior in poultry; and boredom, atypical repetitive behaviors (so-called stereotypies), skins lesions, bone problems, and frustration of nesting behaviors in sows (Gregory and Wilkins, 1989).[5] Management and breeding practices for increased productivity have also significantly exacerbated welfare problems arising from heat stress, lameness, and mastitis in dairy cows (Heringstad, Klemetsdal, and Ruane, 2000). In addition, welfare problems arise from many practices employed to manage the large number of animals on a typical farm, including tail docking of swine, dairy cows and sheep, dehorning and branding of cattle, castration of boars, and debeaking and de-toeing of poultry. These surgeries are typically performed without anesthetic or analgesic (Fraser and Weary, 2008).

---

[4] For a discussion of different ways to justify this claim, see Steinbock and McClamrock (1994); Brock (1995); Rollin (1995).

[5] For further discussion, see Webster (2004).

Because animal welfare issues are so extensive in standard agricultural practices, there is a large potential for improving welfare through animal biotech. A comprehensive literature review performed in 2002 by the National Research Council (NRC) of the National Academies documented several ways that cloning and genetic engineering could be used to improve animal welfare (NRC, 2004, pp. 93–102, 104–107). These include creating animals that are more resistant to parasites and diseases, engineering animals that lack traits that lead to current welfare issues (for example, genetically engineering cattle without horns to avoid dehorning, or pigs that do not require castration to remove "boar taint"), and sex selection to avoid producing animals of an undesired sex that are then killed.

The NRC also documented several ways that cloning and genetic engineering could increase risks to animal welfare. Collecting the genetic materials and embryos necessary for animal biotech often requires that animals undergo painful procedures or be sacrificed. Cloned animals have an increased risk for a variety of problems that are painful and result in abnormal functioning, while the parents of such offspring are at a higher risk of complications during pregnancy. Mutations can result from attempts to create animals with novel genomes and many survivors suffer from problems such as "severe muscle weakness, missing kidneys, seizures, behavioral changes, sterility, disruptions of brain structure, neuronal degeneration, inner ear deformities, and limb deformities" (NRC, 2004, p. 97).

Many of the above welfare issues are unintended and unwanted problems associated with the process of genetically engineering or cloning an animal, and it is important to note that researchers will continue to refine these techniques to minimize their occurrence. However, even if nothing goes wrong with the process of genetically engineering or cloning an animal, the intended effects themselves can have a negative impact on animal welfare. For example, insofar as genetic engineering enables further concentration in animal agriculture – i.e. even greater numbers of animals in even smaller spaces – there is the potential for welfare diminishment. Genetically engineering an animal to be resistant to disease or parasites could be, on balance, bad for animal welfare if it results in animals being subjected to even more severe movement restrictions or otherwise exacerbates the problems described above.

Temple Grandin, a leading expert on animal welfare science, goes so far as to suggest that "the most serious animal welfare problems may be caused by over selection for production traits such as rapid growth, leanness, and high milk yield" (Grandin and Deasing, 1998, p. 219). So although each application of animal biotech should be evaluated on its own individual merits, to the extent that genetic engineering and cloning are used with the same motivations and for the same purposes as traditional selection, it would be naïve and irrational to ignore the presumption that animal biotech, in general, is more likely to exacerbate rather than mitigate animal welfare problems.

2.4 Improving welfare by diminishing animals?

When most people think of using animal biotech to improve animal welfare, they think of enhancing the animal in some way, for example, enhancing its capacity to resist injury or disease. But some have argued that we ought to use animal biotech to disable or diminish animals relative to normally functioning members of their species.

Consider the example of using chemical mutagenesis to produce a line of blind chickens. Chickens are naturally aggressive and have become more aggressive as an unintended byproduct of selective breeding for egg productivity. This aggression is exacerbated by the crowded conditions in which chickens are kept in industrial agriculture settings (Ali and Cheng, 1985, p. 791). Producers typically try to minimize the damage of the resulting attacks by trimming the birds' beaks, combs, and toes, but this is less than ideal. The results of debeaking are described by Ian Duncan: "behavioral changes suggestive of acute pain have been found to occur in the 2 days following surgery. These are followed by chronic pain that lasts at least 5 or 6 weeks after the surgery" (Duncan, 2004, p. 215).

An alternative method became available, though, when researchers used chemical mutagenesis to induce a genetic mutation that resulted in congenital blindness (Cheng et al., 1980). Researchers at the University of British Columbia identified the specific mutation, speculating early on that "because of [blind chickens'] docility and possibly reduced interaction in a social hierarchy, studies of their behavior with relation to growth rate, feed efficiency, and housing density may have some practical application" (Cheng et al., 1980, p. 2182). Follow up studies indicated that flocks of blind chickens use less feed and show increased egg productivity, while also suffering fewer injuries and less feather and skin damage (Ali and Cheng, 1985).

Consider also microencephalic pigs. Researchers identified a gene (Lim1, also referred to as Lhx1) responsible for head morphology in mice (Shawlot and Behringer, 1995). Using genetic engineering to knock out this gene resulted in mice with bodies that were normally formed except for lacking heads. The pups survived in utero about halfway through normal gestation. Such research vividly raises the possibility that highly social and intelligent animals, such as pigs, which suffer greatly in current factory farming conditions, could be engineered or bred to have just enough brain stem to support biological growth, but not enough to support consciousness (Balduini et al., 1986; Bach, 2000). Such animals, lacking the capacity for consciousness, would be incapable of suffering.

Some of these diminished animals are of interest to producers because they use less feed or show higher productivity, but the ethical argument for using them stems from the idea that these animals will be comparatively better off than their non-engineered counterparts. Whether this is so is partly an empirical question and partly a conceptual question. The answer will depend on the empirical facts about the lives of the diminished animals and the lives of their non-diminished counterparts, and it will also depend on the conceptual facts about what constitutes animal welfare.

The traditional Mentalistic view of animal welfare characterizes welfare in terms of the absence of mental states such as pain, suffering, or distress (Dawkins, 1980). A broader, but still Mentalistic, conception of animal welfare also includes the presence of positive mental states such as pleasure, enjoyment, or contentment (McMillan, 2005). If the traditional Mentalistic view is correct, then we need only ask whether the animals we currently use experience more (or more severe) negative mental states than would their diminished counterparts, if we used them instead.

Taking the above examples as philosophical thought experiments, one can simply stipulate that the diminished animals do better in terms of negative mental states than their non-diminished counterparts. But we do not have the liberty to make such stipulations in evaluating whether to actually adopt such technologies. It is also crucial to be clear on the welfare claims being made: even supposing that the traditional Mentalistic view is correct and that a modification alleviates a source of suffering, it is still an open question whether the modification improves the animal's welfare, on balance. After all, it may introduce other sources of negative mental states.

With respect to blind chickens, it isn't clear what the on-balance welfare impact would be. On the one hand, blind chickens would not suffer the well-documented pain of having their beaks, toes, and combs amputated, and there is data to support the claim that they suffer fewer injuries and less feather loss (Ali and Cheng, 1985). On the other hand, there were no significant differences in the two surrogate measures of psychological stress, adrenal weight and plasma corticosterone levels. This suggests that they are just as stressed as sighted birds (Ali and Cheng, 1985). Furthermore even if blind chickens do experience fewer or less intense negative mental states because of their inability to see, sight might still provide a significantly enhanced subjective environment that generates positive mental states. So, even if blindness decreases the number and intensity of negative mental states it still might be worse overall from a broader Mentalistic perspective. It is difficult to even begin conceptualizing how we would measure and compare the negative and positive mental states of individuals of a non-human species where one of them is missing an entire sense modality of such crucial importance as sight.

Even for diminished animals that do fare better than their non-diminished counterparts from a mentalistic perspective, it is still possible that they will do worse in terms of desire-satisfaction, health or natural functioning. Even if blind chickens suffer less, because they are blind, they are still, other things being equal, less healthy and less capable of natural functioning compared to normal members of their species. And the blindness may have other effects that are relevant as well. For example, blind chickens engage in significantly fewer social interactions than do sighted chickens (Ali and Cheng, 1985). Chickens will have desires that will be harder to satisfy if they are blind, and these need to be taken into account if desire satisfaction is a component of welfare. So even if blindness results in a better balance of mentalistic factors, it does not follow that the welfare of blind chickens is better than the welfare of sighted chickens unless these other components are also addressed.

What about microencephalic pigs who lack the capacity for consciousness? If the traditional Mentalistic view of animal welfare were correct, then microencephalic pigs would have a better welfare than normal pigs since microencephalic pigs have no negative mental states. Indeed, if animal welfare were defined solely in terms of the absence of negative mental states, then microencephalic pigs would have a better welfare than any pig, no matter how well treated. But this, of course, demonstrates the implausibility of excluding positive mental states as components of animal welfare.

According to a hybrid Objective List view, positive mental states are components of animal welfare, and whether the welfare of microencephalic pigs is better than their conventional counterparts will largely turn on whether the pigs in current industrial agricultural practices suffer so much that the suffering outweighs any positive mentalistic benefits. That is, in mentalistic terms, are the lives of pigs in industrial agriculture so bad that they would have been better off not having existed at all? If the answer to that question is no, then pigs without the capacity for consciousness do *not* fare better than the animals currently used. Admittedly, the lives of pigs in industrial agriculture fall far below the lives that pigs are capable of in better circumstances, but that isn't the relevant comparison.

3. ANIMAL BIOTECH AND THE ENVIRONMENT

A review by the United Nations of the environmental impact of livestock agriculture concluded that the livestock sector is one of the most significant contributors to environmental problems, both globally and locally (FAO, 2006). In the following sections, we review these impacts. The general ways in which animal biotech might alleviate or exacerbate environmental problems are then discussed, followed by an examination of a specific application of animal biotech: Enviropigs.

3.1 Livestock agriculture and the environment

The most serious global environmental challenge is that of anthropogenic climate change. Costs from climate change come in a variety of forms: health risks, loss of water and land resources, displacement of people (i.e. refugees), biodiversity loss, and the economic costs of adapting to a changing climate, for example (IPCC, 2007; Singer, 2002; Gardiner, 2004; Hulme, 2005; FAO, 2006). According to the United Nations' report "Livestock's Long Shadow," the livestock sector produces significant amounts of three main greenhouse gases: methane, nitrous oxide, and CO_2. These contributions accrue from deforestation to create more land for agricultural use (CO_2), natural digestive processes in agricultural animals (methane), and from the production and use of manure (nitrous oxide).

Animal agriculture also contributes directly to local environmental issues such as water depletion, water degradation, and biodiversity loss. While it is difficult to calculate the full impact of animal agriculture on water resources, the amount used, both directly and indirectly, is substantial. Water is used in animal agriculture to feed and service animals, in processing products after slaughter, and to grow feed crops. The livestock sector makes a major contribution – 93% of withdrawal and 73% of consumption (removal that renders the water unavailable for other uses) – to human water use globally, and this does not include water used in aquaculture (Turner et al., 2004).

In addition to water use, animal agriculture contributes to water pollution. The high number of animals produces large amounts of waste which is often used as fertilizer. Given the scale of agricultural facilities, the amount of waste produced is typically much higher than should be used for fertilization in surrounding fields, leading to excessive applications of waste manure (Hooda et al., 2000). This increases the amount of runoff that contaminates the water system, leading to excessive amounts of phosphorus and nitrogen in the water system.

According to a 2005 Millennium Ecosystems Assessment, the most important drivers of biodiversity loss include climate change, pollution, habitat change, and invasive species (Millennium Ecosystem Assessment, 2005). In addition to animal agriculture's contribution to climate change and pollution discussed above, animal agriculture also drives biodiversity loss through habitat change and the introduction of invasive species (Ilea, 2009; FAO, 2006).

3.2 Potential impact of animal biotech on the environment

Animal biotech has the potential to affect the environmental impact of livestock agriculture in several ways. While using animal biotech to increase individual productivity imposes additional risks to animal welfare, it could have a beneficial effect on the environment by decreasing the number of animals used in agriculture, thereby reducing the amount of feed needed and pollution produced. Increasing feed-conversion efficiency or nutrient utilization would also reduce the amount of feed or additives needed, which would have environmental benefits and might not have any negative impact on animal welfare.

Among the most widely discussed environmental risks of animal biotech is the intentional or unintentional release of genetically engineered animals. Whether the release of a genetically engineered animal results in an environmental harm depends, in part, on what in the environment is valuable and the ways in which it is valuable. Views about the value of the environment can be divided between those views on which the environment has intrinsic value and those on which it has merely instrumental value. To claim that the environment has intrinsic value is to claim that it is valuable for its own sake, independently of whether anyone or anything values it (O'Neill, 2003). To claim that it has merely instrumental value is to claim that it is valuable only because it is instrumentally useful to other entities achieving their worthwhile ends.

Some environmental ethicists argue that the environment has intrinsic value stemming from its naturalness (Elliot, 1982; Rolston III, 1989; Throop, 2000). Naturalness is typically defined as freedom from the influence of humans, and it is understood as coming in degrees. On such views, the novel presence of a genetically engineered animal in the environment would itself detract from the naturalness of the area and so constitute an environmental harm, even if its presence had no further consequences.

Views on which the environment is intrinsically valuable because it is natural are controversial. It is difficult to see why being natural would be a source of value except instrumentally in that we value areas that are free of our impact and that such areas are beneficial to ourselves and other entities. Similar worries arise for views that claim that the intrinsic value of the environment stems from its biodiversity. But while the intrinsic value of the environment is controversial, there is little doubt that the environment is instrumentally valuable. Ecosystems provide services and resources for human and non-humans alike. Whether the release of a genetically engineered animal results in an environmental harm, on instrumental views of environmental value, will depend on (a) the environmental consequences of the release and (b) the moral status of the entities affected.

The environmental risk of the release of a genetically engineered animal depends on several factors, including the animal's ability to initially escape captivity, the ability to travel, the ability to maintain a feral population, and the extent of environmental disruption that might be caused by such a feral population (NRC, 2004). Taking such factors into account, the NRC concluded that fish and shellfish pose the greatest environmental concern, pigs and goats pose a moderate degree of concern, and chickens, cattle, and sheep pose the least concern (NRC, 2004). Concern is increased when the transgene enhances fitness, as would be expected with several traits under study, including salt water tolerance, cold tolerance, increased growth rate, enhanced disease resistance, and improved nutrient utilization.

There are currently two genetically engineered animals that have been seriously considered for commercialization in the United States: Enviropigs and AquAdvantage salmon.[6] In what follows, we discuss the case of Enviropigs to explore the environmental issues raised by animal biotech in more detail.

3.3 Enviropigs

Swine require phosphorus in their diets but the phosphorus in the grain-based foods standard in contemporary agriculture is bound in phytate, which swine are unable to digest (Golovan et al., 2001). To meet the phosphorus requirements of swine, farmers supplement the standard feed with mineral phosphate. This standard practice contributes to a variety of local environmental problems and, in turn, to more global ones. The phosphate must be mined and shipped to farms, where, because it is inexpensive, it is fed to swine in abundance to ensure that their nutritional needs are met (Forsberg, Hilborn, and Hacker, 2003). While the swine are able to digest the mineral phosphate, it increases the amount of phosphorus in their manure. The manure is used as fertilizer and so the amount of phosphorus released into the environment is increased. This causes serious environmental damage, especially when the phosphorus makes its way into the water system where it increases algal bloom and contributes to the death of native species and to drinking water contamination (Carpenter et al., 1998; Jongbloed and Lenis, 1998). Phosphorus pollution is consistently ranked by the Environmental Protection Agency as one of the top causes of water quality problems.

The Enviropig is a transgenic animal, created by scientists at the University of Guelph, which is intended to help alleviate these environmental concerns. The Enviropig has been genetically engineered to produce saliva that contains phytase, an enzyme which allows the pig to digest the plant-based phytate in standard feed, thus reducing or eliminating the need for supplemental phosphorus. Enviropigs secrete up to 75% less phosphorus in their manure than their non-genetically modified counterparts fed the standard diet, and up to 25% less than swine fed supplemental phytase to help digest plant-based phosphorus.

---

6. The Enviropig research program was recently suspended until an industry partner can be identified.

Insofar as the Enviropig merely replaces existing swine without causing further changes to the agricultural system – for example without changing the number of pigs per farm or the number of farms (what we will call *the scenario of mere replacement*) – the use of Enviropigs yields environmental benefits by decreasing the demand for phosphorus mining and reducing phosphorus pollution of ground and surface water. Under the scenario of mere replacement, these environmental benefits provide a reason for replacing existing swine with the Enviropig.

In scenarios where the adoption of Enviropigs results in an increase in the number of swine (*the scenario of growth*), the opposite conclusion may be justified. Environmental pollution serves as one of the main constraints on the growth of the livestock industry (Vestel, 2001). If the Enviropig allows for an increase in the total number of swine used in animal agriculture by reducing the amount of phosphorus pollution per pig, there may be no benefit in terms of aggregate phosphorus reduction. Furthermore, the increase in the number of swine may contribute to other environmental problems: ground compaction, where the number of pigs per unit area increases; deforestation, where the area per farm can be increased; reduction in available water resources; and an increase in other environmental pollutants, such as greenhouse gasses.

Even if Enviropigs do not increase the number of swine, they still pose an environmental risk if they escape into the surrounding environment. As mentioned above, the NRC identified pigs as a species the unintentional release of which poses a moderate degree of environmental risk. Swine have been known to escape and live outside captivity, and feral pigs have been known to cause significant environmental damage. Moreover, Enviropigs, or their descendent feral pigs, could obtain needed phosphorus from plants in the surrounding environment, increasing their pest potential (NRC, 2004, p. 84).

4. CONCLUSION

An all-things-considered judgment about an application of animal biotech must consider its impact on animal welfare and the environment. In this chapter we have presented a general framework for evaluating an application of animal biotech by reviewing its impacts on the animal welfare and environmental problems that plague modern agriculture.

Although genetic engineering and cloning have the potential to mitigate existing animal welfare problems, they continue to be imperfect procedures that often increase the risk of welfare problems compared to the status quo. Even when nothing goes wrong with the procedures, the intended effect can itself exacerbate welfare problems. This will be especially likely when the intended effect is to increase individual productivity, as it often will be. Proposals to improve welfare by using animal biotech to eliminate or reduce livestock's cognitive capacities should be viewed with substantial skepticism.

Animal biotech that reduces the number of animals or increases feed-conversion efficiency or nutrient utilization could reduce the environmental harms of livestock agriculture. Discussions of environmental risk from animal biotech have focused on the effects of releasing genetically engineered animals into the surrounding environment. The proposed benefits of an application of animal biotech should not be evaluated under the assumption that livestock animals will merely be replaced by others that are more environmentally benign. If an application of animal biotech allows for an increase in the number of agricultural animals, there may be no net environmental benefit, or even additional environmental harm.

WORKS CITED

A. Ali, and K. Chang (1985) 'Early Egg Production in Genetically Blind (rc/rc) Chickens in Comparison with Sighted (Rc+/rc) Controls,' *Poultry Science* 4 (5):789–794.

I. Bach, (2000) 'The LIM Domain: Regulation by Association,' *Mechanisms of Development* 91: 5–17.

W. Balduini, M. Cimino, G. Lombardelli, M. Abbracchio, G. Peruzzi, T. Cecchini, G. Gazzanelli, and F. Cattabeni, (1986) 'Microencephalic Rats as a Model for Cognitive Disorders,' *Clinical Neuropharmacology* 9 (3): s8-s18.

D. Brink (1989) *Moral Realism and the Foundations of Ethics* (Cambridge: Cambridge University Press).

D. Brock (1995) 'The Non-Identity Problem and Genetic Harms – The Case of Wrongful Handicaps,' *Bioethics* 9 (3/4): 269–275.

A. Buchanan, D. Brock, N. Daniels, and D. Wikler (2000) *From Chance to Choice: Genetics and Justice* (Cambridge, MA: Cambridge University Press).

S. Carpenter, N. Caracao, D. Correll, R. Howarth, A. Sharpley, and V. Smith, (1998) 'Nonpoint Pollution of Surface Waters with Phosphorus and Nitrogen,' *Ecological Applications* 8 (3): 559–568.

K. Cheng, R. Shoffner, K. Gelatt, G. Gum, J. Otis, and J. Bitgood (1980) 'An Autosomal Recessive Blind Mutant in the Chicken,' *Poultry Science*, 59: 2179–2182.

M. S. Dawkins (1980) *Animal Suffering: The Science of Animal Welfare* (London: Chapman and Hall).

I. Duncan (2004) 'Welfare Problems with Poultry,' in *The Well-Being of Farm Animals: Challenges and Solutions*, (ed.) G. J. Benson and B. Rollin (Iowa: Blackwell Press): 307–323.

R. Elliot (1982) 'Faking Nature,' *Inquiry* 25 (1): 81–93.

C. Forsberg, J. Phillips, S. Golovan, M. Fan, R. Meidinger, A. Ajakaiye, D. Hilborn, and R. Hacker (2003) 'The Enviropig Physiology, Performance, and Contribution to Nutrient Management Advances in a Regulated Environment: The Leading Edge of Change in the Pork Industry,' *Journal of Animal Science* 81(2): e68-e77.

D. Fraser, J. Mench, and S. Millman (2000) 'Farm Animals and Their Welfare in 2000,' in *The State of the Animals: 2001* (ed.) D. Salem, and A. Rowan (Gaithersburg: Humane Society Press) 87–99.

D. Fraser, and D. Weary (2004) 'Quality of Life for Farm Animals: Linking Science, Ethics, and Animal Welfare,' in *The Well-Being of Farm Animals: Challenges and Solutions*, (ed.) G. J. Benson, and B. Rollin, 39–60.

S. Gardiner (2004) 'Ethics and Global Climate Change,' *Ethics* 114: 555–600.

S. Golovan, R. Meidinger, A. Ajakaiye, M. Cottrill, M. Wiederkehr, D. Barney, C. Plante, J. W. Pollard, M. Fan, M. A. Hayes, J. Laursen, J. P. Hjorth, R. H.,

J. Phillips, and C. Forsberg (2001) 'Pigs Expressing Salivary Phytase Produce Low-Phosphorus Manure,' *Nature Biotechnology* 19: 741–745.

T. Grandin, and M. Deasing (1998) 'Genetics and Animal Welfare,' in (ed.) T. Grandin *Genetics and the Behaviour of Domestic Animals*, (San Diego, CA: Academic Press): 319–341.

N. G. Gregory, and L. J. Wilkins (1989) 'Broken Bones in Fowl: Handling and Processing Damage in End-of-Lay Battery Hens,' *British Poultry Science* 30: 555–562.

J. Griffin (1986) *Well-Being: Its Meaning, Measurement, and Moral Importance* (Oxford: Clarendon Press).

B. Heringstad, G. Klemetsdal, and J. Ruane (2000) 'Selection for Mastitis Resistance in Dairy Cattle: A Review with Focus on the Situation in Nordic Countries,' *Livestock Production Science* 64: 95–106.

P. Hooda, A. Edwards, H. Anderson, and A. Miller (2000) 'A Review of Water Quality Concerns in Livestock Farming Areas,' *The Science of the Total Environmental*, 250: 143–167.

P. Hulme (2005) 'Adapting to Climate Change: Is there Scope for Ecological Management in the Face of a Global Threat?' *Journal of Applied Ecology*, 42: 784–794.

R. Ilea (2009) 'Intensive Livestock Farming: Global Trends, Increased Environmental Concerns, and Ethical Solutions,' *Journal of Agricultural and Environmental Ethics*, 22: 153–167.

IPCC (2007) 'Fourth Assessment Report of the Intergovernmental Panel on Climate Change,' (ed.) M. Parry, O. Canziani, J. Palutikof, P. van der Linden, and C. Hanson (Cambridge: Cambridge University Press): 7–22.

A. Jongbloed, and N. Lenis (1998) 'Environmental Concerns about Animal Manure,' *Journal of Animal Science* 76 (10): 2641–2648.

F. D. McMillan (2005) *Mental Health and Well-Being in Animals* (Ames, Iowa: Blackwell Publishing).

Millennium Ecosystem Assessment (2005) *Ecosystems and Human Well-Being: Biodiversity Synthesis* (Washington, DC: World Resources Institute, 2005) (available at http://www.millenniumassessment.org/documents/document.354.aspx.pdf (accessed August 30, 2010).

National Animal Health Monitoring System (NAHMS) (2003) 'Layers '99: Part 1: Reference of 1999 Table Egg Layer Management in the U.S.,' 5 (available at nahms.aphis.usda.gov/poultry/layers99/Layers99_dr_PartII.pdf (accessed August 31, 2010).

National Research Council (NRC) (2004) *Animal Biotechnology: Science-Based Concerns* (Washington, D.C.: National Academies Press): 93–102, 104–107.

J. O'Neill (2003) 'The Varieties of Intrinsic Value,' in *Environmental Ethics: An Anthology*, (ed.) H. Rolston III, and A. Light (Malden: Blackwell Press).

D. Parfit (1984) *Reasons and Persons* (Oxford: Clarendon Press) 351–379.

B. Rollin (1995) *The Frankenstein Syndrome: Ethical and Social Issues in the Genetic Engineering of Animals* (Cambridge: Cambridge University Press).

H. Rolston III (1989) *Philosophy Gone Wild* (Amherst, NY: Prometheus Books).

W. Shawlot, and R. Behringer (1995) 'Requirement for LIM1 in Head-Organizer Function,' *Nature* 374: 425–430.

P. Singer (2002) *One World: The Ethics of Globalization* (New Haven: Yale University Press).

B. Steinbock, and R. McClamrock (1994) 'When is Birth Unfair to the Child,' *Hastings Center Report* 24 (6): 15–21.

R. Streiffer (2008) 'Animal Biotechnology and the Non-Identity Problem,' *American Journal of Bioethics* 8 (6): 47–48.

W. Throop (2000) 'Eradicating the Aliens: Restoration and Exotic Species,' in *Environmental Restoration*, (ed.) W. Throop (Amherst, NY: Humanity Books).

K. Turner, S. Georgiou, R. Clark, R. Brouwer, and J. Burke (2004) 'Economic Valuation of Water Resources in Agriculture: From Sectoral to a Functional Perspective on Natural Resource Management,' *FAO Paper Reports*, 24.

UN FAO (2006) *Livestock's Long Shadow*, Food and Agricultural Organization of the United Nations (available at ftp://ftp.fao.org/docrep/fao/010/a0701e/a0701e.pdf).

L. Vestel (2001) 'The Next Pig Thing,' *Mother Jones*, (available at http://motherjones.com/environment/2001/10/next-pig-thing (accessed August 31, 2010).

A. B. Webster (2004) 'Welfare Implications of Avian Osteoporosis,' *Poultry Science*, 83: 184–192.

DISCUSSION QUESTIONS

1. Do you think that the animal welfare and ecological concerns apply equally to all forms of animal agriculture, or only to particular species or types of agriculture?

2. What is the authors' argument that animals engineered to be ecologically beneficial could in fact turn out to be ecologically detrimental? Did you find their reasoning persuasive?

3. Which theory of animal welfare (Mentalistic, Desire-Satisfaction, or Objective List) do you think is most plausible? Why? What are the implications for animal agriculture?

4. Given that the welfare and ecological problems associated with animal agriculture are so large, is it more justified to try to end (or, at least, reduce) animal agriculture rather than to try to decrease the harms through genetic engineering? Why or why not?

5. Can you think of alternative, less technologically oriented methods to improve animal welfare and reduce the ecological impacts of livestock than genetic engineering?

6. Would you want to eat the products of diminished or engineered animals? Why or why not?

7. Should consumers be informed when animal products (e.g. meat and dairy) are derived from genetically engineered animals – for example, by positive labeling? Why or why not?

8. Do you think there is anything intrinsically ethically problematic about animal biotechnology? Why or why not?

ARTIFICIAL MEAT

Paul Thompson

CHAPTER SUMMARY

Emerging tissue engineering techniques enable growing animal tissue for human consumption *in vitro*, without the animals. Growing meat in this way, rather than "on the hoof," has been promoted as a way to reduce the suffering and eliminate the inefficiencies (and so ecological impacts) associated with animal agriculture. In this chapter, Paul Thompson critically evaluates the case in favor of artificial meat. He argues that the considerations offered in support of it are not decisive, for several reasons – e.g. artificial meat is not likely to displace meat grown on the hoof (for economic and aesthetic reasons), it has not been shown to be more ecological efficient, and it does not provide a substitute for non-meat animal parts such as hides. Moreover, those who are concerned about the industrialization of agriculture are likely to be opposed to artificial meat, since its production will be industrial. Furthermore, the case for synthetic meat involves comparing it to meat produced in concentrated animal feed operations (CAFOs). However, there are other alternatives – less intensive, more humane forms of animal agriculture – that might be preferable to both CAFOs and artificial meat.

RELATED READINGS

Introduction: Innovation Presumption (2.1); Situated Technology (2.2); Form of Life (2.5); Extrinsic Concerns (2.6); Responsible Development (2.8)

Other Chapters: Robert Streiffer and John Basl, *The Ethics of Agricultural Animal Biotechnology* (Ch. 33); Vandana Shiva, *Women and the Gendered Politics of Food* (Ch. 32)

1. INTRODUCTION

Imaginary foods that appear "as if by magic" have been a staple of fantasy literature and science fiction for a long time. People who were born after 1970 may have trouble believing that the ordinary frozen dinner brought to a reasonable semblance of palatable food by a microwave oven actually realizes magical food fantasies from the distant past. But the rice in a Hungry Man™ Mexican Fiesta Style frozen dinner started out in a farmer's field (probably in Texas or California), and more to the point, the beef in the enchiladas started out on the hoof. It came from an animal. Food scientists have scrambled to bring "textured vegetable protein" up to snuff, but the goal of a meat that doesn't come from animals continues to hold allure. Is there any chance that we will get there?

There are, in fact, several possible routes to animal-free meat. One could start by placing a few myocytes (e.g. muscle cells) on a scaffold, and then multiply them using cell culture. Cell culture is, in this case, a process of using chemical and physical cues to promote cell growth in an artificial medium. It has been developed extensively for regenerating plants, for example, and patents currently exist for using it to develop cultured meat. An alternative would begin with an explants – a portion of tissue that contains all the cell-types that make up muscle in their corresponding proportions. The challenge of proliferating explants lies in the need to deliver a constant supply of oxygen and nutrients as the tissue grows, something normally done by the flow of blood. The challenge of making them seem like meat lies in stimulating the cells to achieve the texture that muscle tissue achieves through the alternating tensions of flexion. While a simpler scaffold-based cell culture system might be adequate for producing small meat fibers (something like finely ground hamburger), only the more complex system would produce something that resembled a piece of meat (Edelman et al. 2005). Many of the tissue engineering techniques that would be needed are being developed in connection with medical projects that are trying to derive transplantable tissues from stem cells (Wellin et al. 2012). So perhaps there is reason to think that cultured meat will be carried forward on the wave of biomedical research.

It's hard to know how to begin when it comes to thinking about the ethics of artificial meat. One might start by noting the cacophony of voices critiquing the industrial food system in general and the "factory farming" of food animals, in particular. Some of these voices seem to emerge from a growing recognition of our (meaning human beings) moral responsibilities to other animals and our rather shabby treatment of them in the past. There are really a number of points to be made in connection with factory farming and the moral status of non-humans. One is to overcome a set of philosophical presumptions that mark radical distinctions between humans and other animals: that non-humans are "machines" that neither feel pain nor possess a mental life; that non-humans do not count in our moral considerations because they do not speak, because their mental lives are insufficiently rich, or maybe just because we humans are God's chosen species. Although they continue to be debated, all these presumptions are becoming philosophically out-dated. Yet while the presumption that non-humans are moral sinecures fades into the past, the actual exploitation of them continues and proliferates, and the growth of concentrated animal feeding operations (CAFOs) – factory farms – is put forward as "Exhibit A" to testify to this fact. If one comes out of the recent social movement to recognize the interests of non-human animals – and many philosophers are part it – then putting an end to factory farms is unquestionably a good thing.

Other people are interested in putting an end to factory farms for a very different set of reasons. For them, the concern is that this way of farming is unsustainable. Explaining why is a long story, but one might begin with the burden that concentration of animal manure and ambient emissions place on the environment. CAFOs have played havoc with water quality in some regions, and even when they do not they are ugly. They smell bad and thus damage the local quality of life. People voicing this concern would much rather see smaller herds or flocks dotting an intermittent landscape of pastures and hayfields, an animal agriculture that harkens back to our (that is, again, humans) pastoral and nomadic past. These critics of industrial farming might agree with the pro-animal social movement that CAFOs are hard on the animals that live there, but they don't really envision an end to animal production. They might even argue that a sustainable agriculture is impossible *without* animal production because of the role that properly composted animal manures have traditionally played in sustaining soil quality. For them, sustainability implies a return to a more ecological and natural way of farming.

We thus find two voices starting out in apparent agreement on the evils of industrial livestock production yet who are likely to diverge on the question of artificial meat. Pursuit of the artificial in our food system is what got us into our current mess, if you are of the "sustainability" persuasion. It's quite likely that we (humans again) should eat *less* meat than many of us are eating during the second decade of the third millennium, but it's not like we should give up eating meat in favor of some substitute product that we squirt out of a tube! No. Something like that would probably come from one of those giant food companies, the weasels who gave us "pink slime" (more on this later) and processed chicken parts. Yet if you are among those who want to see the ethically unenlightened past where animals simply did not count disappearing into the vanishing point of our rear-view mirror as we motor comfortably into a high-tech future where non-human animals are our friends, well, for you this so-called sustainable agriculture looks more like the perpetuation of animal suffering and human dominance.

Of course, one might also start in a rather different place altogether. "Emerging technology" is a phrase that is being used to gesture at a fairly large class of tools and techniques that are presumed to be on humanity's horizon, the new new things that will thrill the young and send an older generation reeling into ever greater paroxysms of future shock. Of course, "future shock" is itself rather passé, a bygone notion of pop philosophy proffered by Alvin Toffler during the 1970s. The idea was that while there had certainly been big changes in humanity's past (e.g. planes, trains and automobiles) the pace of change would become so rapid that meaning systems and worldviews – what the elders referred to as culture – would simply be unable to keep up. Ethical norms and our institutions of governance would strain under the pace of technical change. Some would fret over the meaning or acceptability of strange capabilities like "test tube babies"; others would charge ahead, seeking to hang ten on the next wave of new things. Emerging technology has indeed become a fertile ground for philosophical thinking over the last half century, as we have accepted smart phones and genetically engineered drugs, while we have grown queasy over irradiation and genetically engineered food.

From *this* starting point, we note that "artificial meat" is an adaptation of tissue engineering work that is currently being done with biomedical goals in mind. Organs for transplant might one day be grown from a potential recipient's own body tissue, assuring a match to the patient's DNA and reducing the risk of immune system rejection. This line of research draws upon work with stem cells (though the tissue that is eventually transplanted may or may not involve stem cells in its production) and is unequivocally an application of so-called "red" biotechnology. Red means biomedical: manipulation of cells and cellular processes for medical purposes. An adaptation to food production would be an example of "green" or agricultural biotechnology. While red biotechnologies have had high degrees of approval and acceptance, green biotechnologies have suffered mightily from consumer rejection and citizen protest. They are the poster child for a feared and unwanted utilization of technical means. Thus, the emerging technology starting point leads us to suspect that while tissue engineering for organ transplants will be welcomed, tissue engineering as a form of food production will encounter a massive "yuck factor" and will be subjected to organized resistance.

Emerging technology has also been a focal point for social scientists and activists with an interest in participatory governance procedures and deliberative democracy. Part and parcel of the resistance to genetically engineered food, it is claimed, was the way that this product was foisted on an unwary public. We should not let big companies and scientists in government food safety agencies decide what is good for us. We should not let pointy-headed philosophers, self-appointed religious authorities and other gurus decide what is right for us. We humans should develop democratic procedures for making the big social decisions about what kind of world we want, and we should use these procedures to review emerging technologies that have ability to cause dramatic shifts in the way we live our lives. We can call these procedures "anticipatory governance." And then we can have philosophical disagreements on what these procedures should look like, and over whether their primary goals are focused on outcomes or process. With *this* starting point we end up in a very familiar place. The ethics of artificial meat turns out to be another round of a long slugfest between utilitarians, who say we should weigh the costs and benefits of the outcome, and rights theorists, who are all about autonomy and the right to resist having *anything* foisted on anyone. This will be especially good for college professors because these debates over outcome and process will look a lot like *other* debates we've had in ethics and politics over the last 250 years. Thus, you are fine with whatever happens so long as it *either* achieves the greatest good for the greatest number, *or* does not treat someone as means during the social process of technological development and dissemination. We won't have to buy new books, and we won't even have to know very much about the underlying science or engineering of artificial meat.

Another option is to start someplace seemingly far, far from the topic of artificial meat, then circle back to some of these issues as the need arises. Let us consider pink slime.

2. PINK SLIME

Lean finely textured beef is a product developed in the 1980s. It is derived from the flesh of cattle that remains on the bone after the butchering process. The flesh is removed by spinning the carcass in a heated centrifuge. Centrifugal force separates the fleshy bits from bone and presses them through a sieve, where they can be recovered and utilized in food products. The mixture of meat, sinew and fat is further processed by additional heating and treatment with gaseous ammonia or citric acid to eliminate contamination by e-coli. The industrial processing is completed by packaging into blocks, which can then be added to products intended for human consumption. In addition to hotdogs, canned items (such as chili or soups) and other processed foods, lean finely textured beef can constitute up to 15% by weight of ground beef. According to the website of Beef Products, Inc., the use of ammonia gas to kill bacteria was developed in 1994, and its integration into the production of lean beef was accomplished by 1997.

The expression "pink slime" was first used in an e-mail by an USDA scientist who objected to the un-labeled incorporation of lean, finely textured beef into hamburger. Gerald Zirnstein argued that the practice was false and misleading. He coined the expression to underline the fact that he did not consider the product consistent with ordinary consumers' expectations for products being marketed under the name "beef". It was not until seven years later that Zirnstein's comment was picked up by an investigative reporter, and not until five years after that that celebrity chef Jamie Oliver picked up the use of the term in a critique of the food industry. Then everything happened extremely suddenly. Over the course of only a few months, campaigns to eliminate the use of lean, finely textured beef in school lunches were launched, and one major food company after another announced that pink slime would not be used in their processed meat products. Then the news outlets began reporting that the entire industry for producing lean finely textured beef was in collapse, with plant closures and loss of jobs for workers.

Pink slime is an interesting case in its own right, but in the present context it is of interest for two reasons. First, like textured vegetable protein, generally developed from soya, lean finely textured beef is a real-world reference point for the term "artificial meat". Unlike textured vegetable protein, lean finely textured beef was deemed sufficiently similar to "genuine meat" (if we may use this term) to warrant its incorporation into food products without special labeling. This highlights a key ontological question for future forms of artificial meat: Is biophysical make-up at the cellular level the relevant criterion for whether the substance in question is meat or not? Lean finely textured beef is derived from the flesh of once living animals, so in this respect it is quite *unlike* a meat-product developed through tissue engineering. This is a crucial difference for some of the ethical arguments, to be sure, but noting this difference points to a second reason why pink slime is worth a moment or two of reflection: lean finely textured beef illustrates how ontological questions intersect with attitudinal dispositions of producers and consumers, and these attitudinal dispositions, in turn, may have aesthetic, political and ethical significance.

Was there an ethical problem with pink slime? Although some activists raised concerns about the use of ammonia gas, it looks rather like the general distaste over pink slime was the driving factor. Language is part of the story. It seems likely that if processors had been able to prevail in the terminology war, few would have objected to the consumption of lean finely textured beef. Yet no one wants schoolchildren to be fed on pink slime. The initial point to notice is that the public's reaction to this highly processed meat product depends on the way that it is couched within a narrative and rhetorical framework. It seems that Zirnstein's initial judgment has been supported by events that transpired over the first six months of 2012, but it is possible that the counter-narrative of unjustified job loss could play a role in reversing the trend. If people came to see lean finely textured beef as a product that is (a) healthy in virtue of its low fat content and (b) environmentally beneficial in virtue of the way that it wrings more consumable animal protein from each animal, then the collapse of the industry and thousands of newly unemployed wage workers could be the *coup de grâce*. The expression 'pink slime' could be viewed as an irresponsible attempt to taint an ethically laudable food product.

Something similar *could* happen to *in vitro* artificial meat. The developers of lean finely textured beef plausibly thought of themselves as doing something laudable. The product would keep the price of animal protein down, making it more accessible to low-income consumers. The reduction in dietary fat and the load on the environment further support this ethically laudatory viewpoint. Yet the very success of the product and its relative invisibility to consumers left the makers of lean finely textured beef vulnerable to a social movement that began to take shape in 2007. Following on Eric Schlosser's *Fast Food Nation*, Michael Pollan's *An Omnivore's Dilemma*, an international outpouring of anger over genetically engineered crops and popular films such as *Food Inc.*, pink slime was just the next outrage perpetrated by a food industry intent on unrestrained pursuit of profit. While the animal activists who are supporting the development of *in vitro* meats may think of themselves as part and parcel of this food movement, the product they are advocating would be impossible to imagine absent the capabilities of the industrial food system. It seems difficult to picture foodies buying artificial meat from local producers who grew it up in Petri dishes back in their garage. Advocates of artificial meat may picture themselves as moral heroes. As a result they may find it hard to imagine how they could be villainized in a manner comparable to that of the makers of pink slime. Yet elements of the analogy are difficult to deny.

It is also the case that whatever one thinks about pink slime, lean finely textured beef is a product with a stronger ontological claim to being labeled "beef" than would any product derived from tissue engineering. Like any other beef product, pink slime comes from steers and cows raised on the hoof and slaughtered, rendered and processed just like any other piece of meat. It is not reconstituted from some non-animal source, like textured soy-protein, nor is it grown up in a vat (the scientists may prefer to call it a 'rotating bioreactor'), as we may presume artificial meats will be. Zirnstein's seemingly accurate reading of the public mind notwithstanding, the stuff is recognizably *similar* to beef both in biochemical formulation and in the story one would tell about where it comes from. It thus seems that if the authenticity of lean finely textured beef was a source of its difficulty, artificial meat would encounter the problem in spades. Then again, perhaps this would only be the case if there was some attempt to pass off artificial meat as something other than what it is. There are good reasons to doubt that this will happen.

One important disanalogy to pink slime resides in the way that market success for artificial meat will depend on consumers knowing what they are eating. Although no public studies on the economics of meat products derived from tissue engineering are currently available, it is unlikely that the product will be competitive with conventionally produced meat in terms of consumer cost, at least in its early days on supermarket shelves. New technologies typically come on the market when the costs of production are just low enough to insure a market for "luxury" consumption. Only when the specialty market for a product has been proven do manufacturers scale up for mass markets and cheaper production. What is more, all the scientific articles on cultured meat mention cost of production as a barrier to deployment of existing technologies. So while we can expect research and development (R&D) to whittle the costs down, the first products will probably not be cheap. As such, gaining a market will depend on consumers being willing to pay a premium for meat products that have not been produced by slaughtering a live animal. And this means that they will have to *know* that they are consuming the product, something that was not the case for pink slime. This also means that it is very unlikely that *in vitro* meat will appear in school lunches without parents' knowledge and consent. Thus, there is little reason to think that people will feel deceived by the food industry when it comes to artificial meat, and in this an important element in the public's revulsion over the publicity over pink slime is eliminated.

3. CONSUMER ECONOMICS

Although likely market structure for artificial meat illustrates a disanalogy with the case of pink slime, and the disanalogy is a reason to doubt that ethical problems with the deceptive marketing of lean finely textured beef are likely to occur in connection with animal-free meat, this is not an unalloyed bonus for advocates of *in vitro* meats. If artificial meat is significantly more expensive than conventional meat and meat products, that will diffuse the ethical force of the argument for artificial meat to a significant degree. A full appreciation of this point presupposes some understanding of the basic consumer economics of food.

Everyone needs food. People increasingly get the food they need not by producing it themselves, but by purchasing it from supermarkets, co-ops, specialty markets and restaurants. The diversity among retail outlets is enormous, and price is but one feature of importance. Consumers value taste, appearance, convenience and (as the pink slime case illustrates) a host of culturally-based attributes that define some foods as "authentic" or appropriate. Nevertheless, the retail price of food is of significant ethical importance. Not only does everyone need food, everyone needs a minimal amount of food, and they must have access to food on a regular basis. This applies roughly equally to people on the lower end of the income scale (henceforth "the poor") and to people who have significant discretionary income. Although no one likes to see food prices increase, an increase in food prices has a disproportionate impact on the quality of life for the poor, who spend a larger share of their income on food. While the middle-class in industrial societies generally spends between 10% and 20% of their income on food, expenditures for the poor can exceed 50%. What is more, wealthy consumers enjoy considerable discretion *within* their food budget, as they can adjust by spending less on meals outside the home, or by buying a less expensive bottle of wine.

There are thus powerful *prima facie* reasons to resist any increase in the price of food, especially when the increase involves food items that are part of the typical bundle purchased by relatively poor consumers. Animal products are, with very few exceptions, important items in that bundle. The International Food Policy Research Institute has estimated that when the world's poorest people go from making one Euro per day to making two Euros per day, the main thing that they spend their second dollar on is some form of animal protein: meat, milk or eggs.

Even those in extreme poverty (e.g. less than €1 or $1.60 per day) generally know how to utilize animal proteins in their diet. They know how to cook meat, milk and eggs, and these products constitute culturally appropriate components of their diet. The primary exception is Hindus from the Indian subcontinent, but although many Hindus are vegetarian, pure veganism is often more a product of economic necessity than philosophical choice. Thus poor Hindus will incorporate ghee, yogurt and eggs into their diet as their income rises (though not at a rate that supports the "second dollar" generalization stated above).

These general points apply equally well to poor populations in industrialized countries. Although their level of well-being is typically far above that of the World Bank's standard for extreme poverty, poor people in the United States, Canada, Europe, Australia and other countries of the industrialized West spend a much larger percentage of their income on food than do middle-class people. They are disproportionately from aboriginal, Hispanic, black and Muslim cultural origins. All of these cultural groups traditionally consume animal products on a regular (if not necessarily daily) basis, and most tend to associate some form of meat with the centerpiece of an appropriate meal. Poor people in industrial societies also face challenges that limit their opportunity to experiment with alternative diets, even when the sheer cost of food does not. Poor households have disproportionate numbers of adults holding multiple low-paying jobs, maintaining long commutes, and lacking convenient access to retail food outlets with a variety of food options. Thus, even when incomes (sometimes supplemented with government assistance programs) might allow experimentation with alternative diets, the opportunity cost of doing so can be high. What is more, given high rates of obesity and dietary disease among the poor, it is doubtful that a shift to artificial meat would be the most compelling ethical concern on the agenda for changes in poor diets.

The upshot is that some qualifications should be placed upon any argument that would be advanced on the ethical case for artificial meat. Any food ethic which accepts that poor people should not be compelled to adopt vegetarian diets in the *absence* of artificial meat will also need to accept that the ethical case for consuming artificial meats will be significantly constrained, even when and if cultured meat becomes available. If poor people are not currently obligated to practice a vegan diet, the availability of high-priced or inconvenient artificial meat will not alter this situation. The same considerations that relieve them of any duty to eliminate or limit their consumption of animal products under present-day market conditions are very likely to apply when meats from tissue engineering appear as higher priced specialty items. To the extent that poor people view traditional diets and foods as a component of their culture, the reluctance they might have to experiment with high-tech industrial food products of any kind will reinforce the thought that imputed ethical duties to consume artificial meats are just another form of assimilation to the dominant norms of bourgeois society.

Of course, it is possible to develop a philosophically consistent food ethic which holds that desperately poor people in India who cook their curry with a bit of ghee are as culpable as the American businessman who sits down in front of a 32 ounce steak. A view which holds that *all* forms of animal confinement are ethically forbidden might support this judgment, for example. Someone holding such an ethic may support the development of *in vitro* meat for tactical reasons: It becomes a way of easing people who love the taste of meat into a more ethically defensible lifestyle. But this emphasis on taste is itself ethically troubling in the present context. To suggest that a culturally rooted food tradition in which animal products play a prominent role is simply a matter of the way food *tastes* disrespects the autonomy of cultures and of individuals who develop a personality that is deeply informed by identification with a regional, ethnic or religious culture. I do not suggest that philosophers who have defended universalizing arguments for animal rights will find this persuasive; the subjectivity of non-humans will almost certainly be seen as overriding the identities of economically and culturally oppressed groups. Yet many authors in recent decades have struggled to find a voice that speaks for oppressed people and against the tyranny of universalist philosophies that rationalized European expansion during and after the high tide of modern philosophy. Protecting the group rights of racial and ethnic minorities who draw upon food culture to support their sense of self provides yet another source of constraints on the sense in which embracing artificial meat can be viewed as a way to comply with the ethical imperative to give up all forms of consumption (like eating meat) that require the confinement or death of an animal.

4. EXOTIC TASTES

The drive for artificial meat often seems calculated as a riposte to those who say that they would like to practice a vegetarian diet, but who suffer from weakness of the will when it comes to doing so. As mentioned already, the suggestion is that some people just love the taste of meat so much, they cannot be persuaded to give it up, no matter how compelling the ethical argument for doing so might be. Given this starting point, the case for meat products derived from tissue engineering might proceed in any of several directions. First, it might be a *voluntarist* argument: If you want to do what's ethically right, but find it hard to do so, then artificial meat will make it easier for you to do the right thing. Artificial meat looks a lot like nicotine patches that are intended to help people stop smoking. Second, there is a *substitution* argument. One gets all the pleasure of eating meat without any of the guilt. Here artificial meat looks a bit like a 3-D pornographic video game that allows one to satisfy sexual urges without needing the consent of a willing partner.

In both of these cases, the pleasure remains tainted by its unhealthful or immoral associations. There is also the concern that reducing the cost of capitulation reduces the sanction. One is reminded of Immanuel Kant's view on cruelty to animals. It's not that one can harm a non-human in a manner that could violate the categorical imperative, but failing to sanction animal cruelty might inure the offenders and make them more likely to inflict a morally heinous act on other human beings. So perhaps a third type of ethical position is more appealing, one that we might refer to as a *transformational* argument: With *in vitro* meat, a taste that is tainted by moral culpability is technologically transformed into a taste with no moral associations whatsoever (see Pellissier 2012). This is certainly an unusual type of argument, and we would be behooved to inquire further as to whether it is truly plausible.

Carolyn Korsmeyer considers a case that is so similar that she might as well have been writing about artificial meat.

> The chemistry of flavor production is proceeding apace, and it is not out of the question that there will come a time when the morally fastidious can have foie gras made from lentils or gluten. Would this be a taste liking that could be cultivated free of moral compromise? That is, can we with clear conscience look forward to a world where we can have all those delicious flavors free from the moral taint that presently turns them (or perhaps should turn them) to ashes in the mouth? (Korsmeyer 2012 p. 100).

Korsmeyer's larger context is the discussion of aesthetic tastes from the world of art that cannot be produced without committing an immoral act. In considering foods, she notes not only foie gras, which involves force-feeding geese to overtax their livers, but also the ortolan, a tiny bird eaten as a delicacy. To achieve the best flavor the ortolan must be subjected to a violent death by asphyxiation. She is not unaware that animal rights advocates would see *all* meat consumption in a similar manner. Although she restricts the comments above to these extreme (and thus perhaps noncontroversial) cases, her remarks in favor of a flavor chemistry that would transform the pleasure into a morally neutral one nicely summarize the view that seems to underlie the argument for *in vitro* meat.

Yet Korsmeyer balks at endorsing the transformational argument, and asks us to consider "a darker example". "Suppose human flesh tastes delectable. Is it okay to cultivate a taste for faux human being? Isn't there something enduringly terrible about having a taste for human flesh, even if that taste is to be satisfied by means of a substitute?" (Korsmeyer 2012, p. 100). Placed in the present context, Korsmeyer's mischievous thought experiment echoes a theme advanced by Cora Diamond in a very early critique of Peter Singer and Tom Regan. Singer's book *Animal Liberation* stressed the claim that it is an animal's interest in avoiding pain and suffering that demands our moral respect. Regan had argued that this is not enough. We do wrong when any animal that is "the subject of a life" – the center of memories, expectations and a desire to go on living – is made into a mere means for satisfying our own desires. Singer is often cited as having developed a utilitarian approach to animal ethics, while Regan's arguments are said to provide the basis for animal rights.

In a paper entitled "Eating Meat and Eating People," Diamond, who *is* a vegetarian, attempts to show how metaphysical categories run amok in our ordinary language, surfacing in unexpected ways. If *either* Singer *or* Regan had it right, we would have no qualms about eating our dead, presuming perhaps that they "nicked off in some morally inoffensive way," such as an automobile accident. They might be quite tasty, Diamond speculates (Diamond 1978). Her point is not to endorse "yuck factor" reactions in a blanket way. Yet she does want us to be aware of the way that an unreflective but fairly systemic distinction between humans and other species runs through a host of seemingly unrelated phenomena in moral life. *If* the only thing keeping Singer from eating animals is their suffering (and Singer is rather clear in stating that this is the case), he should have no problem with a little cannibalism, so long as the cannibalized suffered no intentionally inflicted or avoidable pain.

Tissue engineering is a technology that really *demands* a reply to these seemingly bizarre and obtuse thought experiments. While it seems unlikely that any flavor laboratory is investing much effort in synthesizing the taste of human flesh, tissue engineers are indeed very hard at work proving the effectiveness of their techniques on human tissues. The driving force behind tissue engineering is medical applications, the creation of human tissues and possibly complete organs or musculature that would be available for transplant or alternative therapeutic purposes. In developing their argument for *in vitro* meat, Stellan Wellin, Julia Gold and Johanna Berlin emphasize the fact that these medical applications focused on human flesh provide the impetus for developing our technical capability to produce artificial meat. This observation is crucial to their claim that artificial meat is a technically feasible project (Wellin, Gold and Berlin 2012). We will thus have advanced capability to produce human flesh from tissue culture methods well before we apply the technology to tissue samples from food animals, and long before we seriously begin to think about how these techniques might be used to produce artificial meats that would be available on a commercial scale.

There is thus an argument that might run like this: Given that we will be able to produce consumable human tissue years before we could seriously produce artificial beef, pork, lamb or poultry, and given that we have no immediate reason to think that there would be any moral problem associated with eating this human tissue, we *should be* pressing for engineered human flesh as the first product in the line of artificial animal proteins that will satisfy our palates over the decades to come. This would certainly be the *quickest* way to achieve the general goals sought by advocates of artificial meat. If shifting to a flesh-like animal protein that does not require the death of an animal is the desired and ethically praiseworthy goal, the species from which cells are derived should be less important than the achievement of that goal.

5. DOWN ON THE FARM

The musings put forward by Korsmeyer and Diamond are founded on the premise that no one would seriously mount such an argument. The philosophical point is to expose the lack of any underlying ethical rationale for distinguishing the human and the non-human when it comes to the consumption of flesh. Diamond is especially clear in stating that the observance of such a distinction is both absolutely necessary and utterly lacking in any rational basis. Although she portrays herself as fully willing to comport herself within the dietary restrictions implied by an ethical vegetarian worldview, her paper can be read as a refutation of arguments that attempt to defend such a worldview on non-human animals' interests in not being harmed in the process of livestock production or slaughter. In other writings she suggests that a pre-philosophical and emotional solidarity with non-human animals – a regard for them as "fellow creatures" – provides a more accurate phenomenology of the relevant intuition (Diamond 2009).

But does a regard for fellow creatures reliably translate into a prohibition against consuming their flesh? The observation that stockpersons choose their vocation because they like animals is a commonplace among farming and ranching communities. Is it simply the height of self-deception, or worse, a canard intended to conceal personalities inured to cruelty? A philosophically adequate defense of the claim that the stockperson's self-image is neither of these things would be considerably beyond the scope of the present argument. A few bald assertions must suffice to put yet another perspective on artificial meat into context. While stockpersons' professed love of animals is undoubtedly inauthentic in some cases, it is equally indubitably genuine in many others. Temple Grandin has become the symbolic expression of the "pro-animal" agriculturalist in the present age. While she does not question the legitimacy of using non-human animals for food and fiber, neither does she tolerate cruel or unnecessary interference with their natures (see Grandin 2009). I refer to her and others like her in making the following claims.

Humans have kept domesticated livestock under their care and supervision for millennia. Although the anthropology of this practice is complex, production for meat has not been the primary purpose until comparatively recent times. Some scholars speculate that the domestication of present-day food animals occurred as a by-product of their use in religious rituals (Bulliet 2005). Whether or not this is so, livestock were historically utilized for transportation and traction and kept for their fleece or the collection of milk, blood or eggs. Livestock manures have come to have an indispensible role in maintaining soil fertility for many farming systems. These uses continue into the present, and the creation of artificial meat will do nothing to affect their utility.

This means that while successful methods for engineering animal tissues for human food production will intersect with livestock production in complex ways, they will not neatly *replace* livestock production. If meat production were to disappear, it would affect industries (including pharmaceuticals) that depend on animal by-products. It would require greater reliance on synthetic fertilizers, or entirely new and expanded methods for producing organic fertilizers from plant-based sources alone. It is unclear what would happen to eggs, milk or wool. These considerations are not, in any direct sense, arguments *against* the development of artificial meat, or the consumption of it on ethical grounds, should it become available. They do explain why some people, including many advocates of more sustainable farming methods, might regard artificial meat as just another unwelcome intrusion of industrial methods into an area where they do not belong.

The belief that some of the more traditional forms of livestock production are compatible with the welfare of livestock species is at least plausible, though in fact some of the changes associated with so-called factory farming have been put in place to counter the harm done by disease, climate and exposure to predation. It is thus reasonable to think that continued attention to the way that livestock actually fare in modern farming practices can address the egregious harms of current practices. Although it is disputed, there is thus a view which holds that elimination of livestock production is inconsistent with the interests of the animals themselves. Donna Haraway has characterized the view that these animals would be better off not to have lived at all as a form of exterminism not unlike that associated with genocide. While she is deeply critical of industrial-scale animal production associated with CAFOs, she finds the kind of philosophical position that would press for a total elimination of livestock deeply problematic from a moral point of view (Haraway 2008).

Yet, like virtually everything that humans seem to do, livestock production is plagued by unintended environmental costs. There is a raging debate as to whether environmental impacts should be addressed by even greater emphasis on efficiency, by a return to more extensive production methods or by eliminating livestock production altogether. At present, there has been no serious attention to the environmental sustainability of artificial meat. Advocates claim that since energy and resources do not go into first the growth and then the disassembly of a whole animal, the process will be more sustainable. But these claims cannot be evaluated until the much more information on the actual process for scaled-up production of cultured meats is at hand. The presumption that non-meat parts of an animal carcass are "wasted" would, in any case, be incorrect, so total sustainability estimates will require some knowledge of where current consumers of hides and animal byproducts would turn, as well. There are, in short, many unanswered questions about the wisdom and future sustainability of livestock production, and equally many parallel questions about a novel food technology that is currently in a totally experimental stage of development. It is at best premature to presume that commercial scale factories churning out thousands of tons of engineered animal tissue for human food consumption would prove superior to reformed methods of livestock production on either animal welfare *or* environmental grounds.

6. SUMMING UP

This rather impressionistic tour of observations and ideas does not provide a knock-down case either for or against the use of tissue engineering to develop commercially available cultured meats. But it does provide a basis for some more limited conclusions about the ethics of artificial meat. First, the moral claims being made on behalf of artificial meat imply that it is an ethically preferable substitute for traditional meats derived from the flesh of living animals. If this were so, we should clamor for it to be used in school lunches, and we should be resisting any labeling policy that could imply some form of taint. No one seriously thinks that artificial meat will enter markets through the school cafeteria, however. Rather, it will be an expensive substitute, not readily available even to those low-income consumers who might be quite inclined to eat it. If the strongly pro-cultured meat position were on firm ethical grounds, this would be a serious problem. As it stands, the likely pattern of market entry removes the practical need for anyone to argue that, like pink slime, artificial meat should be mainstreamed with no thought for labeling or consumer choice. Nonetheless, we should notice that the position being taken by advocates of cultured meat implies exactly that.

Second, if artificial meat really solves the moral problem with meat eating, wouldn't it also solve the moral problem with cannibalism? And why not up the ante of this thought experiment a notch? Why not offer special products that allow us to eat the flesh of especially admired (or especially despised) individuals? If this seems too ridiculous to warrant a response, it nonetheless shows how artificial meat will intersect with arational and deeply emotive impulses in fairly unpredictable ways. It also shows that the *arguments* for cultured meat are rather conveniently ignoring the cultural and emotional underpinnings of the way that we think. And they are ignoring the fact that cultured meat crosses two areas in which our cultural and emotional habits run notoriously at odds with a rationalistic view of right action: animals, on the one hand, and food, on the other. Noting these points proves nothing final, but it does provide some reason to think that the arguments are less easy than their protagonists have led us to think.

And finally, there may be some good reasons to continue producing livestock that are simply not affected by the artificial meat argument. Not only is animal manure a linchpin for contemporary organic farming methods, animal byproducts go into our clothing, our medicines and percolate through our economy in myriad ways. Once again, noting this is not intended as a final rebuttal to the argument for artificial meat. It's rather a rebuttal to the presumption that this argument will be so easy as to be already won. In summation, yuck factors and more politically based forms of resistance provide persuasive reasons for thinking that artificial meat will not be "the greatest thing since sliced bread." Rather *like* sliced bread, some people will want nothing to do with it. As such, the claim that artificial meat is a unilateral moral good in virtue of the fact that it does not require the death of an animal is at the very least much too simple, and assumes much too much. Artificial meat can as easily be regarded as a further extension of the industrial food system, a system that we are coming to see as deeply problematic.

WORKS CITED

Beef Products, Inc. N.D. About. Accessed June 4 2012 at http://www.beefproducts.com/history.php

R. W. Bulliet (2005) *Hunters, Herders and Hamburgers: The Past and Future of Human-Animal Relationships* (New York: Columbia University Press).

C. Choi (2012) 'The making of the term 'pink slime', Associated Press, May 11, 2012. Accessed June 4, 2012 at: http://news.yahoo.com/making-term-pink-slime-173349428.html

C. Diamond (1978) 'Eating Meat and Eating People,' *Philosophy* 53: 465–479.

C. Diamond (2003) 'The Difficulty of Reality and the Difficulty of Philosophy,' *Partial Answers: Journal of Literature and History of Ideas* 1/2: 1–26.

P. D. Edelman, D.C. McFarland, V.A. Mironov, and J.G. Matheny 'In Vitro Cultured Meat Production,' *Tissue Engineering*, May/June 2005, 11(5–6): 659–662

T. Grandin, C. Johnson (2009) *Animals Make Us Human: Creating the Best Life for Animals* (Orlando, FL: Houghton Mifflin Harcourt Publishing).

D. Haraway (2008) *When Species Meet* (St. Paul, MN: University of Minnesota Press).

C. Korsmeyer (2012) 'Ethical Gourmandism,' in *The Philosophy of Food*, D. Kaplan, ed. (Berkeley: University Press of California): 87–102.

H. Pellissier (2012) 'Nine Ways In-Vitro Meat Will Change Our Lives,' Institute for Ethics and Emerging Technologies, Trinity College, Hartford, CT, Accessed May 5, 2012 at http://ieet.org/index.php/IEET/more/5379

M. Pollan (2006) *The Omnivore's Dilemma: A Natural History in Four Meals* (New York: Penguin Books).

E. Schlosser (2001) *Fast Food Nation: The Dark Side of the American Meal* (Boston: Houghton-Mifflin).

A. Toffler (1970) *Future Schock* (New York: Random House).

S. Wellin, J. Gold and J. Berlin (2012) 'In Vitro Meat: What Are the Moral Issues?' In *The Philosophy of Food* D. Kaplan, ed. (Berkeley: University Press of California): 292–304.

DISCUSSION QUESTIONS

1. Do you think that artificial meat would be ethically superior to meat grown "on the hoof"? Why or why not?

2. Would there be anything ethically problematic with incorporating artificial meat into food products without labeling to notify consumers? Why or why not?

3. The author suggests that the ecological benefits of artificial meat are thus far unproven? Can you think of any ecological impacts associated with animal agriculture that artificial meat is certain not to have?

4. The author suggests that the case for artificial meat depends on a false dichotomy – i.e. that we must choose between artificial meat or meat from animals raised in CAFOs, when in fact there are other alternatives. What are those other alternatives? Do you think they are ethically preferable to CAFOs and artificial meat? Why or why not?

5. If eating meat grown "on the hoof" is ethically problematic, is there something ethically (or perhaps aesthetically) troubling about eating it when grown in vitro?

6. Do you find the critique of artificial meat that it is reductionist and would be part of a centralized industrial food system persuasive? Why or why not?

7. Can you think of any considerations in support of artificial meat that were not discussed in the chapter? Or any possible concerns about it that were not discussed?

8. In what respects, if any, would artificial meat production constitute a change in the form of life of agriculture?

SYNTHETIC GENOMICS AND ARTIFICIAL LIFE

We commonly think of living things as being natural rather than artifactual, and typically they are. The vast majority of organisms, from bacteria to mammals, come into existence independent of human intentions and control. Nevertheless, people also often engineer living things for particular purposes, both practical and scientific. Intentional manipulation of organisms has been occurring since at least the beginning of agriculture – through selective breeding, hybridization, and grafting – and recombinant DNA techniques have been used to create transgenic plants and animals by inserting genes from one organism into another for decades now. The organisms that result from these and other modification processes, such as chimeric biomedical research, are, at least partially, artifacts – i.e. the products of human design and creation. In recent years, advances in the tools used to characterize, manipulate and construct genomes have resulted in a substantial increase in the precision, intensity, and comprehensiveness of organism engineering. Researchers have now begun to thoroughly reengineer existing organisms and even create novel organisms from scratch. We are on the cusp of transitioning from creating minimally artifactual organisms to creating highly (or even entirely) artifactual ones. As one leader in the field put it, "We're moving from reading the genetic code to writing it."[1] The goals of synthetic genomics and artificial organism projects are sometimes scientific – e.g. elucidating fundamental "design" principles of life and identifying how life could have emerged. However, they are very often practical – i.e. creating organisms that can perform useful functions, such as coding information, supplementing our immune systems, or "manufacturing" vaccines, chemicals and fuels.

In the first chapter in this section, "Synthetic Biology, Biosecurity, and Biosafety," Michele Garfinkle and Lori Knowles discuss the primary extrinsic or outcome-oriented concerns associated with synthetic genomics – that synthetic

---

[1] J. Craig Venter, in Antonio Regaldo. 2005. "Next Dream for Venter: Create Entire Set of Genes From Scratch." *The Wall Street Journal*, June 29: A1.

organisms could be detrimental to human health and the environment if they were to be released either intentionally or accidentally. They argue that responsible development of synthetic biology requires both hard and soft law approaches to managing the risks associated with the dual-use technology. However, it is also crucial that oversight does not stifle innovation, given the technology's tremendous potential to provide ecological and human health benefits.

Because synthetic biology involves creating novel life forms and combing genomic material from multiple species, it raises several prominent intrinsic ethical concerns or concerns about the nature of the technology itself. In "Evolution and the Deep Past: Responses to Synthetic Biology," Christopher J. Preston critically evaluates several of these. He has reservations about objections to the technology on the grounds that it involves mixing genomic material across species. However, he believes that there may be a compelling intrinsic or deontological argument to be made against synthetic biology on the basis that synthetic organisms, unlike all prior organisms, are not the product of descent with modification from prior organisms. In this way, the connection to historical evolutionary processes is severed.

Synthetic organisms are for the most part engineered using materials and design principles from other biological organisms. However, some researchers are trying to create artificial life forms from simple organic and inorganic materials alone. In the final chapter in this section, "Social and Ethical Implications of Creating Artificial Cells," Mark A. Bedau and Mark Triant address both intrinsic and extrinsic concerns related to this protocell or artificial life research. They argue that each of the intrinsic concerns is misplaced, and that the primary ethical issue really is making decisions about how and whether to proceed with the research under conditions of extreme uncertainty. They advocate what they call a "cautious courage" approach, rather then an approach based on the precautionary principle or the doomsday principle.

More detailed summaries, as well as a listing of related readings, are located at the start of the chapters.

35 SYNTHETIC BIOLOGY, BIOSECURITY, AND BIOSAFETY

Michele Garfinkle and Lori Knowles

CHAPTER SUMMARY

The primary extrinsic concerns regarding synthetic organisms are to do with their human health and environmental risks. One reason for this is that they are (for now, at least) microorganisms, and many microorganisms cause illnesses in humans (and nonhumans) and are ecological disruptive. Another reason is the growth of "DIY (do-it-yourself) biotech" and "garage biotech" – i.e. people who are exploring synthetic biology on their own or outside of institutional structures such as academic, government, and corporate labs. In this chapter, Michele Garfinkle and Lori Knowles present and evaluate governance strategies, both hard law and soft law, for managing the risks associated with synthetic genomics, while also fostering innovation and realizing its potential benefits. They believe that it is possible to develop synthetic biology in biosecure and biosafe ways.

RELATED READINGS

Introduction: Innovation Presumption (2.1); EHS (2.6.1); Dual Use (2.6.5); Responsible Development (2.8)

Other Chapters: Kevin C. Elliott, *Risk, Precaution, and Nanotechnology* (Ch. 27); Mark A. Bedau and Mark Triant, *Societal and Ethical Implications of Creating Artificial Cells* (Ch. 37)

1. INTRODUCTION

Synthetic biology is one of several emerging biotechnologies resulting from increased knowledge about molecular biology, enhanced tools for the computational analysis of DNA, and the ability to synthesize long stretches of DNA routinely and inexpensively. It emerges from the convergence between biology, chemistry, computation, and engineering. Synthetic biology enables the production of copies of existing biological entities, such as bacteria, for use and manipulation in the laboratory, and potentially the modification of plant or animal genomes with enhanced or novel traits.

A subset of synthetic biology known as synthetic genomics involves the construction of genes, which are relatively large stretches of DNA, from smaller pieces of DNA (called oligonucleotides). To put this in context, a gene may be from about 1000 to 10,000 nucleotides long; an oligonucleotide about 45–75 nucleotides long. The nucleotide is the basic building block of DNA.

Whereas traditional genetic engineering requires laborious isolation and recombination of genetic material in a laboratory, if the order of the DNA chemical bases (the sequence) is known, a machine called a synthesizer can make the chemical sequence desired in a fraction of the time. A number of commercial firms with sophisticated gene synthesizers can fill orders for genetic material ranging in size from oligonucleotides to full genomes, quickly and relatively inexpensively. Laboratory work that would take 6 months of full-time effort to combine pieces of DNA can now be essentially dispensed with by placing an order for the precise sequence required. Moreover, the availability of "desktop" oligonucleotide synthesizers makes it possible for individuals to avoid going through a company altogether.

In addition to the speed and power of gene synthesis, another advance that is a hallmark of synthetic biology is the intent to apply engineering principles to biological systems. By focusing on the functional capacities of various genes or gene segments and using design principles derived from engineering, synthetic biology enables researchers to construct novel or enhanced biological systems or potentially even whole organisms. These organisms could then, for example, produce a fuel, or glow a certain color, according to the researcher's design. These two features of synthetic biology, the easy construction of genes or genomes that can be manipulated according to the researcher's plan, and the novel organisms that may be created, give rise to a number of ethical and policy concerns, including concerns about human and environmental health, which are our focus here.

Among the most prominent and pressing concerns regarding synthetic biology are the potential environmental and human health harms that might occur from exposure, either inadvertent or intentional to organisms that are constructed using the tools of gene synthesis. These concerns can be analyzed against the backdrop of existing biosafety and biosecurity laws and guidelines. To be clear, biosafety rules seek to avoid inadvertent harms by keeping scientific personnel safe from accidental exposures to the hazardous biological agents they are working with, and to prevent accidental releases of pathogens from the laboratory that could threaten public health and the environment. Biosecurity measures, in contrast, seek to prevent the deliberate malicious release of synthetic organisms for hostile purposes. Both types of measures are necessary when exploring how to avoid, minimize or manage harms from synthetic biology.

Synthetic biology is an example of a dual-use technology; that is, a technology that is designed to be used in ways that are beneficial, but that can also be used in ways that inflict harm. A classic example of a dual-use technology is nuclear reactors, which can produce electricity to fuel cities, or create weapons to level cities. As with all dual-use technologies, the governance or regulation of synthetic biology (either as a technology or within the realm of research) is a complex task. Overly burdensome regulation could come at the cost of many of the benefits to human health, biomedical research, industrial applications, energy production, or environmental protection that this technology may provide. But inadequate governance could result in significant harms if the technology is deliberately misused or an accident occurs due to lack of oversight. Developing governance strategies that enhance biosecurity and biosafety, protect the environment, minimize the costs and burdens of government and industry while not impeding research is a very tall order (Garfinkel et al. 2007).

Thus, synthetic biology presents policymakers and other stakeholders with a two-step analytic and policy problem. First, are the current regulations or other policies adequate when the process or products of synthetic biology are deemed to be essentially identical to those of other extant biotechnologies? Second, are any new policies warranted to deal with the novel aspects of the technologies underlying synthetic biology or the novel products produced from synthetic biology? The former problem may be especially difficult as it may require confronting inadequacies in current regulations or policies that may have existed for years or decades.

2. ECOLOGICAL, ENVIRONMENTAL, AND HUMAN HEALTH CONCERNS

To date, synthetic biology experiments have largely involved the synthesis and manipulation of microbes (specifically bacteria, fungi, and viruses) to try to produce products or learn more about how organisms function. For example, one research group has modified yeast to create a necessary precursor for an anti-malarial drug that is difficult to derive from naturally-occurring sources (Ro et al., 2006). And a synthetic biology company has begun producing isobutanol from engineered organisms, which is thought to be a superior alternative fuel to ethanol and is also an important manufacturing chemical. The production was to have reached an industrial scale (1 million gallons per month) by the end of 2012 (Noorden, 2012). These synthetic biology products have clear, potential benefits. At the same time, scientists have sometimes caused considerable controversy with synthetic biology, for example, by synthesizing live poliovirus (Cello et al., 2002) and the 1918 influenza virus that killed tens of millions of people almost one hundred years ago (Taubenberger et al., 2005).

As gene synthesis becomes faster and easier, questions about who is using the technology and for what purpose arise. While the individual researchers involved in the virus synthesis experiments described above were from reputable institutions, news of their work raised concerns about synthetic biology being used for bioterrorism by state actors or lone-wolf terrorists. Deliberate misuse is not the only way that biologically engineered organisms could cause harm. The U.S. military has expressed interest in harnessing the power of synthetic biology to create "greener" weapons that rely less on toxic chemicals for production (Hayden, 2011). Equally unnerving is the prospect of the do-it-yourself amateur or recreational biologist who synthesizes something in his or her garage and unwittingly unleashes a destructive microbe into the environment (Wolinsky, 2009).

While malicious use of synthetic microorganisms is a genuine concern, even legitimate scientific uses of biological agents entail both safety and security risks. Concerns about the safety and health of laboratory workers and those living near laboratories working with synthetic genomes also have been raised, in tandem with concerns about environmental health should a pathogenic or environmentally invasive organism escape. The risks, then, that are most often expressed when it comes to synthetic biology are those related to bioterrorism and biosafety.

As such, an understanding of existing regulations in these areas and how those regulations might be enhanced in the case of synthetic biology is helpful when contemplating how to govern this advancing field.

The effective governance of dual-use technologies like synthetic biology requires a multifaceted approach that includes three types of measures: hard law (treaties, statutes, and regulations), soft law (voluntary standards and guidelines), and informal measures (awareness-raising, professional codes of conduct). These three types of governance measures are not mutually exclusive. For example, voluntary standards and guidelines aimed at research oversight or promoting biosafety and biosecurity can be bolstered by criminal laws or tort laws that impose penalties for breaches of the law or the harm caused by accidental or deliberate misuse (Knowles, 2012).

In what follows, we discuss approaches to the governance of synthetic biology to promote safety and security, while minimizing disruptions to the process of scientific discovery.

3. GOVERNANCE OF SYNTHETIC BIOLOGY: BIOSECURITY

Biosecurity, particularly in the United States, is still viewed by most policymakers and law enforcement officials through the filter of the events of September 11, 2001. Thus, much biosecurity governance derives from hard law. In the realm of scientific research, there has been much discussion as to how much law is required, and in fact most research in the United States is governed through hard and soft law approaches. How best to balance the two is an ongoing policy discussion.

In the first post-9/11 experiment that raised widespread public concern, a team from the State University of New York at Stony Brook synthesized a live, infectious poliovirus in 2002 by assembling oligonucleotides ordered from a commercial synthesis company (Cello et al., 2003). In 2003, researchers at the J. Craig Venter Institute synthesized a bacteriophage, a virus that infects bacteria (Smith et al., 2003). While the first experiment took over a year, the second team synthesized its virus in approximately two weeks, illustrating just how fast the technology is evolving.

A concern regarding synthetic biology is that the increasing efficiency and affordability of gene synthesis, together with increased knowledge about viral genomes pose a significant risk to human health. Some worry that the risk may come from the synthesis and release of deadly viruses like the smallpox virus, the sequence of which is known, but to which the public for the most part has no natural or vaccine-induced immunity. Others say that the danger synthetic biology poses is that it offers the ability to make pathogenic viruses more virulent. Recent experiments to make the H5N1 influenza virus (i.e. avian influenza or bird flu) more easily transmissible between laboratory animals, while not using any synthetic biology techniques, point to a situation that could be precipitated by synthetic biology. Moreover, it raises concerns in the public as to what scientists are doing, and which authorities are regulating them (Enserink, 2011).

The question, then, is whether these concerns are well-founded. Specialists in viral microbiology doubt whether it is as easy to synthesize a deadly virus as one might believe (Collett, 2007). In order to synthesize an existing virus, its exact genetic sequence must be known, and to be functional, the sequence must be entirely correct. Some of the viral strains in laboratories are attenuated through spontaneous mutations, and may no longer be transmissible or pathogenic even if they were at the time they were sequenced (Baric, 2007). Moreover, even if a correct sequence for a virus exists, it still requires significant expertise to construct a virus from synthesized DNA and then to express the virus so that it functions as a bioweapon (National Science Advisory Board for Biosecurity, 2006). For the time being at least, it remains the case that it would be easier to steal a sample of a deadly virus from a laboratory or from a viral outbreak in nature (Garfinkel et al., 2007; Epstein, 2008).

Recent experiments on genetically altered or synthesized viruses have raised contentious discussions about possible restrictions on the publication of sensitive information. In 2003, the editors of several major scientific journals issued a joint statement calling for the review of security-sensitive research papers submitted for publication, with the default position favoring public release (Journal Editors and Authors Group, 2003). In response to concerns over dual-use information, the National Science Advisory Board for Biosecurity (NSABB) suggested conducting a risk-benefit analysis before publishing articles that could lead to dual-use risks. Based on biosecurity concerns, the editors can ask the authors to modify an article, delay publication, or reject it entirely. In 2004, an expert panel of the U.S. National Academies considered restrictions on the publication of pathogen genomes and ultimately decided not to endorse such restrictions (National Research Council 2004). A major problem with pre-publication security reviews is that it can be difficult to identify a priori which research findings entail dual-use risks. Critics of such restrictions also argue that scientific freedom and access to information are crucial to technological innovation and that restricting publication would slow the development of medical countermeasures against biological threats (National Academies of Science, 2007).

Given the foregoing assessment of the biosecurity risks of synthetic biology, what are the regulatory options for policy makers and private actors? When sequences can be mail-ordered (as for the 2002 poliovirus experiments), DNA synthesis companies may need to be the gatekeepers that keep synthesized DNA sequences out of the hands of inappropriate or unsupervised parties. Software is available to compare synthetic DNA orders against lists of agents of concern – for example, viruses that could be used in a terrorist attack, or more generally against databases of all known sequences. The US government provides non-binding guidelines for companies as to how to screen these orders. Most companies have already been doing screening of some sort. Consortia of such companies have written their own guidelines, which in some cases may be stricter than government guidelines (IASB, 2009). Almost all of the guidelines by all parties require that the companies be able to trace an order to a person for forensic purposes, and most companies already keep such records as good business practice.

Other proposals for governance have suggested that DNA synthesizers, especially oligonucleotide synthesizers, might be registered, or users could be required to have licenses before they are allowed to buy the chemicals required to make DNA. Such measures would likely be less effective than those interventions aimed at companies making the large pieces of DNA, but could provide some measure of security if implemented properly (Garfinkel et al., 2007; US Department of Health and Human Services, 2010).

Extensive biosecurity regulation, especially in the United States, provides a strong backdrop to the governance of synthetic biology (Tucker, 2003). One very important regulation is the National Select Agent Registry Program (APHIS/CDC, 2011), which was instituted after the 9/11 and the U.S. anthrax attacks as part of the 2001 PATRIOT Act, strengthening an existing Center for Disease Control and Prevention (CDC) registry (United States Government, 2001). These regulations govern the importation and transport of pathogens. The core of the Select Agents program is a list of microbial and other agents (bacteria, viruses, toxins, etc.) that the US government restricts possession of, even for legitimate scientific inquiries. Institutes and personnel must be registered to use these agents, and all movements of the agents (theft or other loss, but also even from one room to another) must be reported. The physical containment of the labs (locks, badges use, etc.) is dictated by the regulations, as are restrictions on who can work with them (e.g., individuals with mental illness, or with links to countries with histories of state-sponsored terrorism are excluded).

Currently, the Select Agents list is reviewed for updates every two years, and has been criticized for focusing on physical agents rather than the DNA sequences that may be more appropriate. It may be replaced by a system for specifying microbial pathogens and toxins based on DNA sequences rather than microbial species, so that it can keep pace with rapid advances in biotechnology (National Research Council, 2010). This would be particularly useful in helping govern illicit use of DNA synthesis, as sequence information continues to expand. Rapidly evolving biotechnologies, such as synthetic biology, require flexible governance strategies, since more rigid measures would soon become obsolete. In part due to the already robust biosecurity regulations described above (as well as an equally established web of biosafety regulations), the US Presidential Commission for the Study of Bioethical Issues recommended that no new regulations were necessary to govern synthetic biology (Presidential Commission for the Study of Bioethical Issues, 2010).

This is not to say that there are no legitimate concerns about harms that synthesized organisms might cause. There are situations that need further exploration and there are governance methods – not necessarily laws and regulations – aimed in particular at the behavior and responsibility of people working with these organisms. For example, one way that the community of amateur or garage biologists has decided to approach biosecurity and biosafety is to set up community laboratories, where many people can come together to work. These laboratories have rules to promote safety and mitigate many of the concerns others might have about people working alone in garages, and they follow all applicable laws. These came about through the community's understanding about the nature of the technology and what individuals wanted to do with it. Similarly, before any guidance was issued, many of the DNA synthesis companies chose to screen both sequences and customers for potential malicious intent, as they recognized the power of the technology and its potential for misuse. It is not a coincidence that these companies also have a profit motive, which can be a powerful and important governance tool on its own.

This section has focused on security issues related to synthetic biology. As we have seen, not all synthetic biology research is done in the same context or by the same type of researchers. Therefore, biosecurity measures need to be tailored to context. We have also seen that there are considerable resources and options for promoting biosecurity for synthetic biology, and that many have already been implemented.

4. GOVERNANCE OF SYNTHETIC BIOLOGY: BIOSAFETY

In addition to concerns about bioterrorism, there are general questions about the unintended impacts of synthetically engineered organisms on humans and the environment. Consider the following three scenarios. In the first, a laboratory worker is accidentally exposed to a pathogenic synthetic organism that results in harm. This scenario does not differ from accidental workplace exposure to any other biological pathogen or toxin, except that it may be difficult to identify the effects of accidental exposure to a novel synthetic organism if its effects have not been documented. In the second scenario, a known pathogen is synthesized and accidentally released into the environment causing unforeseen effects, or harm to humans and the environment. This is a classic biosafety situation that exists for all research involved with pathogens, whether they are plucked from nature, created through genetic engineering, or synthesized. In the third scenario, a synthesized organism with novel characteristics not found in nature escapes from the laboratory. It is this scenario that has been the focus of recent academic and policy discussions regarding synthetic biology.

The escape of novel synthetic microorganisms gives rise to disquiet about the impact they might have on the environment in a number of ways. For example, researchers are currently interested in creating biofuels by synthesizing algae that might have the sole function of turning sugar into ethanol. If these algae were created and became a robust organism, people would be both excited about the possibilities of the alternative fuel and rightly concerned about what might happen if the synthetic algae were accidentally spilled outside the lab. Some have voiced concerns about whether the algae might disrupt natural ecosystems by "taking over" the natural organisms, or invading parts of the ecosystem, such as streams or even oceans (Fehrenbacher, 2011). While this situation might never occur, there is no way of knowing for certain what the effects of such a release might be. It is possible that synthetic organisms might mix with wild relatives of the same organisms and create new genetic variants that could disrupt or change ecosystems in unanticipated ways.

These concerns about escape, gene flow (which happens naturally in the environment between non-engineered species) and disruption have been explored thoroughly in the genetic engineering policy debates. There is no consensus on these issues, but there are international agreements and national regulations on biosafety that are aimed at least in principle at minimizing the negative effects that might arise (Cartagena Protocol, 2000).

However, critics of synthetic biology often respond that because the organisms can be designed to function in completely novel ways, the gene escape and gene flow effects (the movement of DNA between organisms and even between species) are very hard for us to imagine, which makes them difficult to anticipate and manage. Scientists, however, strongly object to this characterization, as there is no evidence that synthetic DNA acts any differently than natural DNA, and gene flow is well understood. Still, gene flow happens, and cannot be mitigated with current technologies. Therefore, scientists have recognized that it might be necessary to construct synthetic organisms that are not fit to live outside the laboratory environment by making them dependent on nutrients that can only be found in the laboratory (a condition called auxotrophy), or perhaps in the extreme make them so that they would self-destruct after a certain time. Other synthetic biologists have suggested that researchers who construct synthetic genes, cells or organisms "sign" their work so that it can be identified if necessary. The addition of identifying sequences of DNA would indicate unambiguously that this microorganism was constructed in a laboratory and did not occur in nature. This could be critical information in the case of an accidental release.

There are a number of biosafety questions that synthetic biology raises, but they are not radically different from those raised by the manipulation of genes that has been occurring for several decades: What types of research should be done? Is there research that should be federally funded and controlled rather than left to the private sector? What types of systems should be in place to evaluate the environmental risks of this research and commercialization? Are new data needed? How do we go about evaluating and managing risks we can neither know nor predict?

This last situation is something that makes many people extremely uncomfortable. The unknowns create tremendous anxiety since the risks are unquantifiable – they might be absolutely minimal but they might be terrible. As humans we have a difficult time knowing how to manage tiny chances of high impact risks: think about the number of people who drive a car while on the phone without a single worry, but are anxious about taking a much safer plane. Some environmental groups have called for a precautionary approach to synthetic biology, insisting on a moratorium on the release and commercialization of the synthetic biology products until their risks are fully known and more regulation is in place (Friends of the Earth et al., 2012).

The truth is that we can never fully know the risks of introducing natural, genetically engineered, hybrid or synthesized organisms into the environment because it is such a complex system about which we know so little. We are right to be concerned, but that concern needs to be weighed carefully against the cost of foregoing the real benefits to humans and the environment that this technology could provide. It is equally important to remember that this technology does not exist in a regulatory vacuum; there are many safeguards that already exist to govern synthetic biology.

For example, the United States has numerous laws, regulations, and guidelines that apply to the safe handling of biological pathogens, such as viral genomes, and are therefore applicable to many of the products of synthetic biology. Many of these regulations relate to health and safety standards for exposure to pathogens, such as those in the Occupational Safety and Health Act of 1970 (United States Government, 1970). Since 1984, the CDC and the National Institutes of Health (NIH) have jointly published a manual that includes a graduated risk-assessment and containment model for work with dangerous pathogens (US Department of Health and Human Services, 2009).The four biosafety levels demand increasingly stringent measures for the handling, containment, and disposal of biohazardous materials, as well as risk-management efforts involving equipment, facilities, and personnel.

Genetic engineering in the United States is governed by the NIH Guidelines on Research involving Recombinant DNA Molecules (National Institutes of Health, 2011). These guidelines outline safe laboratory practices and appropriate levels of physical and biological containment for research with recombinant DNA, including the creation and use of organisms containing foreign genes. Like the graduated biosafety categories listed above, the NIH Guidelines classify research with recombinant microbes into four risk categories based on the pathogenicity of the agents in question in healthy adult humans. The higher the risk category, the more stringent are the safety and oversight precautions. In 2009, in response to developments in synthetic biology, the NIH published proposed changes extending coverage of the Guidelines to molecules constructed outside living cells by joining natural or synthetic DNA segments to DNA molecules that can replicate in a living cell (National Institutes of Health, 2009).

Like the biosafety guidelines, the NIH Guidelines apply to laboratories and institutions that receive federal funding for recombinant DNA research and to other institutions that accept the rules voluntarily. Under the Guidelines, proposed experiments using recombinant DNA must be reviewed by a local Institutional Biosafety Committee (IBC). The IBC assesses the potential harm that might occur to public health and the environment in addition to potential harms to researchers. Based on this risk assessment, the IBC determines the appropriate level of containment, and evaluates the adequacy of training, procedures, and facilities. The IBCs function as sentinels to identify emerging safety (and sometimes policy) issues for consideration by a federal-level body called the Recombinant DNA Advisory Committee (RAC).

The general mechanism that IBCs have used to evaluate experiments has not changed in over thirty years. Members of the IBC, mostly scientists from the institution where the research is taking place, evaluate the experiments involved in the proposed research. As in the scientific system in general, the IBC system is built on the trust that one scientist has in another in the description and evaluation of the experiments to be done. This was fine when the IBC was responsible for thinking about narrow questions regarding the transfer of single pieces of DNA. But now, as a result of a National Academy of Sciences panel that met following 9/11 and a recommendation of the NSABB, IBCs will gradually become responsible for also evaluating the possibility of scientific knowledge being used in a dangerous (dual-use) way. Furthermore, although IBCs are not required to evaluate the broader (i.e. beyond environmental, health and safety) ethical aspects of experiments, they are allowed to take community values into account. As the ethics of experimentation become more widely discussed outside of the laboratory, they may need to become more prominent within IBCs.

In addition to oversight of research, the US and other countries have robust regulatory systems to oversee the safety of products of research. The USDA, FDA, and EPA, for example, all have input into whether products may be made available to the public. The criteria of evaluation used by the FDA are well understood: safety and effectiveness. USDA and EPA both have large roles to play in deciding whether the products of synthetic biology will be allowed on the market; as part of those considerations they will be asking if the deliberate use of products outside the lab (e.g., modified algae in ponds in the desert to make fuel; modified plants for fibers) will negatively impact the environment or agriculture. These agencies already have a significant amount of regulatory authority, but whether these authorities will fully capture the potential problems of synthetic biology-derived products, especially as those problems are outlined by critics, remains to be analyzed, and these analyses have started in earnest (Rodemeyer, 2009).

As with biosecurity, there are significant biosafety issues associated with synthetic biology. However, in most cases the biosafety challenges are not new or unique to synthetic biology, and there are robust regulations and policies in place that apply to synthetic biology (or that can be modified to apply). Moreover, policy makers and researchers are actively working to identify where additional policies may be needed for the any unique biosafety challenges posed by synthetic biology.

5. SOFT LAW AND INFORMAL MEASURES

Researchers and scientists themselves need to be part of the governance of synthetic biology. This includes making difficult decisions about what research ought not to be done, either because the necessary experiments are simply too dangerous or the knowledge that might be produced does not outweigh the real risk of harm that might result. When progress in science is based on both a healthy curiosity and sense of play, and a desire just "to see what might happen" in a particular experiment, it can be difficult to look carefully down the road and anticipate what might result. Additionally, few and far between are the scientists who produce something that works and are content not to publish the results of that experiment when the threats are vague but the outcomes of the experiment are clear. This is especially true where funding, promotions, and recognition are so closely tied with publications of articles of impact. What then needs to happen to guide synthetic biology in a positive way? Regulations and law cannot provide all the safeguards and guidance that is needed, in large part because they are static instruments that do not have the ability to respond dynamically to the rapid advances in science and technology. Moreover, laws are often responsive rather than anticipatory in nature. Consequently, other means of circumscribing problematic behavior or outcomes are needed alongside hard law mechanisms.

A complementary set of governance tools is based on "soft law" measures, including codes of ethics, awareness training, and voluntary guidelines, all of which focus on the people who conduct synthetic biology research (Interacademy Panel on International Issues, 2005). In the biotechnology industry, companies involved in commercial gene synthesis have formed associations to develop guidelines for ethical practice. The International Association Synthetic Biology (IASB, 2009) and the International Gene

Synthesis Consortium (IGSC, no date) have both produced codes of conduct for their member companies that are specifically designed to prevent the exploitation of their products for harmful purposes. These codes include protocols for screening gene synthesis orders for their similarity to the genetic material of pathogens as discussed above.

In the absence of a robust self-governance scheme for synthetic biology, ethics education can help to create a culture of responsibility in the life sciences that can have a significant impact in mitigating risks from research. While some scientific societies have issued codes of ethics that prohibit certain research, like the development of biological weapons (International Union of Microbiological Societies, 2008), codes of conduct have yet to be integrated into science education, professional development, or certification requirements in any concerted and meaningful way (Rappert et al., 2006).

There is also a need for awareness-raising efforts. In 2009, the Federation of American Societies for Experimental Biology (FASEB) issued a statement that "scientists who are educated about the potential dual-use nature of their research will be more mindful of the necessary security controls which strike the balance between preserving public trust and allowing highly beneficial research to continue." Surveys indicate, however, that many life-science researchers lack an awareness of dual-use concerns, including the risk of misuse associated with their own work (Dando, 2009). Overcoming this deficit will require a commitment to ethics education for science and engineering students, as well as training in identifying and managing dual-use risks. This task is daunting because of the difficulty of defining dual-use and the lack of experts in the field (National Science Advisory Board for Biosecurity 2008).

To help contain dual-use risks, ethics education must be combined with mechanisms for reporting risks once they have been identified. Whenever a student or researcher suspects that a colleague is misusing a technology for harmful purposes or is willfully blind to the risks that research might pose, there should be a confidential channel for passing this information to the appropriate authorities so that it can be acted upon.

One interesting experiment in trying to make "soft law" work together in real time along with scientific training is iGEM, the International Genetically Engineered Machine Competition (iGEM, 2012). An annual competition for college level (and starting in 2012, high-school level) synthetic biologists, iGEM brought together over 1000 students at the 2011 competition. The students work in teams to use a system of standard parts to do projects illustrating the applications of synthetic biology to such things as health, food production, and environmental damage mitigation. In addition to promoting good science and engineering, iGEM now includes requirements to ensure that the students understand safety and security (and in some cases, the projects themselves are about safety and security). US law enforcement, representatives of various US government offices, and international biosecurity treaty personnel have attended the competitions to open lines of dialogue between those agencies and the students. While it is unclear whether these idiosyncratic meetings and good practices within the competition carry over to the students' everyday thinking about how to do good science, this will no doubt become clearer over time.

6. CONCLUSION

Unlike medical doctors, scientists are not bound by oaths to "do no harm". As part of the larger "societal contract" they are, however, supposed to work toward improving society in ways ranging from the contribution of knowledge via basic research papers to solving what could be seen as impossibly large problems. Implied in the contract is that scientists will choose their lines of work carefully, and carry them out with an eye toward safety and security. But because those parameters are not defined by law, and because restricting the latitude of scientists to carry out their inquiries unimpeded is more or less universally recognized as bad for society, what those parameters actually are is continually being tested, as seen in the poliovirus, influenza virus, and synthetic cell cases.

Ethics and governance discussions about synthetic biology tend to focus on the negative: biosecurity and biosafety risks for humans and the environment. However, in a world filled with environmental degradation, climate change, population pressures, and shortages of clean water and energy, it is crucial to remember that the risk of doing nothing may be as great or greater than the risk of doing something. Synthetic biology may lead to alternative clean fuels, better vaccine manufacture, production of next-generation medicines, and reductions of carbon dioxide and pesticides in our environment. How to weigh the good a technology could bring against the risks that might come with it is a delicate and difficult quandary, one that requires real data and thoughtful analysis.

WORKS CITED

APHIS/CDC (2011) 'National Select Agent Registry,' http://www.selectagents. gov/, accessed 2 June 2012.

R. S. Baric (2007) 'Synthetic Viral Genomics: Risks and Benefits for Science and Society,' in Garfinkle, Endy, Epstein, Friedman (eds.) *Working Papers for Synthetic Genomics: Risks and Benefits for Science and Society*, 38–51.

Cartagena Protocol (2000) 'Cartagena Protocol on Biosafety to the Convention on Biological Diversity,' *Convention on Biological Diversity* (Montreal).

J. Cello, A. V. Paul, and E. Wimmer (2002) 'Chemical Synthesis of Poliovirus cDNA: Generation of Infectious Virus in the Absence of Natural Template,' *Science* 297: 1016–18.

M. S. Collett (2007) 'Impact of Synthetic Genomics on the Threat of Bioterrorism with Viral Agents,' in *Working Papers for Synthetic Genomics: Risks and Benefits for Science and Society*, 83–103.

M. R. Dando (2009) 'Dual-Use Education for Life Scientists,' *Disarmament Forum: Ideas for Peace and Security*, no. 2, 41–44.

M. Enserink (2011) 'Scientists Brace for Media Storm Around Controversial Flu Studies,' *ScienceInsider*, 23 November, http://news.sciencemag.org/ scienceinsider/2011/11/scientists-brace-for-media-storm.html, accessed 2 June 2012.

G. L. Epstein (2008) 'The challenges of developing synthetic pathogens,' *Bulletin of the Atomic Scientists Web Edition*, 19 May, http://www.thebulletin.org/

web-edition/features/the-challenges-of-developing-synthetic-pathogens, accessed 1 June 2012.

K. Fehrenbacher (2011) 'Craig Venter: Algae Fuel that Can Replace Oil Will Not Come From Nature,' *GigaOM* 23 October http://gigaom.com/cleantech/craig-venter-algae-fuel-that-can-replace-oil-will-not-be-from-nature/ accessed 1 June 2012.

Federation of American Societies for Experimental Biology (2009) 'Statement on Dual Use Education,' http://www.faseb.org/portals/0/pdfs/opa/2009/FASEB_Statement_on_Dual_Use_Education.pdf, accessed 2 June 2012.

Friends of the Earth, International Center for Technology Assessment, and ETC Group (2012), *The Principles for the Oversight of Synthetic Biology,* (Washington DC and Montreal, Quebec).

M. S. Garfinkel, D. Endy, G. L. Epstein, and R. M. Friedman (2007) *Synthetic Genomics: Options for Governance* (Rockville, MD: J Craig Venter Institute, 2007), http://www.jcvi.org/cms/research/projects/syngen-options/overview/, accessed 2 June 2012.

D. G. Gibson, J. I. Glass, C. Lartigue, V. N. Noskov, R.-Y. Chuang, M. A. Algire, G. A. Benders, M. G. Montague, l. Ma, M M. Moodie, C. Merryman, S. Vashee, R. Krishnakumar, N. Assad-Garcia, C Andrews-Pfannkoch, E. A. Denisova, L. Young, Z.-Q. Qi, T. H. Segall-Shapiro, C. H. Calvy, P. P. Parmar, C. A. Hutchison III, H. O. Smith, and J. C. Venter (2010) 'Creation of a Bacterial Cell Controlled by a Chemically Synthesized Genome,' *Science* 329: 52–56.

IASB (2009) 'International Association Synthetic Biology Code of Conduct for Best Practices in Gene Synthesis,' http://www.ia-sb.eu/tasks/sites/synthetic-biology/assets/File/pdf/iasb_code_of_conduct_final.pdf, accessed 2 June 2012.

iGEM (2012) 'Synthetic Biology Based on Standard Parts,' http://igem.org/Main_Page, accessed 5 June 2012.

IGSC (no date) 'International Gene Synthesis Consortium Harmonized Screening Protocol: Gene Sequence & Customer Screening to Promote Biosecurity,' http://www.genesynthesisconsortium.org/wp-content/uploads/2012/02/IGSC-Harmonized-Screening-Protocol1.pdf, accessed 5 June 2012.

Interacademy Panel on International Issues (2005) 'Statement on Biosecurity,' November 7, http://www.interacademies.net/File.aspx?id=5401, accessed 1 June 2012.

International Union of Microbiological Societies (2008) 'IUMS Code of Ethics against Misuse of Scientific Knowledge: Research and Resources,' http://www.iums.org/index.php/code-of-ethics, accessed 5 June 2012.

Journal Editors and Authors Group (2003) 'Statement on Scientific Publication and Security,' *Science* 299, 1149.

L. P. Knowles (2012) 'Existing Dual-Use Governance Measures' in J. Tucker, (ed.) *Innovation, Dual-Use and Biosecurity.*

National Academies of Science, Committee on a New Government-University Partnership for Science and Security, *Science and Security in a Post 9/11 World: A Report Based on Regional Discussions between the Science and Security Communities* (Washington, DC: National Academies Press).

National Institutes of Health (2009) 'Notice of Consideration of a Proposed Action Under the NIH Guidelines,' *Federal Register* 74: 9411–9421 4 March.

National Institutes of Health (2011) 'NIH Guidelines for Research involving Recombinant DNA Molecules (NIH Guidelines),' http://oba.od.nih.gov/oba/rac/Guidelines/NIH_Guidelines.htm, accessed 5 June 2012.

National Research Council (2010) *Sequence-Based Classification of Select Agents: A Brighter Line* (Washington, DC: National Academies Press).

National Research Council, Committee on Genomics Databases for Bioterrorism Threat Agents (2004) *Seeking Security: Pathogens, Open Access, and Genome Databases* (Washington, DC: National Academies Press).

National Science Advisory Board for Biosecurity (2006) *Addressing Biosecurity Concerns Related to the Synthesis of Select Agents* (Bethesda, MD: National Institutes of Health), p. 4.

National Science Advisory Board for Biosecurity (2008) *Strategic Plan for Outreach and Education on Dual Use Issues* (Bethesda, MD: National Institutes of Health).

R. V. Noorden (2012) 'Butanol Hits the Biofuels Big-Time,' *Nature News Blog,* 25 May, http://blogs.nature.com/news/2012/05/butanol-hits-the-biofuels-big-time.html, accessed 5 June 2012.

Presidential Commission for the Study of Bioethical Issues (2010) *New Directions: The Ethics of Synthetic Biology and Emerging Technologies* (PCSBI, Washington, DC).

B. Rappert, M. Chevrier, and M. Dando (2006) 'In-depth Implementation of the BTWC: Education and Outreach,' *Bradford Review Conference Paper,* http://www.brad.ac.uk/acad/sbtwc/briefing/RCP_18.pdf, accessed 2 June 2012.

D-K Ro, E. M. Paradise, M. Oullet, K. J. Fisher, K. L. Newman, J. M. Ndungu, K. A. Ho, R. A. Eachus, T. S. Ham, J. Kirby, M. C. Y. Chang, S. T. Withers, Y. Shiba, R. Sarpong, and J. D. Keasling (2006) 'Production of the Antimalarial Drug Precursor Artemisinic Acid in Engineered Yeast', *Nature* 440, 940–943.

M. Rodemeyer (2009) *New Life, Old Bottles: Regulating First-Generation Products of Synthetic Biology,* Woodrow Wilson International Center for Scholars (Washington, DC).

H. O. Smith, C. A. Hutchinson III, C. Pfannkoch, and J. C. Venter (2003) 'Generating a Synthetic Genome by Whole Genome Assembly: phiX174 Bacteriophage from Synthetic Oligonucleotides,' *Proceedings of the National Academy of Sciences* 100, 15440–15445.

J. B. Tucker (2003) 'Preventing the Misuse of Pathogens: The Need for Global Biosecurity Standards,' *Arms Control Today,* June issue (available at http://www.armscontrol.org/act/2003_06/tucker_june03).

T. M. Tumpey, C. F. Basler, P. V. Aguilar, H. Zeng, A. Solorzano, D. E. Swayne, N. J. Cox, J. M. Katz, J. K. Taubenberger, P. Palese, and A. Garcia-Sastre (2005) 'Characterization of the Reconstructed 1918 Spanish Influenza Pandemic Virus,' *Science* 310: 77–80.

US Department of Health and Human Services (2009) *Biosafety in Microbiological and Biomedical Laboratories, 5th edition,* HHS Publication CDC 21–1112 (Washington, DC).

US Department of Health and Human Services (2010) *Screening Framework Guidance for Providers of Synthetic Double-Stranded DNA* (Assistant Secretary for Preparedness and Response, Washington, DC).

United States Government (1970) *Occupational Safety and Health Act of 1970,* 29 U.S.C. §651 et seq.

United States Government (2001) *Uniting and Strengthening America by Providing Appropriate Tools Required to Intercept and Obstruct Terrorism Act of 2001* (USA PATRIOT Act), Pub. L. No. 107–56, Oct. 12.

H. Wolinsky (2009) 'Kitchen Biology: The Rise of Do-It-Yourself Biology Democratizes Science, but is it Dangerous to Public Health and the Environment?' *EMBO Reports* 10: 683–685.

DISCUSSION QUESTIONS

1. Do the risks associated with garage or do-it-yourself synthetic biology justify hard laws that would curtail the liberty or autonomy of individuals to engage in biotechnology research and development on their own? If so, what are some examples of laws or regulations that might be developed?

2. Are the approaches to responsible development of synthetic biology described in this chapter adequate to protect human and environmental health? Is this a case where proactive responsible development is being accomplished?

3. Given the potential benefits of synthetic biology, should widespread innovation be encouraged? If so, how might this be accomplished?

4. Who – e.g. government officials, researchers, public interest groups, or citizens – do you think should be included on committees, like Institutional Biosafety Committees, that evaluate synthetic biology and make recommendations on governance?

5. What might be some effective ways to increase education about dual-use technologies for synthetic biology researchers?

6. What kinds of soft laws do you think have the best chance of encouraging scientists and others to pay better attention to the risk associated with their research?

7. Under what conditions (if any) is it permissible to prevent publication of scientific research? Does doing so violate individual liberties of speech and expression?

36 EVOLUTION AND THE DEEP PAST: INTRINSIC RESPONSES TO SYNTHETIC BIOLOGY

Christopher J. Preston

CHAPTER SUMMARY

Because synthetic biology involves creating novel forms of life using genomic material drawn from multiply species of organisms it raises several intrinsic ethical objections (i.e. concerns based on the technology itself, rather than its impacts) having to do with crossing species boundaries and "play God." In this chapter, Christopher J. Preston argues that intrinsic concerns regarding synthetic biology should not be treated as lightly as they often are by proponents of the technology, since they are quite widely held. He then discusses several different types of intrinsic objections. He argues that some of them, including those based on the sanctity of species boundaries, are not very well justified. However, he believes that there may be a compelling argument to be made against synthetic organisms based on the fact that they are constructed *de novo*, rather than produced by descent with modification from prior organisms. Therefore, unlike all other organisms, they do not have a continuous causal connection to historical evolutionary processes.

RELATED READINGS

Introduction: Intrinsic Concerns (2.7); Ethics and Public Policy (2.9); Types of Theories (3.2)

Other Chapters: Jason Robert and Francoise Baylis, *Crossing Species Boundaries* (Ch. 10); John Basl, *What to do about Artificial Consciousness* (Ch. 25); Gary Comstock, *Ethics and Genetically Modified Foods* (Ch. 31); Mark A. Bedau and Mark Triant, *Social and Ethical Implications of Creating Artificial Cells* (Ch. 37)

1. INTRODUCTION

Recent innovations in biotechnology challenge some of our most basic understandings of organismic life. Reactions to these technologies run the gamut from an unbridled excitement generated by contemplating the potential goods that might result to a deep pessimism about what this particular breakthrough means for our human future. The technologies falling under the broad label of "synthetic biology" cause responses of both these kinds. On the one hand, there is the promise of massive benefits offered by deliberately designed synthetic bacteria. These synthetic organisms could contribute to new types of bio-based energy, more effective remediation of contaminated sites, powerful and novel pharmaceuticals and delivery methods, innovative bio-based computing, and climate-saving carbon dioxide remediation. On the other hand, there looms the specter of unknown environmental harms, unanticipated synthetic pathogens, devastating terrorist abuses, and startling changes to our Darwinian understanding of life. Underneath both the hype and the hysteria are a set of serious ethical issues that warrant thoughtful discussion before embarking on any large-scale production and use of synthetic organisms.

In the earlier debate over the ethics of genetically modified organisms (GMOs), it was common to make a distinction between 'extrinsic' and 'intrinsic' objections to the technology. Extrinsic objections "focus on the potential harms consequent upon … adoption" (Comstock, 2010, p. 53). A number of possible negative effects, including the disruption of ecosystems, human health impacts, and socio-economic changes were seen as potential results from the widespread use of the GMOs. These types of harms are *consequences* of the deployment of the technology, thereby falling under a broadly utilitarian (or consequentialist) philosophical framework. With extrinsic concerns, careful study and observation can often determine whether these projected harms will in fact materialize. The appearance of any negative consequences then has to be judged in relation to the benefits the technology offers.

Intrinsic objections, by contrast, are based on the assumption that "the process [of their production] … is objectionable *in itself*" (Comstock, 2010, p. 53). This means that, whatever the consequences – good or bad – of the widespread deployment of GMOs, there is said to be an ethical problem inherent in the very act of creating the modified organism. Often this problem is characterized as stemming from the supposed *unnaturalness* of the process, its manufacture involving crossing some intangible line that nature itself is alleged never to cross. Theologically minded critics might call these types of responses "playing God" objections. It is important to notice, however, that no theology is required to critique a technology for departing from anything that had previously appeared in nature. Since these objections usually rest on either a principle-based objection or a worry about the kind of attitude the technology displays, they tend to fall under either a Kantian or a virtue-based philosophical framework in ethics.

Both intrinsic and extrinsic arguments can be found throughout the literature on the ethics of GMOs. Arguably the extrinsic arguments are the most significant when it comes to making the policy that governs the production and use of these organisms. Governments are charged with keeping their citizens safe from harmful consequences. Often agencies exist that are tasked specifically with watching out for human and environmental health, ensuring that a new technology does not disproportionately harm any single group, checking that the technology won't negatively impact the nation's economic structure, or taking preventative measures so that a technology does not fall into the wrong hands. Harm arguments are taken seriously because the stakes directly involve the well-being of flesh and blood individuals. This contrasts with intrinsic arguments where the stakes appear to be merely challenges to certain religious or ethical principles. The resulting temptation to take the intrinsic arguments less seriously has been bolstered by a number of philosophers arguing that the intrinsic arguments don't even stand up to sustained scrutiny (Thompson, 2003; Rollin, 2005). As a result, intrinsic arguments have often remained the province of philosophy classrooms and after-dinner conversations, rather than the stuff of national policy.[1]

There is, however, a strong reason not to simply assume that intrinsic arguments can be ignored (Streiffer and Hedemann, 2005). Intrinsic and principle-based arguments are proven to enjoy widespread public appeal. Social science research on GMO's suggests that intrinsic concerns outweigh extrinsic concerns for large segments of the population. Gaskell (1997) found that "moral doubts act as a veto irrespective of people's views on use and risk" (p. 845). In a similar vein, a PEW study on public attitudes toward biotechnology and food found that, among those uncomfortable with animal cloning, significantly more are uncomfortable based on "religious and ethical" reasons (36%) than based on "safety" reasons (23%) (PEW, 2005, pp. 6–7). A comprehensive US Office of Technology Assessment study likewise found that consequentialist (extrinsic) concerns lagged far behind "tampering with nature" and "playing God" concerns with transgenic organisms in the public's mind (OTA, 1987). In other words, more of the public seem to be bothered by the intrinsic concerns surrounding certain forms of biotechnology than the extrinsic ones. Given that the products of synthetic biology in many cases appear to depart even further from natural processes than the products of traditional biotechnology, one might expect a broadly similar principle-based public reaction to these new technologies.[2]

While an appeal to the power of public sentiment is at best only a political consideration rather than a philosophical argument, one can't help but wonder whether these public reactions are based on legitimate concerns (Deckers, 2005). In the case of synthetic biology, pioneers in the field have fueled the fire behind the intrinsic line of thinking by making some dramatic claims about the new ground they are breaking. Craig Venter, for example, when announcing his lab's fabrication of the world's first entirely synthetic but fully functional bacterial genome, described the resulting organism as "the first self-replicating species we have had on the planet whose parent is a computer," adding that this was as much a philosophical breakthrough as a technical one (Wade, 2010). George Whitesides has talked of the "marvelous challenge" of seeing if humans can "outdesign evolution" (Gibbs, 2004). Talk of "computer based life-forms" and "outdesigning evolution" certainly suggest the idea of some new line being crossed. Rather than allowing practitioners of these technologies to simply dismiss the public's concerns as ill-informed or reactionary,[3] the principle-based objection seem worthy of investigation.

---

[1] Notable exceptions to this generalization are human stem cell research and human cloning. In these two areas, some national (and international) policies have been based on religious or ethical principles rather than projected harms. See the United Nations Declaration on Human Cloning (2005) and U.S. Executive Order 13435 (2007).

[2] In a 2010 report on public perceptions of synthetic biology, respondents were asked to choose their top priority from a list of possible concerns about synthetic biology. The percentage of those who ranked moral – or principle-based – concerns first (25%) exceeded those who were more worried about human health (23%) or environmental safety (13%). Those who prioritized moral concerns fell only just short of those who ranked bioterrorism first (27%) (Hart Research Associates, 2010).

2. SOME BACKGROUND

One of the most noteworthy achievements in synthetic biology to date has been the success of Jay Keasling's Berkeley lab at altering a yeast cell to produce artemisinic acid, a chemical precursor to artemisinin (Ro, 2006). Prior to this breakthrough, artemisinin, an effective drug used in life-saving anti-malarial treatments, had to be painstakingly and expensively derived from extracts of the wormwood plant (*Artemisia annua*). Keasling's lab worked out how to produce small quantities of the enzyme amorphadiene – a precursor to artemisinin – in an *Escherichia coli* bacterium by adding a gene from the wormwood plant. To scale up and make the process more efficient, they recreated a similar pathway in yeast (*Saccharomyces cerevisiae*). By adding the gene from the wormwood plant to the yeast and regulating a number of the yeast's own genes controlling its mevalonate pathway, they were able to dramatically increase the production of amorphadiene over what had been achieved in the *E. coli*. The researchers then went back to the wormwood plant and identified the gene responsible for the enzymes that oxidize amorphadiene into artemesinic acid. They did this by comparing wormwood genes to a gene suspected of performing a similar function in the sunflower and lettuce plants (which belong to the same family). After adding the required gene to the engineered yeast, the yeast started producing artemisinic acid, which could then be retrieved from the cell through washing and purification. The result was more artemisinic acid per unit of mass of the host organism in 4–5 days than the amount produced in several months by wormwood plant, leading to production of artemisinic acid "nearly two orders of magnitude greater than *A. annua*" (Ro 2006, p. 942).

The form of manipulation employed by the Keasling group is known as "metabolic engineering." The artemisinin is only "semi-synthetic" because the engineered yeast does not itself complete the job. After the artemisinic acid has been produced by the synthetically created pathway in the yeast cell, its transformation into useful derivatives takes place through more traditional purification and chemical conversion processes. Even though scaling up the production of artemisinic acid while keeping the costs down continues to be challenging, introduction of the first semi-synthetically produced artemisinin into the supply chain is expected in 2012. Such an improvement in sourcing arteminisic acid promises huge benefits in the global fight against malaria.

The achievements in the Keasling lab go far beyond simply adding a foreign gene to the genome of an existing organism like cotton or soybeans to have them express a desirable trait. For the production of artemisinic acid, a metabolic pathway is engineered within an organism: one type of gene from a different organism is inserted to produce an enzyme, a second gene performing a different function is inserted to oxidize that enzyme. All of this depends on transgenic technology developed in a third organism (the *E. coli*) and comparative genetic analysis performed on two more (the sunflower and the lettuce). The goal of the technology is not simply to alter a trait of an existing organism. It is to construct

---

[3] Such dismissals are (regrettably) all too common among advocates of synthetic biology. Drew Endy, for example, dismissed "playing God" worries as "superficial and embarrassingly simple" (2008).

a biological factory out of genetic parts from various organisms and locate it in a microbial host. Any of the objections to GMOs based on "playing God," "crossing species lines," and "departing from nature" would seem to apply in even more dramatic form to the activities of the Keasling engineers.

A second notable achievement in synthetic biology has been the synthesis of an entire genome entirely from laboratory chemicals. This genome has been inserted into a bacterial host and the new, synthetic genome has proven capable of taking over the command and control mechanisms of that host. Using techniques perfected during the mapping of the human genome, J. Craig Venter and his research team first mapped and then synthesized from scratch the genome of *Mycoplasma genitalium*, a tiny bacterium living in the human urinary tract and possessing the relatively small number of 482 protein-coding genes. The enormous task of stitching together the genetic material in a precise sequence was achieved with the assistance of yeast cells (*Saccharomyces cerevisiae*) acting as temporary hosts to the gene fragments.

The Venter team had earlier mastered the task of inserting a (non-synthetic) genome into a host cell and having the inserted material take over (Latrigue, 2007). To create the first synthetic cell they had to perform the same trick with this newly synthesized genome. The team found themselves forced to switch from the *Mycoplasma genitalium* genome to the larger *Mycoplasma mycoides* genome because of the advantages offered by its much faster reproduction, in the process increasing the gene sequence that had to be synthesized from half a million to a million base pairs. What remained was the challenge of transplanting the synthesized genome into the *M. capricolum* host and "booting up" the cell so that it ran off the inserted DNA.

To protect against undesirable horizontal transfer, numerous restriction systems are present in bacterial cells to fight foreign DNA. After finding ways to circumvent each of the obstacles, success was finally achieved in May of 2010 (Gibson, 2010). The new organism, which Venter called in a subsequent press conference "the world's first synthetic cell" was named by the team *M. mycoides* JCVI-syn1.0. In order that its progeny could be distinguished from naturally occurring *M. mycoides* bacteria, the researchers encoded several 'watermarks' in a non-active part of its genome. They included, with some deliberate flair, a web address for the new organism and the James Joyce quote "[t]o live, to err, to fall, to triumph, to recreate life out of life." Venter claimed that booting up the synthesized genome of *M. mycoides* JCVI-syn1.0 was a "giant philosophical leap" in terms of how we view life.

Like the Keasling achievements, the Venter achievements go far beyond those of traditional forms of genetic modification. In the case of *M. mycoides* JCVI-syn1.0 there is no existing genome that has been "modified." Rather, a genome has been synthesized from scratch using laboratory chemicals (albeit a genome modeled on that of an existing organism). After the successful insertion of the synthesized genetic material into the host cell, an organism exists that had not existed before. Moreover, this species is already reproducing through binary fission in the Venter labs. With further research on how to minimize the size of the genome required, synthesized bacterial organisms such as these might in the future be used as the basis for biological factories or production systems (such as Keasling's), performing highly valuable functions for their human designers.[4]

3. CROSSING SPECIES LINES

The claim there is something inherently wrong with crossing species lines has been widespread and popular in the literature on the ethics of GMOs (Rifkin, 1985; Robert and Bayliss, 2003). It seems probable that a moral case against synthetic biology on intrinsic grounds might begin in a similar fashion. Part of the driving force behind synthetic biology is the idea that biotechnology should be more like engineering with interchangeable functionally useful parts ("biobricks") being plugged into a minimal microbial host chassis. This interchangeability suggests that the technology as a whole pays little regard to existing species lines. Functional gene sequences, discovered through the study of any number of existing species will be mixed and matched to produce desirable properties in synthetically produced bacterial hosts.

While this "crossing species lines" objection has often had a wide *prima facie* appeal, it has been relentlessly attacked by both scientists and ethicists. The challenges are often made on the grounds that species boundaries are inherently dynamic and fluid conventions more designed to serve taxanomic preferences than map any alleged "natural essence" of a species (Rollin 1995). Darwin himself acknowledged "I look at the term species, as one arbitrarily given for the sake of convenience to a set of individuals closely resembling each other" (Darwin, 1898, p. 66). Such an arbitrary tool of convenience seems like an improbable basis for an ethical line in the sand. Furthermore, there is little consensus among biologists about how to define species in the first place, with R.L. Mayden, for example, suggesting there are 22 different species concepts operable among biologists (1995). As a result of this fluidity, contemporary bio-ethicist Gary Comstock insists "[T]o proscribe the crossing of species borders on the grounds that it is unnatural seems scientifically indefensible" (2010, p. 56).

A second challenge to the "crossing species lines" objection seeks to cast doubt on the idea that genetic modification crosses any *new* lines of moral significance. If "going against nature" were a reason to reject GMOs, then many techniques that are already well-accepted would also have to be rejected (Comstock 1998). Opponents of genetic modification sometimes have a tough time showing how this new technology differs from other technologies that appear to meet with public approval. For thousands of years, the biologically inevitable process of descent through modification has been gently nudged in certain directions by industrious humans manipulating reproductive processes for their own agricultural and pecuniary purposes. Plant breeding and animal husbandry long ago showed that natural selection is not the only force giving shape to earth's species. Selection guided by human influence has been responsible for creating numerous

---

[4] Venter has suggested that the world's first trillionaire will be the designer of marketable synthetic organisms. Research into biofuel production using synthetic organisms is currently generating considerable attention.

breeds of domestic animals, flowers, and food crops with characteristics chosen to satisfy human needs and interests. Defenders of genetic modification suggest that artificial selection through recombinant DNA technology differs little from these long accepted practices except perhaps in rate and scope, thereby deflating the power of the objection.

Whether or not these two rejections of the "crossing species lines" worry are completely philosophically sound, as a practical matter, any residual concern about crossing species lines is highly likely to decline as organisms with manipulated genomes proliferate in common usage. Genetically modified plants continue to be grown on ever-increasing acreages. They currently occupy more than 160 million hectares (equivalent to 10% of the world's farmland) (ISAAA, 2011). The horse appears to have already bolted. At some point, the whole line of objection based on crossing species boundaries is likely to become (literally) passé.

4. CAUSAL CONTINUITY WITH THE PAST

A different sort of objection to synthetic biology appeals not to the alleged wrongness of disregarding some scientifically questionable, natural species essence. It appeals instead to a criterion of natural value much more broadly accepted throughout environmental ethics. In a much-read article on the topic of environmental restoration, Robert Elliot suggested that environmental restorations can be questioned on the grounds that they are an attempt to "fake nature" (Elliot, 1982). Making an analogy with forged works of art, Elliot suggested that there is less value in a restoration than the original landscape even if the restoration is functionally equivalent to the previous ecosystem and indistinguishable even to the expert eye. The reason for the diminution of value, according to Elliot, is the break in "causal continuity with the past" (p, 87). One aspect of nature's value that a restoration cannot capture is "to do with its genesis, its history" (p. 87).

Whether or not one finds the analogy between environmental restorations and artistic forgeries compelling, the idea that causal continuity with the historical past confers some form of value onto an organism is an idea deeply embedded into environmental consciousness. Aldo Leopold, for example, often considered the first modern environmental ethicist, thought hard about what lay behind the significance of cranes returning to a crane marsh.

> Their annual return is the ticking of the geologic clock. Upon the place of their return they confer a peculiar distinction. Amid the endless mediocrity of the commonplace, a crane marsh holds a paleontological patent of nobility, won in the march of aeons … (Leopold, 1949, p. 97)

The paleontological patent of nobility is worn by the crane because her species has been performing the same ritual over countless millennia. The forms of many of earth's current biota far predate the arrival of humans and for that reason people often find themselves feeling a sense of awe and respect for them. Many of our own most important characteristics also have roots in this distant evolutionary past. It is the connection to this biological history that made the crane marsh so significant to Leopold. Leopold's insight is the heart of an approach common throughout environmental ethics.

This intuition about the moral significance of a connection to earth's geological and biological history appears in numerous places. In addition to Leopold and Elliot, Holmes Rolston, III, the so-called "father of modern environmental ethics," has spoken lyrically of the significance of geology he witnessed on an Appalachian hike.

> "Out of the lithosphere: atmosphere, hydrosphere, and biosphere. Earth's carbonate and apatite have graced me with the carbon, calcium, and phosphate that support my frame....Those stains of limonite and hematite now coloring this weathered cut will tomorrow be the hemoglobin that flushes my face with red... Here is my cradle. My soul is hidden in the cleft of this rock" (Rolston 1971, pp. 79–80).

In later writing, he suggested that part of the reason we protect wildlands is that they provide "the profoundest historical museum of all, a relic of the way the world was during 99.9% of past time' (Rolston, 1988: 14). "To travel into the wilderness is to go to our aboriginal source... it is by homecoming to enjoy an essential reunion with the earth" (Rolston, 1975, p. 122). For Rolston, as for Leopold and Elliot, it is the causal relationship of wild landscapes to processes embedded deep in earth's history that confers moral significance upon them.

Similarly, Eugene Hargrove, promoting an entirely different kind of environmental ethic from Leopold and Rolston based on the human appreciation of natural beauty, asks that earth's historical geologic and biological processes be left alone to continue creating the forms humans find beautiful. "When we interfere with nature," says Hargrove, "....we create a break in [that] natural history" (1996, p. 195). In yet another kind of environmental ethic, Whitney Stanford and Vandana Shiva recommend an "alternate narrative" in which "evolution offers the roles, narrative and scope to develop sustainable relationships with the biotic community" (2011, p. 204). They quote Leslie Paul Thiel's claim that evolution is "perhaps the only story sufficiently grand is scope, robust in fact, and rich in metaphor to aid us in resolving our ecological concerns today" (p. 204). For all of the above thinkers, an appeal to the significance of the past – and particularly the evolutionary past – appears to play a grounding role in generating an environmental ethic.

5. SYNTHETIC BIOLOGY AND HISTORICAL PROCESSES[5]

Despite the existence of the growing number of modified genomes in agriculture, there is a significant sense in which historical evolutionary processes are uniquely entwined with the genetic identity of every living form. Up to this point in history it has always been possible to say something reassuringly Darwinian about every organism on Earth. Every living thing, from a thermophilic bacterium in a Yellowstone hot spring, to a lately discovered antelope in the Vietnamese forest, to the person reading this chapter, has always had an overwhelming percentage of

---

[5] The argument that follows is developed at greater length in Preston (2008).

their DNA causally connected to the past. All genomes have the majority of their nucleotides handed down by ancestors whose own ancestors stretched back into the recesses of evolutionary time. The three and a half billion years of operation of Darwinian principles have served as a deep time anchor for all of life on earth.

Synthetic biology practiced in the Venter lab for the first time breaks this causal continuity. No part of the genetic sequence in *M. mycoides* JCVI-Syn1.0 is inherited. Though *modeled* on the genetic code of a naturally occurring organism, the synthetic organism has no physical ancestors. The particular genome inserted into the bacterial host has undergone no descent with modification since it was synthesized entirely in a lab from constituent chemicals. There is no physical, causal link between the synthesized genome and any natural genome. Clyde Hutchinson, one of JCVI's researchers, highlighted this difference when he stated in praise of the lab's achievement, "[t]o me the most remarkable thing about our synthetic cell is that its genome was designed in the computer and brought to life through chemical synthesis, without using any pieces of natural DNA" (JVCI, 2010). In the creation of these synthetic genomes, Darwinian processes are playing no role. While the genome of *M. mycoides* JCVI Syn1.0 is *inspired* by a naturally evolved organism, it is not connected to one through any physical, causal history. Organisms built through a combination of a synthetically produced minimal genome and inserted biobricks will take even less inspiration from naturally evolved organisms than *M. mycoides* JCVI Syn1.0. For the first time, Venter says, humans will look out on the living world and find an organism whose DNA is forged "not by Darwinian evolution but created by human intelligence" (JVCI, 2010). This is not a *natural* but a *human* version of biological life. As Wayt Gibbs put it in his Scientific American article in 2004, "[T]hink of it as Life, version 2.0." If one takes the perspective of the environmental thinkers who value causal continuity with the past, this new version is likely to diminish the value of the organism.

6. SYNTHETIC BIOLOGY VS GENETIC MODIFICATION

The difference between this principle-based position and the position of those who oppose genetically modified organisms for crossing-species lines is striking. Genetically modified organisms (GMOs), even though they might cross species lines, retain a continuous genetic causal history connected to the deep past. As the name implies, GMOs contain only modifications of existing genomes, modifications typically affecting far less than 0.1% of the total number of genes in any organism. These modified organisms still contain the majority of their genetic material born entirely of earth's long history, and they have it both in the 99.9% of the genome that has not been modified and in the 0.1% that has, since the foreign gene is itself a product of its own evolutionary history (albeit originally in a different organism). A deletion here, an insertion there, while certainly introducing human design and changing how an organism looks and behaves in sometimes significant ways, does not alter the fundamental fact of genetic inheritance from an evolutionary past. Carefully bred long-haired Pekingese dogs still have progenitors linked to *Canis lupus* and beyond. All of the so-called 'frankenfoods' remain causally connected to the history of life on earth, with the vast majority of its genes, inherited and inserted, traceable to ancient ancestors. Before Venter's type of synthetic biology, every organism had ancestors connecting it to the historical processes environmentalists value.

Synthetic biology does not start with a viable genome and modify it. It starts afresh, engineering life *de novo*. This is something entirely unprecedented. "Products of traditional gene technology are perceived as modifications of their natural ancestors, a version of the natural organism with some altered features or capabilities," states Deplazes-Zemp. "This is different for synthetic organisms" (2012, p. 69). The difference is dramatic. Boldt and Müller have described this change as a move "from manipulation to creation" (2010, p. 387). In synthetic biology, they continue, the goal is "not to amend an organism … [but] to create a new form of life" (p. 388), adding that this creates a "significant ethical concern" distinct from concerns about risks and harms.

It is important to note that this alleged value of non-synthetic organisms over synthetic ones makes no specific mention of species lines. There is no mysterious – and scientifically suspicious – natural essence that requires protecting. There is instead a concrete, historical causal process, one widely viewed as a source of value, being sidestepped. It is the fact that a natural genome has a physical connection to an historical past that confers value upon it. Any person who has traced their own genealogy knows the significance that can be placed in causal genetic connections.

7. A FEW CAVEATS

Even though this criterion of causal continuity with past life does seem to separate many of the products of synthetic biology from other biological achievements on principle, there are a number of considerations to take into account before deciding how this should bear on the synthetic biology industry.

The first consideration is that the idea synthetic organisms lack a certain type of value will obviously only sound plausible to those who share the view that a connection to evolutionary history is important. While this view is widespread throughout the literature in environmental ethics, it could be faulted for having a retrospective, romantic flavor to it. Others would find less significance in this connection to evolutionary history and some ardent futurists might find no significance in the connection at all.

The commitment to the significance of evolutionary processes is, however, not easy to surrender in the context in which many are advocating for synthetic organisms. One of the chief arguments in favor of synthetic organisms is based on their potential environmental benefits. But neither contemporary ecology, nor reflections on our connection to other life forms, nor warnings about climate change and species extinction, nor worries about the future prospects for the shape of ecosystems in an environmentally stressed world make any sense at all without recognition of the significance of evolution to those threatened values. Even a thoroughgoing anthropocentric environmental ethics based solely on human interests is forced to concede that all human values were made possible by our own evolutionary history. Without that history, there would be no human values in the first place. Dismissing the significance of the history is certainly logically possible, but one might argue that it is an attitude that demonstrates a distinct lack of perspective (Rolston, 1988).

A second observation is that even if one admits that the causal connections to the evolutionary past are important, there are other considerations to be invoked. Primary among these are the instrumental values to health, environmental remediation, and energy production that synthetic organisms promise. These are legitimate and important motivations for pursuing the development of synthetic organisms. As Jan Deckers observed in social science research about genetically modified (GM) organisms "GM was perceived as an illegitimate infringement of humans on the natural order, *unless participants thought that serious human needs could be satisfied* by changing the natural order through GM ..." [emphasis added] (Deckers, 2005, p. 469). Certain types of synthetic biology – though probably not all – may provide enough benefits that any moral squeamishness should be overcome.

Furthermore, the argument made above suggests only that a synthetic organism has less natural value than a naturally occurring one but does not suggest that such an organism has zero or negative value. One might be able to find other types of value inhering in synthetic organisms. Sandler and Simon note that synthetic organisms may possess value on the basis that they are creative products of human endeavor and organisms that have a good of their own (2012). Synthetic organisms might possess some value on the grounds that they are "teleological centers of life" (Taylor, 1986, Deplazes-Zemp, 2012). These potential positive values inhering in synthetic organisms will need to be balanced against whatever diminution of value is caused by their being disconnected from evolutionary history.

A fourth observation is that despite the heady talk of "outdesigning evolution" and a "second genesis" in which life is created by humans with their computers and bottles of chemicals, synthetic biology does not mean the death of Darwinism. Neo-Darwinian principles describe how genomes will mutate and organisms will be selected for their fit in a certain environments and their ability to successfully reproduce. These explanations will still be operative in a world filled with synthetic organisms. The fact that synthetic bacteria such as *M. mycoides* JCVI-syn1.0 can reproduce means that mutations will continue to occur and organisms will continue to evolve. If allowed by lab technicians, *M. mycoides* JCVI-syn1.0 will itself evolve according to Darwinian principles.[6]

Other questions crop up. Some suggest the emphasis on Darwinism and the process of descent through modification is misguided given that a great deal of DNA has been transferred between bacterial organisms throughout history by means of horizontal (or lateral) transfer (i.e. without descent).[7] The Keasling achievements may not even depart from the causal connection to evolutionary history

---

[6] This feature of synthetic organisms supplies one of the extrinsic arguments that has been used against the wisdom of their creation, the worry that harmful organisms will escape from the lab and evolve to cause unanticipated harms. This possibility shows that Darwinism is still very much operative in a world of synthetic organisms.

[7] We might note that the horizontal transfer of genetic material between bacteria, however prevalent through evolutionary history, has never resulted in the creation of an organism from scratch.

because – at the present state of the technology – no organism is created entirely *de novo*. Rather a metabolic pathway is synthesized through genetic insertions into a host organism. The type of synthetic biology practiced by Keasling might therefore raise fewer moral issues than the type practiced by Venter (something that appears to correspond with the amount of press coverage each achievement received). One might also note that even the Venter lab has created nothing entirely *de novo* since the synthesized genome, at this stage of the technology, has to be placed in a naturally occurring bacterial cell in order to function.

Given these complexities, it is unlikely that the position articulated here about natural historical value is going to be decisive either for or against proceeding with the large-scale production of useful synthetic organisms. Synthetic biology may yet turn out to be a case like human cloning where intrinsic arguments against the technology are deemed more important than the promised benefits. On the other hand, it may turn out to be like GMOs where the extrinsic arguments slowed the technology down temporarily at the same time as the intrinsic ones were ultimately judged to lack power. More philosophical work, public discussion, and informed political debate awaits.

The work that lies ahead does not, however, diminish the significance of the search for the principles that distinguish synthetic biology from past biotechnologies. The job of ethicists is sometimes simply to provide an alert when a significant line in the sand is being crossed. At times, the line in the sand will turn out to be something worth fighting for. At others, it may only be looked at wistfully as it passes beneath us into history. On occasions its passing may even be regarded as a cause for celebration. What is clear in the case of synthetic biology is that, given the existing weight we tend to place on Darwinism and on the connections between today's world and historical processes originating in deep time, the creation of the first life forms that are causally disconnected from that history is a significant milestone, one that needs to be identified, acknowledged, and scrutinized. We all have a stake in whether to resist, embrace, or simply salute this milestone.

WORKS CITED

J. Boldt and O. Müller (2008) 'Newtons of the Leaves of Grass' *Nature Biotechnology* 26, 4, 387–389.

G. Comstock (1998) 'Is it Unnatural to Genetically Engineer Plants?' *Weed Science* 46, 647–651.

G. Comstock (2010) 'Ethics and Genetically Modified Food' in F. Gottwald, ed. *Food Ethics* (Dordrecht: Springer), pp. 49–66.

C. Darwin (1898) *The Origin of Species, By Means of Natural Selection*, Vol. 1. (New York: Appleton and Company).

J. Deckers (2005) 'Are Scientists Right and Non-scientists wrong? Reflections on Discussions of GM', *Journal of Agricultural and Environmental Ethics*, 18, 451–478

A. Deplazes-Zemp (2012) 'The Moral Impact of Synthesising Living Organisms: Biocentric Views on Synthetic Biology', *Environmental Values*, 21, 63–82

R. Elliot (1982) 'Faking Nature', *Inquiry*, 25, 1, 81–93.

D. Endy (2008, February 19) 'Life, What a Concept (Part III)' *Edge* 237 http://www.edge.org/documents/archive/edge237.html, date accessed 8 March 2012.

G. Gaskell (1997) 'Europe Ambivalent on Biotechnology', *Nature*, vol. 387, 26 June: 845–847.

W. Gibbs (2004, April 26) 'Synthetic Life', *Scientific American*. http://www.scientificamerican.com/article.cfm?id=synthetic-life, date accessed 8 March 2012.

D. Gibson, J. Glass, C. Lartigue, V. Noskov, R. Chuang, M. Algire…. J. Venter (2010) 'Creation of a Bacterial Cell Controlled by a Chemically Synthesized Genome', *Science* 329, 5987, 52–56.

E. Hargrove (1996) *The Foundations of Environmental Ethics* (Denton, TX: Environmental Ethics Books).

Hart Research Associates (2010) *Awareness and Impressions of Synthetic Biology.* http://www.synbioproject.org/process/assets/files/6456/hart_revised_.pdf?, date accessed 14 April 2012.

ISAAA (International Service for the Acquisition of Agri-Biotech Applications) (2011) 'Global Status of the World's Biotech/GM Crops', http://www.isaaa.org/resources/publications/briefs/43/default.asp, date accessed 4 March 2012.

JVCI (J Craig Venter Institute) (2010, May 10) 'First Self-Replicating Synthetic Bacterial Cell', Press Release, http://www.jcvi.org/cms/press/press-releases/full-text/article/first-self-replicating-synthetic-bacterial-cell-constructed-by-j-craig-venter-institute-researcher, date accessed 8 March 2012.

C. Latrigue, J. Glass, N. Alperovich, R. Pieper, P. Parmar, C. Hutchison III, … J. Venter (2007) 'Genome Transplantation in Bacteria: Changing One Species to Another', *Science*, 317, 5838, 632–638.

R. Mayden (1997) 'A hierarchy of Species Concepts: The Denouement in the Saga of the Species Problem' in M. F. Claridge, H. A. Dawah, and M. R. Wilson (eds.) *Species: The Units of Biodiversity*, (London: Chapman and Hall), 381–424.

OTA (Office of Technology Assessment) (1987) 'New Developments in Biotechnology' Background paper: Public perceptions of biotechnology. OTA-BP-BA-45. Washington: GPO.

PEW (Pew Initiative on Food and Biotechnology) (2006). Public Sentiment about Genetically Modified Food, http://www.pewtrusts.org/uploadedFiles/wwwpewtrustsorg/News/Press_Releases/Food_and_Biotechnology/PIFB_Public_Sentiment_GM_Foods2005.pdf, date accessed 9 March 2012

C. Preston (2008) 'Synthetic Biology: Drawing a Line in Darwin's Sand', *Environmental Values*, 17, 1, 23–39.

J. Rifkin (1985) *Declaration of a Heretic* (Boston and London: Routledge and Kegan Paul).

D. Ro, E. Paradise, M. Ouellet, K. Fisher, K. Newman…J. Keasling (2006) "Production of the Antimalarial Drug Precursor Artemisinic Acid in Engineered Yeast" *Nature*, 440, (13 April), 940–943.

J. Robert and F. Bayliss (2003) 'Crossing Species Boundaries' *American Journal of Bioethics*, 3, 3, 1–13.

B.E. Rollin (1995) *The Frankenstein syndrome: Ethical and social issues in the genetic engineering of animals* (Cambridge, U.K.: Cambridge University Press).

H. Rolston, III. (1971) 'Hewn and Cleft from this Rock', *Main Currents in Modern Thought*, 27, 3, 79–83.

H. Rolston, III. (1975) 'Lake Solitude: The Individual in Wildness', *Main Currents in Modern Thought*, 31, 4, 121–126.

H. Rolston, III. (1988) *Environmental Ethics: Duties to and Values in the Natural World* (Philadelphia: Temple University Press).

R. Sandler and L. Simon. (2012) 'The Value of Artefactual Organisms', *Environmental Values*, 21, 43–61.

W. Sanford and V. Shiva (2011) *Growing Stories from India: Religion and the Fate of Agriculture* (Lexington, KY: University Press of Kentucky).

R. Streiffer and T. Hedemann. (2005) 'The Political Import of Intrinsic Objections to Genetically Engineered Food', *Journal of Agricultural and Environmental Ethics*, 18, 191–210

P. Taylor (1986) Respect *for nature: A theory of environmental ethics*. Princeton, NJ: Princeton University Press.

P. Thompson (1997) *Food Biotechnology in Ethical Perspective* (London, U.K.: Blackie Academic & Professional)

N. Wade (2010) 'Researchers Say They Created a Synthetic Cell' *New York Times*, May 10, http://www.nytimes.com/2010/05/21/science/21cell.html, date accessed 4 February 2012

DISCUSSION QUESTIONS

1. Assume that there is a strong intrinsic objection to the creation of synthetic organisms. How should this be taken into account as we deliberate about policies regarding the development of synthetic organisms? Should it be given more or less (or different) weight in deliberations than extrinsic concerns?

2. What does Preston mean when he says that synthetic organisms are *de novo*? Are they really *de novo* in the ways he suggests?

3. Do you believe that historical evolutionary processes give rise to a certain sort of value, natural value, which cannot be replicated and should be respected? Why or why not?

4. Do you agree with the author that synthetic organisms are ethically different from genetically modified organisms? Why or why not?

5. Imagine that complex synthetic organisms are created, for example, synthetic chimpanzees. Do you think these chimpanzees would differ in value or moral status from naturally evolved chimps? Do we have different obligations to them?

6. Does the fact that artifactual organisms lack natural value imply that they should be opposed? That is to ask, is it correct that the existence of organisms that lack natural value is ethically problematic?

SOCIAL AND ETHICAL IMPLICATIONS OF CREATING ARTIFICIAL CELLS[1]

Mark A. Bedau and Mark Triant

CHAPTER SUMMARY

Artificial cells are microscopic self-organizing and self-replicating autonomous entities built from simple organic and inorganic substances. A number of research efforts are currently aimed at creating artificial cells within the next generation. The ability to create artificial cells could have many social and economic benefits, but it would also pose significant risks and raise several ethical concerns. In this chapter, Mark A. Bedau and Mark Triant first respond to the objections that creating artificial cells would be wrong because it is unnatural, it commoditizes life, it fosters reductionism, or it is playing God. Then they consider two principles for acting in the face of uncertain risks – the Doomsday Principle and the Precautionary Principle – and find them wanting. They end by proposing a new method – which they dub the "Cautious Courage" Principle – for deciding whether and how to develop artificial cells. They believe that their conclusions generalize to analogous debates concerning related new technologies, such as genetic engineering and nanotechnology.

RELATED READINGS

Introduction: EHS (2.6.1); Intrinsic Concerns (2.7); Responsible Development (2.8); Types of Theories (3.2)

Other Chapters: Kevin C. Elliott, *Risk, Precaution, and Nanotechnology* (Ch. 27); Michele Garfinkle and Lori Knowles, *Synthetic Biology, Biosecurity, and Biosafety* (Ch. 35); Christopher J. Preston, *Evolution and the Deep Past* (Ch. 36)

1. INTRODUCTION

Artificial cells are microscopic, self-organizing and self-replicating autonomous entities created from simple organic and inorganic substances.

---

[1] Material in this chapter originally appeared in Mark A. Bedau and Mark Triant (2008) 'Social and Ethical Implications of Artificial Cells,' *The Prospect of Protocells: Social and Ethical Implications of the Recreation of Life*, eds. Bedau and Parke (MIT Press). It appears here by permission of MIT Press.

They are artificial, rather than natural, because they come into existence only through the experimental efforts of human scientists and engineers. Artificial cells embody the essential minimal conditions for genuine life. They grow by harvesting nutrients and energy from their environment and converting it into their component molecules. This metabolic process is modulated by genetic information, typically through the production of catalysts that control metabolism. At a specific point during the growth process the cells can undergo division into two or more similar but not identical daughter cells. Because of this variation, populations of artificial cells are subject to selective pressures and can thereby evolve. These are the fundamental properties of the simplest living organisms. Artificial cells are simpler than existing bacteria, and furthermore exist only because of intentional human activity. At least, this is what protocells *will* be like when they exist, for they do not exist today. But artificial cells are closer than many realize. Virtually all of the basic cellular functions that would be required of artificial cells have already been produced in the laboratory (Rasmussen et al., 2009); all that remains is to integrate them all into one chemical system (Deamer 2009).

The race to create artificial cells is already starting to capture public attention. New companies for creating artificial life forms are being founded in the USA and Europe, and an increasing number of multimillion dollar grants are funding research in the USA, Europe, and Japan. Thus it is no surprise that this activity is attracting notice in the popular media. The increasing pace of breakthroughs in artificial cell science will heighten public interest in their broader implications, particularly since the creation of them raises a number of pressing social and ethical issues. There are ample reasons to believe that artificial cells could benefit human health and the environment and lead to new economic opportunities, but they could simultaneously create new risks to human health and the environment and transgress cultural and moral norms. Being living systems created from nonliving matter, they will be unlike any previous technology and their development will lead society into uncharted waters.

When pictures of Dolly, the Scottish sheep cloned from an adult udder cell, splashed across the front pages of newspapers around the world, society was caught unprepared. President Clinton immediately halted all federally funded cloning research in the United States, and polls revealed that ninety percent of the public favored a ban on human cloning (Silver, 1998). To prevent the future announcement of the first artificial cells from provoking similar knee-jerk reactions, we should start thinking through the social and ethical issues today.

This chapter reviews many of the main strategies for deciding whether to create artificial cells. One set of considerations focuses on intrinsic features of artificial cells. These include the suggestions that creating artificial cells is unnatural, that it treats life as a commodity, that it promotes a mistaken reductionism, and that it is playing God. We find all these considerations unconvincing, for reasons we explain below. The alternative strategies focus on weighing the consequences of creating artificial cells. Utilitarianism and decision theory promise scientifically objective and pragmatic methods for deciding what course to chart. Although we agree that consequences have central importance, we are skeptical whether utilitarianism and decision theory can provide much practical help because the consequences of creating artificial cells are so uncertain. The critical problem is to find some method for choosing the best course of action in the face of this uncertainty. In this setting, some people advocate the doomsday principle which, roughly, counsels against any action that might result in catastrophe, but we explain why we find this principle incoherent. An increasing number of decision makers are turning for guidance in such situations to the precautionary principle, but we explain why this principle is also unattractive. We conclude that making decisions about artificial cells requires being courageous about accepting uncertain risks when warranted by the potential gains.

2. THE INTRINSIC VALUE OF LIFE

Arguments about whether it is right or wrong to develop a new technology can take either of two forms. Extrinsic arguments are driven by the technology's consequences. A technology's consequences often depend on how it is implemented, so extrinsic arguments do not usually produce blanket evaluations of all possible implementations of a technology. Presumably, any decision about creating a new technology should weigh its consequences, perhaps along with other considerations. Evaluating extrinsic approaches to decisions about artificial cells is the subject of the two subsequent sections. In this section, we focus on intrinsic arguments for or against a new technology. Such arguments are driven by the nature of the technology itself, yielding conclusions pertinent to any implementation of it. The advances in biochemical pharmacology of the early twentieth century and more recent developments in genetic engineering and cloning have been criticized on intrinsic grounds, for example. Such criticisms include injunctions against playing God, tampering with forces beyond our control, or violating nature's sanctity; the prospect of creating artificial cells raises many of the same kinds of intrinsic concerns. This section addresses arguments about whether creating artificial life forms is intrinsically objectionable.

Reactions to the prospect of synthesizing new forms of life range from fascination to skepticism and even horror. Everyone should agree that the first artificial cell will herald a scientific and cultural event of great significance, one that will force us to reconsider our place in the cosmos. But what some would hail as a technological milestone, others would decry as a new height of scientific hubris. So it is natural to ask whether this big step would cross some forbidden line. In this section, we examine four kinds of intrinsic objections to the creation of artificial cells, all of which frequently arise in debates over genetic engineering and cloning. These arguments all stem from the notion that life has a certain privileged status and should in some respect remain off limits from human intervention and manipulation.

One objection against creating artificial cells is simply that doing so would be unnatural and, hence, unethical. The force of such arguments depends on what is meant by "unnatural," and why increasing unnaturalness is wrong. At one extreme, one could view all human activity and its products as natural since we are part of the natural world. But then creating artificial cells would be natural, and this objection would have no force. At the other extreme, one could consider all human activity and its products as unnatural, defining the natural as what is independent of human influence. But then the objection would deem all human activities to be unethical, which is absurd. So the objection has force only if "natural" is interpreted in such a way that we can engage in both natural and unnatural acts and the unnatural acts are intuitively wrong. But what could that sense of "natural" be? One might consider it "unnatural" to intervene in the workings of other life forms. But then the unnatural is not in general wrong; far from it. For example, it is surely not immoral to hybridize vegetable species or to engage in animal husbandry. And the stricture against interfering in life forms does not arise in the context of interventions concerning humans, for vaccinating one's children is not generally thought to be wrong. So there is no evident sense of "unnatural" in which artificial cells are unnatural and the unnatural is intrinsically wrong.

Another objection is that to create artificial life forms would lead to the commodification of life, which is immoral (Kass, 2002). Underlying this objection is the notion that living things have a certain sanctity or otherwise demand our respect, and that creating them undermines this respect. The commodification of life is seen as analogous to the commodification of persons, a practice most of us would find appalling. By producing living artifacts, one might argue, we would come to regard life forms as one among our many products, and thus valuable only insofar as they are useful to us. This argument is easy to sympathize with, but is implausible when followed to its conclusion. Life is after all one of our most abundant commodities. Produce, livestock, vaccines, and pets are all examples of life forms that are bought and sold every day. Anyone who objects to the commodification of an artificial single-celled organism should also object to the commodification of a tomato. Furthermore, creating, buying, and selling life forms does not prevent one from respecting those life forms. Family farmers, for example, are often among those with the greatest reverence for life.

The commodification argument reflects a commonly held sentiment that life is special somehow, that it is wrong to treat it with no more respect than we treat the rest of the material world. It can be argued that, though it is not inherently wrong to commodify living things, it is still wrong to create life from nonliving matter because doing so would foster a reductionistic attitude toward life, which undermines the sense of awe, reverence, and respect we owe it (Kass, 2002; Dobson, 1995). This objection does not exactly require that biological reductionism be false, but merely that it be bad for us to view life reductionistically. Of course, it seems somewhat absurd to admit the truth of some form of biological reductionism while advocating an antireductionist worldview on moral grounds. If living things are really irreducible to purely physical systems (at least in some minimal sense), then creating life from nonliving chemicals would, presumably, be impossible, so the argument is moot. By the same token, if living things are reducible to physical systems, it is hard to see why fostering reductionistic beliefs would be unethical. It is by no means obvious that life per se is the type of thing that demands the sense of awe and respect this objection is premised on, but even if we grant that life deserves our reverence, there is no reason to assume that this is incompatible with biological reductionism. Many who study the workings of life in a reductionistic framework come away from the experience with a sense of wonder and an enhanced appreciation and respect for their object of study. Life is no less amazing by virtue of being an elaborate chemical process. In fact, only after we began studying life in naturalistic terms have we come to appreciate how staggeringly complex it really is.

Inevitably, the proponents and eventual creators of artificial cells will have to face up to the accusation that what they are doing is *playing God*. The playing-God argument can be fleshed out in two ways: It could be the observation that creating life from scratch brings new dangers that we simply are not prepared to handle, or it could be the claim that, for whatever reason, creating life from scratch crosses a line that humans simply should never cross. The former construal concerns the potential bad consequences of artificial cells, so it will be discussed in a subsequent section.

If creating artificial cells is crossing some line, we must ask exactly where the line is and why we should not cross it. What exactly would be so horrible about creating new forms of life from scratch? If we set aside the *consequences* of doing this, we are left with little to explain why crossing that line should be forbidden.

The term *playing God* was popularized in the early twentieth century by Christian Scientists in reaction to the variety of advances in medical science taking place at the time. With the help of new surgical techniques, vaccines, antibiotics, and other pharmaceuticals, the human lifespan began to extend, and many fatal or otherwise untreatable ailments could be easily and reliably cured. Christian Scientists opted out of medical treatment on the grounds that it is wrong to "play God" – healing the ill was God's business, not ours. Yet if a person living today were to deny her ailing child medical attention on the grounds that medical science is playing God, we would be rightly appalled. So, if saving a life through modern medicine is playing God, then playing God is morally required.

All of the intrinsic objections to the creation of artificial cells canvassed in this section turn out to be unjustified. So, in the next section we examine extrinsic objections that turn on the *consequences* of artificial cells.

3. EVALUATING THE CONSEQUENCES: DECIDING IN THE DARK

New and revolutionary technologies like genetic engineering and nanotechnology typically present us with what we call *decisions in the dark*. The unprecedented nature of these innovations makes their future implications extremely difficult to forecast. The social and economic promise is so huge that many public and private entities have bet vast stakes on the bio-nano future, but at the same time, the imagined risks are generating growing alarm (Joy, 2000). Even though we are substantially ignorant about their likely consequences, we face choices today about whether and how to support, develop, and regulate them. We have to make these decisions in the dark.

The same holds for decisions about artificial cells. We can and should speculate about the possible benefits and risks of artificial cell technology, but the fact remains that we now are substantially ignorant about their consequences. Statistical analyses of probabilities are consequently of little use. So, decisions about artificial cells are typically decisions in the dark. Thus, utilitarianism and decision theory and other algorithmic decision support methods have little if any practical value. Any decision-theoretic calculus we attempt will be limited by our current guesses about the shape of the space of consequences, and in all likelihood our picture of this shape will substantially change as we learn more.

This does not mean that we cannot make wise decisions; rather, it means that deciding will require the exercise of good judgment. Most of us have more or less well-developed abilities to identify relevant factors and take them into account, to discount factors likely to appeal to our own self-interest, and the like. These methods are fallible and inconclusive, but when deciding in the dark they generally are all we have available. It might be nice if we could foist the responsibility for making wise choices onto some decision algorithm such as utilitarianism or decision theory, but that is a vain hope today.

Even though the consequences of creating artificial cells will remain uncertain for some time, scientific leaders and policy makers still will have to face decisions about whether to allow them to be created, and under what circumstances. And as the science and technology behind artificial cells progresses, the range of these decisions will grow. The decisions will include whether to permit various lines of research in the laboratory, where to allow various kinds of field trials, whether to permit development of various commercial applications, how to assign liability for harms of these commercial products, whether to restrict access to research results that could be used for nefarious purposes, and so on. The uncertainty about the possible outcomes of these decisions does not remove the responsibility for taking some course of action, and the stakes involved could be very high. So, how should one meet this responsibility to make decisions about artificial cells in the dark?

When contemplating a course of action that could lead to a catastrophe, many people conclude that it is not worth the risk and instinctively pull back. This form of reasoning illustrates what we call the *doomsday principle*, which says: *Do not pursue a course of action if it might lead to a catastrophe* (Stich, 1978). Many people in the nanotechnology community employ something like this principle. For example, Merkle (1992) thinks that the potential risks posed by nanomachines that replicate themselves and evolve in a natural setting are so great that not only should they not be constructed; they should not even be designed. He concludes that to achieve compliance with this goal will involve enculturating people to the idea that "[t]here are certain things that you just *do not do*" (Merkle, 1992, p. 292). This illustrates doomsday reasoning that absolutely forbids crossing a certain line because it might lead to a disaster.

A little reflection shows that the doomsday principle is implausible as a general rule, because it would generate all sorts of implausible prohibitions. Almost any new technology could, under some circumstances, lead to a catastrophe, but we presumably do not want to ban development of technology in general. To dramatize the point, notice there is always at least some risk that getting out of bed on any given morning *could* lead to a catastrophe. Maybe you will be hit by a truck and thereby be prevented from discovering a critical breakthrough for a cure for cancer; maybe a switch triggering the world's nuclear arsenal has been surreptitiously left beside your bed. These consequences are completely fanciful, of course, but they are still possible. The same kind of consideration shows that virtually every action could lead to a catastrophe and so would be prohibited by the doomsday principle. *Not* creating artificial cells could lead to a disaster because there could be some catastrophic consequence that society could avert only by developing artificial cells, if, for example, artificial cells could be used to cure heart disease. Since the doomsday principle prohibits your action no matter what you do, the principle is incoherent.

The likelihood of triggering a nuclear reaction by getting out of bed is negligible, of course, while the likelihood of self-replicating nanomachines wreaking havoc might be higher. With this in mind, one might try to resuscitate the doomsday principle by modifying it so that it is triggered only when the likelihood of catastrophe is non-negligible. But there are two problems with implementing such a principle. First, the threshold of negligible likelihood is vague and could be applied only after being converted into some precise threshold (for example, probability 0.001). But any such precise threshold would be arbitrary and hard to justify. Second, it will often be impossible to ascertain the probability of an action causing a catastrophe with anything like the requisite precision. For example, we have no way at present of even estimating if the probability of self-replicating nanomachines causing a catastrophe is above or below 0.001. Estimates of human health risks are typically based on three kinds of evidence: toxicological studies of harms to laboratory animals, epidemiological studies of correlations in existing populations and environments, and statistical analyses of morbidity and mortality data (Ropeik and Gray, 2002). We lack even a shred of any of these kinds of evidence concerning self-replicating nanomachines, because they do not yet exist.

When someone proposes to engage in a new kind of activity or to develop a new technology today, typically this is permitted unless and until it has been shown that some serious harm would result. Think of the use of cell phones, the genetic modification of foods, the feeding of offal to cattle. In other words, a new activity is innocent until proven guilty. Anyone engaged in the new activity need not first prove that it is safe; rather, whoever questions its safety bears the burden of proof for showing that it really is unsafe. Furthermore, this burden of proof can be met only with scientifically credible evidence that establishes a causal connection between the new activity and the supposed harm. It is insufficient if someone suspects or worries that there might be such a connection, or even if there is scientific evidence that there *could* be such a connection. The causal connection must be credibly established before the new activity can be curtailed.

This approach to societal decision making has, in the eyes of many, led to serious problems. New activities have sometimes caused great damage to human health or the environment before sufficient evidence of the cause of these damages had accumulated. One notorious case is thalidomide, which was introduced in the 1950s as a sleeping pill and to combat morning sickness, and was withdrawn from the market in the early 1960s when it was discovered to cause severe birth defects (Stephens and Brynner, 2001). Support has been growing for shifting the burden of proof and exercising more caution before new and untested activities are allowed, that is, treating them as guilty until proven innocent. This approach to decision making is now widely known as the *precautionary principle*: *Do not pursue a course of action that might cause significant harm even if it is uncertain whether the risk is genuine*. Different formulations of the precautionary principle can have significantly different pros and cons. Here, we concentrate just on the following elaboration (Geiser, 1999, p. xxiii), which is representative of many of the best known statements of the principle:

The Precautionary Principle asserts that parties should take measures to protect public health and the environment, even in the absence of clear, scientific evidence of harm. It provides for two conditions. First, in the face of scientific uncertainties, parties should refrain from actions that might harm the environment, and, second, that the burden of proof for assuring the safety of an action falls on those who propose it.

The precautionary principle is playing an increasing role in decision making around the world. For example, the contract creating the European Union appeals to the principle, and it governs international legal arrangements such as the United Nations Biosafety Protocol. The precautionary principle is also causing a growing controversy.

We are skeptical of the precautionary principle. It is only common sense to exercise due caution when developing new technologies, but other considerations are also relevant. We find the precautionary principle to be too insensitive to the complexities presented by deciding in the dark. One can think of the precautionary principle as a principle of inaction, recommending that, when in doubt, leave well enough alone. It is sensible to leave well enough alone only if things are well at present and they will remain well by preserving the status quo. But these presumptions are often false, and this causes two problems for the precautionary principle.

Leaving well enough alone might make sense if the world were unchanging, but this is manifestly false. The world's population is continuing to grow, especially in poor and relatively underdeveloped countries, and this is creating problems that will not be solved simply by being ignored. In the developed world, average longevity has been steadily increasing; over the last hundred years in the United States, for example, life expectancy has increased more than 50% (Wilson and Crouch, 2001). Today heart disease and cancer are far and away the two leading causes of death in the United States (Ropeik and Gray, 2002). Pollution of drinking water is another growing problem. There are an estimated hundred thousand leaking underground fuel storage tanks in the United States, and a fifth of these are known to have contaminated the groundwater (Ropeik and Gray, 2002). These few examples illustrate that the key issues that society must confront are continually evolving. The precautionary principle does not require us to stand immobile in the face of such problems. But it prevents us from using a method that has not been shown to be safe. So the precautionary principle ties our hands when we face new challenges.

This leads to a second, deeper problem with the precautionary principle. New procedures and technologies often offer significant benefits to society, many of which are new and unique. Cell phones free long-distance communication from the tether of land lines, and genetic engineering opens the door to biological opportunities that would never occur without human intervention. Whether or not the benefits of these technologies outweigh the risks they pose, they do have benefits. But the precautionary principle ignores such benefits. To forgo these benefits causes harm – what one might call a "harm of inaction." These harms of inaction are opportunity costs, created by the lost opportunities to bring about certain new kinds of benefits. Whether or not these opportunity costs outweigh other considerations, the precautionary principle prevents them from being considered at all. That is a mistake.

These considerations surfaced at the birth of genetic engineering. The biologists who were developing recombinant DNA methods suspected that their new technology might pose various new kinds of risks to society, so the National Academy of Science, U.S.A., empaneled some experts to examine the matter. They quickly published their findings in *Science* in what has come to be known as the *Moratorium letter*, and recommended suspending all recombinant DNA studies "until the potential hazards ... have been better evaluated" (Berg et al., 1974, p. 303). This is an early example of precautionary reasoning. Recombinant DNA studies were suspended even though no specific risks had been scientifically documented. Rather, it was thought that there *might* be such risks, and that was enough to justify a moratorium even on recombinant DNA *research*.

The Moratorium letter provoked the National Academy of Science to organize a conference at Asilomar the following year, with the aim of determining under what conditions various kinds of recombinant DNA research could be safely conducted. James Watson, who signed the Moratorium letter and participated in the Asilomar conference, reports having had serious misgivings about the excessive precaution being advocated, and writes that he "now felt that it was more irresponsible to defer research on the basis of unknown and unquantifiable dangers. There were desperately sick people out there, people with cancer or cystic fibrosis – what gave us the right to deny them perhaps their only hope?" (Watson, 2003, p. 98). Watson is here criticizing excessive precautionary actions because they caused harms of inaction.

Society's initial decisions concerning artificial cells will be made in the dark. The potential benefits of artificial cells seem enormous, but so do their potential harms. Without gathering a lot more basic knowledge, we will remain unable to say anything much more precise about those benefits and risks. We will be unable to determine with any confidence even the main alternative kinds of consequences that might ensue, much less the probabilities of their occurrence. Given this lack of concrete knowledge about risks, an optimist would counsel us to pursue the benefits and have confidence that science will handle any negative consequences that arise. The precautionary principle is a reaction against precisely this kind of blind optimism. Where the optimist sees ignorance about risks as opening the door to action, the precautionary thinker sees that same ignorance as closing the door.

As an alternative to the precautionary principle and to traditional risk assessment, we propose dropping the quest for universal ethical principles, and instead cultivating the virtues we will need for deciding in the dark. Wise decision making will no doubt require balancing a variety of virtues. One obvious virtue is caution. The positive lesson of the precautionary principle is to call attention to this virtue, and its main flaw is ignoring other virtues that could help lead to wise decisions. Another obvious virtue is wisdom; giving proper weight to different kinds of evidence is obviously important when deciding in the dark.

We want to call special attention to a third virtue that is relevant to making proper decisions: *courage*. That is, we advocate carefully but proactively pursuing new courses of action and new technologies when the potential benefits warrant it, even if the risks are uncertain. Deciding and acting virtuously requires more than courage. It also involves the exercise of proper caution; we should pursue new technologies only if we have vigilantly identified and understood the risks involved. But the world is complex and the nature and severity of those risks will remain somewhat uncertain. This is where courage becomes relevant. Uncertainty about outcomes and possible risks should not invariably block action. We should weigh the risks and benefits of various alternative courses of action (including the "action" of doing nothing), and have the courage to make a leap in the dark when on balance that seems most sensible. Not to do so would be cowardly.

We are not saying that courage is an overriding virtue and that other virtues such as caution are secondary. Rather, we are saying that courage is one important virtue for deciding in the dark, and it should be given due weight. Precautionary thinking tends to undervalue courage.

Our exhortation to courage is vague, of course; we provide no mechanical algorithm for generating courageous decisions. We do not view this as a criticism of our counsel for courage. For the reasons outlined in the previous section, we think no sensible mechanical algorithm for deciding in the dark exists. If we are right, then responsible and virtuous agents must exercise judgment, which involves adjudicating conflicting principles and weighing competing interests. Because such decisions are deeply context dependent, any sensible advice will be conditional.

New technologies give us new powers, and these powers make us confront new choices about exercising the powers. The new responsibility to make these choices wisely calls on a variety of virtues, including being courageous when deciding in the dark. We should be prepared to take some risks if the possible benefits are significant enough and the alternatives unattractive enough.

4. CONCLUSIONS

Artificial cells are in our future, and that future could arrive within a decade. By harnessing the automatic regeneration and spontaneous adaptation of life, artificial cells promise society a wide variety of social and economic benefits. But their ability to self-replicate and unpredictably evolve creates unforeseeable risks to human health and the environment. So it behooves us to start thinking through the implications of our impending future with artificial cells now.

From the public discussion on genetic engineering and nanotechnology one can predict the outline of much of the debate that will ensue. One can expect the objections that creating artificial cells is unnatural, that it commodifies life, that it fosters a reductionistic perspective, and that it is playing God; but we have explained why these kinds of considerations are all unpersuasive.

The unknowns involving artificial cells are sufficiently great that we must decided in the dark regarding whether and how to proceed. The precautionary principle is increasingly being applied to important decisions in the dark, but the principle fails to give due weight to potential benefits lost through inaction (what we called "harms of inaction"). We suggest that appropriately balancing the virtues of courage and caution would preserve the attractions of the precautionary principle while avoiding its weaknesses.

WORKS CITED

M. A. Bedau, E. C. Parke, U. Tangen, and B. Hantsche-Tangen (2009) 'Social and Ethical Checkpoints for Bottom-up Synthetic Biology, or Protocells,' *Systems and Synthetic Biology*, 3: 65–75.

P. Berg, D. Baltimore, H. W. Boyer, S. N. Cohen, R. W. Davis, D. S. Hogness, et al (1974) 'Potential Biohazards of Recombinant DNA Molecules,' *Science* 185: 303.

V. Bocci (1992) 'The Neglected Organ: Bacterial Flora has a Crucial Immunostimulatory Role,' *Perspectives in Biology and Medicine*, 35 (2): 251–60.

M. C. Brannigan (2001) 'Introduction,' in M. C. Brannigan (ed.) *Ethical issues in Human Cloning: Cross-disciplinary Perspectives* (New York: Seven Bridges Press).

G. L. Comstock (2000) *Vexing Nature? On the Ethical Case Against Agricultural Biotechnology* (Boston, MA: Kluwer Academic Publishers).

M. Crichton (2002a) 'How Nanotechnology is Changing our World,' *Parade*, November 24: 6–8.

M. Crichton (2002b) *Prey* (New York: HarperCollins).

D. Deamer (2009) 'Experimental Approaches to Fabricating Artificial Cellular Life,' *Protocells: bridging nonliving and living matter* S. Rasmussen, M. A. Bedau, L. Chen, et al (eds.) (Cambridge: MIT Press).

A. Dobson (1995) 'Biocentrism and Genetic Engineering,' *Environmental Values* 4: 227.

F. J. Dyson (2003) 'The Future Needs Us,' *New York Review of Books*, 50 (13 February): 11–13.

M. C. Enright, D. A. Robinson, and G. Randle (2002) 'The Evolutionary History of Methicillin-resistant *Staphylococcus aureus* (MSRA),' *The Proceedings of the National Academy of Sciences of the United States of America* 99 (11): 7687–7692.

K. L. Erickson, and N. E. Hubbard (2000) 'Probiotic Immunomodulation in Health and Disease,' *The Journal of Nutrition* 130 (2): 403S-409S.

ETC Group (2003) *The Big Down: From Genomes to Atoms*. Available online at: http://www.etcgroup.org (accessed June 2003).

K. Geiser (1999) 'Establishing a General Duty of Precaution in Environmental Protection Policies in the United States: A Proposal,' *Protecting Public Health and the Environment: Implementing the Precautionary Principle*. Raffensperger & Tickner (eds.) (Washington, DC: Island Press).

K. Ishikawa, K. Sato, Y. Shima, I. Urabe, and T. Yomo (2004) 'Expression of a Cascading Genetic Network within Liposomes,' *FEBS Letters* 576: 387–390.

B. Joy (2000) 'Why the Future Doesn't Need Us,' *Wired* 8 (April). Available online at: http://www.wired.com/wired/archive/8.04/joy.html (accessed September 2007).

L. R. Kass (2002) *Life, Liberty and the Defense of Dignity: The Challenge for Bioethics* (San Francisco: Encounter Books).

V. Mai, and J. G. Morris Jr. (2004) 'Colonic Bacterial Flora: Changing Understandings in a Molecular Age,' *The Journal of Nutrition* 134 (2): 459–64.

R. Merkle (1992) 'The Risks of Nanotechnology,' *Nanotechnology Research and Perspectives*. B. Crandall, and J. Lewis (eds.) (Cambridge, MA: MIT Press).

K. E. Miller (2000) 'Can Vaginal Lactobacilli Reduce the Risk of STDs?' *American Family Physician* 61 (10): 3139–3140.

P-A. Monnard, and D. W. Deamer (2002) 'Membrane self-assembly processes: Steps toward the first cellular life,' *Anatomical Record*, 268, 196–207.

J. Morris (ed.) (2000) *Rethinking risk and the precautionary principle* (Oxford: Butterworth-Heinemann).

V. Noireaux, and A. Libchaber (2004) 'A vesicle bioreactor as a step toward an artificial cell assembly,' *Proceedings of the National Academy of Science of the United States of America*, 101, 17669–17674.

C. Raffensperger, and J. Tickner (eds.) (1999) *Protecting public health and the environment: Implementing the precautionary principle* (Washington, DC: Island Press).

S. Rasmussen, M. A. Bedau, L. Chen, D. Deamer, D. C. Krakauer, N. H. Packard, and P. F. Stadler (eds.) (2009) *Protocells: bridging nonliving and living matter* (Cambridge: MIT Press)

M. J. Reiss, and R. Straughan (1996) *Improving nature? The science and ethics of genetic engineering* (Cambridge: Cambridge University Press).

M. Resnick (1987) *Choices: An introduction to decision theory* (Minneapolis: University of Minnesota Press).

D. Ropeik, and G. Gray (2002) *Risk: A practical guide to deciding what's really safe and what's really dangerous in the world around you* (Boston: Houghton Mifflin).

L. M. Silver (1998) 'Cloning, ethics, and religion,' *Cambridge Quarterly of Healthcare Ethics*, 7: 168–172. Reprinted in M. C. Brannigan (ed.) (2001) *Ethical issues in human cloning: Cross-disciplinary perspectives* (pp. 100–105) (New York: Seven Bridges Press).

T. D. Stephens, and R. Brynner (2001) *Dark remedy: The impact of thalidomide and its revival as a vital medicine* (New York: Perseus).

S. P. Stich (1978) 'The recombinant DNA debate,' *Philosophy and Public Affairs*, 7: 187–205.

K. Takakura, T. Toyota, and T. Sugawara (2003) 'A novel system of self-reproducing giant vesicles,' *Journal of the American Chemical Society*, 125: 8134–8140.

J. D. Watson (2003) *DNA: The secret of life* (New York: Alfred A. Knopf).

R. Wilson, and E. A. C. Crouch (2001) *Risk-benefit analysis* (Cambridge, MA: Harvard University Press).

World Health Organization (2000) 'World Health Organization report on infectious diseases 2000: Overcoming antimicrobial resistance,' Available online at: http://www.who.int/infectious-disease-report/2000/ (accessed May 2004).

DISCUSSION QUESTIONS

1. Do you agree with the authors that there is no compelling intrinsic argument against the creation of protocells? Why or why not?
2. Does the creation of artificial cells pose any extrinsic risks that other technologies, such as synthetic biology or nanotechnology, do not?

3. The authors discuss various virtues to employ when making decisions in the dark; what are some vices that we should avoid in making such decisions?

4. What measures could policy makers, researchers, and citizens put in place to minimize risks when making decisions in the dark?

5. Do you agree with the authors that decision procedures, such as the precautionary principle and the doomsday principle, are little or no help in cases of extreme uncertainty? Why or why not?

6. Are there any technologies that we should never develop? What grounds or bases justify relinquishing or prohibiting a technology?

7. Are there any ethical concerns associated with creating living things that are not also associated with the creation of nonliving artifacts?

8. Does the creation of protocells establish decisively that a reductionist or materialist view about life is the correct one, since it appears possible to create life from chemical materials alone?

INDEX

Printed and bound in the United States of America